T0155837

Multifunctional Nanostructured Metal Oxides for Energy Harvesting and Storage Devices

Multifunctional Nanostructured Metal Oxides for Energy Harvesting and Storage Devices

Edited by
Vijay B. Pawade
Paresh H. Salame
Bharat A. Bhanvase

CRC Press is an imprint of the
Taylor & Francis Group, an informa business

First edition published 2020
by CRC Press
6000 Broken Sound Parkway NW, Suite 300, Boca Raton, FL 33487-2742

and by CRC Press
2 Park Square, Milton Park, Abingdon, Oxon, OX14 4RN

CRC Press is an imprint of Taylor & Francis Group, LLC

ISBN: 978-0-367-27547-1 (hbk)
ISBN: 978-0-367-49858-0 (pbk)
ISBN: 978-0-429-29687-1 (ebk)

Typeset in Times
by Lumina Datamatics Limited

Contents

Editors

Dr. Vijay B. Pawade is an Assistant Professor in the Department of Applied Physics at the Laxminarayan Institute of Technology, R.T.M. Nagpur University, Nagpur, India. His research is mainly focused on rare-earth-doped oxide materials and their applications in LEDs and solar cell devices. He has published 33 research papers in referred international journals and has served as a referee for many peer-reviewed journals published by Elsevier, Springer, Willey, Taylor & Francis, RSC, and ACS. He is the author of *Phosphor for Energy Saving and Conversion Technology* (CRC Press, 2018). He also is an editor of the two books: *Nanomaterials for Green Energy* (Elsevier, 2018), contributing four chapters, and *Spectroscopy of Lanthanide Doped Oxide Materials* (Woodhead Publishing, 2019). Email: vijaypawade003@gmail.com

Dr. Paresh H. Salame is an Assistant Professor in the Department of Physics, Institute of Chemical Technology, Mumbai, India. His research is focused on nanostructured cathode materials for energy storage and conversion, colossal dielectric constant materials, and multiferroics. He has written eight international papers in peer-reviewed journals and also has authored two book chapters in reputed books published by Elsevier 2018. He is also a recipient of the Early Career Research Award (previously called the Young Scientist Research Award) by the Science and Engineering Research Board (SERB), Government of India for his project on Cathode materials for Rechargeable Na-ion Batteries. Email: paresh.salame@gmail.com

Dr. Bharat A. Bhanvase is currently working as an Professor and Head in the Department of Chemical Engineering at the Laxminarayan Institute of Technology, R. T. M. Nagpur University, Nagpur, Maharashtra, India. His research interests are focused on wastewater treatment, cavitation-based nanomaterials and nanocomposites, process intensification, microfluidics, and nanofluids. He earned his PhD in chemical engineering from the University of Pune, Pune, India. He has published several research articles in both national and international journals and presented several papers at both national and international conferences. He has written several book chapters in internationally renowned books, and has edited the books published *Process Modeling, Simulation, and Environmental Applications in Chemical Engineering Hardcover* (Apple Academic Press, 2016), *Novel Water Treatment and Separation Methods: Simulation of Chemical Processes* (Apple Academic Press, 2017), *Nanomaterials for Green Energy* (Elsevier, 2018) and *Encapsulation of Active Molecules and their Delivery System* (Elsevier, 2020). He is the recipient of the Young Scientist Research Award start-up research grant awarded by the Science and Engineering Research Board, New Delhi (India) in 2015. He received the Best Scientist Award from Rashtrasant Tukadoji Maharaj Nagpur University, Nagpur in 2017. He has more than 16 years of teaching experience. He serves as a reviewer for several International journals. Email: bharatbhanvase@gmail.com

Contributors

M. V. Bagal
Department of Chemical Engineering
Bharati Vidyapeeth College of Engineering
Navi Mumbai, India

Deepa B. Bailmare
Energy Materials and Devices Laboratory
Department of Physics
R. T. M. Nagpur University
Nagpur, India

Sagar D. Balgude
Department of Engineering Science
DY Patil College of Engineering SP Pune University
Pune, India

Shrikant S. Barkade
Department of Chemical Engineering
Sinhgad College of Engineering
Pune, India

Bharat A. Bhanvase
Department of Chemical Engineering
Laxminarayan Institute of Technology
R. T. M. Nagpur University
Nagpur, India

Jaykumar B. Bhasarkar
Department of Chemical Technology (Pulp and Paper Technology)
Laxminarayan Institute of Technology
R. T. M. Nagpur University
Nagpur, India

S. S. Bhoga
Department of Physics
R. T. M. Nagpur University
Nagpur, India

Abhay D. Deshmukh
Energy Materials and Devices Laboratory
Department of Physics
R. T. M. Nagpur University
Nagpur, India

S. J. Dhoble
Department of Physics
R. T. M. Nagpur University
Nagpur, India

D. Haranath
Department of Physics
National Institute Technology
Warangal, India

A. P. Khandale
Indian Institute of Information Technology, Design and Manufacturing
Kancheepuram, Chennai, India

Jayant Kolte
School of Physics and Materials Science
Thapar Institute of Engineering and Technology
Patiala, India

Shrikaant Kulkarni
Department of Chemical Engineering
Vishwakarma Institute of Technology
Pune, India

Satish P. Mardikar
Department of Chemistry
Smt. RS College, SGB Amravati University
Amravati, India

Vijay B. Pawade
Department of Applied Physics
Laxminarayan Institute of Technology
R. T. M. Nagpur University
Nagpur, India

R. Rakesh Kumar
Department of Physics
National Institute Technology
Warangal, India

S. Raut-Jadhav
Department of Chemical Engineering
Bharati Vidyapeeth (Deemed to be University) College of Engineering
Pune, India

Paresh H. Salame
Department of Physics
Institute of Chemical Technology Mumbai
Mumbai, India

K. N. Shinde
Department of Physics
NS Science & Arts College
Bhadrawati, India

Rajan Singh
Key Laboratory of Advanced Materials (MOE)
School of Materials Science and Engineering
Tsinghua University
Beijing, China

and

Beijing National Center for Electron Microscopy
Tsinghua University
Beijing, China

and

Department of Electronics and Communication Engineering
School of Engineering and Technology
Central University of Rajasthan
Ajmer, India

Chandrasekar M. Subramaniyam
Department of Fibre and Polymer Technology
KTH Royal Institute of Technology
Stockholm, Sweden

N. Thejo Kalyani
Department of Applied Physics
Laxminarayan Institute of Technology
Nagpur, India

K. Uday Kumar
Department of Physics
National Institute Technology
Warangal, India

1 Synthesis, Properties, and Applications of Transition Metal Oxide Nanomaterials

R. Rakesh Kumar, K. Uday Kumar, and D. Haranath

CONTENTS

1.1 INTRODUCTION

Transition metal oxide nanomaterials (TMONMs) in the form of nanowires, nanoparticles (NPs), nanosheets, nanoflowers, nanoribbons, nanobelts, 3D networks, and hierarchical nanostructures have attracted a lot of attention from the last decade due to their multifunctional properties. The unusual electronic structure of the base transition metal and the bonding with oxide makes TMONMs a fascinating class of materials. The partially filled *d* orbital is the basis for a wide range of oxides with unique physical and chemical properties. This characteristic feature brings these materials with unique and exceptional reactive electronic transitions, high dielectric constants, high density, tunable band gap, and morphologies controlled on the nanodimension. Therefore, TMONMs are considered to be one of the fascinating functional materials due to tunable physical and chemical properties with a wide range of applications that include energy storage, energy harvesting, photocatalysis, sensors, electrochromic devices, wastewater treatment, and microelectronics.

A variety of synthesis methods have been employed for the synthesis of TMONMs both in the vapor phase and the solution phase at higher temperatures and lower temperatures, respectively. Vapor phase methods include thermal evaporation, electron beam evaporation, pulsed-laser deposition, and chemical vapor deposition. Morphology in vapor phase methods can be controlled by changing the parameters such as growth temperature, catalyst, substrate, pre- and post-treatment, and oxygen pressure. In the vapor phase method, morphologies such as wires, rods, needles, tubes, and belts will be obtained. The growth mechanism can be easily summarized by the vapor-solid and vapor-liquid-solid (VLS) mechanism. In the liquid-phase method, a greater variety of nanostructures can be synthesized than in the vapor-phase method. Hydrothermal growth, electrochemical deposition, and template-directed synthesis are more popular methods in the liquid phase. Morphology in the liquid-phase synthesis can be controlled by growth temperature, pressure, time, and reaction medium.

TMONMs of WO_3, Fe_2O_3, ZnO, TiO_2, V_2O_5, MnO, CoO, and SnO_2 are currently playing a major role in various applications. TMONMs have been extensively investigated as electrode material for Li-ion batteries for high-energy density as well as a long life cycle. Hierarchically nanostructured transition metal oxides (TMOs) have become a hot research area in the field of batteries. Hierarchical nanostructures provide more accessible electroactive sites for redox reactions, shorten the diffusion length of Li-ion, and also accommodate a large volume expansion during cycling. TMONMs also play a major role in supercapacitor applications for high-energy storage and harvesting energy in the form of solar cells, nanogenerators. TMONMs of Fe_2O_3, Fe_3O_4, CoO, CO_3O_4, NiO, Mn_2O_3, TiO_2, Nb_2O_5, V_2O_5, and WO_3 are extensively studied for energy storage and conservation applications. Recently, synthesis of bimetallic TMO nanostructures such as $NiCo_2O_4$, $MnCo_2O_4$, $ZnFe_2O_4$, and $ZnCo_2O_4$ also tested for energy-storage applications.

Another useful application of TMONMs is photocatalytic activity. Single TMOs such as TiO_2, Fe_2O_3, V_2O_5, ZnO, and TiO_2 loaded with various transition metals (Co, Cr, Fe, Mo, V, and W) have shown excellent photocatalytic activity. Photocatalytic activity finds applications in the disinfection of both air and water. TMONM-coated surfaces were developed as self-disinfecting materials. Photocatalytic activity has potential application in environmental health, biological, medical, hospitals, food, and pharmaceutical applications.

The first part of this chapter discusses the different synthesis methods for TMONMs, and in the second part, different applications of TMONMs are presented.

1.2 SYNTHESIS METHODS OF TMONMS

1.2.1 Zinc Oxide (ZnO) Nanomaterials

1.2.1.1 Zero-Dimensional ZnO Nanomaterials

Zero-dimensional (0D) nanostructures such as NPs, quantum dots, and hollow spheres of ZnO have attracted a lot of attention due to their potential applications in various fields [1–4]. Strong UV-absorption properties of ZnO are used in personal care products such as cosmetics and sunscreens. In addition to the above, ZnO NPs have superior antibacterial and antimicrobial properties, which makes them useful in the textile industry. ZnO hollow spheres are used in catalysis and chemical and biological sensors. Similarly, ZnO quantum dots are useful in drug delivery, optical imaging, cancer cell sensing, and DNA detection.

ZnO NPs were mainly prepared by the chemical precipitation method shown in Figure 1.1 [5], sol-gel method [6], solid-state pyrolytic method [7], and other biologic

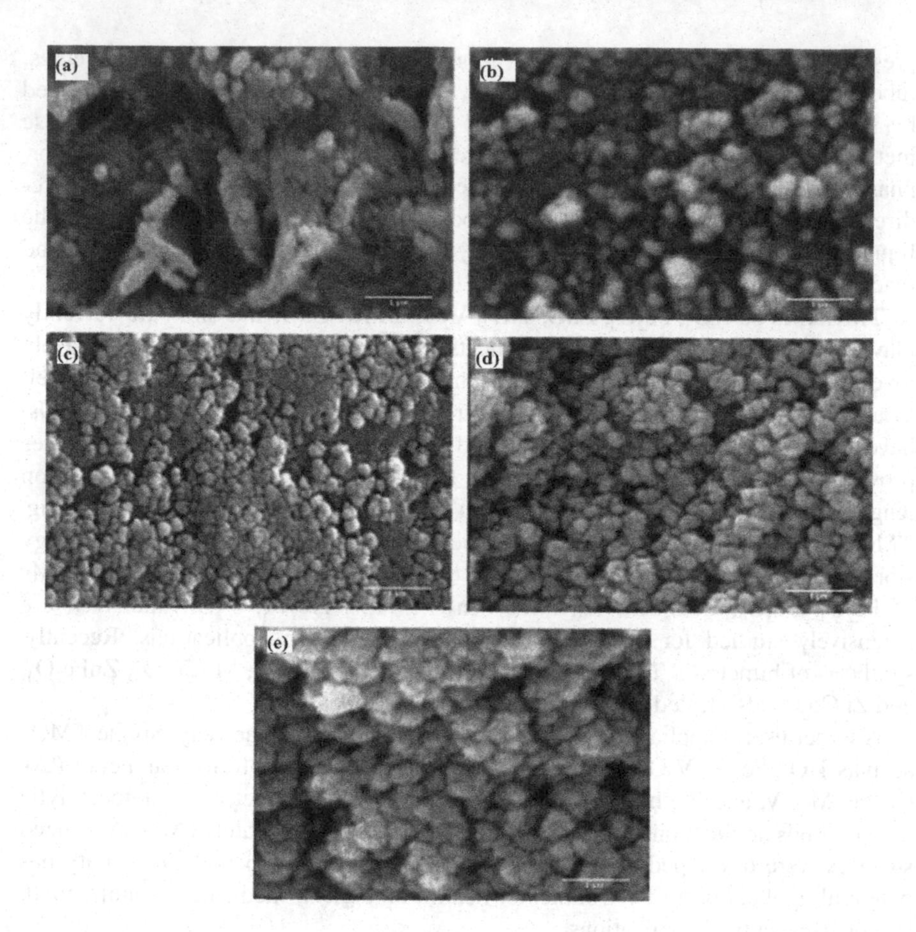

FIGURE 1.1 SEM images of the ZnO nanoparticle formation by chemical precipitation at different annealing temperatures of (a) as-obtained precursor, (b) 250°C, (c) 350°C, (d) 450°C, and (e) 550°C. (Reproduced with permission from Raoufi, D., *Renew. Energy*, 50, 932–937, 2013. With permission. Copyright 2013.)

methods [8]. The chemical precipitation method involves two reaction reagents, zinc acetate ($Zn(CH_3COO)_2 2H_2O$) or zinc nitrate ($Zn(NO_3)_2$) or zinc sulfate ($ZnSO_4$), and a solution precipitator such as sodium hydroxide (NaOH) or ammonium hydroxide (NH_3H_2O). In a typical process, the precipitator is added to the dissolved zinc precursor in dropwise until the pH levels reach the desired number and then the resulting solution is stirred to get a white intermediate zinc hydroxide ($Zn(OH)_2$). $Zn(OH)_2$ is further converted into ZnO after sintering at higher temperatures. The concentration of zinc precursor and precipitator, molar ratio of two solutions, and the annealing temperature are controlled parameters for the ZnO NP production. The advantages of this method are simple and easily controlled, and the disadvantage of this method is an agglomeration of the obtained NPs.

Raoufi et al. prepared ZnO NPs by this method using $Zn(NO_3)_2$ and $(NH_4)_2 CO_3$ in a molar ratio of 1:1.5, and intermediate products were annealed in a muffle furnace in

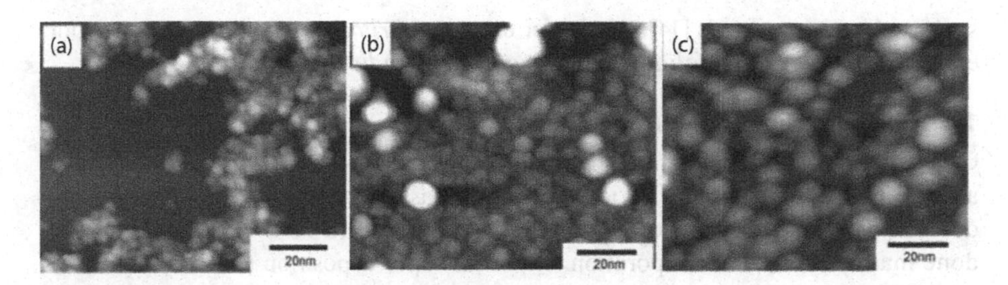

FIGURE 1.2 AFM images of the ZnO nanoparticles formed by the chemical precipitation method at different pH values (a) 10, (b) 12, and (c) 14. (Reproduced from Patra, M.K. et al., *J. Lumin.*, 129, 320–324, 2009. With permission. Copyright© 2009.)

the range of 250°C–550°C for 4 hours to obtain the ZnO NPs [5]. Sizes of the obtained ZnO NPs were in the range of 8–27 nm depending on the annealing temperature.

The chemical precipitation method was also used for the production of ZnO quantum dots (shown in Figure 1.2). Patra et al. prepared ZnO quantum dots [9]. $Zn(CH_3COO)_2$ was dissolved in methanol and a potassium hydroxide (KOH) solution was added to zinc precursor solution dropwise with constant stirring. The final solution pH was maintained at ~10. The resulting solution shows bright bluish-green emission under UV excitation, indicating the formation of ZnO quantum dots. Further, a tetraethylorthosillicate solution was added to the ZnO quantum dot solution to control the particle growth. ZnO quantum dots of different sizes 4, 5.5 and 8 nm were prepared with different pH values of ~10, 12, and 14, respectively [9].

The second most important synthesis method for the ZnO NPs is the sol-gel method. It has several advantages such as simple method, low cost, mild reaction conditions of synthesis. It was first proposed by Spanhel et al. in 1991 [10]. This method contains three important steps: (1) preparation of the zinc precursor, (2) preparation of ZnO cluster from precursor solution, and (3) crystal growth.

Spanhel et al. prepared ZnO colloids (crystallite size 3–6 nm) by the sol-gel method [10]. A zinc precursor was prepared by dissolving $Zn(CH_3COO)_2$ $2H_2O$ in methanol, and the solution was boiled at 80°C to get the hygroscopic mixture. LiOH H_2O powder was added to the above hygroscopic mixture and kept in the ultrasonic bath to get a transparent solution. Crystal growth takes in the last step at room temperature. The amount of LiOH strongly affects the pH of the solution, which in turn affects the crystal growth rate, shape, and size of ZnO NPs.

Rani et al. synthesized ZnO NPs using NaOH, and the obtained ZnO NPs are near 14 nm [11]. Vafaee et al. also reported the synthesis of ZnO NPs of size 3–4 nm using triethanolamine as a surfactant [12].

The sol-gel method was also used for the production of ZnO quantum dots. Bera et al. synthesized 3–8 nm size ZnO quantum dots using $Zn(CH_3COO)_2$ and NaOH [13].

Green chemistry has attracted more and more attention because of it being environmentally friendly. Green synthesis methods are low cost, non-toxic, and biocompatible. Abroad, a variety of plant extracts are used for ZnO NPs synthesis, such as *Aloe barbadensis* [14], the flower extract of *Trifolium pretense* [15], the root extract of *Polygala tenuifolia* [16] and *Azadirachta india* [17].

1.2.1.2 One-Dimensional ZnO Nanomaterials

One-dimensional (1D) ZnO nanomaterials such as rods, wires, belts, and tubes have attracted a lot of attention in various fields, such as photonics, sensors, photovoltaic devices, and field-emission devices [18–20]. Synthesis of 1D ZnO nanomaterials was done mainly by thermal evaporation, chemical vapor deposition (CVD), hydrothermal growth, and pulsed-laser deposition.

In a thermal evaporation method, material to be deposited is put in a graphite crucible with the appropriate addition of catalyst material placed in the high-temperature zone, and the substrate to be coated is placed slightly at a low temperature zone in a tube furnace. The tube furnace has provisions to send the carrier gas and other reactive gases. ZnO nanostructures are easily prepared by this method. Zinc powder is evaporated by thermal heating in a tube furnace; evaporated atoms are transported by an argon (Ar) carrier gas along the reactive oxygen gas. The substrates are placed in a low-temperature zone with or without catalyst particles coated. The substrate, which has catalyst particles, produces the 1D nanorods/wires via VLS growth (shown in Figure 1.3) [21]. The substrate that does not have catalyst particles produces 1D nanocones by VS mechanism [22].

Huang et al. reported the catalytic growth of ZnO nanowires by the VLS growth method [21]. In this report, zinc vapor was produced by heating the mixture of ZnO and graphite powder in an alumina boat placed in a tube furnace at a temperature

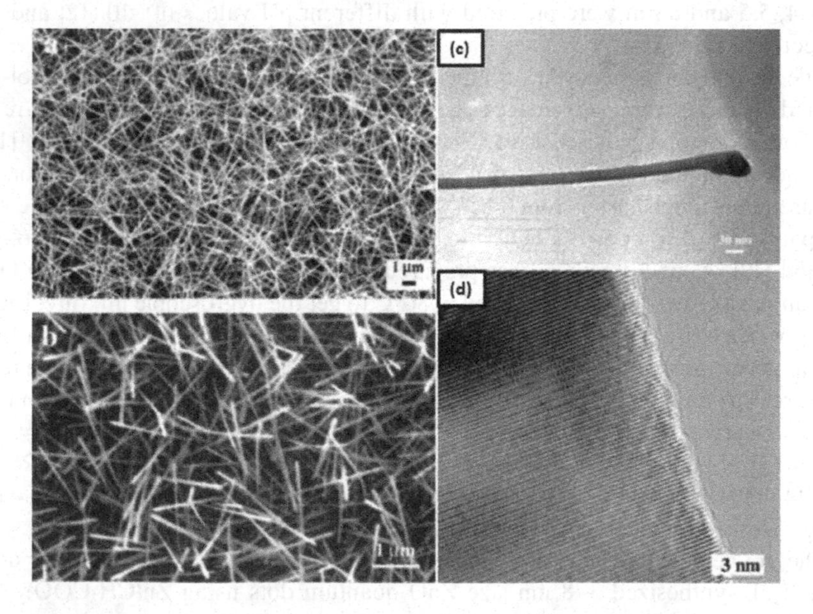

FIGURE 1.3 ZnO nanowire formation by the thermal evaporation method via VLS growth (a,b) SEM images of the ZnO nanowires with different catalyst layer thickness, (c) TEM image of the single nanowire, and (d) HRTEM recorded on the nanowire. (Reproduced from Huang, M.H. et al., *Adv. Mater.*, 13, 113–116, 2001. With permission. Copyright© 2001.)

of 950°C. The evaporated zinc atoms are deposited on the gold catalyst-coated substrate placed 10 cm away from the alumina boat in the tube furnace. Nanowires have the diameter in the range of 20–30 nm and length 5–10 mm. A transition electron microscopy (TEM) study of a single nanowire confirmed the gold-catalyst-assisted VLS growth of the ZnO nanowire. Fan et al. reported the patterned growth of ZnO nanowires by catalyst-assisted VLS growth by thermal evaporation [23]. Umar et al. reported the growth of hexagonal-shaped ZnO nanorods with Zn-terminated (0001) facets on nickel-coated Si substrates via thermal evaporation of zinc powder in a tube furnace under oxygen ambient [24].

ZnO nanowires, nanobelts, nanorings, and nanohelix were reported without aid any catalyst by the thermal evaporation method through the vapor-solid (VS) growth mechanism. Pan et al. reported the growth of ZnO nanobelts of width 50–300 nm and length of a few tens of microns [25]. In this report, ZnO powder was heated at 1400°C for 2 hours, and a white wool-like product deposited on the alumina substrate. TEM investigations suggested that ZnO nanobelts were formed through two growth directions (0001) and (011'0). The ZnO crystal structure consists of tetrahedrally coordinated Zn^{2+} and O^{2-} ions stacked along the *c*-axis. The polar surfaces of O^{2-} (0001') and Zn^{2+} (0001) produce divergence of surface energy for polar surfaces. The other nonpolar planes (011'0) and (21'1'0) have low surface energy, and this leads to the anisotropic growth of ZnO. Wang's research group further synthesized ZnO nanonanorings, nanohelixes, and nanospirals, and the representative scanning electron microscopy (SEM) image is shown in Figure 1.4 [26,27]. In these structures, positive and negative polarized surfaces would give rise to spontaneous polarization across the nanobelt thickness.

Rusli et al. reported the growth of a ZnO nanorod porous silicon substrate at a growth temperature of 600°C–1000°C by simple thermal evaporation of zinc powder in the presence of oxygen gas without any catalyst [22]. In this report, substrates

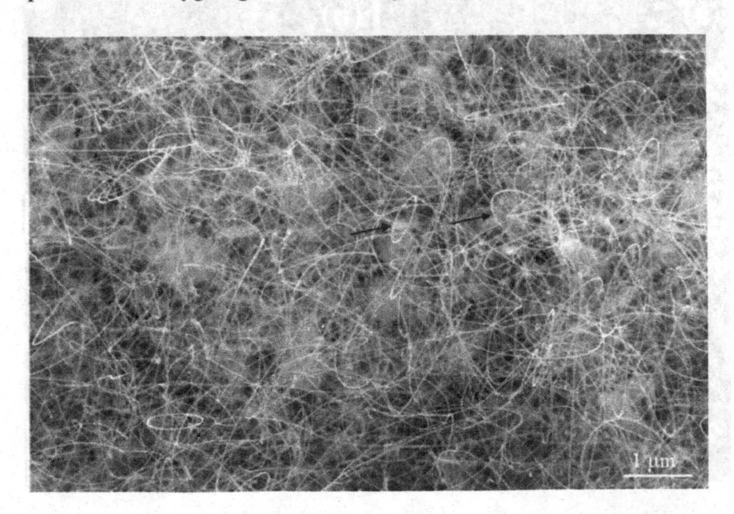

FIGURE 1.4 SEM images of the ZnO nanobelts, nanorings, and helical structures synthesized by the thermal evaporation method via the VS growth mechanism. (Reproduced from Kong, X.Y., *Nano Letters*, 3, 1625–1631, 2003. With permission. Copyright© 2003.)

were placed directly over the evaporation boat itself. The growth of ZnO nanorods took place via self-catalyzed VLS and VS growth mechanisms.

1D ZnO materials were also synthesized using CVD methods, such as CVD [28], Metal Organic Chemical Vapor Deposition (MOCVD) [29], Metal Organic Vapour-Phase Epitaxy (MOVPE) [30]. The CVD process was usually carried out in high temperatures in order to form a Zn vapor. Further low temperature CVD methods were developed for ZnO nanowires synthesis. Wu et al. reported the growth of ZnO nanorods at a low growth temperature of 500°C [28]. In this report, zinc acetylacetonatehydrate was vaporized at 140°C in a furnace, and the vapor was carried out by N_2/O_2 flow into the high-temperature zone of a furnace where substrates were located. In MOVPE, no catalyst was involved in the synthesis. Park et al. reported the growth of ZnO nanorods on an Al_2O_3 substrate at 400°C using diethylzinc and oxygen as reactants and argon as a carrier gas [30]. Further Plasma Enhanced Chemical Vapor Deposition (PECVD), Atmospheric Pressure Chemical Vapor Deposition (APCVD), MOCVD methods were developed for the synthesis ZnO nanowires. Liu et al. reported doping of ZnO nanowires by CVD methods [31].

Recently, solution-phase synthesis methods have attracted much attention due their low growth temperature, low cost, good yield of nanomaterial, any kind of substrate, and large-scale production. Vayssieres et al. reported the growth of ZnO nanorods by hydrothermal route at low growth temperatures [32]. This report provided a new route for the synthesis of 1D ZnO nanomaterials. Our research group also reported the growth of ZnO nanorods on an alloy substrate at a low growth temperature of 60°C–90°C (shown in Figure 1.5) [33]. In this report, equal molar $Zn(CH_3COO)_2$ and Hexamethylenetetramine (HMTA) solutions were mixed, and the resulting solution was used for the synthesis of ZnO nanorods. Seed-layer-coated

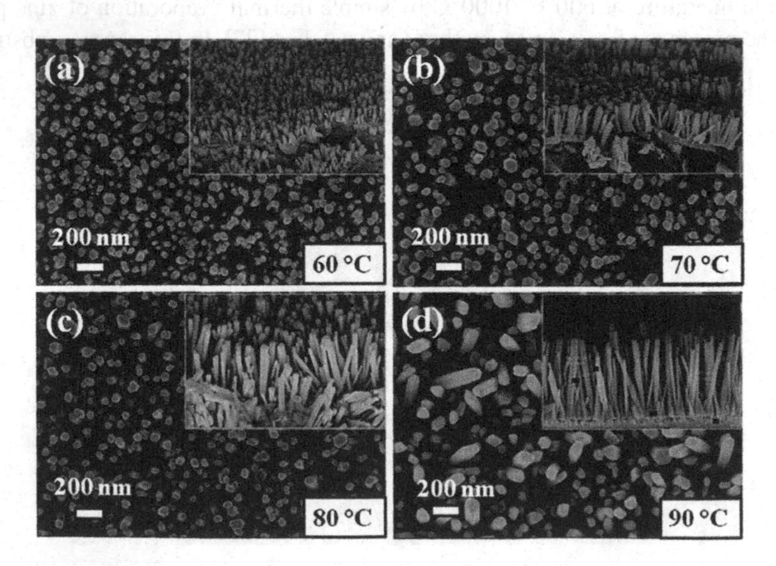

FIGURE 1.5 ZnO nanorods synthesized by hydrothermal growth with ZnO seed layer at different growth temperatures as mentioned in (a–d). (Reproduced from Gaddam, V. et al., *RSC Adv.*, 5, 89985–89992, 2015. With permission. Copyright© 2015.)

substrates were placed over the solution, and the seed layer faced the solution in a screw reagent bottle. The reagent bottle along with the substrates was placed inside a hot air oven at different growth temperatures of 60°C–90°C. A further hydrothermal method was used for the synthesis of ZnO nanowires and ZnO nanotubes [19,34].

1.2.1.3 Two-Dimensional ZnO Nanomaterials

Two-Dimensional (2D) ZnO nanomaterials such as nanosheets, nanoplates, nanowalls, and nanodisks have attracted a lot of attention due to their application in nanogenerators, gas sensors, dye sensitive solar cells, and Li-ion batteries.

The hydrothermal method is the most popular method for the synthesis of ZnO nanosheets. Our research group reported the growth of ZnO nanosheets on aluminum foils at a low growth temperature of 60°C–90°C without any catalyst (shown in Figure 1.6) [35]. In this report, equimolar zinc and HMTA solutions were mixed, and the resulting solution was used for the synthesis of ZnO nanosheets. Aluminum substrates were placed over the mixed solution and placed inside the oven at a growth temperature of 60°C–90°C for a growth duration of 4 hours. Gupta et al. also reported the growth of pure and vanadium-doped ZnO nanosheets by the hydrothermal method [36]. Cao et al. reported the growth of hexagonal-shaped ZnO nanosheets by this method [37]. In this report, 0.1M $Zn(NO_3)_2$ and 0.05M dimethylamine borane solutions were prepared and mixed. The mixed solution was put in a glass jar, and substrates were placed inside the mixed solution and placed inside the oven at a temperature of 70°C for a growth duration of 3 hours.

The electrochemical deposition method has been widely adapted for the growth of 2D ZnO structures. In the electrodeposition method, 2D nanostructures such as

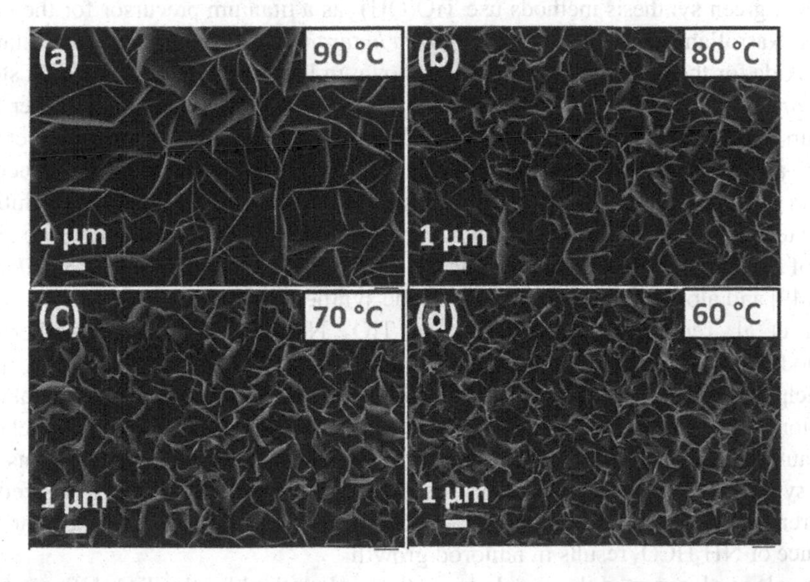

FIGURE 1.6 ZnO nanosheets synthesized by hydrothermal growth without seed layer at different growth temperatures as mentioned in (a–d). (Reproduced from Gaddam, V. et al., *RSC Adv.*, 5, 13519–13524, 2015. With permission. Copyright© 2015.)

nanosheets [38], flake-like nanostructures, ZnO plate structures, and ZnO nanowalls have been produced [39]. Yang et al. reported a simple electrochemical synthesis of ZnO nanosheets [38]. In this report, zinc chloride or $Zn(NO_3)_2$ is dissolved in the deionized (DI) water, and a KCl solution is added to increase the conductivity of the above solution. The above solution is used for the synthesis of ZnO nanosheets in two/three electrode configuration. Indium Tin Oxide (ITO) is considered a working electrode, platinum as a counterelectrode, and the third electrode Ag/AgCl as a reference electrode. A potential is applied between the electrodes, and the cell is heated up to a temperature ranging between 70°C and 85°C. With the increase of electrodepositing time, the size of the nanosheets increased.

Other methods such as thermal evaporation [40], sol-gel [41], MOCVD [42], and Pulsed Laser Deposition (PLD) [43] were also used for the synthesis of 2D ZnO nanomaterials.

1.2.2 Titanium Dioxide (TiO_2) Nanomaterials

TiO_2 is widely used as multifunctional material due to its potential application in sunscreens, Li-ion batteries, gas sensors, photocatalysis, and solar cells [44–46].

1.2.2.1 Zero-Dimensional TiO_2 Nanomaterials

TiO_2 NPs were mostly synthesized by liquid-phase methods such as the hydro/solvothermal method, green synthesis methods, and sol-gel. Green synthesis methods are the most favorable methods because they involve nontoxic plant extracts, microorganisms, and other natural reagents for the synthesis of NPs. Plant extract includes leaves, stems, flowers, and roots.

Most green synthesis methods use $TiO(OH)_2$ as a titanium precursor for the synthesis. Nasrollahzadeh et al. used *Euphorbia heteradena* jaub root extract and titanyl hydroxide for the synthesis of TiO_2 NPs (shown in Figure 1.7) [47]. In the first step, *Euphorbia heteradena* jaub root powder refluxed at 70°C with distilled water for 2 hours and cooled to room temperature. The aqueous extract of the root was centrifuged, and the supernatant was filtered. Further, $TiO(OH)_2$ was added to the aqueous extract of the root with constant stirring for 2 hours at 60°C. Change in the solution indicated the formation of NPs. A change in the color was associated with the excitation of surface plasmon resonance. Further, *Mangifera indica* extract [48] *S. trilobatum* [49] and aloe vera [50] were used for the synthesis of TiO_2 NPs.

Li et al. reported the monodisperse TiO_2 NPs using simple solvothermal method [51]. NH_4HCO_3 and LA, triethylamine, and cyclohexane were mixed with the help of stirring at room temperature. $Ti(OBu)_4$ was added dropwise to the above solution, and the resulting solution was transferred to a Teflon®-lined autoclave. The autoclave was heated at different temperatures and different time durations for NPs synthesis. Figure 1.8 shows the TEM images of the TiO_2 NPs synthesized at different temperatures of 150 and 180°C, respectively. It was also found that the absence of NH_4HCO_3 results in nanorod growth.

Navale et al. reported the simple hydrothermal method for the TiO_2 NPs synthesis [52]. Titanium glycolate was used as a precursor; this precursor avoids the annealing treatment at elevated temperatures.

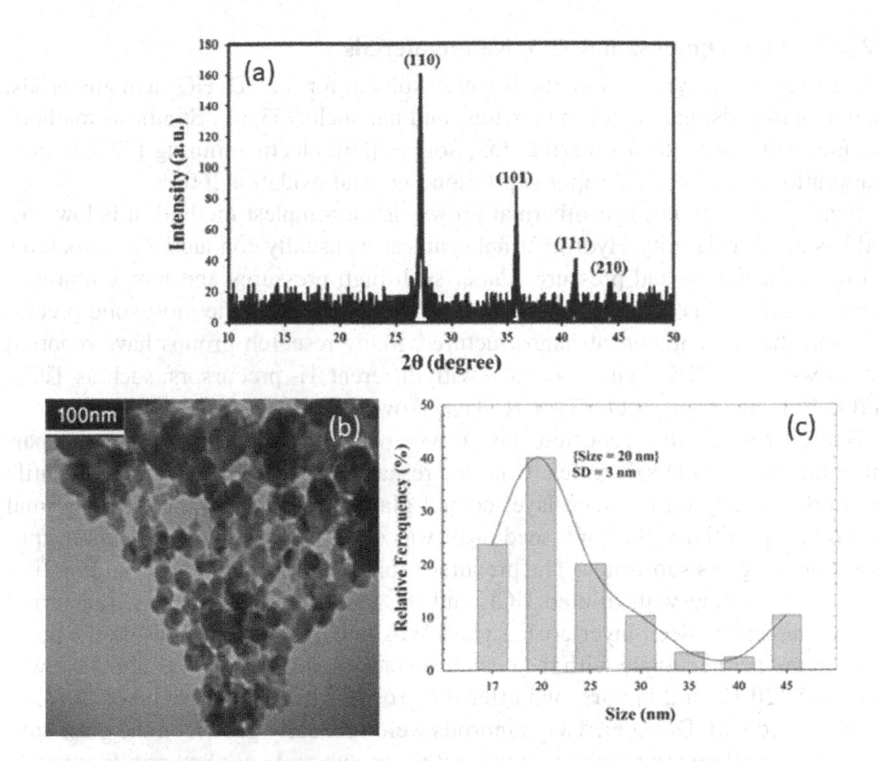

FIGURE 1.7 (a) XRD diffraction pattern, (b) TEM image of the green synthesized TiO_2, and (c) NPs and particle size distribution. (With kind permission from Springer Science+Business Media: *Ceram. Int.*, Synthesis and characterization of titanium dioxide nanoparticles using euphorbia heteradena jaub root extract and evaluation of their stability, 41, 2015, 14435–14439, Nasrollahzadeh, M., and Sajadi, S. M.)

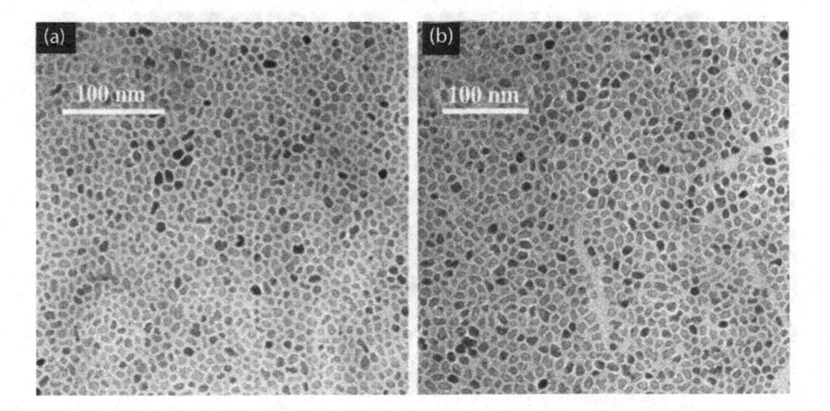

FIGURE 1.8 TEM images of the TiO_2 nanoparticles prepared by hydrothermal growth at different temperatures of (a) 150°C and (b) 180°C. (With kind permission from Springer Science+Business Media: *Chem.: A Eur. J.*, Near monodisperse TiO_2 nanoparticles and nanorods, 12, 2006, 2383–2391, Li, X.-L. et al.)

1.2.2.2 One-Dimensional TiO_2 Nanomaterials

A large variety of synthesis methods were explored for the 1D TiO_2 nanomaterials, such as nanorods, nanotubes, nanowires, and nanobelts [53,54]. Synthesis methods include hydrothermal/solvothermal [55] sol-gel [56], electrospinning [57], thermal evaporation [58], chemical vapor deposition [59], and oxidation [60].

Among all methods, hydrothermal growth is the simplest method; it is low cost and has good scalability. Hydrothermal synthesis is usually conducted in autoclaves at high temperature and pressure. Under such high pressures and temperatures, a supercritical fluid is formed, and this allows the solubility of the most solid precursors and the precipitation of nanostructures. Many research groups have reported the growth of 1D TiO_2 nanostructures with different Ti- precursors, such as $TiCl_4$, Ti $(Oc_3H_7)_4$, and $C_6H_{22}N_2O_8$ Ti for 1D TiO_2 growth.

Boercker et al. first reported the growth of TiO_2 nanorods on the transparent conducting oxide substrates [61]. Our research group synthesized TiO_2 rutile nanorods directly on the seed-layer-coated glass substrates by the hydrothermal method [62]. Initially, the TiO_2 seed layer was deposited by electron beam evaporation on the glass substrates. The precursor solution was prepared by adding titanium isopropoxide with diluted HCl, and the resulting solution was transferred into an autoclave. Seed-layer-coated glass substrates were placed inside the autoclave in an inclined angle with the coated surface facing down. The autoclave was heated at 120°C for 20 hours, and after the growth substrates were taken out they were cleaned with DI water. TiO_2 nanorods were vertically grown on the glass substrate with random alignment with respect to the substrate as shown in Figure 1.9.

Further, Boercker et al. reported the growth of TiO_2 nanowires directly on titanium foil [61]. In this report, TiO_2 nanowires were synthesized in a three-step procedure. In the first step, sodium titanate nanowires were grown on the Ti foil by the alkali hydrothermal growth process. In the second step, obtained nanowires were converted

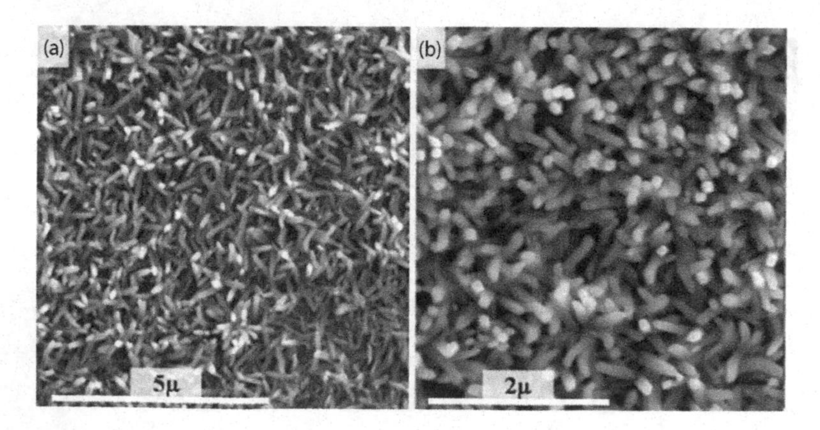

FIGURE 1.9 SEM images of TiO_2 nanorod growth by hydrothermal growth at different magnifications as mentioned in (a–b). (Reprinted from *Appl. Surf. Sci.*, 258, Tamilselvan, V. et al., Growth of rutile TiO_2 nanorods on TiO_2 seed layer deposited by electron beam evaporation, 4283–4287, Copyright 2012, with permission from Elsevier.)

into protonated bititanate nanowires through an ion exchange reaction, and finally, bititanate nanowires were converted into TiO_2 nanowires by a calcination step.

Kasuga et al. first reported the growth of TiO_2 nanotubes by the hydrothermal method [63]. In this report, TiO_2 nanotubes were formed when sol-gel-derived TiO_2-based powders were treated chemically with 5–10 M NaOH. The formation mechanism of TiO_2 nanotubes was reported by Yao et al. [64] TEM investigations revealed that raw material TiO_2 powder was exfoliated into layered crystalline sheets (shown in Figure 1.10). These sheets are unstable because the surface-to-volume ratio has more and more dangling bands at a growth temperature of more than 90°C. Rolling of the sheets can reduce the number of surface dangling bonds that result in lowering the system energy.

Direct oxidation is an easy method for the production of 1D TiO_2 nanomaterial. TiO_2 nanorods and nanowires were directly formed by oxidizing the Ti foils in the literature [60,65]. TiO_2 nanorods were obtained by direct oxidation of Ti foil with H_2O_2 solution at 353 K for 72 hours. Similarly, TiO_2 nanotubes were also prepared by anodic oxidation of Ti foil [66].

Wu et al. reported the growth of TiO_2 nanowires by the two-step thermal evaporation method [58]. TiO_2 seeds were formed on the substrate in the first step by reacting to the TiO_2 powder at 1050°C with the substrate. In the second step, TiO_2 powder was evaporated in a tube furnace, and evaporated atoms condensed on the substrate containing TiO_2 seeds and grown as nanowires by the VS mechanism. Pradhan et al. also reported the growth of TiO_2 nanorods using the MOCVD method [67].

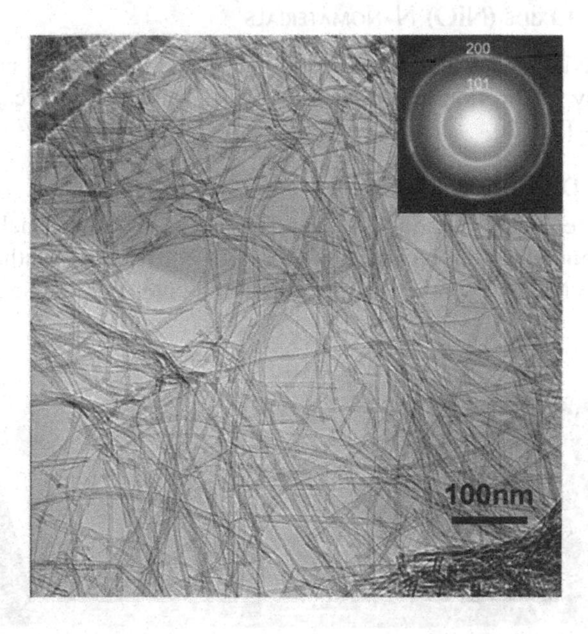

FIGURE 1.10 TEM images of as prepared TiO_2 nanotube growth by the hydrothermal method. The inset is the selected area electron diffraction pattern. (Reprinted with permission from Yao, B.D. et al., *Appl. Phys. Lett.*, 82, 281–283, 2003. Copyright© 2003 by the American Institute of Physics.)

1.2.2.3 Two-Dimensional TiO_2 Nanomaterials

Two-dimensional TiO_2 nanomaterials such as sheets and flowers are mostly synthesized by hydro/solvothermal methods [68–70]. Sun et al. reported a simple synthesis of TiO_2 nanosheets by scalable solvothermal synthesis [71]. In a typical synthesis, 6 mL of Hydrofluoric acid (HF) was added to 50 mL of titanium butoxide dropwise under continuous stirring. The resulting solution was placed in an autoclave and heated to 190°C for 24 hours. The obtained white gel-like precipitate was washed several times with ethanol and dried at 60°C to get the TiO_2 nanosheets.

Reactivity of the (001) surface of anatase TiO_2 is much more than the stable (101) surface. More reactivity of the (001) surface provides active sites for various applications [70]. Therefore, research is more focused on the growth of (001) facet TiO_2 nanosheets.

Wen et al. first reported the anatase phase TiO_2 nanosheets with a large percentage of reactive facets (47%) by the hydrothermal method by HF as a capping agent [72]. The same research group further modified the synthesis procedure to obtain the 64% off (001) facet TiO_2 single crystal nanosheets. In this new synthesis, 2-proponal was used as a synergistic capping agent and reaction medium together with HF [70]. In a typical synthesis, 14.5 mL of TiF4 solution (2.76 mM), 13.38 mL of 2-propanal, and 0.5 mL of HF were added together. The resulting solution was placed in a Teflon autoclave and heated at 180°C for different time durations. Figure 1.11 shows the SEM images of the obtained TiO_2 sheets for 11 hours of growth.

1.2.3 Nickel Oxide (NiO) Nanomaterials

Nickel oxide (NiO) is an abundant and important TMO, and its nanostructures are found many applications, including catalysis, batteries, supercapacitors, and electrochromics [73–75].

1.2.3.1 Zero-Dimensional NiO Nanomaterials

NiO NPs are prepared by thermal decomposition [76], hydrothermal [77], sol-gel methods [78], and chemical precipitation [79]. Recently, green synthesis was also reported for the NiO NP synthesis using Ageratum conyzoides L. leaf extract and Ni $(NO_3)_2$ [80].

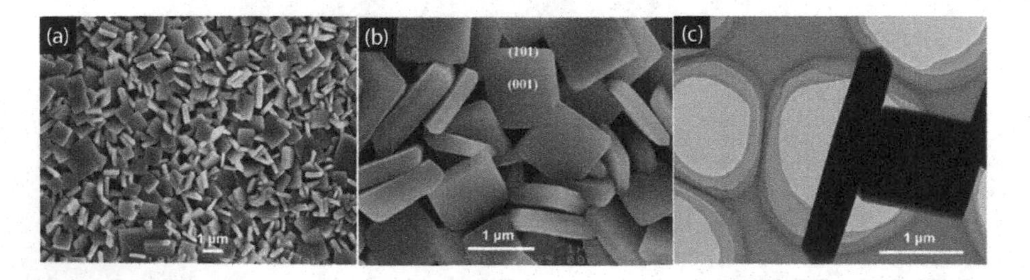

FIGURE 1.11 Anatase phase TiO_2 sheets by hydrothermal growth (a,b) SEM images and (c) TEM image. (Reprinted with Yang, H.G. et al., *J. Am. Chem. Soc.*, 131, 4078–4083, 2009. With permission. Copyright© 2009 American Chemical Society.)

The thermal decomposition method was reported an efficient and easy method for the production of NiO NPs. Different types of precursors such as $Ni(Phen)_2$, Ni $(octa)_2$, nickel-salicylate, $[Ni (en)_3](NO_3)_2$, Ni (II) complex, and Ni (salen) were used to obtain the NiO NPs by this method.

Farhadi et al. presented a simple thermal decomposition route for the synthesis uniform-size NiO NPs with weak agglomeration and ferromagnetic property [76]. To prepare NiO NPs, $[Ni (en)_3]$ $(NO_3)_2$ complex was prepared, and the complex was decomposed at various temperatures for 1 hour under ambient air conditions. It was observed that uniform NPs were obtained at a decomposition temperature of 400°C. NiO was formed via explosive decomposition of [Ni(en)3] $(NO_3)_2$ complex due to the redox reaction between reductants and oxidants in the process (shown in Figure 1.12).

Jafari et al. reported the synthesis of NiO NPs using sol-gel method [78]. In this report, $Ni(NO_3)_2$ $6H_2O$ and gelatin were dissolved in DI water separately and stirred for 30 minutes. Both the solutions were added and heated in a water bath at 80°C under continues stirring for 12 hours to obtain a green gel. This gel was transferred into the crucible and placed inside the tube furnace, which was heated to different temperatures under atmosphere conditions to get the NiO NPs. The NP sizes varied from 9 to 75 nm with different heating temperatures of 400°C–800°C in the furnace.

Fominykh et al. reported the synthesis ultra-small, crystalline, and dispersible NiO NPs for the first time by the hydrothermal method (shown in Figure 1.13) [77]. Nickel acetylacetonate was added to 14 mL of tetrabutanol at ambient conditions, which resulted in a light green suspension. The reaction mixture was transferred to Teflon autoclave, sealed tightly, and kept it in the oven for different reaction times at 200°C. The obtained powders further dried at 60°C in the oven. It was also found that the size

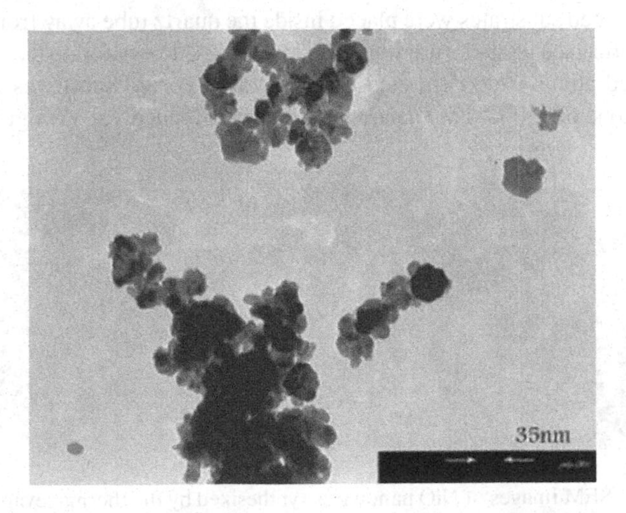

FIGURE 1.12 TEM images of the NiO NPs synthesized by the thermal decomposition method. (Reproduced from Farhadi, S. et al., *Polyhedron*, 30, 971–975, 2011. With permission. Copyright© 2011.)

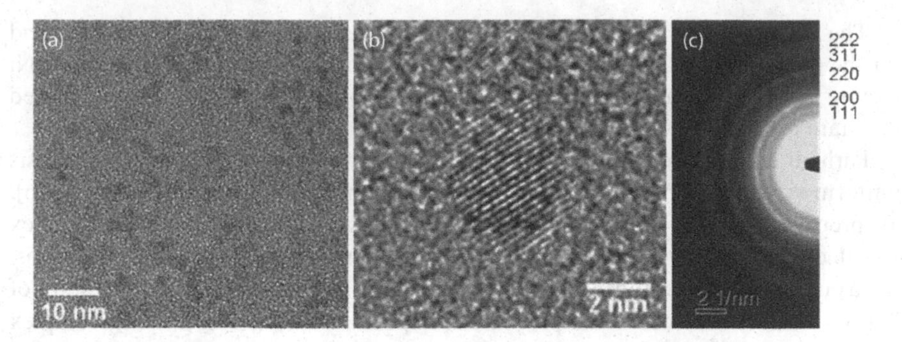

FIGURE 1.13 (a) TEM images of the ultra-small crystalline NiO NPs synthesized by the hydrothermal method, (b) HRTEM of the single nanoparticle, and (c) selected area electron diffraction of NPs. (Fominykh, K. et al.: Ultrasmall Dispersible Crystalline Nickel Oxide Nanoparticles as High-Performance Catalysts for Electrochemical Water Splitting. *Adv. Funct. Mater.* 2014. 24. 3123–3129. Copyright Wiley-VCH Verlag GmbH & Co. KGaA. Reproduced with permission.)

of the obtained NPs increased with the reaction time. Particle sizes varied from 2.5, 3.3, 3.8, and 4.8 nm for 16-, 17-, 24-, and 33-hour reaction times, respectively.

1.2.3.2 One-Dimensional NiO Nanomaterials

One-dimensional NiO nanomaterials such as nanowires and nanofibers were reported in the literature using different methods such as thermal evaporation [81], thermal oxidation [82], electrospinning [83], and sol-gel [84].

Kaur et al. reported the growth of NiO nanowires by catalyst-assisted VLS growth through the thermal evaporation method [81]. For the growth of NiO nanowires (shown in Figure 1.14), NiO powder was taken as source material in a horizontal tube furnace, and catalyst-coated substrates were placed inside the quartz tube away from the source material. The furnace temperature was raised to 1400°C to evaporate the NiO powder, and evaporated atoms were deposited on the catalyst-coated substrates in the lower temperature zone of 930°C. NiO nanowire growth proceeded via VLS growth.

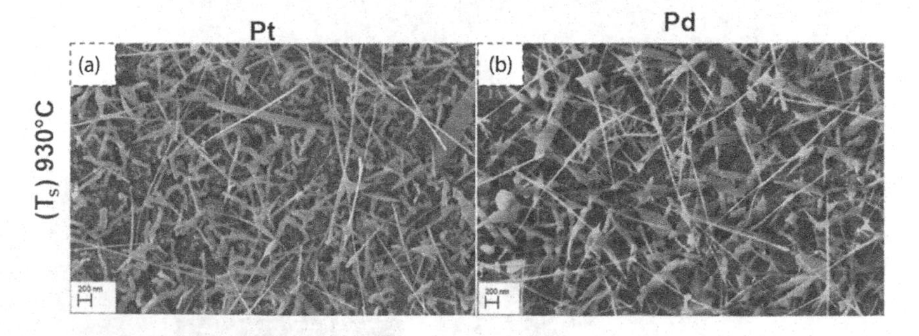

FIGURE 1.14 SEM images of NiO nanowires synthesized by the thermal evaporation using (a) Pt as a catalyst and (b) Pd as a catalyst. (Reprinted with permission from Kaur, N. et al., Nickel Oxide Nanowires: Vapor Liquid Solid Synthesis and Integration into a Gas Sensing Device, *Nanotechnology*, 27(20), 205701, 2016. Copyright 2016, Institute of Physics.)

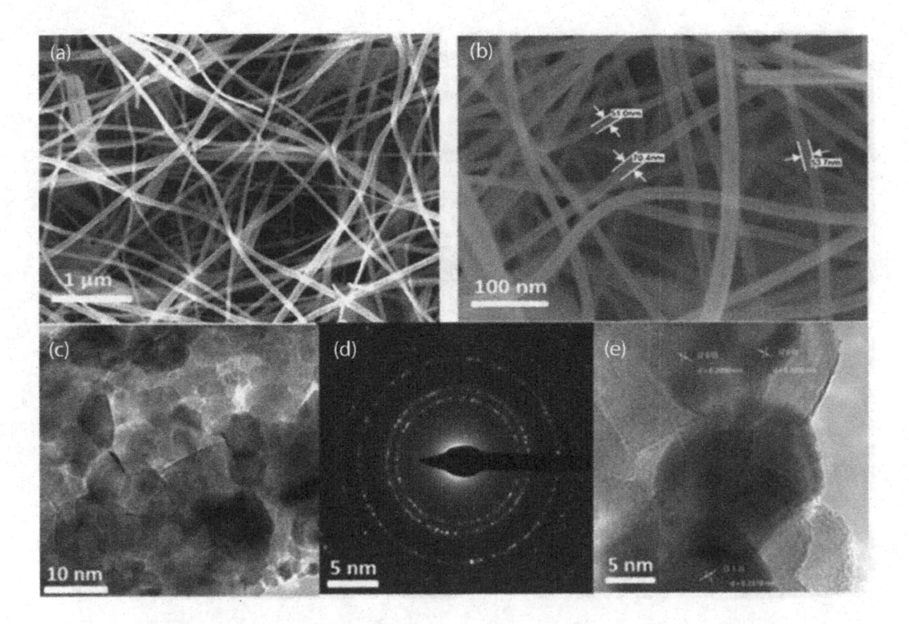

FIGURE 1.15 (a,b) the FESEM images of calcined NiO nanowires at different magnifications, (c) bright field image of a typical nanowire, (d) selected area electron diffraction pattern, and (e) high-resolution lattice image of a typical particle in the TEM sample. (Reprinted from *J. Alloys Compd.*, 610, Vidhyadharan, B. et al., High Performance Supercapacitor Electrodes from Electrospun Nickel Oxide Nanowires, 143–150, Copyright 2014, with permission from Elsevier.)

Vidhyadharan et al. reported NiO nanofibers by electrospinning method (shown in Figure 1.15) [83]. In the electrospinning method, a precursor solution is allowed to pass through a narrow syringe via a syringe pump, and a high potential difference will be applied between the needle and collector. The collector is usually the substrate on which nanofibers will deposit. For NiO nanofibers synthesis, nickel acetate tetrahydrate is dissolved in a Polyvinyl alcohol (PVA) solution and stirred for 20 hours for the homogenous solution. The prepared solution was electrospun onto the aluminum foil collector using the electrospinning machine. The potential difference of 24 kV was applied between the needle and collector, and the separation between them was 17 cm. Khalil et al. also reported the NiO nanofibers with the same precursors [85].

1.2.3.3 Two-Dimensional NiO Nanomaterials

Two-dimensional NiO nanosheets/nanoflakes were mainly synthesized by liquid-phase methods. NiO nanosheets were reported using hydrothermal [86], microwave-assisted hydrothermal [87], and solution-immersed hydrothermal methods [88].

Direct growth of NiO nanosheets on carbon cloth was reported by Long et al. by the hydrothermal method [89]. In this report, the appropriate amount of $NiCl_2$ $6H_2O$ and $CO(NH_2)_2$, NH_4F were dissolved in DI water and stirred to get the homogenous solution. This solution was transferred into the autoclave and a carbon cloth placed vertically in the solution. Autoclave with the solution and carbon cloth was

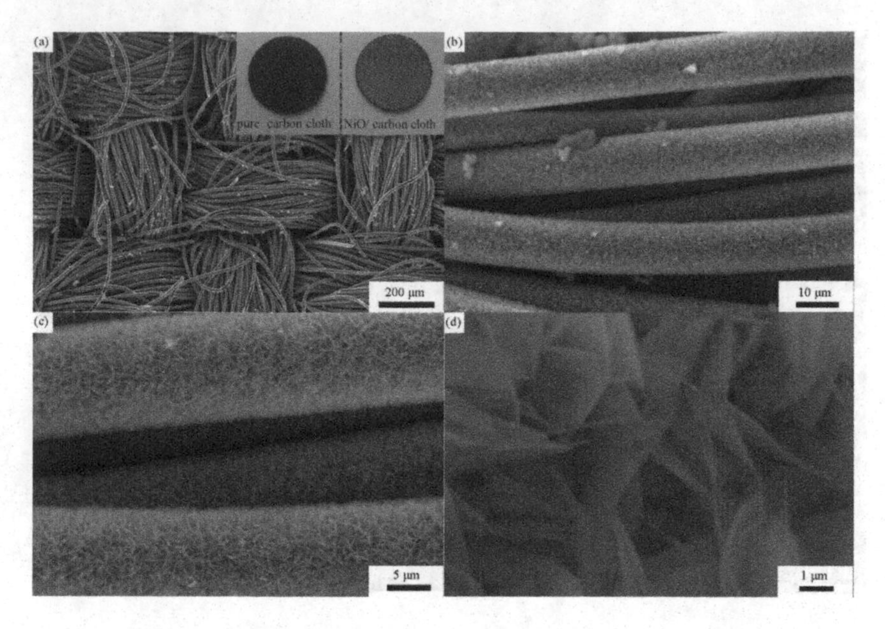

FIGURE 1.16 (a–d) NiO nanosheet growth on carbon cloth by hydrothermal method. (Reprinted by permission from Macmillan Publishers Ltd. *Sci. Rep.*, Long et al., 2015, copyright 2015.)

put in an oven at 120°C for 7 hours. The NiO-coated carbon cloth was taken from the autoclave and rinsed several times and dried in an oven. Finally, the samples were annealed at 350°C for 2 hours. The carbon cloth color changed from a deep black color to green after NiO sheets growth on the carbon cloth. Uniform nanosheet growth was observed throughout the carbon cloth (shown in Figure 1.16). Similarly, Ahamad et al. synthesized NiO nanosheets using Ni $(NO_3)_2$ $6H_2O$ and NaOH as precursor solutions [90]. NiO nanosheets were also synthesized by microwave-assisted hydrothermal growth reported by Zhu et al. [91].

Abdullah et al. reported the direct growth of NiO nanosheets on an ITO substrate by the solution immersion method without any seed layer [88]. For the synthesis of NiO nanosheets, Ni $(NO_3)_2$ 6 H_2O and HMTA were dissolved in DI water and stirred to obtain a homogenous solution. This solution was transferred into the Schott bottle, and the ITO substrate placed at the bottom and sealed tightly. The bottle was immersed in a water bath at 95°C for different growth durations for the growth of NiO nanosheets on the ITO. Finally, samples were annealed for the improvement of crystallinity of the films (shown in Figure 1.17).

1.2.4 Tungsten Oxide (WO_x) Nanomaterials

Tungsten oxides (WO_x $2 < x < 3$) nanostructures have attracted a lot of attention due to their outstanding applications, such as electrochromic devices, super capacitors, photocatalysis, gas sensors, and anode materials for Li-ion batteries [92–94].

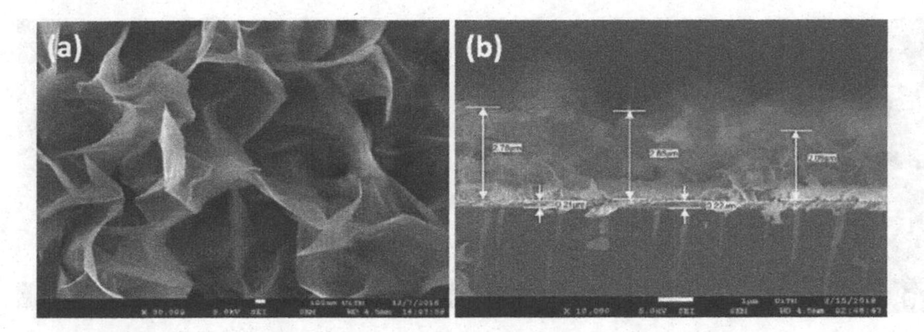

FIGURE 1.17 NiO nanosheet growth directly on ITO substrate by hydrothermal method for the growth duration of 3 hours viz. top view (a) and cross sectional view (b). (Reprinted from *Mater. Lett.*, 236, Abdullah, M.A.R. et al., Direct and seedless growth of nickel oxide nanosheet architectures on ITO using a novel solution immersion method, 460–464, Copyright 2019, with permission from Elsevier.)

1.2.4.1 Zero-Dimensional WO_x Nanomaterials

WO_3 crystals are generally formed by corner and edge sharing of WO_6 octahedra. The following phases are obtained by corner sharing: E-Wo_3, delta-WO_3, Y-WO_3, beta-WO_3, alpha-WO_3, and cubic WO_3. Zero-dimensional WO_x nanomaterials such as quantum dots and NPs were mostly synthesized by liquid-phase methods. Liquid-phase methods include hydrothermal/solvothermal methods and precipitation methods. The precipitation or acid precipitation method was the most reported method for WO_x NP synthesis. Rezaee et al. reported the synthesis of tungsten oxide NPs in the size range of 70–100 nm at the synthesis temperature of 500°C by the precipitation method (shown in Figure 1.18) [95]. In this method, the aqueous solution of sodium tungstate $Na_2WO_42H_2O$ was acidified with HCl till a yellow precipitate was obtained. The formed yellow precipitate was washed several times with DI water and heated at 90°C to remove solvents and form the gel. The resulting gel was calcinated at different temperatures of 500 and 600°C. Acid precipitation method for the synthesis of WO_X NPs was reported many reports in the literature [96] Ahmadi et al. reported the growth of tungsten oxide NPs using the hydrothermal method [97]. Movlaee et al. reported the growth of tungsten oxide NPs using the microwave-assisted hydrothermal method [98].

For the synthesis of WO_x quantum dots, template-directed synthesis is the mostly used method. Watanabl et al. synthesized WO_3 quantum dots using newly synthesized subnanoporous silica as a template. Band-gap engineering of WO_3 QDS was achieved to 3.7 eV from a 2.6 eV bulk energy gap [99].

Y-WO_3 QDs with size 1.6 nm were produced by a two-step procedure reported by Cong et al. In the first step, tungsten chloride reacted with phenol to produce a tungsten aryloxide precursor. In the second step, octylamine reacted with a tungsten aryloxide precursor under N_2 flow at 180°C in order to produce the well-dispersed quantum dots (shown in Figure 1.19) [100].

1.2.4.2 One-Dimensional WO_x Nanomaterials

One-dimensional WO_X nanomaterials are mostly produced by vapor phase and solution phase methods and other methods such as thermal oxidation and aerosol-assisted CVD.

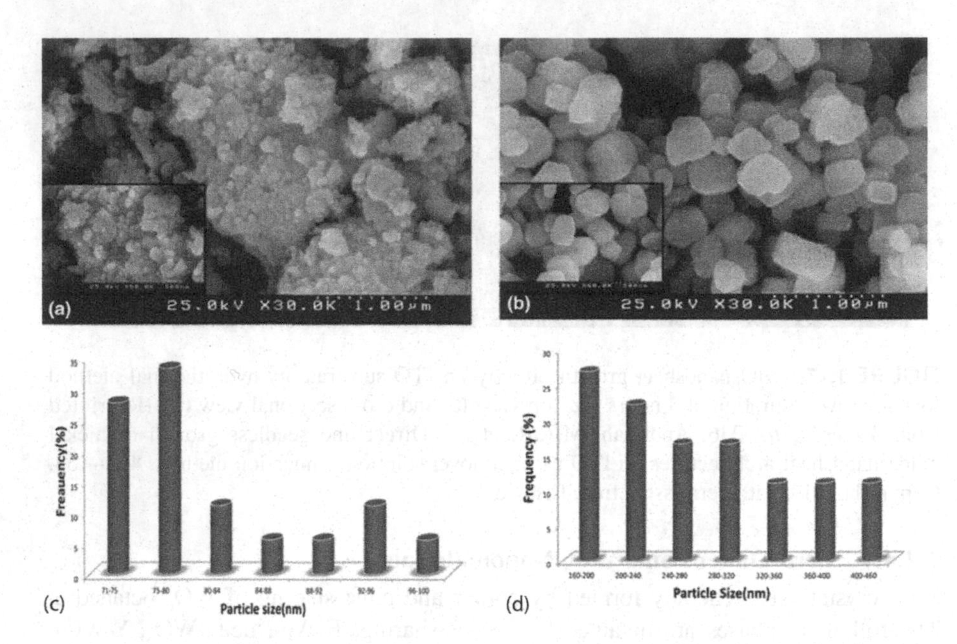

FIGURE 1.18 SEM images (a,b) and particle size distribution (c,d) of WO_X NP synthesized by the precipitation method and calcinated at 500°C and 600°C, respectively. (With kind permission from Springer Science+Business Media: *J. Sol-Gel Sci. Technol.*, Precipitation synthesis of tungsten oxide nanoparticles: X-Ray line broadening analysis and photocatalytic efficiency study, 80, 2016, 109–118, Rezaee, O. et al.)

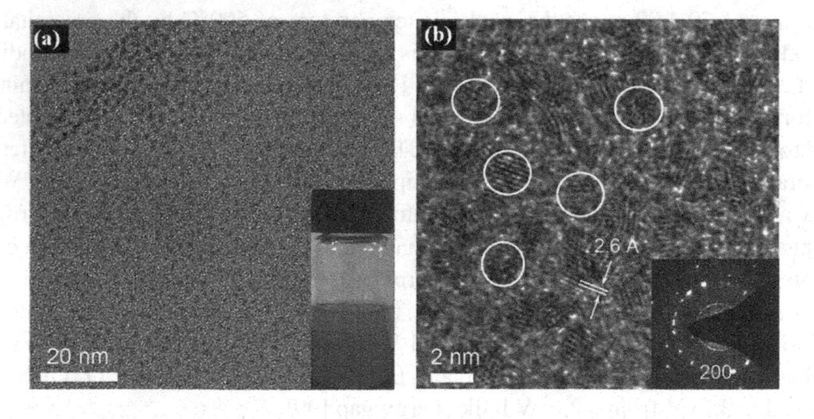

FIGURE 1.19 Low- and high-magnification TEM images of the WO_X NP synthesis by the template method. (From Cong, S. et al., Single-Crystalline Tungsten Oxide Quantum Dots for Fast Pseudocapacitor and Electrochromic Applications. *Adv. Mater.* 2014. 26. 4260–4267. Copyright Wiley-VCH Verlag GmbH & Co. KGaA. Reproduced with permission.)

Baek et al. reported the growth of WO_3 nanowires on a tungsten substrate by thermal evaporation of WO_3 powder at higher temperatures in a tube furnace without any catalyst [101]. Initially, WO_3 powder was put in the alumina boat, and the tungsten substrate was placed over the boat. The entire boat with WO_3 powder and substrate was placed insider a horizontal tube furnace and maintained at a temperature of 900°C–1000°C. After the sufficient time of heating, the furnace was cooled down to room temperature and a white wool-like product was on the substrate. SEM confirms the wool-like product consisting WO_3 nanowires (shown in Figure 1.20). Park et al. reported the growth of WO_3 nanoneedles by the thermal evaporation method [102]. In this report, WO_3:graphite powder (1:1) was placed inside the tube furnace, and five Si substrates were placed at different temperature zones of the furnace (450°C–930°C). The furnace temperature was increased to 1050°C and maintained at that temperature under constant Ar and O_2 flow for 1 hour for the growth of WO_x nanowires on the Si substrates. Needle-like nanowires were obtained on the Si substrates placed in Zone II where the growth temperature ~650°C.

Gu et al. reported a simple thermal oxidation method for the production of WO_3 nanowires (shown in Figure 1.21) [103]. In this report, tungsten oxide nanowires were directly grown on the tungsten metal tip. Initially, tungsten tips were prepared by electrochemical etching of tungsten wires. Tungsten tips were immediately transferred to the tube furnace, and furnace temperature increased to 700°C under Ar flow for 10 minutes growth. Lengths of the nanowires were small in the case of the 10-minutes growth duration, and it was increased to 1–2 μm with the increase of the growth duration. Many researchers reported the thermal oxidation method for the production of pure and doped WO_x nanowires [104,105].

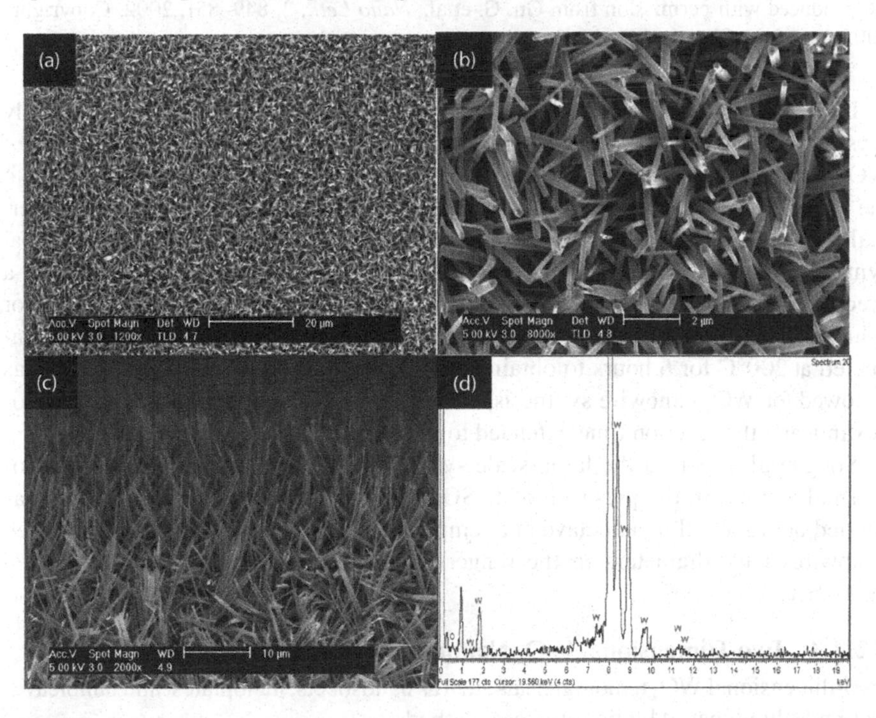

FIGURE 1.20 SEM images at different magnifications (a–c) of the WO_3 nanowire synthesized by thermal evaporation method on a tungsten substrate and a representative energy dispersive spectroscopy (EDS) analysis (d) of WO_3 nanowires. (From Baek, Y., and Yong, K., *J. Phys. Chem. C*, 111, 1213– 1218, 2007. Reproduced by permission of The Royal Society of Chemistry.)

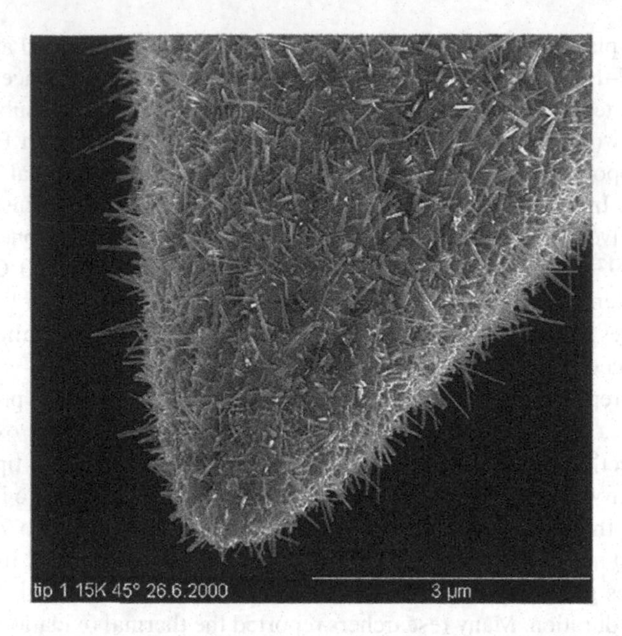

FIGURE 1.21 WO_3 nanowire synthesis on the tungsten tip by the thermal oxidation method. (Reproduced with permission from Gu, G. et al., *Nano Lett.*, 2, 849–851, 2002. Copyright© 2002 American Chemical Society.)

Lee et al. first reported the solution-based preparation of soluble and highly crystalline Tungsten Oxide nanorods [106]. The one-dimensional nanostructure of WO_X with a variety of morphologies were synthesized in liquid phase compared to the vapor-phase methods (shown in Figure 1.22). Qin et al. reported the solvothermal method for the production of WO_x nanowires and nanorods [107]. In a typical synthesis, tungsten hexachloride (WCl_6) was dissolved in ethanol, which acts as a precursor for WO_x nanowire/rod growth. Cyclohexanol was added to the precursor solution and transferred to the Teflon autoclave and a solvothermal reaction conducted at 200°C for 6 hours to obtain the WO_X nanorods. A similar procedure was followed for WO_x nanowire synthesis, except 1-proponal was used instead of cyclohexanol and the reaction time extended to 9 hours.

Song et al. reported the large-scale synthesis of WO_3 nanowires by the hydrothermal method in the presence of K_2SO_4 [108]. The hydrothermal synthesis was carried out in a Teflon autoclave at a temperature of 180°C for 12 hours. Obtained nanowires have diameters in the range of 10–20 nm and lengths up to several micrometers.

1.2.4.3 Two-Dimensional WO_x Nanomaterials

Two-dimensional WO_x nanomaterials such as nanosheets, nanoplates, and nanoleaves were mostly prepared by liquid-phase methods.

Wei et al. reported the synthesis of WO_3 nanosheets by a simple hydrothermal method with the help of structure-directing agents [109]. In the synthesis procedure, ethanol and distilled water was added together to form the solvent, and 0.2 g

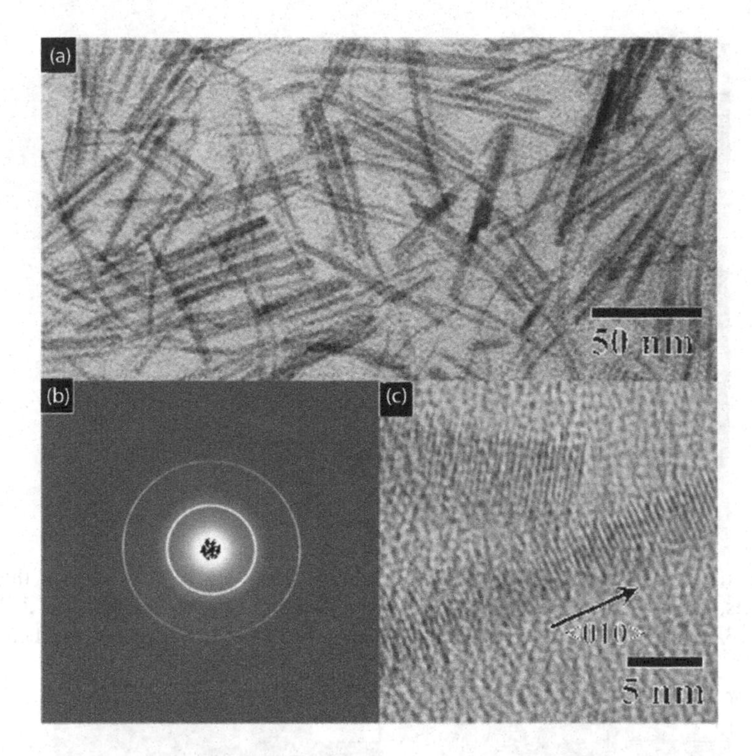

FIGURE 1.22 TEM (a), Selected Area Electron Diffraction (SAED) pattern (b) and high-resolution TEM (HRTEM) image (c) of single WO_3 nanorod synthesized by solution growth method. (Reprinted with permission from Lee, K. et al., *J. Am. Chem. Soc.*, 125, 3408–3409, 2003. Copyright 2003 American Chemical Society.)

of P123 was added to form the transparent solution. WCl_4 was added to the above transparent solution till a yellowish solution obtained. The color of the solution changed gradually, and finally a cloudy suspension was obtained. This suspension was put in an autoclave and heated at 110°C for a different time duration in order to obtain WO_3 nanosheets. Nanosheet agglomeration was observed at higher growth temperatures and high-growth durations (shown in Figure 1.23).

Ming et al. and Yan et al. reported the two-step procedure for the synthesis of WO_3-defective nanosheets using tungstic acid [110,111]. In the first step, exfoliation of layered tungstic acid to tungsten oxide nanosheets was performed, and in the second step, oxygen vacancies were introduced in the nanosheets.

Chen et al. reported the synthesis of WO_x nanoplates [112]. In this report, WO_x nanoplates were synthesized via a three-step procedure. Step 1 involves the synthesis of tungsten-based inorganic hybrid nanobelts via reactions of microscale $H_2W_2O_7 \times H_2O$ powder and n-octylamine. In Step 2, oxidizing acids were used to remove the selective organic species from the above nanobelts to synthesize H_2WO_4 nanoplates. In Step 3, there was a conversion of H_2WO_4 nanoplates into WO_3 nanoplates via a shape- and size-controlled condensation process (shown in Figure 1.24).

FIGURE 1.23 (a,b) SEM images of the WO_3 nanosheets, (c) TEM images of the WO_3 nanosheets, and (d) synthesis by the hydrothermal method. (From Wei, J. et al., *J. Mater. Chem. C*, 3, 7597–7603, 2015. Reproduced by permission of The Royal Society of Chemistry.)

FIGURE 1.24 (a,b) TEM images of the WO_3 nanoplates, (c) SAED pattern along the [001] zone axis, and (d) HRTEM image. (Reproduced Chen, D. et al., *Small*, 4, 1813–1822, 2008. With permission. Copyright© 2008.)

1.2.5 Vanadium Pentoxide (V_2O_5) Nanomaterials

Vanadium pentoxide (V_2O_5) nanomaterials find many applications, such as electrodes for Li-ion batteries, electrochromic devices, photocatalysis, and sensors [113–116]. V_2O_5 nanomaterials can be synthesized by a variety of methods that include, hydrothermal/solvothermal, thermal evaporation, sol-gel, template-based methods, and electrochemical deposition. Among all these methods, hydrothermal synthesis is considered the easiest and most effective way.

1.2.5.1 Zero-Dimensional V_2O_5 Nanomaterials

NPs of V_2O_5were synthesized by mostly liquid-based methods [117–120]. Menezes et al. synthesized V_2O_5 NPs from thermal treatment synthetic bariadite-like vanadium oxide [117]. Initially, bariadite-like vanadium oxide ($V_{10}O_{24}$ $9H_2O$) was synthesized the by sol-gel method. This dark green $V_{10}O_{24}$ $9H_2O$ was heated in the air at 400°C for 2 hours, resulting in an orange powder. An electron microscope study of the powder revealed that the formation of uniform NPs with size in the range of ~15 nm (shown in Figure 1.25).

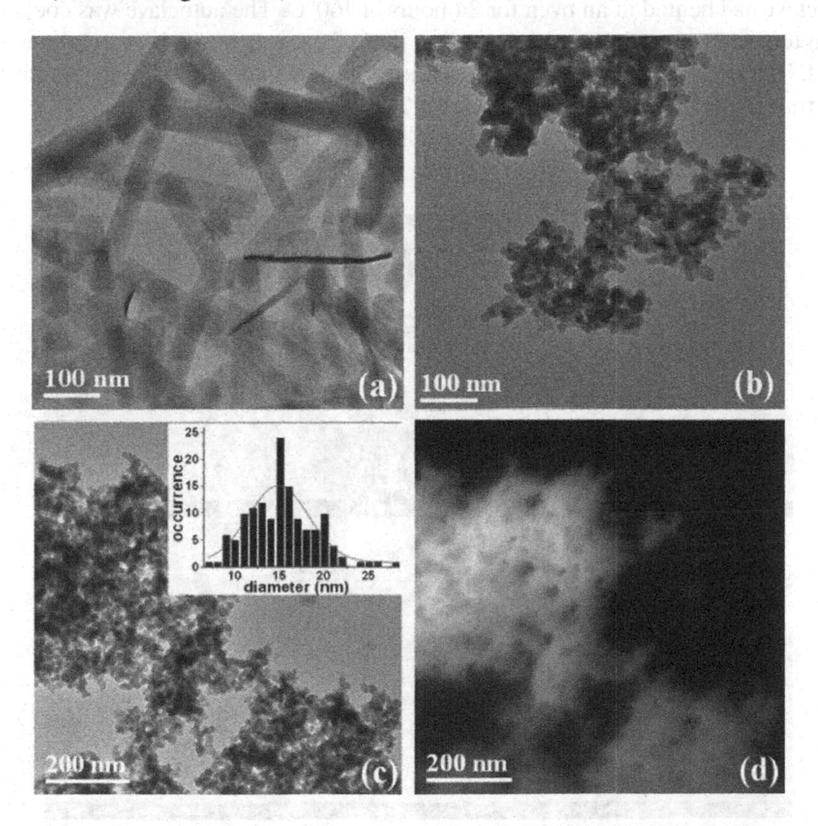

FIGURE 1.25 (a) TEM images of the V_2O_5 bariandite-like $V_{10}O_{24}9H_2O$ NPs, (b,c) bright field image, and (d) dark field image of the $V_{10}O_{24}9H_2O$ after the thermal treatment at 400°C. Inset shows the diameter histogram of V_2O_5 NP. (Reproduced from Menezes, W.G. et al., *J. Colloid Interface Sci.*, 337, 586–593, 2009. With permission. Copyright© 2019.)

Alghool et al. introduced the eco-friendly synthesis of V_2O_5 NPs using cotton fibers as an effective catalytic degradation material [119]. In this method, NH_4VO_3 was first dissolved in DI water, and HCl was added to the solution. Dried and loose cotton fibers were infiltrated with the prepared solution and after dried at 80°C for 24 hours. Dried fibers were put in a ceramic crucible and sintered at 500°C to remove the cotton template and allow the V_2O_5 particle formation.

1.2.5.2 One-Dimensional V_2O_5 Nanomaterials

One-dimensional V_2O_5 nanomaterials were synthesized in the form nanowires, nanotubes, nanofibers, nanobelts, and nanorods in the literature by hydrothermal, electrospinning, electrochemical, and thermal evaporation methods [121–124].

Zhou et al. synthesized ultra-long orthorhombic V_2O_5 nanowires by the simple hydrothermal method without the use of any template [121]. In a typical procedure, $VOSO_4\ xH_2O$ (10 mmol) and $KBRO_3$ (5 mmol) were dissolved in distilled water and stirred to obtain good homogeneity. Nitric acid was added dropwise to the above solution till the pH of the solution became 1–2. The resulting solution was put in an autoclave and heated in an oven for 24 hours at 160°C. The autoclave was cooled to room temperature, and the resulting yellow precipitate was washed several times and dried. The length of the nanowires extended to a few hundred micrometers for the first time in this report (shown in Figure 1.26).

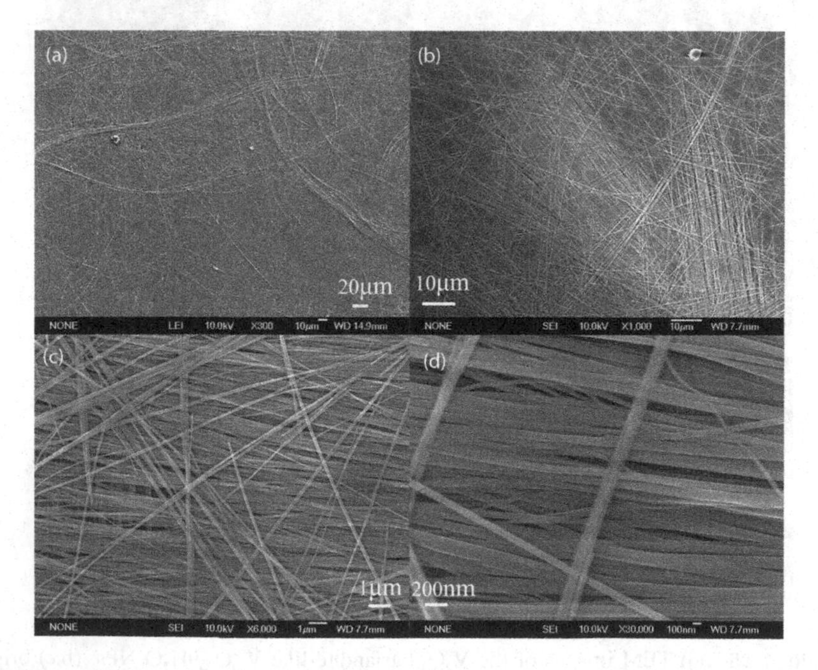

FIGURE 1.26 SEM images at different magnifications (a–d) of ultra-long orthorhombic V_2O_5 nanowires grown by simple hydrothermal method. (Reproduced from Zhou, F. et al., *Cryst. Growth Des.*, 8, 723–727, 2008. With permission. Copyright© 2008.)

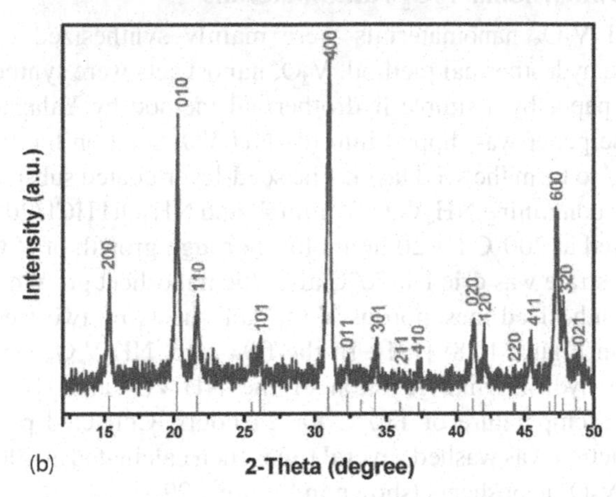

FIGURE 1.27 (a) SEM image of the centimeter-long V_2O_5 nanowires and (b) XRD pattern of V_0O_5 nanowires. Inset of (a) shows the photograph of centimeter-long nanowires. (From Zhai, T. et al., Centimeter-Long V_2O_5 Nanowires: From Synthesis to Field-Emission, Electrochemical, Electrical Transport, and Photoconductive Properties. *Adv. Mater.*, 2010. 22. 2547–2552. Copyright Wiley-VCH Verlag GmbH & Co. KGaA. Reproduced with permission.)

Zhai et al. synthesized centimeter-long V_2O_5 nanowires for the first time by simple hydrothermal synthesis (shown in Figure 1.27) [125]. The aspect ratio of length to the diameter of the nanowires was in the range of 10^5 to 10^6. The synthesis involves the hydrothermal reaction of V_2O_5 and H_2O_2 solutions at 205°C for 4 days. Further V_2O_5 nanorods and nanobelts were reported by the hydrothermal method in the literature [126,127].

Cheah et al. prepared V_2O_5 nanofibers by the electrospinning method [122]. A precursor solution for the nanofibers was prepared to mix the polyvinylpyrrolidone, vanadyl acetylacetonate, and acetic acid in ethanol and stirred for 12 hours. The resulting solution was electrospun at DC voltage of 10 kV, the needle tip to the collector a distance of 10 cm, and at a flow rate of 2 mL/h. The spun nanofibers had a smooth surfaces for each fiber, and the annealed fibers had a porous structure. Annealing of the as-prepared fibers led to crystallization of V_2O_5 nanocrystallites and removal of the polymer.

The electrodeposition method was also used for the synthesis of V_2O_5 nanorods and nanotubes with the help of templates [128,129]. A radiation track-etched polycarbonate membrane was used for the production of nanorods and nanotubes.

The thermal evaporation method for the synthesis of V_2O_5 nanowires was reported by Velazquez et al. [124]. V_2O_5 nanowires were synthesized by evaporating V_2O_5 powder in a tube furnace at 900°C and transporting the evaporated material onto the catalyst-coated substrates for the growth of the nanowires in the pure argon environment. In another method, VO_2 powder was evaporated in the presence of O_2 and Ar gas in a tube furnace, and the evaporated material reacted with oxygen and deposited on the catalyst-coated substrates and grew as nanowires.

1.2.5.3 Two-Dimensional V_2O_5 Nanomaterials

Two-dimensional V_2O_5 nanomaterials were mainly synthesized in the form of nanosheets by the hydrothermal method. V_2O_5 nanosheets were synthesized directly on the cellulose paper by a simple hydrothermal method by Yalagala et al. [130]. Initially, cellulose paper was dipped into the NH_4VO_3 solution for 1 hour followed by drying at 70°C to form the seed layer. The seed-layer-coated substrate was dipped into the autoclave containing NH_4VO_3 (30 mmol) and NH_2OH HCl (20 mmol) growth solution and heated at 200°C for 20 hours for thorough growth of V_2O_5 nanosheets. The cellulose substrate was dried at 70°C after the nanosheet growth.

Song et al. synthesized mesoporous V_2O_5 nanosheets by two step liquid phase method (shown in Figure 1.28) [131]. In the first step, $NH_4V_4O_{10}$ nanosheets were produced by the hydrothermal reaction of the NH_3VO_3 and $H_2C_2O_4$ 2 H_2O in an autoclave at a temperature of 180°C for 24 hours. Collected powder from the hydrothermal reaction was washed several times then calcinated at 400°C to produce the mesoporous V_2O_5 nanosheets (shown in Figure 1.29).

Rui et al. prepared ultrathin V_2O_5 nanosheets by a simple and scalable liquid exfoliation method (shown in Figure 1.30) [132]. In this method, bulk V_2O_5 was swelled by the intercalation of formamide molecules into the interlayer space. This intercalation process weakens the interlayer attraction. Ultrasonication of the solution results in the exfoliation of the V_2O_5 layers. A similar kind of synthesis of V_2O_5 nanosheets was also reported by Nagaraju et al. [133]

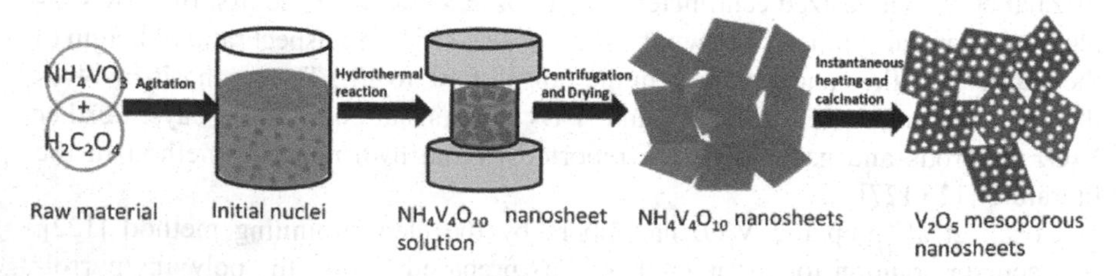

FIGURE 1.28 Schematic of synthesis procedure of mesoporous V_2O_5 nanosheets by the two-step liquid-phase method. (Reproduced from Song, H. et al., *J. Power Sources*, 294, 1–7, 2015. With permission. Copyright© 2015.)

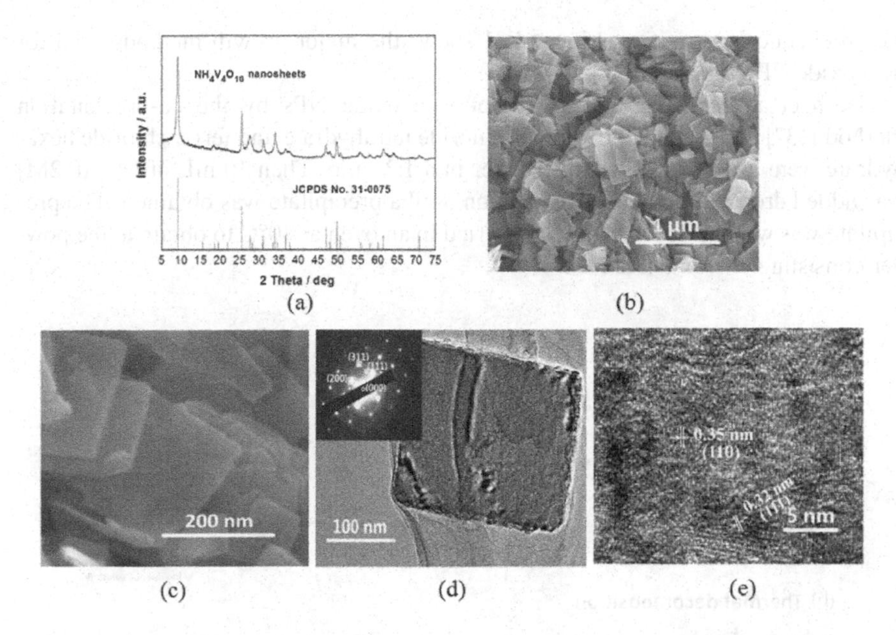

FIGURE 1.29 (a) XRD pattern of V_2O_5 nanosheets, (b,c) SEM image of the mesoporous V_2O_5 nanosheets, and (d,e) TEM images of V_2O_5 nanosheets. (Reproduced with permission from Song, H. et al., *J. Power Sources*, 294, 1–7, 2015.)

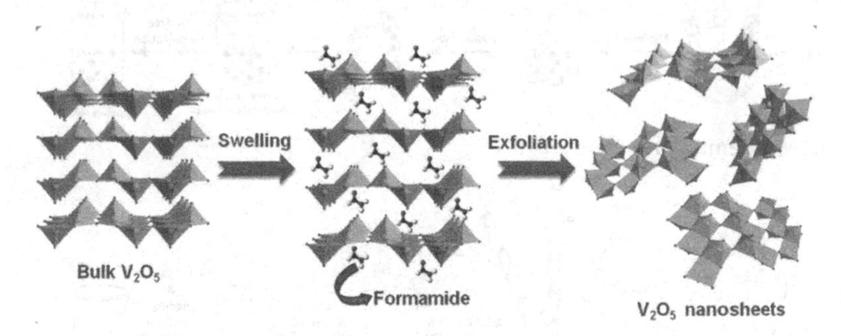

FIGURE 1.30 Schematic of the liquid exfoliation method for V_2O_5 nanosheet synthesis. (Reproduced from Rui, X. et al., *Nanoscale*, 5, 556–560, 2013. With permission. Copyright© 2013.)

1.2.6 Iron Oxide Fe_2O_3 Nanomaterials

1.2.6.1 Zero-Dimensional Fe_2O_3 Nanomaterials

Iron oxide NPs belong to the ferrimagnetic class of magnetic materials and are used for a wide variety of biomedical, energy storage, and environmental applications [134]. Superparamagnetic iron oxide NPs can be employed for magnetic resonance imaging, biosensing, tissue repair, targeted delivery of drugs, proteins, antibodies, and hyperthermia [135,136].

Iron oxide NPs were mainly prepared by mainly co-precipitation, sol-gel, thermal decomposition, microemulsion, hydrothermal, sonochemical, green synthesis, and electrochemical deposition. Figure 1.31 shows the major growth methods used for iron oxide NP synthesis [135].

Saqib et al. reported the growth of iron oxide NPs by the co-precipitation method [137]. In this method, ferrous chloride tetrahydrate and ferric chloride hexahydrate were mixed in a 100 mL beaker in a 1:2 ratio. Then 10 mL of HCl (0.2M) was added dropwise to the above solution until a precipitate was obtained. The precipitate was washed several times and dried in an oven at 40°C to obtain a fine powder consisting of iron oxide NPs.

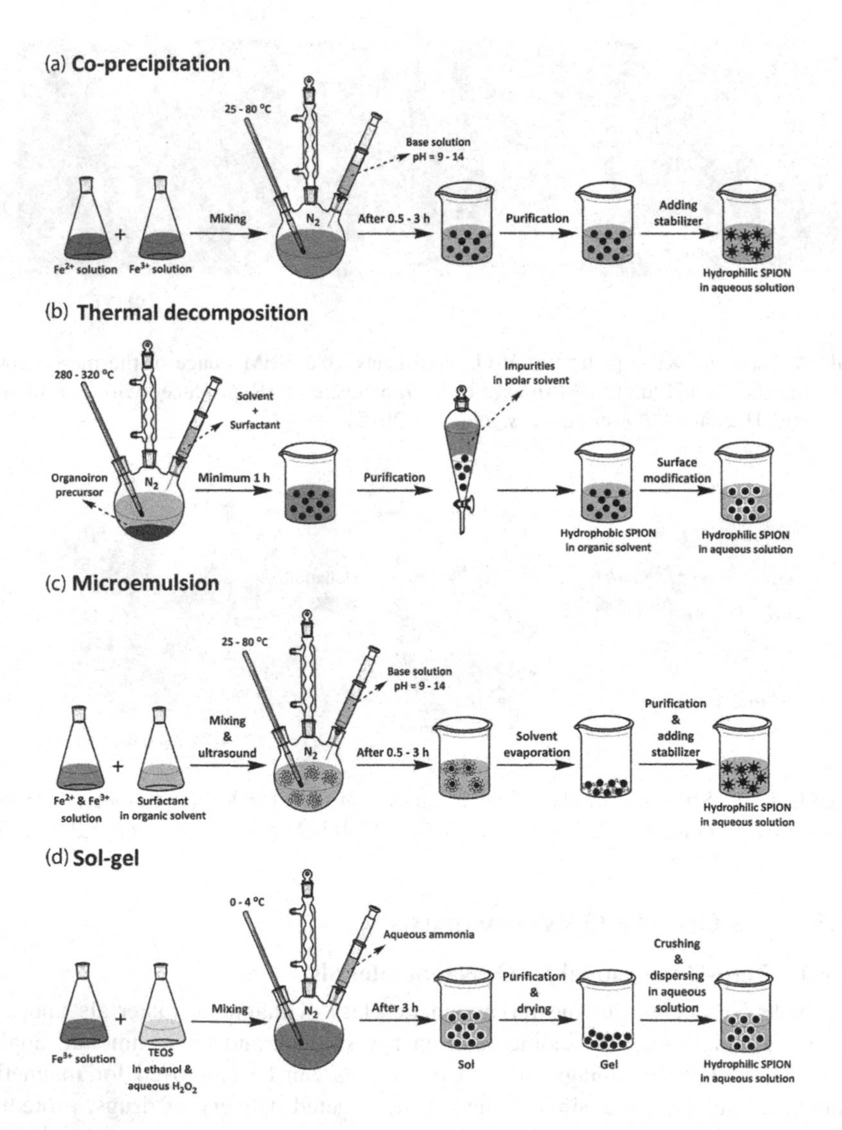

FIGURE 1.31 (a–d) Different synthesis schemes/methods for Fe_2O_3 NPs. (Reproduced from Dadfar, S.M. et al., *Adv. Drug Deliv. Rev.*, 138, 302–325, 2019. With permission. Copyright© 2019.)

Hufschmid et al. reported the phase pure and monodisperse iron oxide NPs by thermal decomposition [138]. In this method, thermal decomposition of organic iron precursors, such as iron(III) oleate, iron pentacarbonyl, and iron oxyhydroxide, in high-boiling-point organic solvents in the presence of stabilizing surfactants. This method has some disadvantages, such as not being environmentally friendly and requiring additional surface modification to get the water-dispersible NPs.

Gao et al. established a novel hydrothermal route by the dephosphorylation approach to synthesize ultra-small superparamagnetic γ-Fe_2O_3 NPs [139]. NPs obtained by this method exhibit high thermal stability and water dispersibility. Similarly, Shen et al. synthesized highly water-dispersible iron oxide NPs with tailored sizes (1.9 nm, 3.1 nm, or 4.2 nm) by a one-pot method [140].

Microemulsion systems are thermodynamically stable isotropic dispersions of two immiscible liquids. Microemulsions can essentially be subdivided into two categories: oil-in-water and water-in-oil. Several kinds of amphiphilic surfactants, such as dioctyl sodium sulfosuccinate, cetyltrimethylammonium bromide (CTAB), and sodium dodecylsulfate, are used for the formation of micellar microemulsion systems [141]. The main advantage of these methods is that the NP size can be controlled by varying the size of the micelles.

The sol-gel technique is also commonly employed for the production of silica-coated iron oxide NPs. Hydrolysis and condensation of tetraethylorthosilicate in ethanol and 30% aqueous H_2O_2 with Fe (III) solutions results in the formation of colloidal sols [142]. The sol is then gelled by chemical reaction or solvent removal to obtain a 3D iron oxide network. The formed gel requires an additional crushing step after drying and solvent removal to get the iron oxide NPs.

1.2.6.2 One-Dimensional Fe_2O_3 Nanomaterials

One-dimensional Fe_2O_3 nanomaterials, such as nanowires, nanobelts, nanoneedles, and nanotubes, were synthesized in the literature. Prepared Fe_2O_3 nanowires had shown excellent gas-sensing properties and superior storage in the Li-ion batteries. The most used synthesis methods for the production of Fe_2O_3 nanowires are thermal oxidation [143], hydrothermal method [144], electrochemical anodization [145], and electrospinning [146].

The thermal oxidation method is the simple method to produce the Fe_2O_3 nanowires and nanobelts [143,147]. Fe_2O_3 nanowires were obtained by oxidizing the pure iron under the reaction atmosphere, temperature, and required time. Fu et al. reported the growth of Fe_2O_3 nanowires by the reacting surface of the iron substrate with oxygen at a growth temperature of 550°C–650°C (shown in Figure 1.32) [147].

Yuan et al. reported the growth of nanobelts and nanowires under controlled oxidation of pure iron foils [148]. Nanowire growth was observed at 200 torr oxygen gas pressure, and nanotube growth was observed 0.1 torr oxygen gas pressure.

The resistive heating method is the simple, inexpensive method for the synthesis of iron oxide nanowires under ambient conditions without any extra equipment other than a small DC power supply. Nasibulin et al. reported the synthesis of Fe_2O_3 nanowires by simple resistive heating of an iron wire [149]. In this method, growth of iron oxide nanowires were observed as fast as a few seconds. In a typical synthesis, 2.5 A (2V–7V) current was allowed to pass through the iron wire, resulting in heating of

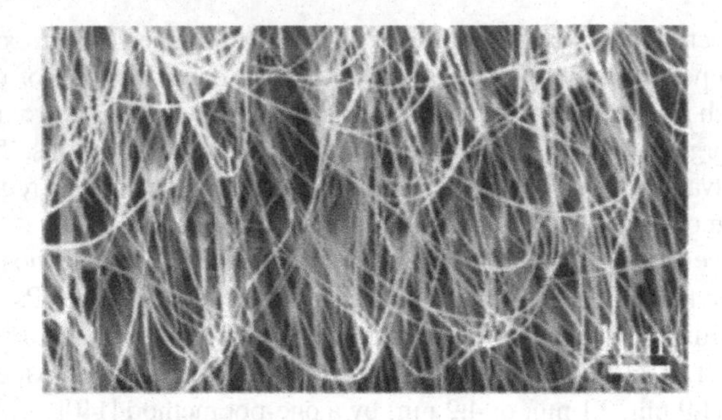

FIGURE 1.32 Fe_2O_3 nanowires prepared by the thermal oxidation method. (Reproduced Fu, Y. et al., *Chem. Phys. Lett.*, 350, 491–494, 2001. With permission. Copyright 2001.)

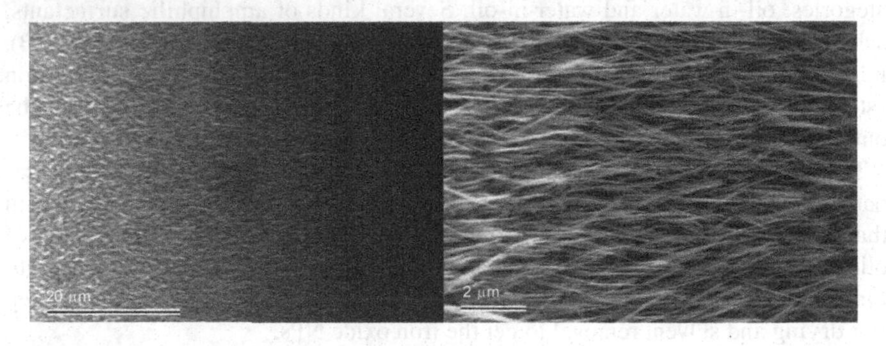

FIGURE 1.33 Fe_2O_3 nanowires formed on the iron wire by resistive heating. (Reproduced from Nasibulin, A.G. et al., *Nano Res.*, 2, 373–379, 2009. With permission. Copyright 2009.)

the wire and heated wire react with ambient oxygen, producing iron oxide nanowires (shown in Figure 1.33). Nanowire formation on the iron wire can be easily identified by observing the color of the wire.

Wang et al. reported a two-step method for the synthesis of all TMO nanowires [144]. In this method, the solvothermal method and organic chains as a mediating template were combined for preparing the Fe_2O_3 nanowires in large quantities. In a typical synthesis, an $FeCl_3$ aqueous solution with isopropanol and nitrotriacetic acid was added and stirred for uniform mixing. The resulting solution was treated at 180°C for 24 hours in a Teflon-lined autoclave. White floccules were collected from the autoclave, further sintered at 350°C for 1 hour to form Fe_2O_3 nanowires.

Rangaraju et al. reported a room-temperature synthesis of Fe_2O_3 nanotubes on a pure iron substrate by an electrochemical anodization method [145]. Zhao et al. reported a novel method for the synthesis of Fe_2O_3 nanowires and nanobelts by a simple iron and water reaction at a temperature of 350°C–450°C [150].

1.2.6.3 Two-Dimensional Fe_2O_3 Nanomaterials

Two-dimensional Fe_2O_3 nanomaterials include mainly nanosheets and nanoflowers. The preparation methods of 2D Fe_2O_3 include microwave synthesis [151], electrospinning [152], and electrodeposition [153].

Zhao et al. reported the electrodeposition of Fe_2O_3 nanosheets on carbon fabric [153]. The Fe_2O_3 nanosheets were electrodeposited on Carbon Fabric (CF) through a two-electrode system by using a piece of as-pretreated CF as an anode and Pt-coated titanium mesh as a cathode. The electrolyte solution was prepared using ammonium sulfate hexahydrate, aqueous ammonia, and reduced iron powder. The electrodepositing process was performed under an anode current density of 12.5 mA cm^{-2} for various reaction times of 1, 5, 10, 15 minutes. Finally, the samples were heat-treated at 350°C for 60 minutes in a high-purity N_2 atmosphere, converting FeOOH into Fe_2O_3

Yang et al. synthesized mesoporous Fe_2O_3 nanosheets by microwave-assisted synthesis followed by annealing (shown in Figure 1.34) [151]. In a typical synthesis, $FeCl_3{\cdot}6H_2O$ and urea were dissolved in ethylene glycol, and then isopropyl alcohol was added, followed by vigorous stirring. The final solution was heated in the microwave oven under the power of 700 W at 180°C for 20 minutes. The final product after the microwave was washed several times and calcinated at 500°C for 2 hours.

Liu et al. used the co-precipitation method for the synthesis of Fe_2O_3 nanosheets with the assistance of Polyethylene Glycol (PEG) through the self-assembly behavior of PEG-coated iron oxide NPs [154]. Zhang et al. produced 2D porous Fe_2O_3 nanosheets using the electrospinning method [152]. Initially, polyvinylpyrrolidone was dissolved in a mixed solution of N, N-dimethyl formamide and ethanol. Iron nitrate nonahydrate ($Fe(NO_3)_3{\cdot}9H_2O$) was then added into the solution. The final solution was loaded into the electrospinning unit, and a DC voltage of 18 KV at 15 cm distance was kept between the needle tip and collector for the formation of Fe_2O_3 nanosheets.

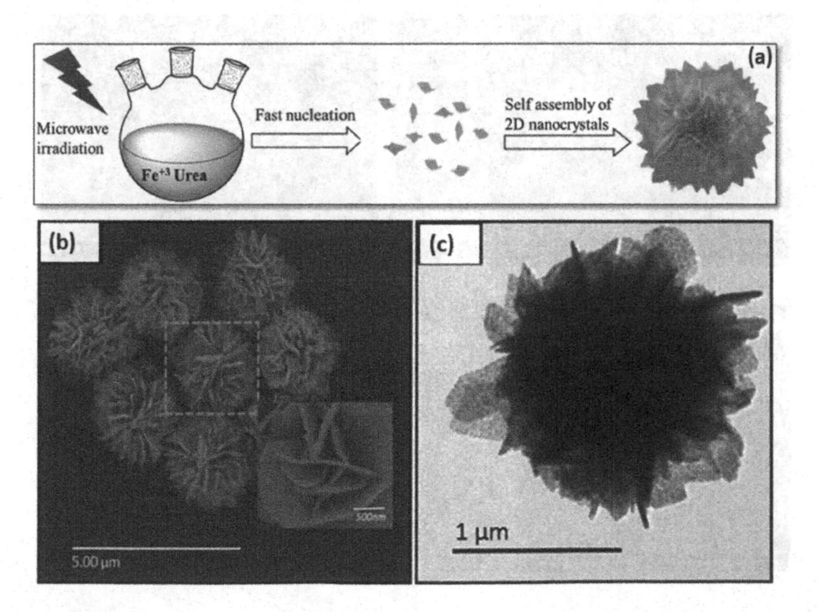

FIGURE 1.34 (a–c) Scheme of the synthesis and flower-like Fe_2O_3 mesoporous structures. (Reproduced from Yang, L. et al., *J. Energy Storage*, 23, 363–370, 2019. With permission. Copyright 2019.)

1.2.7 Tin Oxide SnO_2 Nanomaterials

SnO_2 is an n-type direct wide band-gap semiconductor material that is transparent in the visible region. SnO_2 nanomaterials are useful in optoelectronic devices, transparent conducting electrodes, Li-ion batteries, antireflection coatings, and gas sensors [155–157].

1.2.7.1 Zero-Dimensional SnO_2 Nanomaterials

Zero-dimensional SnO_2 nanomaterials were mostly prepared by solution-based methods. Aziz et al. prepared tetragonal-phase SnO_2 NPs by the sol-gel method [158]. In a typical procedure, $SnCl_2$ $6H_2O$ was dissolved in pure ethanol under constant stirring, and 5 mL of acetyl acetone was added dropwise for the hydrolysis of SnO_2. The resulting solution was refluxed at 80°C for 5 hours to form the SnO_2 sol solution. Then 1 mL of PEG was added to the SnO_2 sol solution and aged for 72 hours at 30°C. Further, sol was dried and calcinated at temperatures of 450 and 600°C for obtaining SnO_2 NPs (Figure 1.35).

SnO_2 NPs also synthesized by the simple hydrothermal method reported by Chiu et al. [159]. The particle size obtained was 3–4 nm, and they showed a good response for ethanol sensing. The hydrothermal method is an easy way to produce SnO_2 NPs. For the preparation of SnO_2 NPs, $SnCl_4$ 5 H_2O was added to the mixed solution of 2-proponal with distilled water at a ratio of 4:1. NaOH was added to the above solution to adjust the pH, and the resulting solution was transferred to an autoclave and heated at 150°C for 24 hours for the production of NPs.

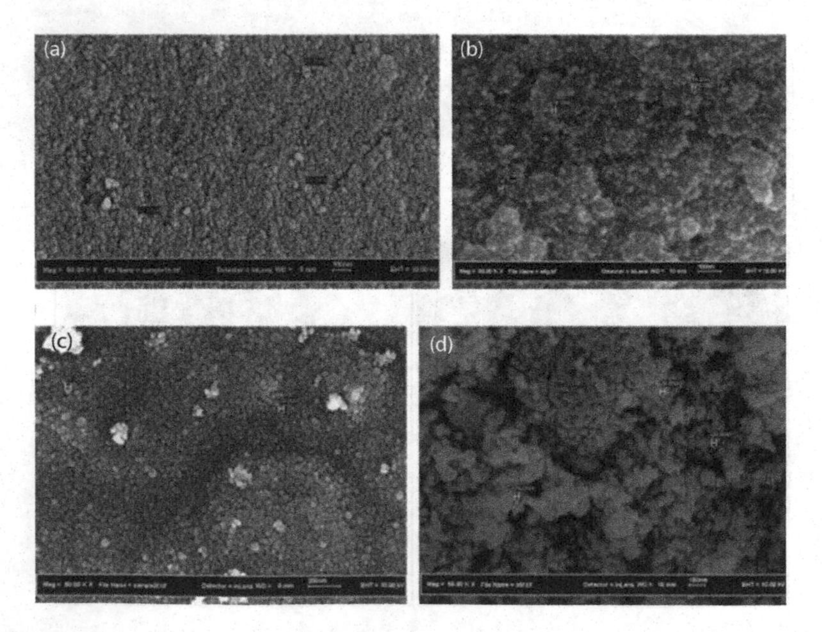

FIGURE 1.35 SnO_2 NPs formed at different calcination temperatures (a,b) 450°C and (c,d) 600°C. (Reprinted from *Mater. Lett.*, 91, Aziz, M. et al., Size-controlled synthesis of SnO_2 nanoparticles by Sol-gel method, 31–34, Copyright 2013, with permission from Elsevier.)

Bhattacharjee et al. reported the simple chemical precipitation method for the production of SnO_2 NPs [160]. The size of the NPs was in the range of approximately 45, 10, and 30 nm for different calcination temperatures of 200, 400, 600°C, respectively, using the amino acid glycine as a complexing agent and sodium dodecyl as a stabilizing agent.

1.2.7.2 One-Dimensional SnO_2 Nanomaterials

One-dimensional SnO_2 was synthesized in the form of nanowires, nanotubes, nanobelts, and nanoribbons [155]. One-dimensional SnO_2 nanomaterials were synthesized by thermal and electron beam evaporation [161,162], laser ablation [163], electrospinning [164], solvothermal method [165], and chemical vapor deposition [166]. The thermal evaporation method was a more popular method for the production of SnO_2 nanowires.

Our research group reported SnO_2 nanowires relatively at lower temperatures of ~450°C growth by simple techniques such as resistive and electron beam evaporation [161,162]. In both the methods, SnO_2 nanowire growth was achieved via gold catalyst-assisted VLS growth (shown in Figure 1.36). In these methods, cleaned Si substrates were coated with a thin layer of gold film, and the film was annealed at 450°C for the formation of catalyst particles on the substrates under high vacuum conditions.

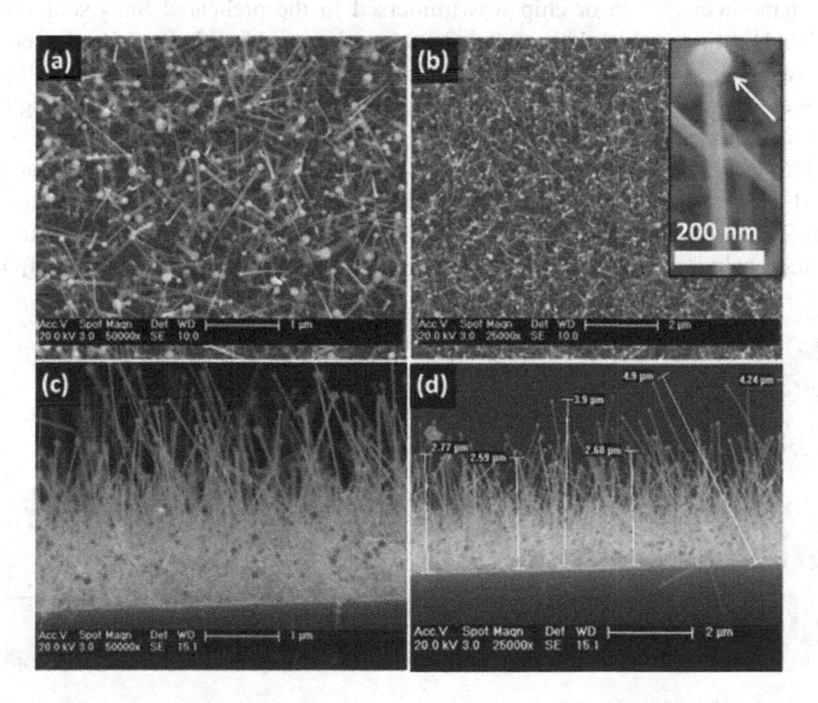

FIGURE 1.36 (a–d) SnO_2 nanowire growth by Au-assisted VLS growth by the electron beam evaporation method at 450°C. (Reprinted from *Mater. Res. Bull.*, 48, Kumar, R.R. et al., Low temperature growth of SnO_2 nanowires by electron beam evaporation and their application in UV light detection, 1545–1552, Copyright 2013, with permission from Elsevier.)

In electron beam and resistive evaporation methods, Sn material was evaporated with the assistance of e-beam heating or filament/boat heating, and generated vapor will condense catalyst formed substrate at the desired temperature. Sn material formed an alloy with gold (Au) and further reacted with the surrounding oxygen and precipitated SnO_2 from the Sn/Au alloy particle upon supersaturation.

Luo et al. reported that Sn catalyzed the growth of SnO_2 nanowires by simple evaporation of SnO_2 powder in a tube furnace [167]. In this report, SnO_2 powder was heated in a tube furnace at a temperature of 1050°C in the presence of Ar and H_2 gas, and evaporated material condensed on the substrates placed 17 cm distance from the source material. From TEM studies, it was concluded that growth was preceded via the VLS growth mechanism with Sn acting as a catalyst for the growth (shown in Figure 1.37). Dai et al. reported the growth of SnO_2 nanotubes, nanowires, and nanoribbons by the high-temperature thermal oxide synthesis method [168].

1.2.7.3 Two-Dimensional SnO_2 Nanomaterials

Two-dimensional SnO_2 nanomaterials were mostly prepared by solution-based methods, such as the hydrothermal method [169], sonochemical method [170], microwave [171], and chemical precipitation method [172].

Choi et al. reported the growth SnO_2 nanosheets directly on the sensor chip via the hydrothermal method (shown in Figure 1.38) [173]. For the synthesis of SnO_2 nanosheets, a sensor chip was immersed in the preheated SnF_2 solution at 90°C and heating of the solution carried out for 6 hours. After the growth, the sensor chip was washed and dried for further studies. Obtained SnO_2 nanosheets had shown an excellent gas-sensing response toward alkene gases, which have high Highest Occupied Molecular Orbital (HOMO) energy values.

Umar et al. reported the growth of 2D SnO_2 nanodisks by hydrothermal growth [174]. Stannous chloride dehydrate ($SnCl_2$ $2H_2O$) was used as a precursor. $SnCl_2$ 2 H_2O (0.05M) and HMTA (0.05M) were dissolved in DI water to make the precursor solution. A pH value of 8.7 was maintained by the addition of NH_4OH.

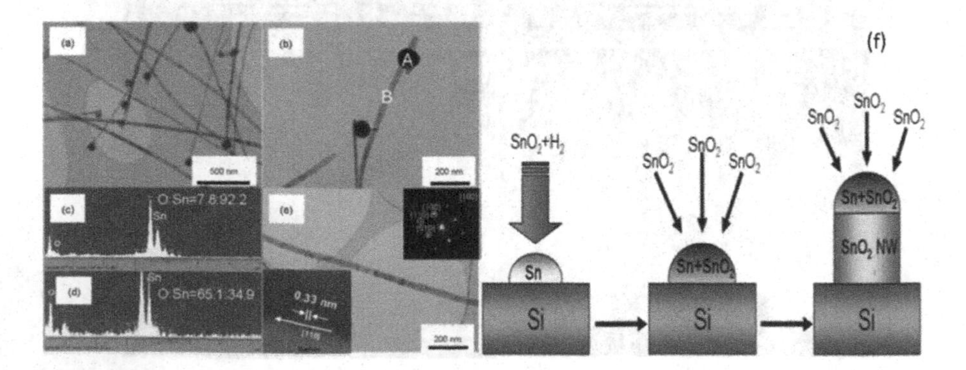

FIGURE 1.37 (a–e) TEM images of the SnO_2 nanowire growth by the self-catalytic VLS growth and (f) growth mechanism schemes. (Reprinted with permission from Luo, L. et al., Sn-Catalyzed Synthesis of SnO_2 Nanowires and Their Optoelectronic Characteristics. *Nanotechnology*, 22(48), 485701, 2011. Copyright 2011, Institute of Physics.)

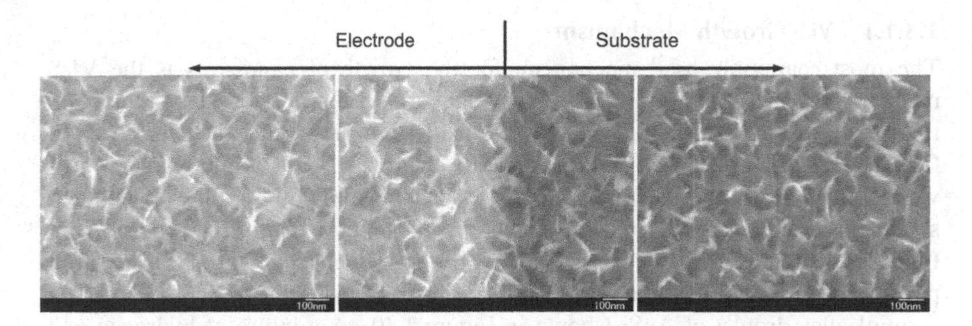

FIGURE 1.38 SnO_2 nanosheets formed on the sensor chip after 6 hours of growth. (Reprinted with permission from Choi, P.G. et al., *ACS Appl. Nano Mater.*, 2, 1820–1827, 2019.)

FIGURE 1.39 (a,b) 2D SnO_2 nanodisks synthesized by the solution growth method. (Reproduced from Umar, A. et al., *Int. J. Hydrogen Energy*, 2019. With permission. Copyright© 2019.)

The final solution was heated in an autoclave at 160°C for 12 hours, and the obtained powder over post growth was annealed at 450°C for 2 hours. An electron microscope study of the powder confirms the formation of SnO_2 nanodisks, such as shown in Figure 1.39. Obtained SnO_2 nanodisks have shown a better response toward H_2 compared to the other gases.

1.3 GROWTH MECHANISMS OF TMO NANOMATERIALS

1.3.1 One-Dimensional Nanomaterial Growth Mechanisms

Understanding the growth mechanism plays an important role in controlling the morphology of 1D nanomaterial. Many different growth mechanisms have been used to grow 1D nanomaterials. Each growth mechanism has its own defining characteristics and is often coupled with a specific deposition method. Among all growth mechanisms, the VLS mechanism and VS growth are mostly accepted mechanisms for 1D oxide nanostructures. A few of the widely accepted growth mechanisms will be discussed here in this section.

1.3.1.1 VLS Growth Mechanism

The most commonly used mechanism for the growth of nanowires is the VLS mechanism. This was first proposed by Wagner and Ellis in 1964 [175] to explain the growth of semiconducting Si whiskers using impurity metal particles (Au, Ni, Cu). The as-grown whiskers had diameters in the range of a few microns, and they were single crystalline structures without dislocations. Therefore, the conventional screw dislocation crystal growth mechanism cannot explain the 1D growth. Figure 1.40 schematically describes the VLS growth mechanism [175]. A small particle of Au is placed on the surface of a Si substrate and heated to 950°C, forming a small alloy droplet of AuSi (shown in Figure 1.40). A mixture of hydrogen and $SiCl_4$ is introduced in the reaction chamber. The Si atoms enter the liquid alloy particle and dissolve in it until the supersaturation stage is reached. After reaching the supersaturation, the alloy particle precipitates Si in order to re-establish the stable composition of Si and Au in the binary liquid alloy. By a continuation of this process, the alloy droplet becomes displaced from the substrate crystal and rides on top of the growing whiskers. The whisker grows in length by this mechanism until the Au is consumed or until the growth conditions are changed. According to this work, the growth mechanism is characterized by three important features: (1) Si whiskers do not contain an axial screw dislocation (which in itself is a growth mechanism); (2) an impurity is essential for whisker growth; and (3) a small globule is present at the tip of the whisker during the growth.

The process is named VLS because it describes a system involving a vapor-phase precursor ($SiCl_4$), a liquid alloying as the mediating material (AuSi), and a solid 1D whisker that is formed (Si). Wu et al. provided direct evidence for the VLS growth of nanowires by in-situ TEM growth of nanowires [176].

Recently, the VLS mechanism has been applied for the synthesis of TMO nanowires, such as SnO_2 [162], ZnO [177], Fe_2O_3 [178], NiO [179], WO_3 [180], and TiO_2 [181]. The VLS mechanism can be easily used with a variety of deposition techniques. The main advantages of the VLS mechanism are control of the diameter,

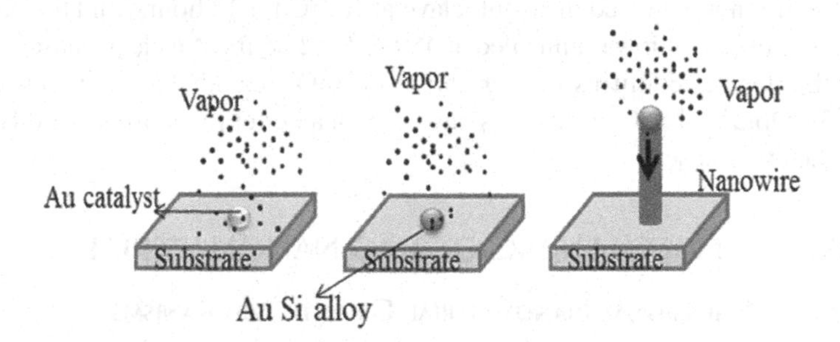

FIGURE 1.40 Schematic illustration of the growth of a silicon nanowire by the VLS growth mechanism. (Reproduced from Wagner, R.S. et al., *Appl. Phys. Lett.*, 4, 89–90, 1964. With permission. Copyright© 1964.)

length, and doping of the nanowires, and growth can be achieved at low growth temperatures. Nanowires can grow selectively over an area by patterning the catalyst particles there.

1.3.1.2 Self-Catalytic VLS Growth Mechanism

The self-catalytic VLS mechanism is a variation of the standard VLS growth mechanism. This mechanism works with a liquid catalyst that is formed in an in-situ mode. When nanowires are made up of more than one type of element, one of these elements must act as a catalyst. For example, it was found that GaN nanowires can be grown without a metal catalyst [182]. The decomposition of the GaN solid leads to the formation of Ga liquid NPs, and the formed Ga liquid NPs act as a catalyst for GaN nanowires. For metal oxides, nitrides, sulfides, etc., it is possible that the compound will decompose at an elevated temperature and then form a metal element. The formed metal element acts as a catalyst for the corresponding oxide, nitride, and sulfide nanowire growth. The major advantage of self-catalytic VLS growth is that nanowires cannot be contaminated by unwanted metals, unlike VLS growth methods. The self-catalytic VLS mechanism was observed for SnO_2 [167] and ZnO [183].

1.3.1.3 VS Growth Mechanism

The VS growth process is a catalyst-free process in which the atoms and molecules evaporated from the source condense on the substrates, which are kept in a different temperature zone and grow into nanostructures. The high substrate temperature and the anisotropic nature of the crystal help in the formation of nanostructures. The growth of nanostructures by the VS mechanism was mainly due to internal anisotropic surfaces, crystal defects, and self-catalytic growth. This mechanism has been used to explain the growth of metal oxide nanowires, nanobelts, nanotubes, and nanorods [184–188]. It is easily differentiated from VLS by the absence of metal particles at the tip of the nanowire.

1.3.2 Two-Dimensional Nanomaterial Growth Mechanisms

There are different mechanisms that could lead to 2D crystal growth, including oriented attachment [189], screw dislocation-driven growth [190], surfactant-assisted synthesis [191], and solvothermal synthesis [192]. These mechanisms are often operative simultaneously to produce the 2D nanomaterials. For example, adsorption of surfactants often plays a critical role in directing the dimension of oriented attachment and in modifying the growth rate at screw dislocation steps.

1.3.2.1 Screw Dislocation Growth Mechanism

Morin et al. elaborately explained the formation of 2D nanoplates by screw dislocation and provided a general mechanism for 2D nanomaterial synthesis (shown in Figure 1.41) [190]. The proposed growth mechanism by Morin et al. considered the relative step velocities of at the dislocation core versus the outer edges of the growth spiral under various supersaturation to explain the dislocation-driven morphologies,

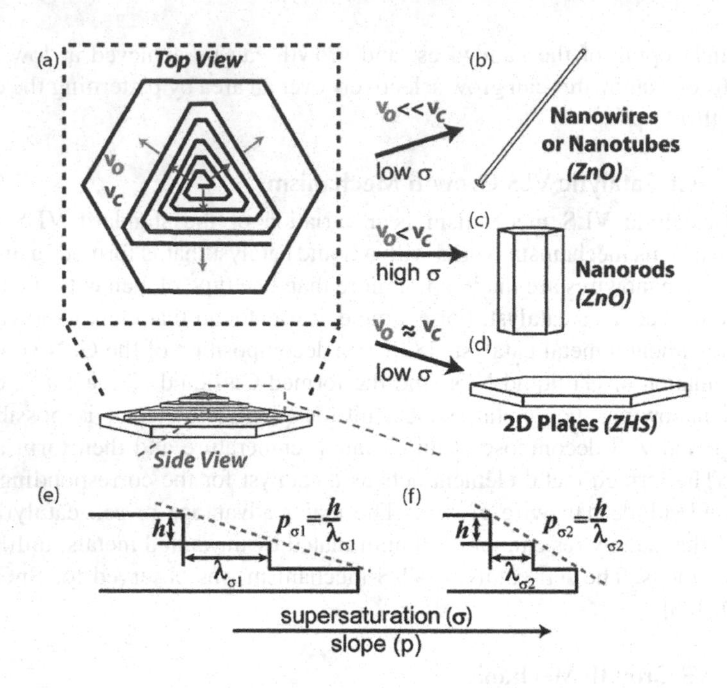

FIGURE 1.41 Dislocation-based growth mechanism of nanomaterials as mentioned in (a–f). (Reproduced from Morin, S. A. et al., *Nano Lett.*, 11, 4449–4455, 2011. With permission. Copyright© 2011.)

such as nanowires, nanotubes, nanorods, and nanoplates. Nanoplates will be formed if the velocity of the steps at the core (v_c) is nearly equal to those of the outer edges of the dislocation (v_o) as shown in Figure 1.41.

1.3.2.2 Surfactant-Assisted Growth Mechanism

Leng et al. reported the growth of TiO_2 nanosheets via the surfactant-assisted hydrothermal method (shown in Figure 1.42) [191]. Tetrabutylammonium hydroxide (TBAOH) was used as a surfactant to produce the TiO_2 nanosheets by the exfoliation method. When TiO_2 crystals are treated with NaOH solution at elevated temperatures it results in layered titanates. Layered titanates can be further exfoliated into a single layer of titanate nanosheets. These nanosheets tend to form nanotubes due to the dangling bonds, such as the negatively charged Ti-O and the positively charged Ti at the sides of the nanosheets. Conversion of nanosheets into nanotubes can be avoided by reducing the surface energy of the nanosheets. This can be done by adding suitable surfactant; TBAOH surfactant has four butyl groups. The TBA+ ion can attach to the negatively charge Ti-O bond on both sides of the nanosheets, thus making the nanosheets stable in the solution.

1.3.2.3 Oriented Attachment Growth Mechanism

Zhang et al. reported of growth of ZnO nanoplates and nanowires via oriented attachment [189]. In this report, the growth solution containing a zinc source was heated in a Teflon-lined autoclave at ~120°C, and reaction products were analyzed by microscope

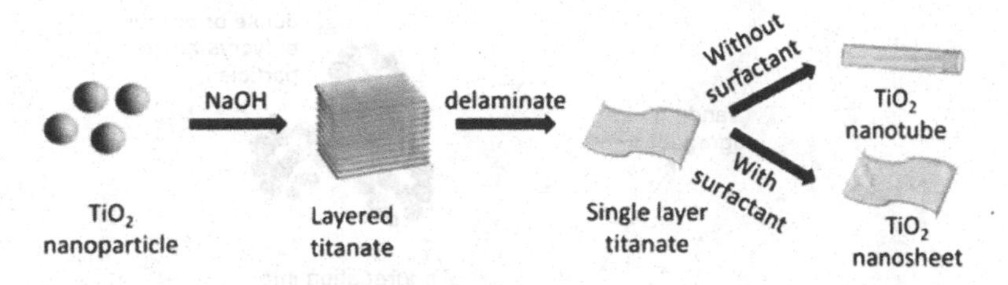

FIGURE 1.42 Surfactant-assisted synthesis of TiO_2 nanosheets. (Reproduced Leng, M. et al., *Nanoscale*, 6, 8531–8534, 2014. With permission. Copyright© 2014.)

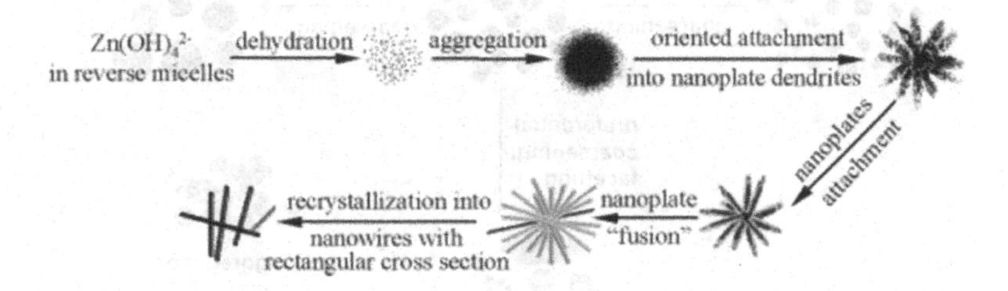

FIGURE 1.43 Growth mechanism of nanosheets and nanowires by oriented attachment. (Reproduced Zhang, D.-F. et al., *J. Phys. Chem. B*, 109, 8786–8790, 2005. With permission. Copyright© 2005.)

at different stages of the growth (shown in Figure 1.43). It was found that ZnO NP clusters formed initially and later converted into nanosheets by the oriented attachment of ZnO NPs. Increasing the reaction time leads to the formation of ZnO nanowires by oriented attachment. Due to the intrinsic anisotropic property of hexagonal ZnO, the aggregated NPs oriented and attach each other to decrease the energy of the system. Due to the Brownian motion of the particles in the hydrothermal system, attachment of NPs was not limited to the [0001] plane but was also favorably perpendicular to the *c*-axis. Thus, nanoplates growing along [0002] in length and [112′0] in width were obtained.

1.3.3 Zero-Dimensional Nanomaterial Growth Mechanisms

Zero-dimensional nanomaterial growth was achieved mostly by wet chemical synthesis methods in the literature. Wet chemical synthesis of NP formation mechanism depends on many factors, such as reaction temperature, supersaturation, time, surfactant, and the surface energy [193]. Different nucleation and growth mechanism can occur within the solution, such as the Lamer nucleation [194], Fineke-Watzky two step mechanism Ostwald ripening [195], and coalescence and oriented attachment growth [196].

1.3.3.1 Ostwald Ripening Growth Mechanism

Classical growth mechanism (Ostwald ripening) states that NPs originate through the formation of small crystalline nuclei in a supersaturated solution followed by particle growth [195]. Bigger NPs will grow at the cost of the small ones to some extent. This mechanism is generally believed to be the main path of crystal growth in synthetic reaction systems.

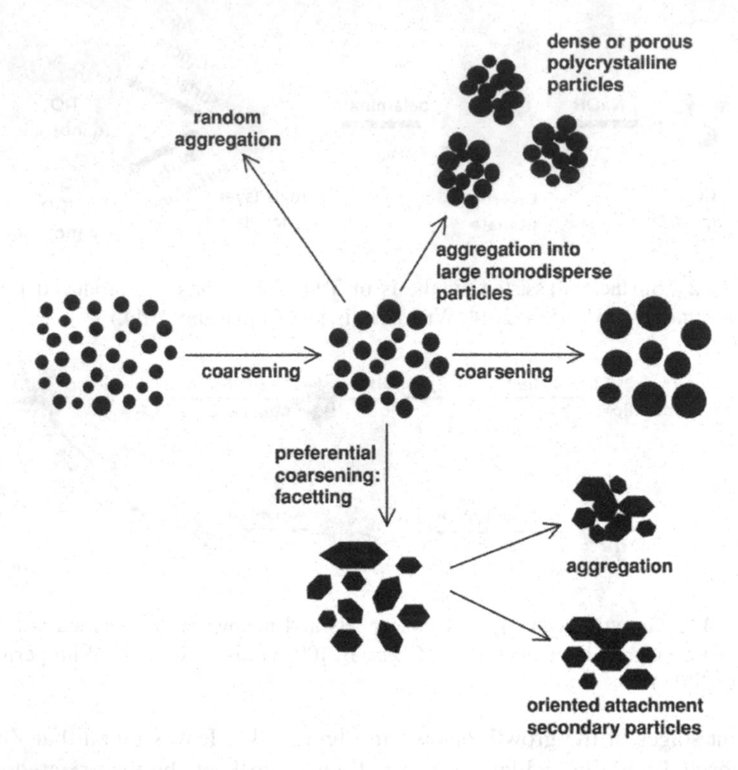

FIGURE 1.44 Schematic illustration of aging processes in the synthesis of NPs. (Reproduced from Oskam, G., *J. Sol-Gel Sci. Technol.*, 37, 161–164, 2006. With permission. Copyright© 2006.)

Oskam et al. reported the synthesis from the growth solution divided into four steps: precursor formation, nucleation, growth, and aging [197]. The aging process is shown in Figure 1.44. After the nucleation and growth, NPs have formed with a different size distribution. Further size of the particle and particle morphology and size distribution occur by aging.

1.3.3.2 Oriented Attachment Growth Mechanism

Typically, in oriented attachment growth proceeds by repeating attachment events of merging particles on lattice-matched crystal facets. Many researchers tried to identify the complete process of oriented attachment, as well as the growth kinetics of NPs. No comprehensive theory or clear conclusions were reported so far [198,199].

Cao et al. reported the growth of ZnO NPs with high dispensability based on the oriented attachment mechanism [199]. The possible growth mechanism of ZnO NPs is depicted in Figure 1.45. It starts with classical nucleation and crystal growth of particles, and the NP surface structure becomes rough. Highly oriented aggregation between the NPs takes place finally through the oriented attachment process as shown in Figure 1.45.

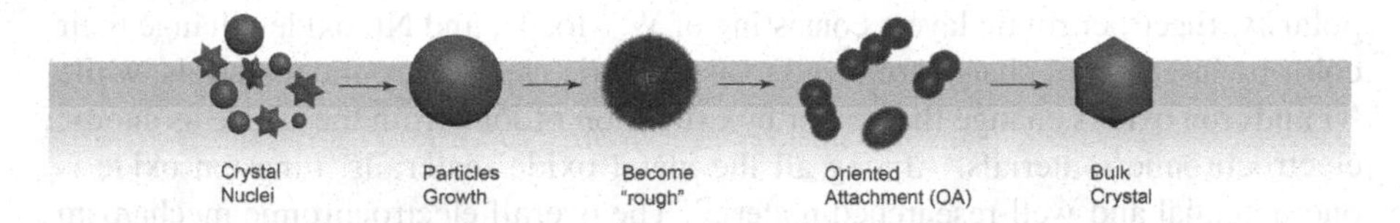

FIGURE 1.45 Growth mechanism of ZnO NPs by the oriented attachment growth mechanism. (Reproduced from Cao, D. et al., *Nanoscale Res. Lett.*, 14, 210, 2019. With permission. Copyright© 2019.)

1.4 APPLICATIONS OF TRANSITION METAL OXIDE NANOMATERIALS

1.4.1 ELECTROCHROMICS

TMOs are the most attracted and potential materials for electrochromic applications [200,201]. TMO applications range in many technologies, including smart windows [201,202], rearview mirrors [203], electrochromic displays [204], and sensors [205]. Electrochromism refers to a reversible change in color state to the transparent state under the insertion or extraction of charge by application of small voltages.

Typical electrochromic devices (ECDs) consist of five superimposed layers on a single substrate or sandwiched between two substrates (Figure 1.46). These five layers consist of two transparent conducting layers, electrochromic active layer, and electrolyte and ion storage layer. When the potential is applied between two transparent conducting oxides, the mobile ions from the ion-conducting layer will move between electrochromic layer and ion storage layer depending upon the

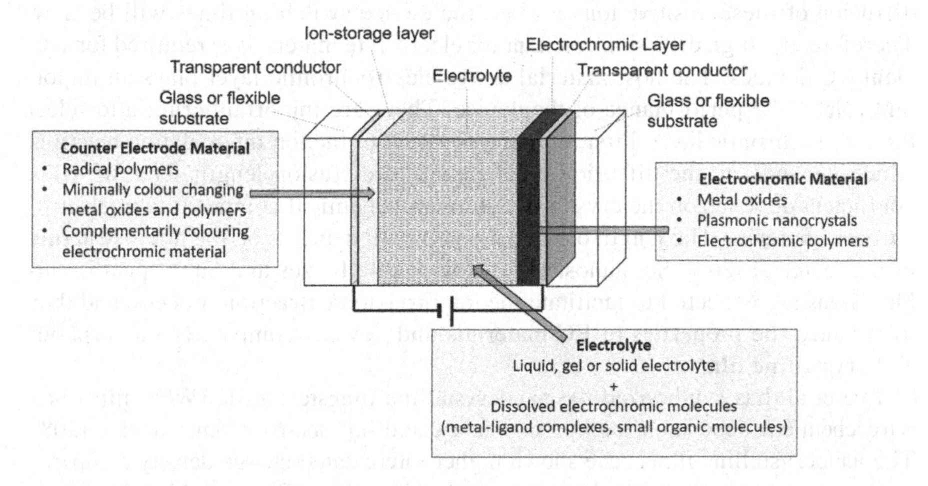

FIGURE 1.46 Schematic representation of an ECD composed of electrochromic material, electrolytes, and counterelectrode material sandwiched between transparent conductive substrates. (From Li, X. et al., *J. Mater. Chem. C*, 7, 12761–12789, 2019. Reproduced by permission of The Royal Society of Chemistry.)

polarity. Electrochromic layers consisting of W, Mo, Ti, and Nb oxides change their color by insertion of charges referred to as cathodic electrochromic materials, while Ni and iron oxides change their color by extraction of ions from the lattice as anodic electrochromic materials. Among all the metal oxide materials, tungsten oxide is one potential and well-researched material. The overall electrochromic mechanism can be expressed with the following expression:

$$MO_y + xA^+ + xe^- \leftrightarrow A_xMO_y(Colored)(A:H,Li,Na..)$$
$$(Transparent)$$

The device performance is evaluated by the parameters such as high coloration efficiency, high optical contrast, long-term redox and photostability, and fast-switching kinetics. Among them, coloration efficiency (CE) is defined as the ratio between optical contrast per inserted charge per area [207].

$$\mathrm{CE} = \frac{1}{\left(Q/A\right)} \ln\left(\frac{T_b}{T_c}\right)$$

where Q is the intercalated charge, A is the area of the film, and T_b and T_c refer to the optical transmission at bleached and colored states, respectively.

The coloration and bleaching times (switching kinetics) can be defined as the times required to achieve 90% of the maximum transmittance change.

The operation of an electrochromic device depends on the double injection of small metal alkali ions (Na^+, Li^+, and H^+) and electrons into the host lattice. However, if these positive metal alkali ions are very slow in nature and if the diffusion of these positive ions is slow, the device switching times will be slow. Therefore, the high diffusion constant for electrolyte materials is required for efficient EC devices. The host material active electrochromic layer plays an important role in the performance of the device. There are important criteria to select the electrochromic layer. First, the time constant of the ion intercalation reaction, which depends on the diffusion coefficient and diffusion length. The diffusion coefficient depends on the crystal structure and chemical composition of electrochromic material. The length of the diffusion path depends on the microstructure of the material [208]. So, nanostructures with small sizes and large specific surface areas are expected to facilitate the ion insertion/extraction process, and then to enhance the properties of EC materials and devices compared to amorphous and crystalline films.

Lee et al. has synthesized the nanocrystalline tungsten oxide (WO_3) films hot-wire chemical deposition method and fabricated an electrochromic device [208]. The nanocrystalline films have shown higher intercalated charge density compared to the crystalline and amorphous tungsten oxide films (Figure 1.47a). The intercalated charge densities for nanocrystalline films are 32 mC/cm^2, whereas for crystalline films 3 mc/cm^2 and amorphous films 9 mc/cm^2. The improved hydrogen-insertion ability of the NP film is attributed to its low density (2.5 g cm^{-3}) and high

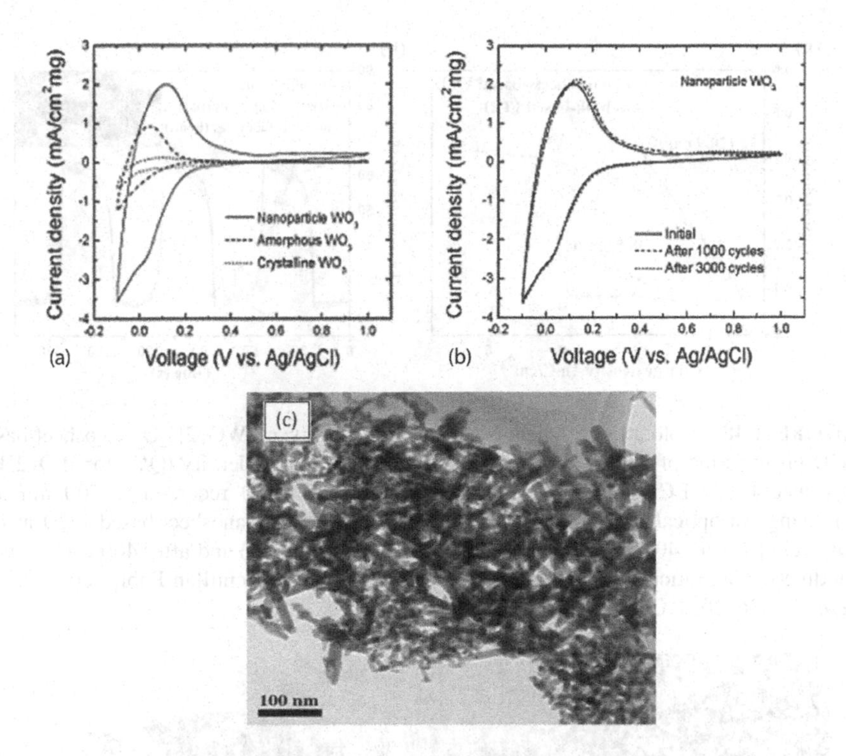

FIGURE 1.47 Cyclic voltammetry curves (a,b) of the EC WO_3 NPs and nanorods prepared by hot-wire chemical vapor deposition, and (c) TEM image of nanocrystalline WO_3 films. (From Lee, S.-H. et al., Crystalline WO_3 Nanoparticles for Highly Improved Electrochromic Applications. *Adv. Mater.* 2006. 18. 763–766. Copyright Wiley-VCH Verlag GmbH & Co. KGaA. Reproduced with permission.)

active surface area, compared with amorphous and crystalline films (5.5 and 6.4 g cm^{-3}, respectively). The nanocrystalline films have shown very good cycling stability (Figure 1.47b).

Zheng et al. prepared WO_3 nanorods (1D nanorods) by the DC magnetron sputtering method with coloration efficiency of 50% at 600 nm wavelength [209]. Azam et al. prepared 2D WO_3 nanosheets by the direct exfoliation method [210]. The large 2D WO_3 nanosheets up to 20 μm were obtained from WS_2-layered network material. These 2D WO_3 nanosheets exhibited transmittance change at 700 nm with enhanced EC performance more than 243% compared to bulk WO_3 prepared by the same method.

Liang et al. synthesized the ultra-nanosheet WO_3: $2H_2O$ films with thickness about 1.4 nm via ultrasonic exfoliation method. Compared to bulk, ultra-thin WO_3: $2H_2O$ nanosheets are flexible in nature, have better interface contact with the substrate, and high Li^+ diffusion constants were observed (shown in Figure 1.48) [211].

Sun and Wang fabricated an electrochromic display device based on ZnO nanowire array (shown in Figure 1.49). The device responds times are remarkably well compared to other EC devices based on bulk and organic molecules such as viologen [212].

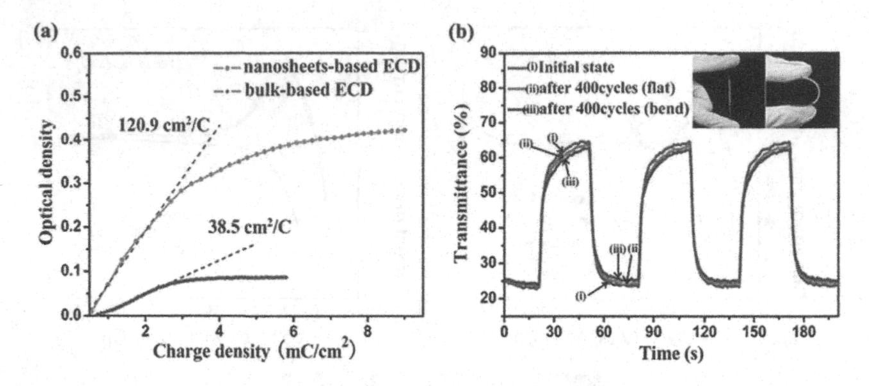

FIGURE 1.48 Coloration efficiency and cyclic stability of $WO_3{\cdot}2H_2O$ nanosheet-based ECD (a) variation of the in-situ optical density vs. the charge density (Q/A) for $WO_3{\cdot}2H_2O$ nanosheet-based ECD and the comparative bulk-based ECD recorded at 700 nm and (b) change in optical transmittance vs. time for $WO_3{\cdot}2H_2O$ nanosheet-based ECD at initial cycle (i), after 400 cycles under extending configuration (ii) and after 400 cycles under bending configuration (iii). (Reprinted by permission from Macmillan Publishers Ltd. *Sci. Rep.*, *3*, 1936, 2013, Copyright 2013.)

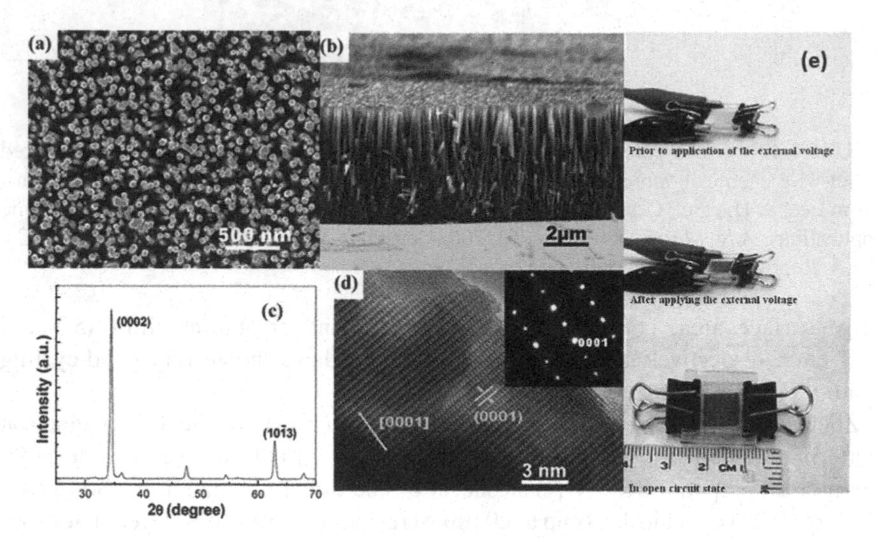

FIGURE 1.49 (a,b) SEM images, (c) X-ray powder diffraction pattern, (d) high-resolution TEM image of the ZnO nanowire array hydrothermally grown on ITO-coated glass, and (e) photos of the EC device at different voltages. (Reproduced Sun, X. W. et al., *Nano Lett.*, 8, 1884–1889, 2008. With permission. Copyright© 2008.)

1.4.2 Lithium Ion Battery

Li-ion batteries are the most advanced rechargeable batteries that attract several industries and researchers for the last few decades. These advanced batteries are now dominant power sources for mobile phones, laptops, and other electronic devices. The Li-ion battery market is now gaining importance in the field of electric vehicles, which requires high power, high capacity, high charging rate, and long life with high safety performance and low cost. The Li-ion battery consists of an anode, cathode, electrolyte, and separator. The anode and cathode electrodes are isolated by a separator. The separator usually has a microporous polymer membrane that allows the exchange of lithium ions between the two electrodes but not electrons. The chemical energy is converted into electrical energy by the process of migration of Li ions between anode and cathode electrolytes. The performance of the Li-ion battery depends on the inherent properties of the electrode materials. Nanomaterials are promising in terms of the performance. The nanostructure of the electrode materials greatly influences the kinetics of the device because of the reduction in the size from micrometer to the nano range. The smaller scale reduces the diffusion length. TMOs have become promising electrode materials for Li-ion batteries because of their multiple chemical valence states and diverse morphological characteristics (shown in Figure 1.50) [213].

Chan's group demonstrated the first Si-based nanowire material for the energy storage battery. The device has shown a 10 times increment in the specific capacity than the graphite anode materials [214]. The small size of the electrode has also enhanced the surface area with more active sites and facilitates the ion exchange with a faster rate between electrode and electrolyte interface. Zhang et al. prepared Fe_2O_3 NPs with 5 nm domains in a diameter assembled into a mesoporous network. The mesoporous Fe_2O_3 NPs exhibit excellent cycling performance (1009 mA h g^{-1} at 100 mA g^{-1} up to 230 cycles) and rate capability (reversible charging capacity of 420 mA h g^{-1} at

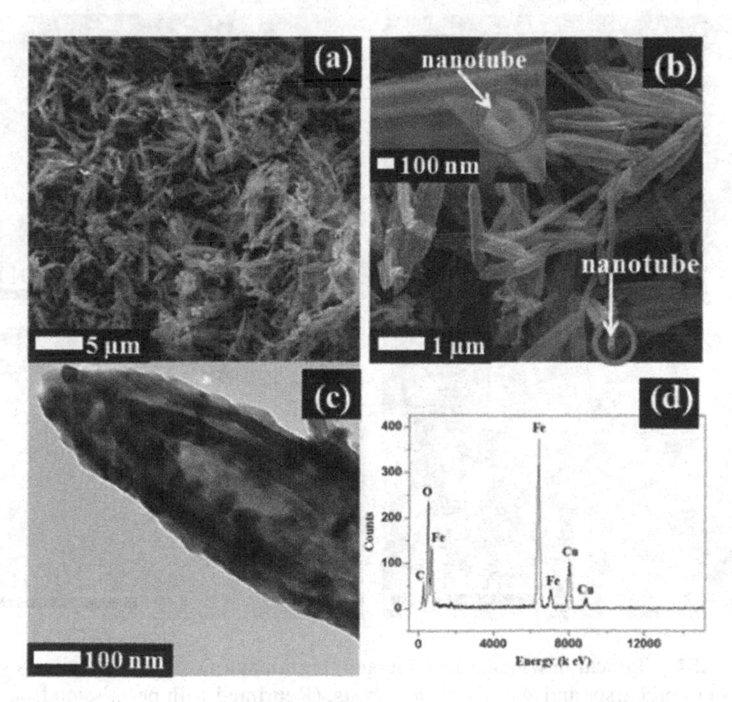

FIGURE 1.50 (a–d) Fe_2O_3 nanotube anodes for high capacity. (Reproduced from Sun, M. et al., *Ceram. Int.*, 1, 363–367, 2017. With permission. Copyright© 2017.)

1000 mA g^{-1} during 230 cycles) [215]. Sun et al. synthesized 1D porous Fe_2O_3 nanotubes with 2 μm length and 220 nm outer diameter and 65 nm wall thickness by low temperature hydrothermal method followed by thermal treatment. Fe_2O_3 nanotubes exhibit enhanced electrochemical properties in terms of lithium storage capacity (1050.1 mA h g^{-1} at 100 mA g^{-1} rate), initial columbic efficiency (78.4%), cycle performances (90.6% capacity retention at 50th cycle), and rate capability (613.7 mA h g^{-1} at 1000 mA g^{-1} rate) [213]. Huang et al. prepared 1D mesoporous single-crystalline Co_3O_4 nanobelt by the hydrothermal method followed by the calcination treatment [216]. The size of the nanobelts is in the range of 100–300 nm in width and several micrometers in length (shown in Figure 1.51). The specific capacity of Co_3O_4 nanobelts could remain over 614 mA h g^{-1} at a current density of 1 A g^{-1} after 60 cycles. Even at a high-current density of 3 A g^{-1}, these Co_3O_4 nanobelts still could deliver a remarkable discharge capacity of 605 mA h g^{-1} with good cycling stability.

Zhang et al. reported, Porous MnO_2 nanoplates were prepared by the agile polyol solution method combined with a simple post-annealing process [217]. MnO_2 nanoplates exhibited excellent cyclic retention and a specific capacity of 813.7 mAhg^{-1} at a current density of 100 mAg^{-1} after 50 cycles.

Nana Wang et al. reported Hierarchically porous NiO microtubes by simple precipitation method (shown in Figure 1.52). The porous NiO microtubes as an anode material for lithium-ion batteries exhibit excellent performances, ~640 mA h g^{-1} after 200 cycles at 1 A g^{-1} [218].

FIGURE 1.51 Typical low-magnification and high-magnification SEM images of as-prepared (a,b) precursor and (c,d) Co_3O_4 nanobelts. (Reprinted with permission from Huang, H. et al., *ACS Appl. Mater. Interfaces*, 4, 2012, 5974–5980. Copyright 2012 American Chemical Society.)

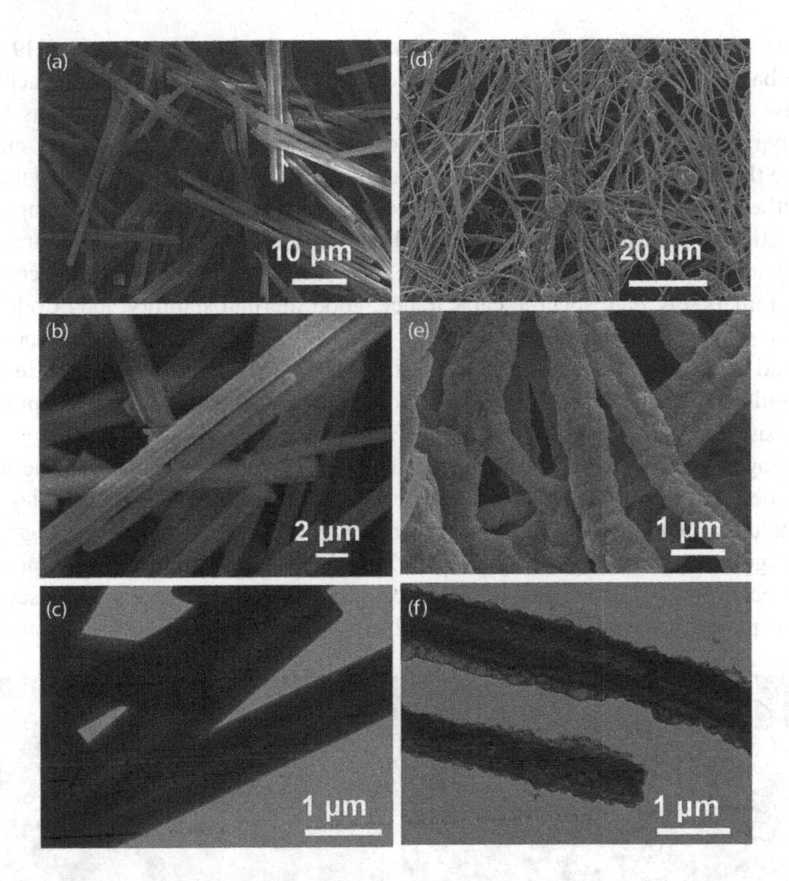

FIGURE 1.52 SEM and TEM images of (a–c) Ni(dmg)$_2$ and (d–f) porous NiO microtubes. (From Wang, N. et al., *J. Mater. Chem. A*, 2, 16847–16850, 2014. Reproduced by permission of The Royal Society of Chemistry.)

1.4.3 Supercapacitors

Supercapacitors are high-performance charge storage devices through adsorption or desorption process, which is currently an alternative to the conventional batteries to fulfill the high demand for the charge storage devices. Supercapacitors have the advantage over batteries because of high power densities, long cycle life, and rapid recharge capability. Because of these superior functionalities, supercapacitors are now devoted to meeting the ever-growing consumption demands and alleviating the energy crisis. Depending on the charge storage mechanism, supercapacitors are categorized into two distinctive classes: electrochemical double-layer capacitors and pseudocapacitors. The electrode materials play an important role in the performance of the supercapacitor. Usually, there are three categories of electrode materials: carbon material, conducting polymers, and TMOs. Carbon electrode materials are high-conducting materials with large surface areas and ideal electrochemical double-layer capacitors, but they suffer from low specific capacitance. Whereas conducting polymers have very high specific capacitance values but poor cycling life because of substantial extraction and contraction during the charging and discharging process [219,220]. TMO-based supercapacitors are considered to be much higher specific capacitance because of multiple oxidation states that are in favor of fast redox reactions [221]. In general, metal oxide nanomaterials provide higher energy densities for energy storage than carbon materials and better stability compared to conducting polymers.

Ruthenium oxide (RuO_2) is one of the well-studied TMOs for supercapacitor application with a wide potential window, highly reversible redox reaction, three distinct oxidation states accessible within a 1.2 V voltage window, high proton conductivity, remarkably high specific capacitance, good thermal stability, long cycle life, metallic-type conductivity, and high rate capability [222]. Unfortunately, the high cost and environmental toxicity of RuO_2 limits the large-scale application. Hu et al. reported nanotubular-arrayed RuO_2 electrode material by the membrane-templated synthesis route (shown in Figure 1.53) [223].

Zhang et al. reported the various porous nanostructured NiO films by the facile hydrothermal process. The NiO nanostructures include nanoslices, nanoplates, and nanocolumns with specific capacitances at about 176, 285, and 390 F A g^{-1} at a discharge current of 5 A g^{-1}, respectively [224]. Yang et al. prepared NiO nanowires by sol-gel synthesis [225]. The sol-gel process, in which the citric acid was used as a chelate, played a crucial role in the formation of NiO nanowires. NiO nanomaterial is

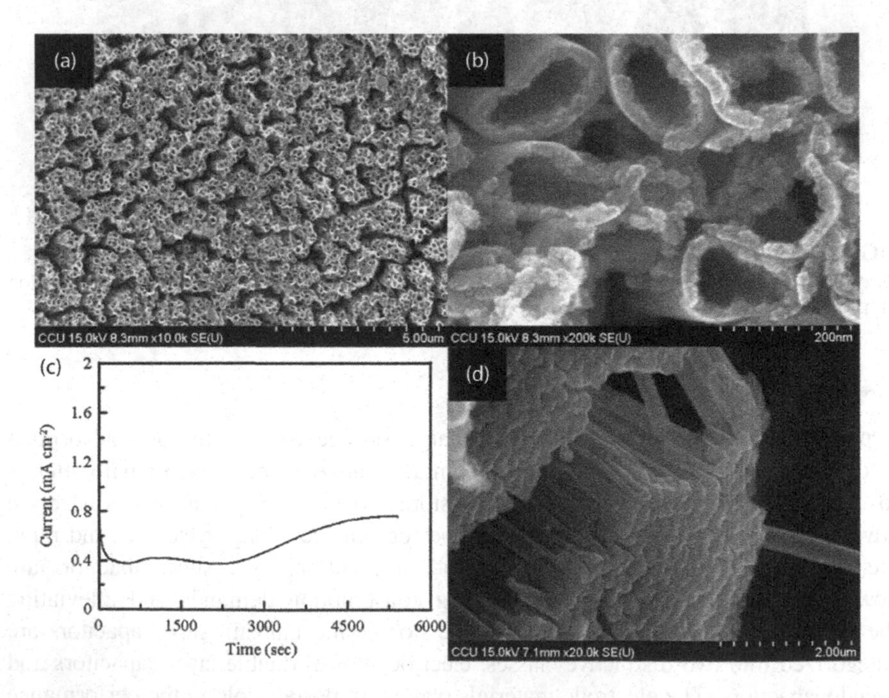

FIGURE 1.53 (a,b) SEM top images of an $RuO_2.xH_2O$ nanotube-arrayed electrode, (c) the *i-t* curve of anodic deposition measured at 1.0 V from a 10 mM $RuCl_3.xH_2O$ + 0.1 M CH_3COONa solution, and (d) SEM image showing the onset-overfilled morphology of a $RuO_2.xH_2O$ NT arrayed electrode. All nanotubes were deposited at 1.0 V from a 10 mM $RuCl_3.xH_2O$ + 0.1 M CH3COONa solution. (Reprinted with permission from Hu, C. C. et al., *Nano Lett.*, 6, 2690–2695, 2006. Copyright 2006 American Chemical Society.)

considered to be one of the alternative electrode materials in alkaline electrolytes with high specific capacitance (theoretical value up to 3750 F/g), being environmentally friendly, and low cost, but suffers from poor cycling and high resistance. Yeager et al. prepared NiO NPs with an average diameter 5.3 ± 1.3 nm as an electrode material with a maximum specific capacitance of 243 F/g [226]. Vijayakumar et al. reported nanoflake-like NiO electrodes synthesized at different calcination temperatures using CTAB as a surfactant via the microwave method [227]. The NiO nanoflakes exhibited maximum specific capacitance 401 Fg^{-1} at a current density of 0.5 mA cm^{-1}. Chen et al. prepared Al-doped NiO nanosheets by the facile hydrothermal method followed by a calcination process (shown in Figure 1.54). The reported values are high specific capacitance 192 ± 23 F g^{-1} at 0.4 A g^{-1} with a high-energy density of 215 ± 15 Wh kg^{-1} and power density of 21.6 kW kg^{-1} [228].

MnO_2 is another TMO promising an alternative to RuO_2 because of its low cost, relatively low toxicity, and high theoretical capacitance (1370 F g^{-1}). Wang et al. synthesized nanoporous MnO_2 electrodes for a supercapacitor by the facile sonochemical method. The porous MnO_2 NPs in the range 20–50 nm nanorods were obtained with 2 mm diameter and length of 4–8 nm. At a current density of 820 mA g^{-1}, a high value of 196.7 F g^{-1} in 9M KOH electrolyte was obtained [229]. Wu et al. prepared MnO_2 nanowires by the electrodeposition technique. By varying the applied potential from 0.1 to 0.6 V with saturated calomel, the electrode morphology of the electrode material changed from nanorod (15–35 nm in diameter) to nanowire structure (8–16 nm in diameter). The specific capacitance increased with an increased potential range from 243 to 350 F. g^{-1}[230].

1.4.4 Photocatalysis

In recent years, there has been a great deal of concern over environmental pollution and water pollution. Therefore, to develop high efficiency, clean and sustainable technology for clean energy have become the need of the hour. The photocatalytic technology of water splitting into hydrogen and oxygen using solar energy is considered to be one of the efficient green technologies of the last few decades [231]. One of the major advantages with this technology is the removal of organic contaminants from water, and air steams are considered to be the most advanced oxidation process. For the electrochemical decomposition of water, a potential difference of minimum 1.23 V needs to be applied between two electrodes [232]. This potential difference is equal to the wavelength of 1000 nm approximately. Therefore, if the energy of light is used effectively in an electrochemical system, it should be possible to decompose water with visible light. The first photocatalysts were suitable for water splitting, or for activating hydrogen production from carbohydrate compounds made by plants from water and carbon dioxide [233]. But these catalysts operate with UV light, which is just 4% of the whole solar energy spectrum. Therefore, numerous efforts are attempted to produce nanomaterials that are capable of utilizing the visible light spectrum, which accounts for nearly 43% to decompose the water into hydrogen and oxygen.

The photocatalytic mechanism can be simply described as a change in the rate of a chemical transformation under the action of light in the presence of a catalyst that absorbs light and is involved in the chemical reaction (shown in Figure 1.55) [234].

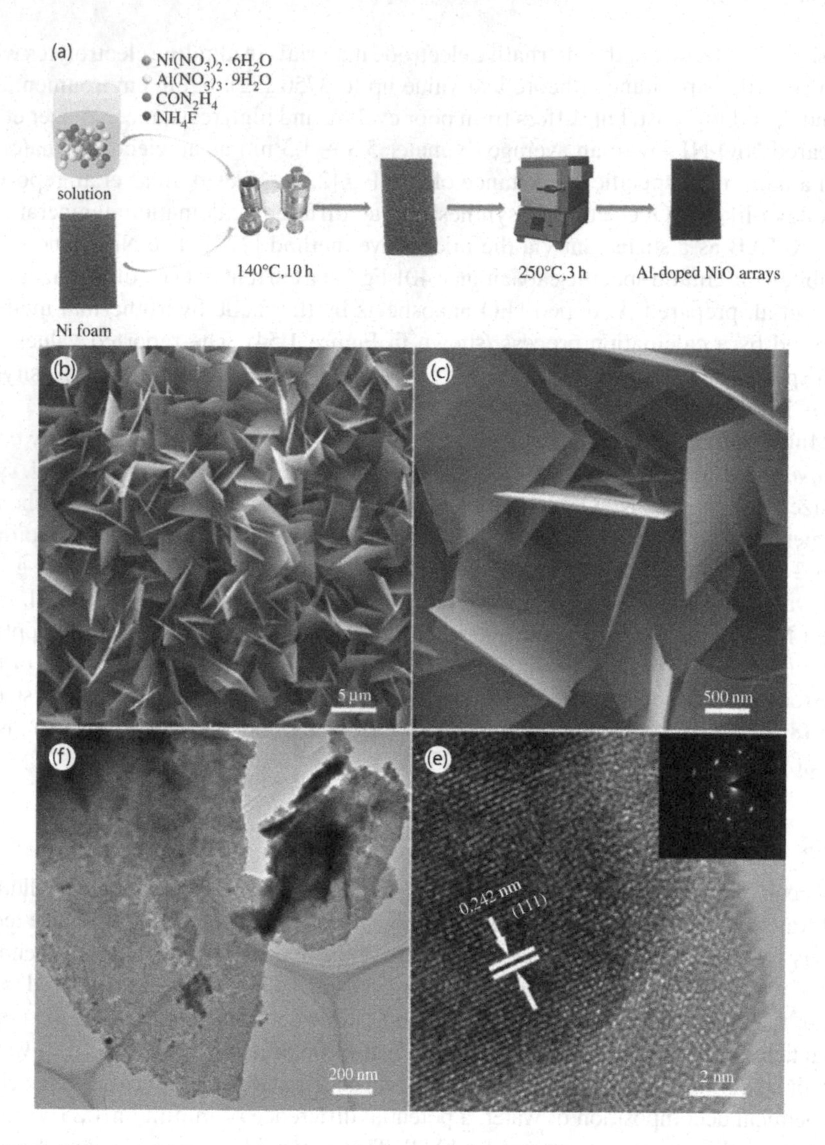

FIGURE 1.54 (a) Methods for preparing Al-doped NiO catalyst, (b,c) low- and high-magnification SEM images of Al-doped NiO nanosheet arrays, and (d,e) TEM and high-resolution TEM images of the nanosheets. The inset in (e) is an Selected Area Electron Diffraction (SAED) pattern corresponding to the high-resolution TEM image. (Reproduced from Chen, J. et al., *R. Soc. Open Sci.*, 5, 180842, 2018. With permission. Copyright© 2018.)

The photocatalytic technique has a wide range of applications, such as degradation of heavy metals, water disinfection, hydrogen generation, carbon dioxide photoreduction, photoelectric sensing, and photodynamic therapy [235]. Figure 1.55 describes the photocatalysis mechanism. The absorption of a photon ($h\nu \geq E_g$) by the semiconductor causes the electron-hole pair generation. The electron-hole pairs can migrate separately to the surface of the semiconductor and participate in the redox reactions

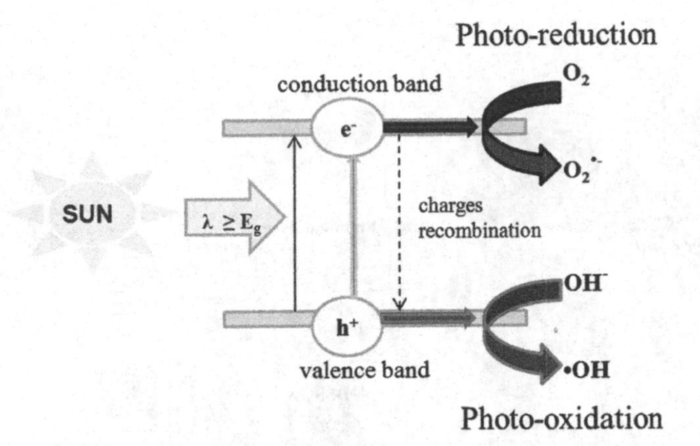

FIGURE 1.55 Principle mechanism of photocatalysis. (Reproduced from Colmenares, J.C., and Luque, R., *Chem. Soc. Rev.*, 43, 765–778, 2014. With permission. Copyright© 2014.)

in order to degrade or to reduce the adsorbed molecules [236]. Among all TMOs, TiO_2 is considered well-researched material for photocatalytic application because of its oxidizing ability, chemical stability, and low cost. Out of all the phases, the anatase phase is most effective and shows high photocatalytic activity. However, fast electron-hole pair recombination limits the expected photocatalytic activity and can be excited at only 388 nm below UV light. Therefore, several attempts were made to improve the visible light absorbance (which covers 43% of solar energy spectrum) and increase the lifetime of carriers.

Nanomaterials having the large surface area and smaller size compared to bulk materials show an increase in the photocatalytic activity because the size reduction results in a large number of atoms being present at the surface, resulting in an enhancement in the number of active centers at surface [237]. Photocatalytic activity depends on how efficiently it separates the photogenerated electron-hole pair. Ramakrishna et al. synthesized size-controlled 0D TiO_2 NPs by the solvothermal method (shown in Figure 1.56). They obtained 5 nm spherical shape TiO_2 NPs with the anatase phase. The photodegradation efficiency of methyl orange and methylene blue dyes under the influence of UV-A light was 96% and 97%, respectively [238].

Tian et al. [239] reported use of 1D TiO_2 nanostructured surface heterostructure to improve the photocatalytic activity. It includes an increase of broadening of light absorption from UV to visible and near-infrared region and effective separation of electron-hole pair, thus increasing the life of the carriers. Zhou et al. reported preparation of single and double heterostructure TiO_2 nanobelts by the hydrothermal method (shown in Figure 1.57). The photocatalytic results show that the double heterostructure TiO_2 nanobelts exhibit a much higher photocatalytic activity than common variety TiO_2 nanobelts and single heterostructure TiO_2 nanobelts. The corresponding decomposition ratios are 48%, 79.4%, and 100% in 30 minutes, respectively [240].

Wang et al. synthesized single crystal TiO_2 with 2D and 3D mesoporous structures by the solvothermal method. The 3D-ordered TiO_2 structures were shown with better photocatalytic activity than the similar structure of polycrystalline TiO_2 under UV light absorption [241].

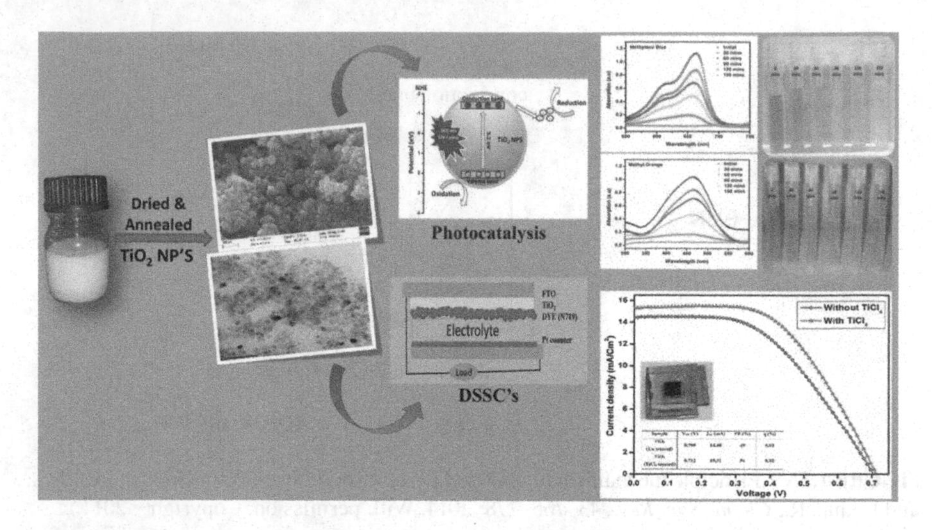

FIGURE 1.56 Photocatalytic activity of solvothermally prepared TiO_2 NPs. (Reprinted from *Mater. Res. Bull.*, 97, Ramakrishnan, V.M. et al., Size controlled synthesis of TiO_2 nanoparticles by modified solvothermal method towards effective photocatalytic and photovoltaic applications, 351–360, Copyright 2018, with permission from Elsevier.)

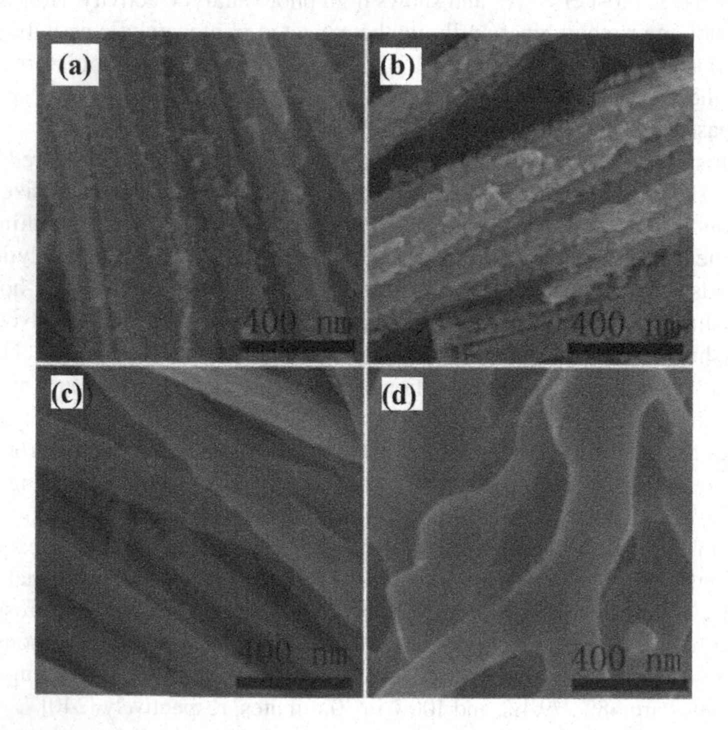

FIGURE 1.57 Typical SEM images (a–d), of the as-synthesized TiO_2-P/TiO_2-B heterostructures obtained with different temperatures: (a) 400°C, (b) 600°C, (c) 800°C, and (d) 1000°C. (From Zhou, W. et al., *J. Mater. Chem.*, 21, 7937, 2011. Reproduced by permission of The Royal Society of Chemistry.)

The photocatalytic activity reported in a variety of other TMONMs such as V_2O_5 [242], MoO_3 [243], WO_3 [244], MnO_2 [245], and Fe_2O_3 [246].

1.4.5 Energy-Harvesting Applications: Solar Cells and Nanogenerators

TMO nanomaterials play important roles in energy the harvesting, which includes ZnO-based nanogenerators [247,248], ZnO and TiO_2-based solar cells [249,250] and thermal electric generators [251].

Solar cells basically harvest energy by utilizing sunlight. TMO nanostructures can act as light absorbers, transparent electrodes, transport layers, and also some perform unique functionalities. The TMOs, namely ZnO, SnO_2, NiO, and TiO_2, were used extensively in solar cells [252–255].

TiO_2 is mostly used as anode in dye-sensitized solar cells because of the large band gap, suitable band edge levels for charge injection and extraction, long lifespan of excited electrons, exceptional resistance to photo corrosion, nontoxicity, and low cost. It has made TiO_2 a popular material for solar energy applications. Among many different kinds of TiO_2 nanostructures, such as rice, star, and flower-like structures, the star-like TiO_2 photo-anode-based dye-sensitized solar cell (DSSC) exhibits the highest photovoltaic conversion efficiency (PCE) of 9.56% [256].

Zhang et al. reported the facile hydrothermal synthesis of rutile TiO_2 nanorod arrays on FTO substrates without the use of acids and used for perovskite solar cells. Perovskite solar cells showed power conversion efficiencies up to 11.1% [257].

Wu et al. reported the vertically oriented rutile TiO_2 nanorod array because an efficient electron transport layer has been used in the perovskite solar cells (PSCs), and its microstructure has a great impact on the corresponding PCE. The PCE of the optimized nanorod array film-based PSC can be further increased to 14.3% (shown in Figure 1.58]) [258].

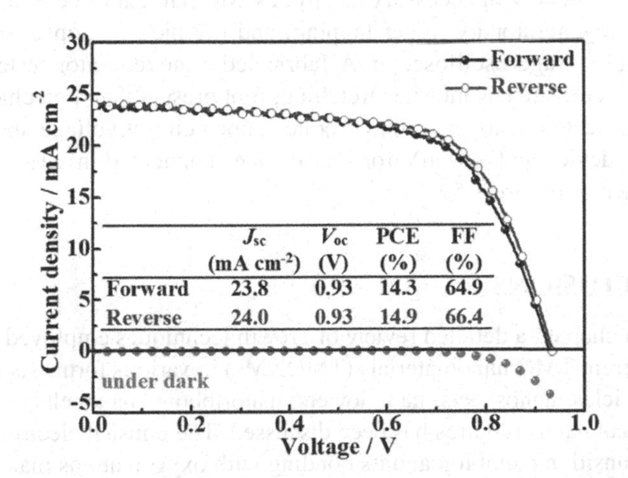

FIGURE 1.58 J–V curves of the PSCs fabricated with optimized nanorod films. (Reprinted with permission from Wu, S. et al., *ACS Appl. Energy Mater.*, 1, 1649–1657, 2018. Copyright 2018 American Chemical Society.)

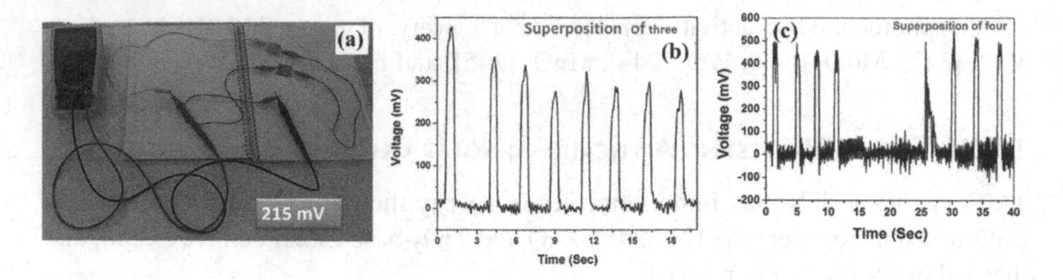

FIGURE 1.59 (a) A photograph of the series-connected two nanogenerators and their response to finger tapping in the multimeter display, (b) piezoresponse voltage of the series-connected three nanogenerators, and (c) piezoresponse voltage of the series-connected of four nanogenerators. (Reproduced from Manjula, Y. et al., *Chem. Phys.*, 533, 110699, 2020. With permission. Copyright© 2020.)

TMO, which exhibits the piezoelectric property such as ZnO, is used for harvesting energy by utilizing easily available waste mechanical energy. Wang et al. extensively worked on the piezoelectric property of ZnO nanostructures and fabricated devices (nanogenerators) for mechanical energy harvesting [259]. Initially, the piezoelectric property of the single ZnO nanowire was measured by an Atomic Force Microscope (AFM) tip and found good piezoelectric property [260]. Later, piezoelectric properties of the array of nanorods, nanotubes, and nanosheets were studied with proper electrode attachment [261–263]. Our reach group also prepared simple and flexible nanogenerators based on ZnO nanosheet networks. ZnO nanosheets are synthesized by simple, cost-effective, and single-step hydrothermal methods on flexible conducting aluminum substrates at a low growth temperature of 80°C. The nanogenerator is fabricated using ZnO nanosheet networks as an active piezoelectric element with necessary electrodes. Mechanical force is applied on the fabricated nanogenerator by finger tapping, and the piezoresponse was recorded using a digital storage oscilloscope. A fabricated nanogenerator tested real-time mechanical forces, such as muscle stretching, foot pressure, and mechanical bending [264]. Fabricated nanogenerators produced open circuit voltage above 100 mV for the single device and 400 mV for four devices connected in series upon finger tapping (shown in Figure 1.59).

1.5 CONCLUSIONS

In the current chapter, a detailed review of growth techniques employed for the synthesis of different TMO nanomaterials (TMONMs) in various forms such as nanowires, nanoparticles, nanosheets, nanoflowers, nanoribbons, nanobelts, 3D networks, and hierarchical nanostructures has been discussed. The unusual electronic structure of the base transition metal ion and its bonding with oxygen atoms make TMONMs a very interesting class of materials. Moreover, unique physical and chemical properties are aroused due to the partial filling of *d* orbitals. It is concluded that the morphology of the TMO is controlled by the synthesis parameters such as temperature, pressure, substrate condition, growth duration, and growth environment. The applications discussed in this chapter are only a few out of many; there could also be many more applications in the areas of gas sensors, photodetectors, drug delivery, magnetic fluid hypothermia, transparent conducting oxides, optical antennas, and optoelectronics. It could be concluded that TMONMs have many impending applications of the future for the humankind.

REFERENCES

1. Theerthagiri, J.; Salla, S.; Senthil, R. A.; Nithyadharseni, P.; Madankumar, A.; Arunachalam, P.; Maiyalagan, T.; Kim, H.-S. A Review on ZnO Nanostructured Materials: Energy, Environmental and Biological Applications. *Nanotechnology*, **2019**, *30* (39), 392001. https://doi.org/10.1088/1361-6528/ab268a.
2. Wang, Z. L. Zinc Oxide Nanostructures: Growth, Properties and Applications. *J. Phys. Condens. Matter*, **2004**, *16* (25), R829–R858. https://doi.org/10.1088/0953-8984/16/25/R01.
3. Kołodziejczak-Radzimska, A.; Jesionowski, T. Zinc Oxide—From Synthesis to Application: A Review. *Materials (Basel).*, **2014**, *7* (4), 2833–2881. https://doi.org/10.3390/ma7042833.
4. Rong, P.; Ren, S.; Yu, Q. Fabrications and Applications of ZnO Nanomaterials in Flexible Functional Devices: A Review. *Crit. Rev. Anal. Chem.*, **2019**, *49* (4), 336–349. https://doi.org/10.1080/10408347.2018.1531691.
5. Raoufi, D. Synthesis and Microstructural Properties of ZnO Nanoparticles Prepared by Precipitation Method. *Renew. Energy*, **2013**, *50*, 932–937. https://doi.org/10.1016/j.renene.2012.08.076.
6. Rani, S.; Suri, P.; Shishodia, P. K.; Mehra, R. M. Synthesis of Nanocrystalline ZnO Powder via Sol-Gel Route for Dye-Sensitized Solar Cells. *Sol. Energy Mater. Sol. Cells*, **2008**, *92* (12), 1639–1645. https://doi.org/10.1016/J.SOLMAT.2008.07.015.
7. Elumalai, K.; Velmurugan, S. Green Synthesis, Characterization and Antimicrobial Activities of Zinc Oxide Nanoparticles from the Leaf Extract of Azadirachta Indica (L.). *Appl. Surf. Sci.*, **2015**, *345*, 329–336. https://doi.org/10.1016/j.apsusc.2015.03.176.
8. Ali, K.; Dwivedi, S.; Azam, A.; Saquib, Q.; Al-Said, M. S.; Alkhedhairy, A. A.; Musarrat, J. Aloe Vera Extract Functionalized Zinc Oxide Nanoparticles as Nanoantibiotics against Multi-Drug Resistant Clinical Bacterial Isolates. *J.Colloid Interface Sci.*, **2016**, *472*, 145–156. https://doi.org/10.1016/j.jcis.2016.03.021.
9. Patra, M. K.; Manoth, M.; Singh, V. K.; Siddaramana Gowd, G.; Choudhry, V. S.; Vadera, S. R.; Kumar, N. Synthesis of Stable Dispersion of ZnO Quantum Dots in Aqueous Medium Showing Visible Emission from Bluish Green to Yellow. *J. Lumin.*, **2009**, *129* (3), 320–324. https://doi.org/10.1016/j.jlumin.2008.10.014.
10. Spanhel, L.; Anderson, M. A. Semiconductor Clusters in the Sol-Gel Process: Quantized Aggregation, Gelation, and Crystal Growth in Concentrated Zinc Oxide Colloids. *J. Am. Chem. Soc.*, **1991**, *113* (8), 2826–2833. https://doi.org/10.1021/ja00008a004.
11. Rani, S.; Suri, P.; Shishodia, P. K.; Mehra, R. M. Synthesis of Nanocrystalline ZnO Powder via Sol-Gel Route for Dye-Sensitized Solar Cells. *Sol. Energy Mater. Sol. Cells*, **2008**, *92* (12), 1639–1645. https://doi.org/10.1016/J.SOLMAT.2008.07.015.
12. Vafaee, M.; Ghamsari, M. S. Preparation and Characterization of ZnO Nanoparticles by a Novel Sol-Gel Route. *Mater. Lett.*, **2007**, *61* (14–15), 3265–3268. https://doi.org/10.1016/J.MATLET.2006.11.089.
13. Bera, D.; Qian, L.; Sabui, S.; Santra, S.; Holloway, P. H. Photoluminescence of ZnO Quantum Dots Produced by a Sol-Gel Process. *Opt. Mater. (Amst).*, **2008**, *30* (8), 1233–1239. https://doi.org/10.1016/J.OPTMAT.2007.06.001.

14. Mahendiran, D.; Subash, G.; Arumai Selvan, D.; Rehana, D.; Senthil Kumar, R.; Kalilur Rahiman, A. Biosynthesis of Zinc Oxide Nanoparticles Using Plant Extracts of Aloe Vera and Hibiscus Sabdariffa: Phytochemical, Antibacterial, Antioxidant and Anti-Proliferative Studies. *Bionanoscience*, **2017**, *7* (3), 530–545. https://doi.org/10.1007/s12668-017-0418-y.
15. Dobrucka, R.; Długaszewska, J. Biosynthesis and Antibacterial Activity of ZnO Nanoparticles Using Trifolium Pratense Flower Extract. *Saudi J. Biol. Sci.*, **2016**, *23* (4), 517–523. https://doi.org/10.1016/j.sjbs.2015.05.016.
16. Nagajyothi, P. C.; Cha, S. J.; Yang, I. J.; Sreekanth, T. V. M.; Kim, K. J.; Shin, H. M. Antioxidant and Anti-Inflammatory Activities of Zinc Oxide Nanoparticles Synthesized Using Polygala Tenuifolia Root Extract. *J. Photochem. Photobiol. B Biol.*, **2015**, *146*, 10–17. https://doi.org/10.1016/J.JPHOTOBIOL.2015.02.008.
17. Elumalai, K.; Velmurugan, S. Green Synthesis, Characterization and Antimicrobial Activities of Zinc Oxide Nanoparticles from the Leaf Extract of Azadirachta Indica (L.). *Appl. Surf. Sci.*, **2015**, *345*, 329–336. https://doi.org/10.1016/J.APSUSC.2015.03.176.
18. Zhang, Y.; Ram, M. K.; Stefanakos, E. K.; Goswami, D. Y. Synthesis, Characterization, and Applications of ZnO Nanowires. *J. Nanomater.*, **2012**, *2012*, 1–22. https://doi.org/10.1155/2012/624520.
19. Bagga, S.; Akhtar, J.; Mishra, S. Synthesis and Applications of ZnO Nanowire: A Review. In *AIP Conference Proceedings*; AIP Publishing LLC, 2018; Vol. 1989, p 020004. https://doi.org/10.1063/1.5047680.
20. Rackauskas, S.; Barbero, N.; Barolo, C.; Viscardi, G. ZnO Nanowire Application in Chemoresistive Sensing: A Review. *Nanomaterials*, **2017**, *7* (11), 381. https://doi.org/10.3390/nano7110381.
21. Huang, M. H.; Wu, Y.; Feick, H.; Tran, N.; Weber, E.; Yang, P. Catalytic Growth of Zinc Oxide Nanowires by Vapor Transport. *Adv. Mater.*, **2001**, *13* (2), 113–116. https://doi.org/10.1002/1521-4095(200101)13:2<113::AID-ADMA113>3.0.CO;2-H.
22. Rusli, N.; Tanikawa, M.; Mahmood, M.; Yasui, K.; Hashim, A. Growth of High-Density Zinc Oxide Nanorods on Porous Silicon by Thermal Evaporation. *Materials (Basel).*, **2012**, *5* (12), 2817–2832. https://doi.org/10.3390/ma5122817.
23. Fan, H. J.; Fuhrmann, B.; Scholz, R.; Syrowatka, F.; Dadgar, A.; Krost, A.; Zacharias, M. Well-Ordered ZnO Nanowire Arrays on GaN Substrate Fabricated via Nanosphere Lithography. *J. Cryst. Growth*, **2006**, *287* (1), 34–38. https://doi.org/10.1016/j.jcrysgro.2005.10.038.
24. Umar, A.; Karunagaran, B.; Suh, E.-K.; Hahn, Y. B. Structural and Optical Properties of Single-Crystalline ZnO Nanorods Grown on Silicon by Thermal Evaporation. *Nanotechnology*, **2006**, *17* (16), 4072–4077. https://doi.org/10.1088/0957-4484/17/16/013.
25. Pan, Z. W. Nanobelts of Semiconducting Oxides. *Science*, **2001**, *291* (5510), 1947–1949. https://doi.org/10.1126/science.1058120.
26. Kong, X. Y.; Wang, Z. L. Spontaneous Polarization-Induced Nanohelixes, Nanosprings, and Nanorings of Piezoelectric Nanobelts. **2003**. https://doi.org/10.1021/NL034463P.
27. Gao, P. X.; Ding, Y.; Mai, W.; Hughes, W. L.; Lao, C.; Wang, Z. L. Conversion of Zinc Oxide Nanobelts into Superlattice-Structured Nanohelices. *Science*, **2005**, *309* (5741), 1700–1704. https://doi.org/10.1126/science.1116495.
28. Wu, J.-J.; Liu, S.-C. Low-Temperature Growth of Well-Aligned ZnO Nanorods by Chemical Vapor Deposition. *Adv. Mater.*, **2002**, *14* (3), 215–218. https://doi.org/10.1002/1521-4095(20020205)14:3<215::AID-ADMA215>3.0.CO;2-J.
29. Juárez, G, J. M.; Andraca, A, J.; Cuenca, A, R.; Jaramillo, V, D.; Juárez, L, F. MOCVD Growth of ZnO Nanowires Through Colloidal and Sputtered Au Seed Via Zn [TMHD] 2 Precursor. *Phys. Procedia*, **2012**, *25*, 278–286. https://doi.org/10.1016/j.phpro.2012.03.084.

30. Park, W. I.; Kim, D. H.; Jung, S.-W.; Yi, G.-C. Metalorganic Vapor-Phase Epitaxial Growth of Vertically Well-Aligned ZnO Nanorods. *Appl. Phys. Lett.*, **2002**, *80* (22), 4232–4234. https://doi.org/10.1063/1.1482800.
31. Liu, J. J.; Yu, M. H.; Zhou, W. L. Well-Aligned Mn-Doped ZnO Nanowires Synthesized by a Chemical Vapor Deposition Method. *Appl. Phys. Lett.*, **2005**, *87* (17), 172505. https://doi.org/10.1063/1.2084321.
32. Vayssieres, L. Growth of Arrayed Nanorods and Nanowires of ZnO from Aqueous Solutions. *Adv. Mater.*, **2003**, *15* (5), 464–466. https://doi.org/10.1002/adma.200390108.
33. Gaddam, V.; Kumar, R. R.; Parmar, M.; Nayak, M. M.; Rajanna, K. Synthesis of ZnO Nanorods on a Flexible Phynox Alloy Substrate: Influence of Growth Temperature on Their Properties. *RSC Adv.*, **2015**, *5* (109), 89985–89992. https://doi.org/10.1039/C5RA12773D.
34. Xu, S.; Wang, Z. L. One-Dimensional ZnO Nanostructures: Solution Growth and Functional Properties. *Nano Res.*, **2011**, *4* (11), 1013–1098. https://doi.org/10.1007/s12274-011-0160-7.
35. Gaddam, V.; Kumar, R. R.; Parmar, M.; Yaddanapudi, G. R. K.; Nayak, M. M.; Rajanna, K. Morphology Controlled Synthesis of Al Doped ZnO Nanosheets on Al Alloy Substrate by Low-Temperature Solution Growth Method. *RSC Adv.*, **2015**, *5* (18), 13519–13524. https://doi.org/10.1039/C4RA14049D.
36. Gupta, M. K.; Lee, J.-H.; Lee, K. Y.; Kim, S.-W. Two-Dimensional Vanadium-Doped ZnO Nanosheet-Based Flexible Direct Current Nanogenerator. *ACS Nano*, **2013**, *7* (10), 8932–8939. https://doi.org/10.1021/nn403428m.
37. Cao, B.; Cai, W.; Li, Y.; Sun, F.; Zhang, L. Ultraviolet-Light-Emitting ZnO Nanosheets Prepared by a Chemical Bath Deposition Method. *Nanotechnology*, **2005**, *16* (9), 1734–1738. https://doi.org/10.1088/0957-4484/16/9/054.
38. Yang, J.; Wang, Y.; Kong, J.; Jia, H.; Wang, Z. Synthesis of ZnO Nanosheets via Electrodeposition Method and Their Optical Properties, Growth Mechanism. *Opt. Mater. (Amst).*, **2015**, *46*, 179–185. https://doi.org/10.1016/J.OPTMAT.2015.04.016.
39. Taleatu, B. A.; Fasasi, A. Y.; Di Santo, G.; Bernstorff, S.; Goldoni, A.; Fanetti, M.; Floreano, L.; Borghetti, P.; Casalis, L.; Sanavio, B.; et al. Electro-Chemical Deposition of Zinc Oxide Nanostructures by Using Two Electrodes. *AIP Adv.*, **2011**, *1* (3), 032147. https://doi.org/10.1063/1.3633476.
40. Umar, A.; Hahn, Y. B. ZnO Nanosheet Networks and Hexagonal Nanodiscs Grown on Silicon Substrate: Growth Mechanism and Structural and Optical Properties. *Nanotechnology*, **2006**, *17* (9), 2174–2180. https://doi.org/10.1088/0957-4484/17/9/016.
41. Tseng, Y.-K.; Chuang, M.-H.; Chen, Y.-C.; Wu, C.-H. Synthesis of 1D, 2D, and 3D ZnO Polycrystalline Nanostructures Using the Sol-Gel Method. *J. Nanotechnol.*, **2012**, *2012*, 1–8. https://doi.org/10.1155/2012/712850.
42. Shi, Z.-F.; Zhang, Y.-T.; Zhuang, S.-W.; Yan, L.; Wu, B.; Cui, X.-J.; Huang, Z.; Dong, X.; Zhang, B.-L.; Du, G.-T. Vertically Aligned Two-Dimensional ZnO Nanowall Networks: Controllable Catalyst-Free Growth and Optical Properties. *J. Alloys Compd.*, **2015**, *620*, 299–307. https://doi.org/10.1016/J.JALLCOM.2014.09.149.
43. Liu, B.; Hu, Z.; Che, Y.; Allenic, A.; Sun, K.; Pan, X. Growth of ZnO Nanoparticles and Nanorods with Ultrafast Pulsed Laser Deposition. *Appl. Phys. A*, **2008**, *93* (3), 813–818. https://doi.org/10.1007/s00339-008-4754-1.
44. Wang, X.; Li, Z.; Shi, J.; Yu, Y. One-Dimensional Titanium Dioxide Nanomaterials: Nanowires, Nanorods, and Nanobelts. *Chem. Rev.*, **2014**, *114* (19), 9346–9384. https://doi.org/10.1021/cr400633s.
45. Rahimi, N.; Pax, R. A.; Gray, E. M. Review of Functional Titanium Oxides. I: TiO_2 and Its Modifications. *Prog. Solid State Chem.*, **2016**, *44* (3), 86–105. https://doi.org/10.1016/J.PROGSOLIDSTCHEM.2016.07.002.

46. Shen, S.; Chen, J.; Wang, M.; Sheng, X.; Chen, X.; Feng, X.; Mao, S. S. Titanium Dioxide Nanostructures for Photoelectrochemical Applications. *Prog. Mater. Sci.*, **2018**, *98*, 299–385. https://doi.org/10.1016/j.pmatsci.2018.07.006.
47. Nasrollahzadeh, M.; Sajadi, S. M. Synthesis and Characterization of Titanium Dioxide Nanoparticles Using Euphorbia Heteradena Jaub Root Extract and Evaluation of Their Stability. *Ceram. Int.*, **2015**, *41* (10), 14435–14439. https://doi.org/10.1016/J.CERAMINT.2015.07.079.
48. Rajakumar, G.; Rahuman, A. A.; Roopan, S. M.; Chung, I.-M.; Anbarasan, K.; Karthikeyan, V. Efficacy of Larvicidal Activity of Green Synthesized Titanium Dioxide Nanoparticles Using Mangifera Indica Extract against Blood-Feeding Parasites. *Parasitol. Res.*, **2015**, *114* (2), 571–581. https://doi.org/10.1007/s00436-014-4219-8.
49. Rajakumar, G.; Rahuman, A. A.; Jayaseelan, C.; Santhoshkumar, T.; Marimuthu, S.; Kamaraj, C.; Bagavan, A.; Zahir, A. A.; Kirthi, A. V.; Elango, G.; et al. Solanum Trilobatum Extract-Mediated Synthesis of Titanium Dioxide Nanoparticles to Control Pediculus Humanus Capitis, Hyalomma Anatolicum and Anopheles Subpictus. *Parasitol. Res.*, **2014**, *113* (2), 469–479. https://doi.org/10.1007/s00436-013-3676-9.
50. Nadeem, M.; Tungmunnithum, D.; Hano, C.; Abbasi, B. H.; Hashmi, S. S.; Ahmad, W.; Zahir, A. The Current Trends in the Green Syntheses of Titanium Oxide Nanoparticles and Their Applications. *Green Chem. Lett. Rev.*, **2018**, *11* (4), 492–502. https://doi.org/10.1080/17518253.2018.1538430.
51. Li, X.-L.; Peng, Q.; Yi, J.-X.; Wang, X.; Li, Y. Near Monodisperse TiO_2 Nanoparticles and Nanorods. *Chem.: A Eur. J.*, **2006**, *12* (8), 2383–2391. https://doi.org/10.1002/chem.200500893.
52. Navale, S. T.; Yang, Z. B.; Liu, C.; Cao, P. J.; Patil, V. B.; Ramgir, N. S.; Mane, R. S.; Stadler, F. J. Enhanced Acetone Sensing Properties of Titanium Dioxide Nanoparticles with a Sub-ppm Detection Limit. *Sensors Actuators B Chem.*, **2018**, *255*, 1701–1710. https://doi.org/10.1016/j.snb.2017.08.186.
53. Ge, M.; Cao, C.; Huang, J.; Li, S.; Chen, Z.; Zhang, K.-Q.; Al-Deyab, S. S.; Lai, Y. A Review of One-Dimensional TiO_2 Nanostructured Materials for Environmental and Energy Applications. *J. Mater. Chem. A*, **2016**, *4* (18), 6772–6801. https://doi.org/10.1039/C5TA09323F.
54. Maheswari, D.; Sreenivasan, D. Review of TiO_2 Nanowires in Dye Sensitized Solar Cell. *Appl. Sol. Energy*, **2015**, *51* (2), 112–116. https://doi.org/10.3103/S0003701 × 15020085.
55. Liu, B.; Boercker, J. E.; Aydil, E. S. Oriented Single Crystalline Titanium Dioxide Nanowires. *Nanotechnology*, **2008**, *19* (50), 505604. https://doi.org/10.1088/0957-4484/19/50/505604.
56. Hayle, S. T.; Gonfa, G. G. Synthesis and Characterization of Titanium Oxide Nanomaterials Using Sol-Gel Method. *Am. J. Nanosci. Nanotechnol.*, **2014**, *2* (1), 1. https://doi.org/10.11648/j.nano.20140201.11.
57. Krishnamoorthy, T.; Thavasi, V.; Subodh, G. M.; Ramakrishna, S. A First Report on the Fabrication of Vertically Aligned Anatase TiO_2 Nanowires by Electrospinning: Preferred Architecture for Nanostructured Solar Cells. *Energy Environ. Sci.*, **2011**, *4* (8), 2807. https://doi.org/10.1039/c1ee01315g.
58. Wu, J.-M.; Shih, H. C.; Wu, W.-T.; Tseng, Y.-K.; Chen, I.-C. Thermal Evaporation Growth and the Luminescence Property of TiO_2 Nanowires. *J. Cryst. Growth*, **2005**, *281* (2–4), 384–390. https://doi.org/10.1016/j.jcrysgro.2005.04.018.
59. Li, Z.; Wang, F.; Kvit, A.; Wang, X. Nitrogen Doped 3D Titanium Dioxide Nanorods Architecture with Significantly Enhanced Visible Light Photoactivity. *J. Phys. Chem. C*, **2015**, *119* (8), 4397–4405. https://doi.org/10.1021/jp512622j.
60. Rahmat, S. T.; Tan, W. K.; Kawamura, G.; Matsuda, A.; Lockman, Z. Synthesis of Rutile TiO_2 Nanowires by Thermal Oxidation of Titanium in the Presence of KOH and Their Ability to Photoreduce Cr(VI) Ions. *J. Alloys Compd.*, **2020**, *812*, 152094. https://doi.org/10.1016/J.JALLCOM.2019.152094.

61. Boercker, J. E.; Enache-Pommer, E.; Aydil, E. S. Growth Mechanism of Titanium Dioxide Nanowires for Dye-Sensitized Solar Cells. *Nanotechnology*, **2008**, *19* (9), 095604. https://doi.org/10.1088/0957-4484/19/9/095604.
62. Tamilselvan, V.; Yuvaraj, D.; Rakesh Kumar, R.; Narasimha Rao, K.; Tamilselvan, V.; Yuvaraj, D.; Rakesh Kumar, R.; Narasimha Rao, K. Growth of Rutile TiO_2 Nanorods on TiO_2 Seed Layer Deposited by Electron Beam Evaporation. *Appl. Surf. Sci.*, **2012**, *258* (10), 4283–4287. https://doi.org/10.1016/j.apsusc.2011.12.079.
63. Kasuga, T.; Hiramatsu, M.; Hoson, A.; Sekino, T.; Niihara, K. Formation of Titanium Oxide Nanotube. *Langmuir*, **1998**, *14* (12), 3160–3163. https://doi.org/10.1021/la9713816.
64. Yao, B. D.; Chan, Y. F.; Zhang, X. Y.; Zhang, W. F.; Yang, Z. Y.; Wang, N. Formation Mechanism of TiO_2 Nanotubes. *Appl. Phys. Lett.*, **2003**, *82* (2), 281–283. https://doi.org/10.1063/1.1537518.
65. Peng, X.; Chen, A. Aligned TiO_2 Nanorod Arrays Synthesized by Oxidizing Titanium with Acetone. *J. Mater. Chem.*, **2004**, *14* (16), 2542. https://doi.org/10.1039/b404750h.
66. Mor, G. K.; Carvalho, M. A.; Varghese, O. K.; Pishko, M. V.; Grimes, C. A. A Room-Temperature TiO_2 -Nanotube Hydrogen Sensor Able to Self-Clean Photoactively from Environmental Contamination. *J. Mater. Res.*, **2004**, *19* (2), 628–634. https://doi.org/10.1557/jmr.2004.19.2.628.
67. Pradhan, S. K.; Reucroft, P. J.; Yang, F.; Dozier, A. Growth of TiO_2 Nanorods by Metalorganic Chemical Vapor Deposition. *J. Cryst. Growth*, **2003**, *256* (1–2), 83–88. https://doi.org/10.1016/S0022-0248(03)01339-3.
68. Jiang, Z.; Miao, W.; Zhu, X.; Yang, G.; Yuan, Z.; Chen, J.; Ji, X.; Kong, F.; Huang, B. Modifying Lewis Base on TiO_2 Nanosheets for Enhancing CO_2 Adsorption and the Separation of Photogenerated Charge Carriers. *Appl. Catal. B Environ.*, **2019**, *256*, 117881. https://doi.org/10.1016/J.APCATB.2019.117881.
69. Yang, H. G.; Sun, C. H.; Qiao, S. Z.; Zou, J.; Liu, G.; Smith, S. C.; Cheng, H. M.; Lu, G. Q. Anatase TiO_2 Single Crystals with a Large Percentage of Reactive Facets. *Nature*, **2008**, *453* (7195), 638–641. https://doi.org/10.1038/nature06964.
70. Yang, H. G.; Liu, G.; Qiao, S. Z.; Sun, C. H.; Jin, Y. G.; Smith, S. C.; Zou, J.; Cheng, H. M.; Lu, G. Q. (Max). Solvothermal Synthesis and Photoreactivity of Anatase TiO_2 Nanosheets with Dominant {001} Facets. *J. Am. Chem. Soc.*, **2009**, *131* (11), 4078–4083. https://doi.org/10.1021/ja808790p.
71. Sun, L.; Xie, J.; Li, Q.; Wang, F.; Xi, X.; Li, L.; Wu, J.; Shao, R.; Chen, Z. Facile Synthesis of Thin Black $TiO_2 - x$ Nanosheets with Enhanced Lithium-Storage Capacity and Visible Light Photocatalytic Hydrogen Production. *J. Solid State Electrochem.*, **2019**, *23* (3), 803–810. https://doi.org/10.1007/s10008-018-04180-7.
72. Wen, C. Z.; Jiang, H. B.; Qiao, S. Z.; Yang, H. G.; Lu, G. Q. (Max). Synthesis of High-Reactive Facets Dominated Anatase TiO_2. *J. Mater. Chem.*, **2011**, *21* (20), 7052. https://doi.org/10.1039/c1jm00068c.
73. Zhang, H.; Wang, R.; Lin, D.; Zeng, Y.; Lu, X. Ni-Based Nanostructures as High-Performance Cathodes for Rechargeable Ni–Zn Battery. *ChemNanoMat*, **2018**, *4* (6), 525–536. https://doi.org/10.1002/cnma.201800078.
74. Wang, X.; Chen, L.; Li, F.; Zhang, S.; Chen, X.; Yin, J. Synthesis of Hollow NiO Nanostructures and Their Application for Supercapacitor Electrode. *Ionics (Kiel).*, **2019**, *25* (2), 697–705. https://doi.org/10.1007/s11581-018-2771-4.
75. Paulose, R.; Mohan, R.; Parihar, V. Nanostructured Nickel Oxide and Its Electrochemical Behaviour: A Brief Review. *Nano-Structures & Nano-Objects*, **2017**, *11*, 102–111. https://doi.org/10.1016/J. NANOSO.2017.07.003.
76. Farhadi, S.; Roostaei-Zaniyani, Z. Preparation and Characterization of NiO Nanoparticles from Thermal Decomposition of the [Ni(En)3](NO3)2 Complex: A Facile and Low-Temperature Route. *Polyhedron*, **2011**, *30* (6), 971–975. https://doi.org/10.1016/J.POLY.2010.12.044.

77. Fominykh, K.; Feckl, J. M.; Sicklinger, J.; Döblinger, M.; Böcklein, S.; Ziegler, J.; Peter, L.; Rathousky, J.; Scheidt, E.-W.; Bein, T.; et al. Ultrasmall Dispersible Crystalline Nickel Oxide Nanoparticles as High-Performance Catalysts for Electrochemical Water Splitting. *Adv. Funct. Mater.*, **2014**, *24* (21), 3123–3129. https://doi.org/10.1002/adfm.201303600.
78. Jafari, A.; Pilban Jahromi, S.; Boustani, K.; Goh, B. T.; Huang, N. M. Evolution of Structural and Magnetic Properties of Nickel Oxide Nanoparticles: Influence of Annealing Ambient and Temperature. *J. Magn. Magn. Mater.*, **2019**, *469*, 383–390. https://doi.org/10.1016/J. JMMM.2018.08.005.
79. Bahari Molla Mahaleh, Y.; Sadrnezhaad, S. K.; Hosseini, D. NiO Nanoparticles Synthesis by Chemical Precipitation and Effect of Applied Surfactant on Distribution of Particle Size. *J. Nanomater.*, **2008**, *2008*, 1–4. https://doi.org/10.1155/2008/470595.
80. Angel Ezhilarasi, A.; Judith Vijaya, J.; Kaviyarasu, K.; John Kennedy, L.; Ramalingam, R. J.; Al-Lohedan, H. A. Green Synthesis of NiO Nanoparticles Using Aegle Marmelos Leaf Extract for the Evaluation of In-Vitro Cytotoxicity, Antibacterial and Photocatalytic Properties. *J. Photochem. Photobiol. B Biol.*, **2018**, *180*, 39–50. https://doi.org/10.1016/j.jphotobiol.2018.01.023.
81. Kaur, N.; Comini, E.; Zappa, D.; Poli, N.; Sberveglieri, G. Nickel Oxide Nanowires: Vapor Liquid Solid Synthesis and Integration into a Gas Sensing Device. *Nanotechnology*, **2016**, *27* (20), 205701. https://doi.org/10.1088/0957-4484/27/20/205701.
82. Xiang, W.; Dong, Z.; Luo, Y.; Zhao, J.; Wang, J.; Ibrahim, K.; Zhan, H.; Yue, W.; Guo, H. Synthesis of NiO Nanotubes via a Dynamic Thermal Oxidation Process. *Materials (Basel).*, **2019**, *12* (5), 805. https://doi.org/10.3390/ma12050805.
83. Vidhyadharan, B.; Zain, N. K. M.; Misnon, I. I.; Aziz, R. A.; Ismail, J.; Yusoff, M. M.; Jose, R. High Performance Supercapacitor Electrodes from Electrospun Nickel Oxide Nanowires. *J. Alloys Compd.*, **2014**, *610*, 143–150. https://doi.org/10.1016/J.JALLCOM.2014.04.211.
84. Yang, Q.; Sha, J.; Ma, X.; Yang, D. Synthesis of NiO Nanowires by a Sol-Gel Process. *Mater. Lett.*, **2005**, *59* (14–15), 1967–1970. https://doi.org/10.1016/J.MATLET.2005.02.037.
85. Khalil, A.; Hashaikeh, R. Electrospun Nickel Oxide Nanofibers: Microstructure and Surface Evolution. *Appl. Surf. Sci.*, **2015**, *357*, 1333–1342. https://doi.org/10.1016/J.APSUSC.2015.09.250.
86. Sun, W.; Rui, X.; Zhu, J.; Yu, L.; Zhang, Y.; Xu, Z.; Madhavi, S.; Yan, Q. Ultrathin Nickel Oxide Nanosheets for Enhanced Sodium and Lithium Storage. *J. Power Sources*, **2015**, *274*, 755–761. https://doi.org/10.1016/J.JPOWSOUR.2014.10.105.
87. Zhu, Y.; Cao, C.; Tao, S.; Chu, W.; Wu, Z.; Li, Y. Ultrathin Nickel Hydroxide and Oxide Nanosheets: Synthesis, Characterizations and Excellent Supercapacitor Performances. *Sci. Rep.*, **2014**, *4* (1), 5787. https://doi.org/10.1038/srep05787.
88. Abdullah, M. A. R.; Mamat, M. H.; Ismail, A. S.; Malek, M. F.; Suriani, A. B.; Ahmad, M. K.; Shameem Banu, I. B.; Amiruddin, R.; Rusop, M. Direct and Seedless Growth of Nickel Oxide Nanosheet Architectures on ITO Using a Novel Solution Immersion Method. *Mater. Lett.*, **2019**, *236*, 460–464. https://doi.org/10.1016/J.MATLET.2018.10.163.
89. Long, H.; Shi, T.; Hu, H.; Jiang, S.; Xi, S.; Tang, Z. Growth of Hierarchal Mesoporous NiO Nanosheets on Carbon Cloth as Binder-Free Anodes for High-Performance Flexible Lithium-Ion Batteries. *Sci. Rep.*, **2015**, *4* (1), 7413. https://doi.org/10.1038/srep07413.
90. Ahmad, R.; Bedük, T.; Majhi, S. M.; Salama, K. N. One-Step Synthesis and Decoration of Nickel Oxide Nanosheets with Gold Nanoparticles by Reduction Method for Hydrazine Sensing Application. *Sensors Actuators B Chem.*, **2019**, *286*, 139–147. https://doi.org/10.1016/j.snb.2019.01.132.

91. Zhu, Y.; Cao, C.; Tao, S.; Chu, W.; Wu, Z.; Li, Y. Ultrathin Nickel Hydroxide and Oxide Nanosheets: Synthesis, Characterizations and Excellent Supercapacitor Performances. *Sci. Rep.*, **2015**, *4* (1), 5787. https://doi.org/10.1038/srep05787.
92. Wu, C. M.; Naseem, S.; Chou, M. H.; Wang, J. H.; Jian, Y. Q. Recent Advances in Tungsten-Oxide-Based Materials and Their Applications. *Front. Mater.*, **2019**, *4*, 49. https://doi.org/10.3389/fmats.2019.00049.
93. Cong, S.; Geng, F.; Zhao, Z. Tungsten Oxide Materials for Optoelectronic Applications. *Adv. Mater.*, **2016**, *28* (47), 10518–10528. https://doi.org/10.1002/adma.201601109.
94. Zheng, M.; Tang, H.; Hu, Q.; Zheng, S.; Li, L.; Xu, J.; Pang, H. Tungsten-Based Materials for Lithium-Ion Batteries. *Adv. Funct. Mater.*, **2018**, *28* (20), 1–26. https://doi.org/10.1002/adfm.201707500.
95. Rezaee, O.; Mahmoudi Chenari, H.; Ghodsi, F. E. Precipitation Synthesis of Tungsten Oxide Nanoparticles: X-Ray Line Broadening Analysis and Photocatalytic Efficiency Study. *J. Sol-Gel Sci. Technol.*, **2016**, *80* (1), 109–118. https://doi.org/10.1007/s10971-016-4073-5.
96. Supothina, S.; Seeharaj, P.; Yoriya, S.; Sriyudthsak, M. Synthesis of Tungsten Oxide Nanoparticles by Acid Precipitation Method. *Ceram. Int.*, **2007**, *33* (6), 931–936. https://doi.org/10.1016/J.CERAMINT.2006.02.007.
97. Ahmadi, M.; Younesi, R.; Guinel, M. J.-F. Synthesis of Tungsten Oxide Nanoparticles Using a Hydrothermal Method at Ambient Pressure. *J. Mater. Res.*, **2014**, *29* (13), 1424–1430. https://doi.org/10.1557/jmr.2014.155.
98. Movlaee, K.; Periasamy, P.; Krishnakumar, T.; Ganjali, M. R.; Leonardi, S. G.; Neri, G.; Chavali, M.; Siril, P. F.; Devarajan, V. P. Microwave-Assisted Synthesis and Characterization of WO_x Nanostructures for Gas Sensor Application. *J. Alloys Compd.*, **2018**, *762*, 745–753. https://doi.org/10.1016/J.JALLCOM.2018.05.189.
99. Watanabe, H.; Fujikata, K.; Oaki, Y.; Imai, H. Band-Gap Expansion of Tungsten Oxide Quantum Dots Synthesized in Sub-Nano Porous Silica. *Chem. Commun.*, **2013**, *49* (76), 8477. https://doi.org/10.1039/c3cc44264k.
100. Cong, S.; Tian, Y.; Li, Q.; Zhao, Z.; Geng, F. Single-Crystalline Tungsten Oxide Quantum Dots for Fast Pseudocapacitor and Electrochromic Applications. *Adv. Mater.*, **2014**, *26* (25), 4260–4267. https://doi.org/10.1002/adma.201400447.
101. Baek, Y.; Yong, K. Controlled Growth and Characterization of Tungsten Oxide Nanowires Using Thermal Evaporation of WO_3 Powder. *J. Phys. Chem. C*, **2007**, *111* (3), 1213–1218. https://doi.org/10.1021/jp0659857.
102. Park, S.; Kim, H.; Jin, C.; Lee, C. Intense Ultraviolet Emission from Needle-like WO_3 Nanostructures Synthesized by Noncatalytic Thermal Evaporation. *Nanoscale Res. Lett.*, **2011**, *6* (1), 451. https://doi.org/10.1186/1556-276X-6-451.
103. Gu, G.; Zheng, B.; Han, W. Q.; Roth, S.; Liu, J. Tungsten Oxide Nanowires on Tungsten Substrates. *Nano Lett.*, **2002**, *2* (8), 849–851. https://doi.org/10.1021/nl025618g.
104. Li, L.; Zhang, Y.; Fang, X.; Zhai, T.; Liao, M.; Sun, X.; Koide, Y.; Bando, Y.; Golberg, D. WO_3 Nanowires on Carbon Papers: Electronic Transport, Improved Ultraviolet-Light Photodetectors and Excellent Field Emitters. *J. Mater. Chem.*, **2011**, *21* (18), 6525. https://doi.org/10.1039/c0jm04557h.
105. Qi, H.; Wang, C.; Liu, J. A Simple Method for the Synthesis of Highly Oriented Potassium-Doped Tungsten Oxide Nanowires. *Adv. Mater.*, **2003**, *15* (5), 411–414. https://doi.org/10.1002/adma.200390094.
106. Lee, K.; Seo, W. S.; Park, J. T. Synthesis and Optical Properties of Colloidal Tungsten Oxide Nanorods. *J. Am. Chem. Soc.*, **2003**, *125* (12), 3408–3409. https://doi.org/10.1021/ja034011e.
107. Qin, Y.; Li, X.; Wang, F.; Hu, M. Solvothermally Synthesized Tungsten Oxide Nanowires/Nanorods for NO_2 Gas Sensor Applications. *J. Alloys Compd.*, **2011**, *509* (33), 8401–8406. https://doi.org/10.1016/j.jallcom.2011.05.100.

108. Song, X. C.; Zheng, Y. F.; Yang, E.; Wang, Y. Large-Scale Hydrothermal Synthesis of WO_3 Nanowires in the Presence of K_2SO_4. *Mater. Lett.*, **2007**, *61* (18), 3904–3908. https://doi.org/10.1016/J.MATLET.2006.12.055.
109. Wei, J.; Jiao, X.; Wang, T.; Chen, D. The Fast and Reversible Intrinsic Photochromic Response of Hydrated Tungsten Oxide Nanosheets. *J. Mater. Chem. C*, **2015**, *3* (29), 7597–7603. https://doi.org/10.1039/C5TC01350J.
110. Ming, X.; Guo, A.; Wang, G.; Wang, X. Two-Dimensional Defective Tungsten Oxide Nanosheets as High Performance Photo-Absorbers for Efficient Solar Steam Generation. *Sol. Energy Mater. Sol. Cells*, **2018**, *185*, 333–341. https://doi.org/10.1016/J.SOLMAT.2018.05.049.
111. Yan, J.; Wang, T.; Wu, G.; Dai, W.; Guan, N.; Li, L.; Gong, J. Tungsten Oxide Single Crystal Nanosheets for Enhanced Multichannel Solar Light Harvesting. *Adv. Mater.*, **2015**, *27* (9), 1580–1586. https://doi.org/10.1002/adma.201404792.
112. Chen, D.; Gao, L.; Yasumori, A.; Kuroda, K.; Sugahara, Y. Size- and Shape-Controlled Conversion of Tungstate-Based Inorganic-Organic Hybrid Belts to WO_3 Nanoplates with High Specific Surface Areas. *Small*, **2008**, *4* (10), 1813–1822. https://doi.org/10.1002/smll.200800205.
113. Liu, M.; Su, B.; Tang, Y.; Jiang, X.; Yu, A. Recent Advances in Nanostructured Vanadium Oxides and Composites for Energy Conversion. *Adv. Energy Mater.*, **2017**, *7* (23), 1700885. https://doi.org/10.1002/aenm.201700885.
114. Diaz, C.; Barrera, G.; Segovia, M.; Valenzuela, M. L.; Osiak, M.; O'Dwyer, C. Crystallizing Vanadium Pentoxide Nanostructures in the Solid-State Using Modified Block Copolymer and Chitosan Complexes. *J. Nanomater.*, **2015**, *2015*, 1–13. https://doi.org/10.1155/2015/105157.
115. Kianfar, E. Recent Advances in Synthesis, Properties, and Applications of Vanadium Oxide Nanotube. *Microchem. J.*, **2019**, *145*, 966–978. https://doi.org/10.1016/J.MICROC.2018.12.008.
116. Liu, X.; Zeng, J.; Yang, H.; Zhou, K.; Pan, D. V_2O_5-Based Nanomaterials: Synthesis and Their Applications. *RSC Adv.*, **2018**, *8* (8), 4014–4031. https://doi.org/10.1039/C7RA12523B.
117. Menezes, W. G.; Reis, D. M.; Benedetti, T. M.; Oliveira, M. M.; Soares, J. F.; Torresi, R. M.; Zarbin, A. J. G. V_2O_5 Nanoparticles Obtained from a Synthetic Bariandite-like Vanadium Oxide: Synthesis, Characterization and Electrochemical Behavior in an Ionic Liquid. *J. Colloid Interface Sci.*, **2009**, *337* (2), 586–593. https://doi.org/10.1016/j.jcis.2009.05.050.
118. Raj, A. T.; Ramanujan, K.; Thangavel, S.; Gopalakrishan, S.; Raghavan, N.; Venugopal, G. Facile Synthesis of Vanadium-Pentoxide Nanoparticles and Study on Their Electrochemical, Photocatalytic Properties. *J. Nanosci. Nanotechnol.*, **2015**, *15* (5), 3802–3808. https://doi.org/10.1166/jnn.2015.9543.
119. Alghool, S.; Abd El-Halim, H. F.; Mostafa, A. M. An Eco-Friendly Synthesis of V_2O_5 Nanoparticles and Their Catalytic Activity for the Degradation of 4-Nitrophrnol. *J. Inorg. Organomet. Polym. Mater.*, **2019**, *29* (4), 1324–1330. https://doi.org/10.1007/s10904-019-01096-1.
120. Zhu, K.; Meng, Y.; Qiu, H.; Gao, Y.; Wang, C.; Du, F.; Wei, Y.; Chen, G.; Wang, C.; Chen, G. Facile Synthesis of V_2O_5 Nanoparticles as a Capable Cathode for High Energy Lithium-Ion Batteries. *J. Alloys Compd.*, **2015**, *650*, 370–373. https://doi.org/10.1016/J.JALLCOM.2015.07.122.
121. Zhou, F.; Zhao, X.; Yuan, C.; Li, L. Vanadium Pentoxide Nanowires: Hydrothermal Synthesis, Formation Mechanism, and Phase Control Parameters. *Cryst. Growth Des.*, **2008**, *8* (2), 723–727. https://doi.org/10.1021/cg060816x.
122. Cheah, Y. L.; Gupta, N.; Pramana, S. S.; Aravindan, V.; Wee, G.; Srinivasan, M. Morphology, Structure and Electrochemical Properties of Single Phase Electrospun Vanadium Pentoxide Nanofibers for Lithium Ion Batteries. *J. Power Sources*, **2011**, *196* (15), 6465–6472. https://doi.org/10.1016/J.JPOWSOUR.2011.03.039.

123. Takahashi, K.; Limmer, S. J.; Wang,Y.; Cao, G. Synthesis and Electrochemical Properties of Single-Crystal V_2O_5 Nanorod Arrays by Template-Based Electrodeposition. **2004**. https://doi.org/10.1021/JP0491820.
124. Velazquez, J. M.; Banerjee, S. Catalytic Growth of Single-Crystalline V_2O_5 Nanowire Arrays. *Small*, **2009**, *5* (9), 1025–1029. https://doi.org/10.1002/smll.200801278.
125. Zhai, T.; Liu, H.; Li, H.; Fang, X.; Liao, M.; Li, L.; Zhou, H.; Koide, Y.; Bando, Y.; Golberg, D. Centimeter-Long V2O5 Nanowires: From Synthesis to Field-Emission, Electrochemical, Electrical Transport, and Photoconductive Properties. *Adv. Mater.*, **2010**, *22* (23), 2547–2552. https://doi.org/10.1002/adma.200903586.
126. Navyashree, G. R.; Hareesh, K.; Nagabhushana, H.; Nagaraju, G.; Sunitha, D. V. Vanadium Pentoxide Nanorods in Latent Finger Print Detection. *Mater. Res. Express*, **2019**, *6* (8), 084003. https://doi.org/10.1088/2053-1591/ab1949.
127. Yu, Y.; Li, J.; Wang, X.; Chang, B.; Wang, J.; Ahmad, M.; Sun, H. Oxygen Vacancies Enhance Lithium Storage Performance in Ultralong Vanadium Pentoxide Nanobelt Cathodes. *J. Colloid Interface Sci.*, **2019**, *539*, 118–125. https://doi.org/10.1016/J.JCIS.2018.12.046.
128. Takahashi, K.; Limmer, S. J.; Wang, Y.; Cao, G. Synthesis and Electrochemical Properties of Single-Crystal V_2O_5 Nanorod Arrays by Template-Based Electrodeposition. *J. Phys. Chem. B*, **2004**, *108* (28), 9795–9800. https://doi.org/10.1021/jp0491820.
129. Wang, Y.; Takahashi, K.; Shang, H.; Cao, G. Synthesis and Electrochemical Properties of Vanadium Pentoxide Nanotube Arrays. *J. Phys. Chem. B*, **2005**, *109* (8), 3085–3088. https://doi.org/10.1021/jp044286w.
130. Yalagala, B. P.; Sahatiya, P.; Kolli, C. sekhar R.; Khandelwal, S.; Mattela, V.; Badhulika, S. V_2O_5 Nanosheets for Flexible Memristors and Broadband Photodetectors. *ACS Appl. Nano Mater.*, **2019**, *2* (2), 937–947. https://doi.org/10.1021/acsanm.8b02233.
131. Song, H.; Zhang, C.; Liu, Y.; Liu, C.; Nan, X.; Cao, G. Facile Synthesis of Mesoporous V_2O_5 Nanosheets with Superior Rate Capability and Excellent Cycling Stability for Lithium Ion Batteries. *J. Power Sources*, **2015**, *294*, 1–7. https://doi.org/10.1016/J.JPOWSOUR.2015.06.055.
132. Rui, X.; Lu, Z.; Yu, H.; Yang, D.; Hng, H. H.; Lim, T. M.; Yan, Q. Ultrathin V_2O_5 Nanosheet Cathodes: Realizing Ultrafast Reversible Lithium Storage. *Nanoscale*, **2013**, *5* (2), 556–560. https://doi.org/10.1039/C2NR33422D.
133. Nagaraju, D. H.; Wang, Q.; Beaujuge, P.; Alshareef, H. N. Two-Dimensional Heterostructures of V_2O_5 and Reduced Graphene Oxide as Electrodes for High Energy Density Asymmetric Supercapacitors. *J. Mater. Chem. A*, **2014**, *2* (40), 17146–17152. https://doi.org/10.1039/C4TA03731F.
134. Tanaka, S.; Kaneti, Y. V.; Septiani, N. L. W.; Dou, S. X.; Bando, Y.; Hossain, M. S. A.; Kim, J.; Yamauchi, Y. A Review on Iron Oxide-Based Nanoarchitectures for Biomedical, Energy Storage, and Environmental Applications. *Small Methods*, **2019**, *3* (5), 1800512. https://doi.org/10.1002/smtd.201800512.
135. Dadfar, S. M.; Roemhild, K.; Drude, N. I.; von Stillfried, S.; Knüchel, R.; Kiessling, F.; Lammers, T. Iron Oxide Nanoparticles: Diagnostic, Therapeutic and Theranostic Applications. *Adv. Drug Deliv. Rev.*, **2019**, *138*, 302–325. https://doi.org/10.1016/j. addr.2019.01.005.
136. Song, C.; Sun, W.; Xiao, Y.; Shi, X. Ultrasmall Iron Oxide Nanoparticles: Synthesis, Surface Modification, Assembly, and Biomedical Applications. *Drug Discov. Today*, **2019**, *24* (3), 835–844. https://doi.org/10.1016/j.drudis.2019.01.001.
137. Saqib, S.; Munis, M. F. H.; Zaman, W.; Ullah, F.; Shah, S. N.; Ayaz, A.; Farooq, M.; Bahadur, S. Synthesis, Characterization and Use of Iron Oxide Nano Particles for Antibacterial Activity. *Microsc. Res. Tech.*, **2019**, *82* (4), 415–420. https://doi. org/10.1002/jemt.23182.
138. Hufschmid, R.; Arami, H.; Ferguson, R. M.; Gonzales, M.; Teeman, E.; Brush, L. N.; Browning, N. D.; Krishnan, K. M. Synthesis of Phase-Pure and Monodisperse Iron Oxide Nanoparticles by Thermal Decomposition. *Nanoscale*, **2015**, 7 (25), 11142–11154. https://doi.org/10.1039/C5NR01651G.

139. Gao, G.; Wu, H.; Zhang, Y.; Luo, T.; Feng, L.; Huang, P.; He, M.; Cui, D. Synthesis of Ultrasmall Nucleotide-Functionalized Superparamagnetic γ-Fe_2O_3 Nanoparticles. *CrystEngComm*, **2011**, *13* (15), 4810. https://doi.org/10.1039/c1ce05371j.
140. Shen, L.; Bao, J.; Wang, D.; Wang, Y.; Chen, Z.; Ren, L.; Zhou, X.; Ke, X.; Chen, M.; Yang, A. One-Step Synthesis of Monodisperse, Water-Soluble Ultra-Small Fe_3O_4 Nanoparticles for Potential Bio-Application. *Nanoscale*, **2013**, *5* (5), 2133. https://doi.org/10.1039/c2nr33840h.
141. Darbandi, M.; Stromberg, F.; Landers, J.; Reckers, N.; Sanyal, B.; Keune, W.; Wende, H. Nanoscale Size Effect on Surface Spin Canting in Iron Oxide Nanoparticles Synthesized by the Microemulsion Method. *J. Phys. D. Appl. Phys.*, **2012**, *45* (19), 195001. https://doi.org/10.1088/0022-3727/45/19/195001.
142. Tadić, M.; Kusigerski, V.; Marković, D.; Panjan, M.; Milošević, I.; Spasojević, V. Highly Crystalline Superparamagnetic Iron Oxide Nanoparticles (SPION) in a Silica Matrix. *J. Alloys Compd.*, **2012**, *525*, 28–33. https://doi.org/10.1016/J.JALLCOM.2012.02.056.
143. Fu, Y..; Wang, R..; Xu, J.; Chen, J.; Yan, Y.; Narlikar, A..; Zhang, H. Synthesis of Large Arrays of Aligned α-Fe_2O_3 Nanowires. *Chem. Phys. Lett.*, **2003**, *379* (3–4), 373–379. https://doi.org/10.1016/J.CPLETT.2003.08.061.
144. Wang, G.; Gou, X.; Horvat, J.; Park, J. Facile Synthesis and Characterization of Iron Oxide Semiconductor Nanowires for Gas Sensing Application. *J. Phys. Chem. C*, **2008**, *112* (39), 15220–15225. https://doi.org/10.1021/jp803869e.
145. Rangaraju, R. R.; Raja, K. S.; Panday, A.; Misra, M. An Investigation on Room Temperature Synthesis of Vertically Oriented Arrays of Iron Oxide Nanotubes by Anodization of Iron. *Electrochim. Acta*, **2010**, *55* (3), 785–793. https://doi.org/10.1016/j.electacta.2009.07.012.
146. Deng, J.; Liu, J.; Dai, H.; Wang, W. Preparation of α-Fe_2O_3 Nanowires through Electrospinning and Their Ag_3PO_4 Heterojunction Composites with Enhanced Visible Light Photocatalytic Activity. *Ferroelectrics*, **2018**, *528* (1), 58–65. https://doi.org/10.1080/00150193.2018.1448625.
147. Fu, Y.; Chen, J.; Zhang, H. Synthesis of Fe_2O_3 Nanowires by Oxidation of Iron. *Chem. Phys. Lett.*, **2001**, *350* (5–6), 491–494. https://doi.org/10.1016/S0009-2614(01)01352-5.
148. Yuan, L.; Jiang, Q.; Wang, J.; Zhou, G. The Growth of Hematite Nanobelts and Nanowires—tune the Shape via Oxygen Gas Pressure. *J. Mater. Res.*, **2012**, *27* (7), 1014–1021. https://doi.org/10.1557/jmr.2012.19.
149. Nasibulin, A. G.; Rackauskas, S.; Jiang, H.; Tian, Y.; Mudimela, P. R.; Shandakov, S. D.; Nasibulina, L. I.; Jani, S.; Kauppinen, E. I. Simple and Rapid Synthesis of α-Fe_2O_3 Nanowires under Ambient Conditions. *Nano Res.*, **2009**, *2* (5), 373–379. https://doi.org/10.1007/s12274-009-9036-5.
150. Zhao, Y. M.; Li, Y.-H.; Ma, R. Z.; Roe, M. J.; McCartney, D. G.; Zhu, Y. Q. Growth and Characterization of Iron Oxide Nanorods/Nanobelts Prepared by a Simple Iron–Water Reaction. *Small*, **2006**, *2* (3), 422–427. https://doi.org/10.1002/smll.200500347.
151. Yang, L.; Wu, Y.; Wu, Y.; Younas, W.; Jia, J.; Cao, C. Hierarchical Flower-like Fe_2O_3 Mesoporous Nanosheets with Superior Electrochemical Lithium Storage Performance. *J. Energy Storage*, **2019**, *23*, 363–370. https://doi.org/10.1016/J.EST.2019.04.003.
152. Zhang, Z.; Li, Z.; Zhang, J.; Bian, H.; Wang, T.; Gao, J.; Li, J. Structural and Magnetic Properties of Porous Fe_xO_y Nanosheets and Nanotubes Fabricated by Electrospinning. *Ceram. Int.*, **2019**, *45* (1), 457–461. https://doi.org/10.1016/j.ceramint.2018.09.189.
153. Zhao, P.; Wang, N.; Hu, W.; Komarneni, S. Anode Electrodeposition of 3D Mesoporous Fe_2O_3 Nanosheets on Carbon Fabric for Flexible Solid-State Asymmetric Supercapacitor. *Ceram. Int.*, **2019**, *45* (8), 10420–10428. https://doi.org/10.1016/j.ceramint.2019.02.101.

154. Liu, Y.; Xu, J.; Zhao, C.; Li, Y.; Rao, Z.; Zhao, S.; Yang, J.; Zhu, H.; Shuai, S.; Hao, J. Preparation and Properties of Hierarchically Self-Assembled Iron Oxide Nanosheets with Specific Removal Performance of Lead Ions. *Mater. Res. Express*, **2019**, *6* (4), 045020. https://doi.org/10.1088/2053-1591/aaf81c.
155. Pan, J.; Shen, H.; Mathur, S. One-Dimensional SnO_2 Nanostructures: Synthesis and Applications. *J. Nanotechnol.*, **2012**, *2012*, 1–12. https://doi.org/10.1155/2012/917320.
156. Zhao, Q.; Ma, L.; Zhang, Q.; Wang, C.; Xu, X. SnO_2 -Based Nanomaterials: Synthesis and Application in Lithium-Ion Batteries and Supercapacitors. *J. Nanomater.*, **2015**, *2015*, 1–15. https://doi.org/10.1155/2015/850147.
157. Das, S.; Jayaraman, V. SnO_2: A Comprehensive Review on Structures and Gas Sensors. *Prog. Mater. Sci.*, **2014**, *66*, 112–255. https://doi.org/10.1016/j.pmatsci.2014.06.003.
158. Aziz, M.; Saber Abbas, S.; Wan Baharom, W. R. Size-Controlled Synthesis of SnO_2 Nanoparticles by Sol-Gel Method. *Mater. Lett.*, **2013**, *91*, 31–34. https://doi.org/10.1016/J.MATLET.2012.09.079.
159. Chiu, H. C.; Yeh, C. S. Hydrothermal Synthesis of SnO_2 Nanoparticles and Their Gas-Sensing of Alcohol. *J. Phys. Chem. C*, **2007**, *111* (20), 7256–7259. https://doi.org/10.1021/jp0688355.
160. Bhattacharjee, A.; Ahmaruzzaman, M.; Sinha, T. A Novel Approach for the Synthesis of SnO_2 Nanoparticles and Its Application as a Catalyst in the Reduction and Photodegradation of Organic Compounds. *Spectrochim. Acta. A. Mol. Biomol. Spectrosc.*, **2015**, *136 Pt B*, 751–760. https://doi.org/10.1016/j.saa.2014.09.092.
161. Kumar, R. R.; Parmar, M.; Narasimha Rao, K.; Rajanna, K.; Phani, A. R. Novel Low-Temperature Growth of SnO_2 Nanowires and Their Gas-Sensing Properties. *Scr. Mater.*, **2013**, *68* (6), 408–411. https://doi.org/10.1016/j.scriptamat.2012.11.002.
162. Kumar, R. R.; Rao, K. N.; Rajanna, K.; Phani, A. R. Low Temperature Growth of SnO_2 Nanowires by Electron Beam Evaporation and Their Application in UV Light Detection. *Mater. Res. Bull.*, **2013**, *48* (4), 1545–1552. https://doi.org/10.1016/j.materresbull.2012.12.050.
163. Liu, Z.; Zhang, D.; Han, S.; Li, C.; Tang, T.; Jin, W.; Liu, X.; Lei, B.; Zhou, C. Laser Ablation Synthesis and Electron Transport Studies of Tin Oxide Nanowires. *Adv. Mater.*, **2003**, *15* (20), 1754–1757. https://doi.org/10.1002/adma.200305439.
164. Krishnamoorthy, T.; Tang, M. Z.; Verma, A.; Nair, A. S.; Pliszka, D.; Mhaisalkar, S. G.; Ramakrishna, S. A Facile Route to Vertically Aligned Electrospun SnO_2 Nanowires on a Transparent Conducting Oxide Substrate for Dye-Sensitized Solar Cells. *J. Mater. Chem.*, **2012**, *22* (5), 2166–2172. https://doi.org/10.1039/C1JM15047B.
165. Lupan, O.; Chow, L.; Chai, G.; Schulte, A.; Park, S.; Heinrich, H. A Rapid Hydrothermal Synthesis of Rutile SnO_2 Nanowires. *Mater. Sci. Eng. B*, **2009**, *157* (1–3), 101–104. https://doi.org/10.1016/J.MSEB.2008.12.035.
166. Jiang, J.; Heck, F.; Hofmann, D. M.; Eickhoff, M. Synthesis of SnO_2 Nanowires Using SnI_2 as Precursor and Their Application as High-Performance Self-Powered Ultraviolet Photodetectors. *Phys. status solidi*, **2018**, *255* (3), 1700426. https://doi.org/10.1002/pssb.201700426.
167. Luo, L.; Liang, F.; Jie, J. Sn-Catalyzed Synthesis of SnO_2 Nanowires and Their Optoelectronic Characteristics. *Nanotechnology*, **2011**, *22* (48), 485701. https://doi.org/10.1088/0957-4484/22/48/485701.
168. Dai, Z. R.; Gole, J. L.; Stout, J. D.; Wang, Z. L. Tin Oxide Nanowires, Nanoribbons, and Nanotubes. *J. Phys. Chem. B*, **2002**, *106* (6), 1274–1279. https://doi.org/10.1021/jp013214r.
169. Xu, R.; Zhang, L.-X.; Li, M.-W.; Yin, Y.-Y.; Yin, J.; Zhu, M.-Y.; Chen, J.-J.; Wang, Y.; Bie, L.-J. Ultrathin SnO_2 Nanosheets with Dominant High-Energy {001} Facets for Low Temperature Formaldehyde Gas Sensor. *Sensors Actuators B Chem.*, **2019**, *289*, 186–194. https://doi.org/10.1016/j.snb.2019.03.012.

170. Zhu, J.; Lu, Z; Aruna, S. T.; Aurbach, Doron.; Gedanken, A. Sonochemical Synthesis of SnO_2 Nanoparticles and Their Preliminary Study as Li Insertion Electrodes. **2000**. https://doi.org/10.1021/CM990683L.
171. Wang, Y.; Tian, J.; Fei, C.; Lv, L.; Liu, X.; Zhao, Z.; Cao, G. Microwave-Assisted Synthesis of SnO_2 Nanosheets Photoanodes for Dye-Sensitized Solar Cells. *J. Phys. Chem. C*, **2014**, *118* (45), 25931–25938. https://doi.org/10.1021/jp5089146.
172. Wei, W.; Du, P.; Liu, D.; Wang, H.; Liu, P. Facile Mass Production of Nanoporous SnO_2 Nanosheets as Anode Materials for High Performance Lithium-Ion Batteries. *J. Colloid Interface Sci.*, **2017**, *503*, 205–213. https://doi.org/10.1016/j.jcis.2017.05.017.
173. Choi, P. G.; Izu, N.; Shirahata, N.; Masuda, Y. SnO_2 Nanosheets for Selective Alkene Gas Sensing. *ACS Appl. Nano Mater.*, **2019**, *2* (4), 1820–1827. https://doi.org/10.1021/acsanm.8b01945.
174. Umar, A.; Ammar, H. Y.; Kumar, R.; Almas, T.; Ibrahim, A. A.; AlAssiri, M. S.; Abaker, M.; Baskoutas, S. Efficient H_2 Gas Sensor Based on 2D SnO_2 Disks: Experimental and Theoretical Studies. *Int. J. Hydrogen Energy*, **2019**. https://doi.org/10.1016/j.ijhydene.2019.04.269.
175. Wagner, R. S.; Ellis, W. C. VAPOR-LIQUID-SOLID MECHANISM OF SINGLE CRYSTAL GROWTH. *Appl. Phys. Lett.*, **1964**, *4* (5), 89–90. https://doi.org/10.1063/1.1753975.
176. Wu, Y.; Yang, P. Direct Observation of Vapor-Liquid-Solid Nanowire Growth. *J. Am. Chem. Soc.*, **2001**, 3165–3166. https://doi.org/10.1021/ja0059084.
177. Simon, H.; Krekeler, T.; Schaan, G.; Mader, W. Metal-Seeded Growth Mechanism of ZnO Nanowires. *Cryst. Growth Des.*, **2013**, *13* (2), 572–580. https://doi.org/10.1021/cg301640v.
178. Morber, J. R.; Ding, Y.; Haluska, M. S.; Li, Y.; Liu, J. P.; Wang, Z. L.; Snyder, R. L. PLD-Assisted VLS Growth of Aligned Ferrite Nanorods, Nanowires, and Nanobelts-Synthesis, and Properties. *J. Phys. Chem. B*, **2006**, *110* (43), 21672–21679. https://doi.org/10.1021/jp064484i.
179. Kaur, N.; Comini, E.; Poli, N.; Zappa, D.; Sberveglieri, G. Nickel Oxide Nanowires Growth by VLS Technique for Gas Sensing Application. *Procedia Eng.*, **2015**, *120*, 760–763. https://doi.org/10.1016/J.PROENG.2015.08.805.
180. Kaur, N.; Zappa, D.; Poli, N.; Comini, E. Integration of VLS-Grown WO_3 Nanowires into Sensing Devices for the Detection of H_2S and O_3. *ACS Omega*, **2019**, *4* (15), 16336–16343. https://doi.org/10.1021/acsomega.9b01792.
181. Lee, J.-C.; Park, K.-S.; Kim, T.-G.; Choi, H.-J.; Sung, Y.-M. Controlled Growth of High-Quality TiO_2 Nanowires on Sapphire and Silica. *Nanotechnology*, **2006**, *17* (17), 4317–4321. https://doi.org/10.1088/0957-4484/17/17/006.
182. Stach, E. A.; Pauzauskie, P. J.; Kuykendall, T.; Goldberger, J.; He, R.; Yang, P. Watching GaN Nanowires Grow. *Nano Lett.*, **2003**, *3* (6), 867–869. https://doi.org/10.1021/nl034222h.
183. Wei, M.; Zhi, D.; MacManus-Driscoll, J. L. Self-Catalysed Growth of Zinc Oxide Nanowires. *Nanotechnology*, **2005**, *16* (8), 1364–1368. https://doi.org/10.1088/0957-4484/16/8/064.
184. Umar, A.; Kim, S. H.; Lee, Y.-S.; Nahm, K. S.; Hahn, Y. B. Catalyst-Free Large-Quantity Synthesis of ZnO Nanorods by a Vapor–Solid Growth Mechanism: Structural and Optical Properties. *J. Cryst. Growth*, **2005**, *282* (1–2), 131–136. https://doi.org/10.1016/J.JCRYSGRO.2005.04.095.
185. Zhang, Z.; Wang, Y.; Li, H.; Yuan, W.; Zhang, X.; Sun, C.; Zhang, Z. Atomic-Scale Observation of Vapor–Solid Nanowire Growth via Oscillatory Mass Transport. *ACS Nano*, **2016**, *10* (1), 763–769. https://doi.org/10.1021/acsnano.5b05851.

186. Chueh, Y.-L.; Lai, M.-W.; Liang, J.-Q.; Chou, L.-J.; Wang, Z. L. Systematic Study of the Growth of Aligned Arrays of α-Fe_2O_3 and Fe_3O_4 Nanowires by a Vapor–Solid Process. *Adv. Funct. Mater.*, **2006**, *16* (17), 2243–2251. https://doi.org/10.1002/adfm.200600499.
187. Wei, Y.; Zheng, H.; Hu, S.; Pu, S.; Peng, H.; Li, L.; Sheng, H.; Zhou, S.; Wang, J.; Jia, S. Controllable Synthesis of Single-Crystal SnO_2 Nanowires and Tri-Crystal SnO_2 Nanobelts. *CrystEngComm*, **2018**, *20* (44), 7114–7119. https://doi.org/10.1039/C8CE01507D.
188. Lee, T.-Y.; Lee, C.-Y.; Chiu, H.-T. Vapor–Solid Reaction Growth of Rutile TiO_2 Nanorods and Nanowires for Li-Ion-Battery Electrodes. *ACS Omega*, **2019**, *4* (14), 16217–16225. https://doi.org/10.1021/acsomega.9b02453.
189. Zhang, D.-F.; Sun, L.-D.; Yin, J.-L.; Yan, C.-H.; Wang, R.-M. Attachment-Driven Morphology Evolvement of Rectangular ZnO Nanowires. *J. Phys. Chem. B*, **2005**, *109* (18), 8786–8790. https://doi.org/10.1021/jp050631l.
190. Morin, S. A.; Forticaux, A.; Bierman, M. J.; Jin, S. Screw Dislocation-Driven Growth of Two-Dimensional Nanoplates. *Nano Lett.*, **2011**, *11* (10), 4449–4455. https://doi.org/10.1021/nl202689m.
191. Leng, M.; Chen, Y.; Xue, J. Synthesis of TiO_2 Nanosheets via an Exfoliation Route Assisted by a Surfactant. *Nanoscale*, **2014**, *6* (15), 8531–8534. https://doi.org/10.1039/C4NR00946K.
192. Sun, Z.; Liao, T.; Dou, Y.; Hwang, S. M.; Park, M.-S.; Jiang, L.; Kim, J. H.; Dou, S. X. Generalized Self-Assembly of Scalable Two-Dimensional Transition Metal Oxide Nanosheets. *Nat. Commun.*, **2014**, *5* (1), 3813. https://doi.org/10.1038/ncomms4813.
193. Thanh, N. T. K.; Maclean, N.; Mahiddine, S. Mechanisms of Nucleation and Growth of Nanoparticles in Solution. *Chem. Rev.*, **2014**, *114* (15), 7610–7630. https://doi.org/10.1021/cr400544s.
194. LaMer, V. K.; Dinegar, R. H. Theory, Production and Mechanism of Formation of Monodispersed Hydrosols. *J. Am. Chem. Soc.*, **1950**, *72* (11), 4847–4854. https://doi.org/10.1021/ja01167a001.
195. Wilson, G. J.; Matijasevich, A. S.; Mitchell, D. R. G.; Schulz, J. C.; Will, G. D. Modification of TiO_2 for Enhanced Surface Properties: Finite Ostwald Ripening by a Microwave Hydrothermal Process. **2006**. https://doi.org/10.1021/LA052716J.
196. Li, D.; Nielsen, M. H.; Lee, J. R. I.; Frandsen, C.; Banfield, J. F.; De Yoreo, J. J. Direction-Specific Interactions Control Crystal Growth by Oriented Attachment. *Science (80-.)*., **2012**, *336* (6084), 1014–1018. https://doi.org/10.1126/science.1219643.
197. Oskam, G. Metal Oxide Nanoparticles: Synthesis, Characterization and Application. *J. Sol-Gel Sci. Technol.*, **2006**, *37* (3), 161–164. https://doi.org/10.1007/s10971-005-6621-2.
198. Hu, X.; Gong, J.; Zhang, L.; Yu, J. C. Continuous Size Tuning of Monodisperse ZnO Colloidal Nanocrystal Clusters by a Microwave-Polyol Process and Their Application for Humidity Sensing. *Adv. Mater.*, **2008**, *20* (24), 4845–4850. https://doi.org/10.1002/adma.200801433.
199. Cao, D.; Gong, S.; Shu, X.; Zhu, D.; Liang, S. Preparation of ZnO Nanoparticles with High Dispersibility Based on Oriented Attachment (OA) Process. *Nanoscale Res. Lett.*, **2019**, *14* (1), 210. https://doi.org/10.1186/s11671-019-3038-3.
200. Granqvist, C. G. *Handbook of Inorganic Electrochromic Materials*; Elsevier, 1995.
201. Runnerstrom, E. L.; Llordés, A.; Lounis, S. D.; Milliron, D. J. Nanostructured Electrochromic Smart Windows: Traditional Materials and NIR-Selective Plasmonic Nanocrystals. *Chem. Commun.*, **2014**, *50* (74), 10555–10572. https://doi.org/10.1039/c4cc03109a.
202. Kumar, K. U.; Subrahmanyam, A. Electrochromic Properties of Tungsten Oxide (WO_3) Thin Films on Lexan (Polycarbonate) Substrates Prepared with Neon as Sputter Gas. *Mater. Res. Express*, **2019**, *6* (6), 065502. https://doi.org/10.1088/2053-1591/ab093f.

203. Lynam, N. R. Electrochromic Automotive Day/Night Mirrors. *SAE Transactions*. SAE International 1987, pp. 891–899. https://doi.org/10.2307/44470742.
204. Nanophase Metal Oxide Materials for Electrochromic Displays. In *Handbook of Nanophase and Nanostructured Materials*; Kluwer Academic Publishers: Dordrecht, 2003; pp. 1380–1414. https://doi.org/10.1007/0-387-23814-X_39.
205. Granqvist, C. G.; Azens, A.; Heszler, P.; Kish, L. B.; Österlund, L. Nanomaterials for Benign Indoor Environments: Electrochromics for "Smart Windows," Sensors for Air Quality, and Photo-Catalysts for Air Cleaning. *Sol. Energy Mater. Sol. Cells*, **2007**, *91* (4), 355–365. https://doi.org/10.1016/J.SOLMAT.2006.10.011.
206. Li, X.; Perera, K.; He, J.; Gumyusenge, A.; Mei, J. Solution-Processable Electrochromic Materials and Devices: Roadblocks and Strategies towards Large-Scale Applications. *J. Mater. Chem. C*, **2019**, *7* (41), 12761–12789. https://doi.org/10.1039/C9TC02861G.
207. Uday Kumar, K.; Murali, D. S.; Subrahmanyam, A. Flexible Electrochromics: Magnetron Sputtered Tungsten Oxide (WO_{3-x}) Thin Films on Lexan (Optically Transparent Polycarbonate) Substrates. *J. Phys. D. Appl. Phys.*, **2015**, *48* (25), 255101. https://doi.org/10.1088/0022-3727/48/25/255101.
208. Lee, S.-H.; Deshpande, R.; Parilla, P. A.; Jones, K. M.; To, B.; Mahan, A. H.; Dillon, A. C. Crystalline WO3 Nanoparticles for Highly Improved Electrochromic Applications. *Adv. Mater.*, **2006**, *18* (6), 763–766. https://doi.org/10.1002/adma.200501953.
209. Zheng, H. J.; Wang, X. D.; Gu, Z. H. Preparation and Characterization of Highly Ordered WO_3 Nanorod Arrays. *Wuli Huaxue Xuebao/ Acta Phys. - Chim. Sin.*, **2009**, *25* (8), 1650–1654. https://doi.org/10.3866/PKU.WHXB20090801.
210. Azam, A.; Kim, J.; Park, J.; Novak, T. G.; Tiwari, A. P.; Song, S. H.; Kim, B.; Jeon, S. Two-Dimensional WO_3 Nanosheets Chemically Converted from Layered WS_2 for High-Performance Electrochromic Devices. *Nano Lett.*, **2018**, *18* (9), 5646–5651. https://doi.org/10.1021/acs.nanolett.8b02150.
211. Liang, L.; Zhang, J.; Zhou, Y.; Xie, J.; Zhang, X.; Guan, M.; Pan, B.; Xie, Y. High-Performance Flexible Electrochromic Device Based on Facile Semiconductor-to-Metal Transition Realized by $WO_3 \cdot 2H_2O$ Ultrathin Nanosheets. *Sci. Rep.*, **2013**, *3* (1), 1936. https://doi.org/10.1038/srep01936.
212. Sun, X. W.; Wang, J. X. Fast Switching Electrochromic Display Using a Viologen-Modified ZnO Nanowire Array Electrode. *Nano Lett.*, **2008**, *8* (7), 1884–1889. https://doi.org/10.1021/nl0804856.
213. Sun, M.; Sun, M.; Yang, H.; Song, W.; Nie, Y.; Sun, S. Porous Fe_2O_3 Nanotubes as Advanced Anode for High Performance Lithium Ion Batteries. *Ceram. Int.*, **2017**, *43* (1), 363–367. https://doi.org/10.1016/J.CERAMINT.2016.09.166.
214. Chan, C. K.; Peng, H.; Liu, G.; McIlwrath, K.; Zhang, X. F.; Huggins, R. A.; Cui, Y. High-Performance Lithium Battery Anodes Using Silicon Nanowires. *Nat. Nanotechnol.*, **2008**, *3* (1), 31–35. https://doi.org/10.1038/nnano.2007.411.
215. Zhang, J.; Huang, T.; Liu, Z.; Yu, A. Mesoporous Fe_2O_3 Nanoparticles as High Performance Anode Materials for Lithium-Ion Batteries. *Electrochem. Commun.*, **2013**, *29*, 17–20. https://doi.org/10.1016/J.ELECOM.2013.01.002.
216. Huang, H.; Zhu, W.; Tao, X.; Xia, Y.; Yu, Z.; Fang, J.; Gan, Y.; Zhang, W. Nanocrystal-Constructed Mesoporous Single-Crystalline Co_3O_4 Nanobelts with Superior Rate Capability for Advanced Lithium-Ion Batteries. *ACS Appl. Mater. Interfaces*, **2012**, *4* (11), 5974–5980. https://doi.org/10.1021/am301641y.
217. Zhang, Y.; Yan, Y.; Wang, X.; Li, G.; Deng, D.; Jiang, L.; Shu, C.; Wang, C. Facile Synthesis of Porous Mn_2O_3 Nanoplates and Their Electrochemical Behavior as Anode Materials for Lithium Ion Batteries. *Chem.: A Eur. J.*, **2014**, *20* (20), 6126–6130. https://doi.org/10.1002/chem.201304935.

218. Wang, N.; Chen, L.; Ma, X.; Yue, J.; Niu, F.; Xu, H.; Yang, J.; Qian, Y. Facile Synthesis of Hierarchically Porous NiO Micro-Tubes as Advanced Anode Materials for Lithium-Ion Batteries. *J. Mater. Chem. A*, **2014**, *2* (40), 16847–16850. https://doi.org/10.1039/C4TA04321A.
219. Snook, G. A.; Kao, P.; Best, A. S. Conducting-Polymer-Based Supercapacitor Devices and Electrodes. *J. Power Sources*, **2011**, *196* (1), 1–12. https://doi.org/10.1016/J.JPOWSOUR.2010.06.084.
220. Snook, G. A.; Chen, G. Z. The Measurement of Specific Capacitances of Conducting Polymers Using the Quartz Crystal Microbalance. *J. Electroanal. Chem.*, **2008**, *612* (1), 140–146. https://doi.org/10.1016/J.JELECHEM.2007.08.024.
221. Yuan, C.; Wu, H. Bin; Xie, Y.; Lou, X. W. D. Mixed Transition-Metal Oxides: Design, Synthesis, and Energy-Related Applications. *Angew. Chemie Int. Ed.*, **2014**, *53* (6), 1488–1504. https://doi.org/10.1002/anie.201303971.
222. Lee, H.; Cho, M. S.; Kim, I. H.; Nam, J. Do; Lee, Y. RuOx/Polypyrrole Nanocomposite Electrode for Electrochemical Capacitors. *Synth. Met.*, **2010**, *160* (9–10), 1055–1059. https://doi. org/10.1016/J.SYNTHMET.2010.02.026.
223. Hu, C. C.; Chang, K. H.; Lin, M. C.; Wu, Y. T. Design and Tailoring of the Nanotubular Arrayed Architecture of Hydrous RuO_2 for next Generation Supercapacitors. *Nano Lett.*, **2006**, *6* (12), 2690–2695. https://doi.org/10.1021/nl061576a.
224. Zhang, X.; Shi, W.; Zhu, J.; Zhao, W.; Ma, J.; Mhaisalkar, S.; Maria, T. L.; Yang, Y.; Zhang, H.; Hng, H. H.; et al. Synthesis of Porous NiO Nanocrystals with Controllable Surface Area and Their Application as Supercapacitor Electrodes. *Nano Res.*, **2010**, *3* (9), 643–652. https://doi.org/10.1007/s12274-010-0024-6.
225. Yang, Q.; Sha, J.; Ma, X.; Yang, D. Synthesis of NiO Nanowires by a Sol-Gel Process. *Mater. Lett.*, **2005**, *59* (14–15), 1967–1970. https://doi.org/10.1016/J. MATLET.2005.02.037.
226. Yeager, M. P.; Su, D.; Marinković, N. S.; Teng, X. Pseudocapacitive NiO Fine Nanoparticles for Supercapacitor Reactions. *J. Electrochem. Soc.*, **2012**, *159* (10), A1598–A1603. https://doi.org/10.1149/2.025210jes.
227. Vijayakumar, S.; Nagamuthu, S.; Muralidharan, G. Supercapacitor Studies on NiO Nanoflakes Synthesized Through a Microwave Route. *ACS Appl. Mater. Interfaces*, **2013**, *5* (6), 2188–2196. https://doi.org/10.1021/am400012h.
228. Chen, J.; Peng, X.; Song, L.; Zhang, L.; Liu, X.; Luo, J. Facile Synthesis of Al-Doped NiO Nanosheet Arrays for High-Performance Supercapacitors. *R. Soc. Open Sci.*, **2018**, *5* (11), 180842. https://doi.org/10.1098/rsos.180842.
229. Wang, H.-Q.; Yang, G.; Li, Q.-Y.; Zhong, X.-X.; Wang, F.-P.; Li, Z.-S.; Li, Y. Porous Nano-MnO_2: Large Scale Synthesis via a Facile Quick-Redox Procedure and Application in a Supercapacitor. *New J. Chem.*, **2011**, *35* (2), 469–475. https://doi.org/10.1039/C0NJ00712A.
230. Wu, M.-S. Electrochemical Capacitance from Manganese Oxide Nanowire Structure Synthesized by Cyclic Voltammetric Electrodeposition. *Appl. Phys. Lett.*, **2005**, *87* (15), 153102. https://doi.org/10.1063/1.2089169.
231. Zou, Z.; Ye, J.; Sayama, K.; Arakawa, H. Direct Splitting of Water Under Visible Light Irradiation with an Oxide Semiconductor Photocatalyst. *Nature*, **2001**, *414* (6864), 625–627. https://doi.org/10.1038/414625a.
232. Fujishima, A.; Honda, K. Electrochemical Photolysis of Water at a Semiconductor Electrode. *Nature*, **1972**, *238* (5358), 37–38. https://doi.org/10.1038/238037a0.
233. Kawai, T.; Sakata, T. Conversion of Carbohydrate into Hydrogen Fuel by a Photocatalytic Process. *Nature*, **1980**, *286* (5772), 474–476. https://doi.org/10.1038/286474a0.
234. Colmenares, J. C.; Luque, R. Heterogeneous Photocatalytic Nanomaterials: Prospects and Challenges in Selective Transformations of Biomass-Derived Compounds. *Chem. Soc. Rev.*, **2014**, *43* (3), 765–778. https://doi.org/10.1039/C3CS60262A.

235. Patil, S. M.; Dhodamani, A. G.; Vanalakar, S. A.; Deshmukh, S. P.; Delekar, S. D. Multi-Applicative Tetragonal TiO_2/SnO_2 Nanocomposites for Photocatalysis and Gas Sensing. *J. Phys. Chem. Solids*, **2018**, *115*, 127–136. https://doi.org/10.1016/J.JPCS.2017.12.020.
236. Linsebigler, A. L.; Lu, G.; Yates, J. T. Photocatalysis on TiO_2 Surfaces: Principles, Mechanisms, and Selected Results. *Chem. Rev.*, **1995**. https://doi.org/10.1021/cr00035a013.
237. Cernuto, G.; Masciocchi, N.; Cervellino, A.; Colonna, G. M.; Guagliardi, A. Size and Shape Dependence of the Photocatalytic Activity of TiO_2 Nanocrystals: A Total Scattering Debye Function Study. *J. Am. Chem. Soc.*, **2011**, *133* (9), 3114–3119. https://doi.org/10.1021/ja110225n.
238. Ramakrishnan, V. M.; Natarajan, M.; Santhanam, A.; Asokan, V.; Velauthapillai, D. Size Controlled Synthesis of TiO_2 Nanoparticles by Modified Solvothermal Method towards Effective Photo Catalytic and Photovoltaic Applications. *Mater. Res. Bull.*, **2018**, *97*, 351–360. https://doi.org/10.1016/J.MATERRESBULL.2017.09.017.
239. Tian, J.; Zhao, Z.; Kumar, A.; Boughton, R. I.; Liu, H. Recent Progress in Design, Synthesis, and Applications of One-Dimensional TiO_2 Nanostructured Surface Heterostructures: A Review. *Chem. Soc. Rev.*, **2014**, *43* (20), 6920–6937. https://doi.org/10.1039/C4CS00180J.
240. Zhou, W.; Du, G.; Hu, P.; Li, G.; Wang, D.; Liu, H.; Wang, J.; Boughton, R. I.; Liu, D.; Jiang, H. Nanoheterostructures on TiO_2 Nanobelts Achieved by Acid Hydrothermal Method with Enhanced Photocatalytic and Gas Sensitive Performance. *J. Mater. Chem.*, **2011**, *21* (22), 7937. https://doi.org/10.1039/c1jm10588d.
241. Wang, J.; Bian, Z.; Zhu, J.; Li, H. Ordered Mesoporous TiO_2 with Exposed (001) Facets and Enhanced Activity in Photocatalytic Selective Oxidation of Alcohols. *J. Mater. Chem. A*, **2013**, *1* (4), 1296–1302. https://doi.org/10.1039/C2TA00035K.
242. Kajita, S.; Yoshida, T.; Ohno, N.; Ichino, Y.; Yoshida, N. Fabrication of Photocatalytically Active Vanadium Oxide Nanostructures via Plasma Route. *J. Phys. D. Appl. Phys.*, **2018**, *51* (21), 215201. https://doi.org/10.1088/1361-6463/aabe44.
243. Luo, Z.; Miao, R.; Huan, T. D.; Mosa, I. M.; Poyraz, A. S.; Zhong, W.; Cloud, J. E.; Kriz, D. A.; Thanneeru, S.; He, J.; et al. Mesoporous MoO_{3-x} Material as an Efficient Electrocatalyst for Hydrogen Evolution Reactions. *Adv. Energy Mater.*, **2016**, *6* (16), 1600528. https://doi.org/10.1002/aenm.201600528.
244. Lai, C. W. Photocatalysis and Photoelectrochemical Properties of Tungsten Trioxide Nanostructured Films. *ScientificWorldJournal.*, **2014**, *2014*, 843587. https://doi.org/10.1155/2014/843587.
245. Menezes, P. W.; Indra, A.; Littlewood, P.; Schwarze, M.; Göbel, C.; Schomäcker, R.; Driess, M. Nanostructured Manganese Oxides as Highly Active Water Oxidation Catalysts: A Boost from Manganese Precursor Chemistry. *ChemSusChem*, **2014**, *7* (8), 2202–2211. https://doi.org/10.1002/cssc.201402169.
246. Mishra, M.; Chun, D.-M. α-Fe_2O_3 as a Photocatalytic Material: A Review. *Appl. Catal. A Gen.*, **2015**, *498*, 126–141. https://doi.org/10.1016/j.apcata.2015.03.023.
247. Briscoe, J.; Dunn, S. Piezoelectric Nanogenerators: A Review of Nanostructured Piezoelectric Energy Harvesters. *Nano Energy*, **2015**, *14*, 15–29. https://doi.org/10.1016/J.NANOEN.2014.11.059.
248. Zhu, G.; Yang, R.; Wang, S.; Wang, Z. L. Flexible High-Output Nanogenerator Based on Lateral ZnO Nanowire Array. *Nano Lett.*, **2010**, *10* (8), 3151–3155. https://doi.org/10.1021/nl101973h.
249. Vittal, R.; Ho, K.-C. Zinc Oxide Based Dye-Sensitized Solar Cells: A Review. *Renew. Sustain. Energy Rev.*, **2017**, *70*, 920–935. https://doi.org/10.1016/j.rser.2016.11.273.
250. Bai, Y.; Mora-Seró, I.; De Angelis, F.; Bisquert, J.; Wang, P. Titanium Dioxide Nanomaterials for Photovoltaic Applications. *Chem. Rev.*, **2014**, *114* (19), 10095–10130. https://doi.org/10.1021/cr400606n.

251. Norouzi, M.; Kolahdouz, M.; Ebrahimi, P.; Ganjian, M.; Soleimanzadeh, R.; Narimani, K.; Radamson, H. Thermoelectric Energy Harvesting Using Array of Vertically Aligned Al-Doped ZnO Nanorods. *Thin Solid Films*, **2016**, *619*, 41–47. https://doi.org/10.1016/j.tsf.2016.10.041.
252. Zhang, P.; Yang, F.; Kapil, G.; Shen, Q.; Toyoda, T.; Yoshino, K.; Minemoto, T.; Pandey, S. S.; Ma, T.; Hayase, S. Enhanced Performance of ZnO Based Perovskite Solar Cells by Nb_2O_5 Surface Passivation. *Org. Electron.*, **2018**, *62*, 615–620. https://doi.org/10.1016/j.orgel.2018.06.038.
253. Dong, Q.; Shi, Y.; Wang, K.; Li, Y.; Wang, S.; Zhang, H.; Xing, Y.; Du, Y.; Bai, X.; Ma, T. Insight into Perovskite Solar Cells Based on SnO_2 Compact Electron-Selective Layer. *J. Phys. Chem. C*, **2015**, *119* (19), 10212–10217. https://doi.org/10.1021/acs.jpcc.5b00541.
254. Gibson, E. A.; Smeigh, A. L.; Le Pleux, L.; Fortage, J.; Boschloo, G.; Blart, E.; Pellegrin, Y.; Odobel, F.; Hagfeldt, A.; Hammarström, L. A P-Type NiO-Based Dye-Sensitized Solar Cell with an Open-Circuit Voltage of 0.35 V. *Angew. Chemie Int. Ed.*, **2009**, *48* (24), 4402–4405. https://doi.org/10.1002/anie.200900423.
255. Chen, J.; Li, B.; Zheng, J.; Zhao, J.; Zhu, Z. Role of Carbon Nanotubes in Dye-Sensitized TiO_2 -Based Solar Cells. *J. Phys. Chem. C*, **2012**, *116* (28), 14848–14856. https://doi.org/10.1021/jp304845t.
256. Lekphet, W.; Ke, T.-C.; Su, C.; Kathirvel, S.; Sireesha, P.; Akula, S. B.; Li, W.-R. Morphology Control Studies of TiO_2 Microstructures via Surfactant-Assisted Hydrothermal Process for Dye-Sensitized Solar Cell Applications. *Appl. Surf. Sci.*, **2016**, *382*, 15–26. https://doi.org/10.1016/J.APSUSC.2016.04.115.
257. Cai, B.; Zhong, D.; Yang, Z.; Huang, B.; Miao, S.; Zhang, W.-H.; Qiu, J.; Li, C. An Acid-Free Medium Growth of Rutile TiO_2 Nanorods Arrays and Their Application in Perovskite Solar Cells. *J. Mater. Chem. C*, **2015**, *3* (4), 729–733. https://doi.org/10.1039/C4TC02249A.
258. Wu, S.; Chen, C.; Wang, J.; Xiao, J.; Peng, T. Controllable Preparation of Rutile TiO_2 Nanorod Array for Enhanced Photovoltaic Performance of Perovskite Solar Cells. *ACS Appl. Energy Mater.*, **2018**, *1* (4), 1649–1657. https://doi.org/10.1021/acsaem.8b00106.
259. Wang, Z. L. Piezo-Phototronic Effect on Electrochemical Processes and Energy Storage; 2012; pp. 223–236. https://doi.org/10.1007/978-3-642-34237-0_11.
260. Wang, Z. L.; Song, J. Piezoelectric Nanogenerators Based on Zinc Oxide Nanowire Arrays. *Science*, **2006**, *312* (5771), 242–246. https://doi.org/10.1126/science.1124005.
261. Kim, D. Y.; Lee, S.; Lin, Z.-H.; Choi, K. H.; Doo, S. G.; Chang, H.; Leem, J.-Y.; Wang, Z. L.; Kim, S.-O. High Temperature Processed ZnO Nanorods Using Flexible and Transparent Mica Substrates for Dye-Sensitized Solar Cells and Piezoelectric Nanogenerators. *Nano Energy*, **2014**, *9*, 101–111. https://doi.org/10.1016/j.nanoen.2014.07.004.
262. Gupta, M. K.; Lee, J.-H.; Lee, K. Y.; Kim, S.-W. Two-Dimensional Vanadium-Doped ZnO Nanosheet-Based Flexible Direct Current Nanogenerator. *ACS Nano*, **2013**, *7* (10), 8932–8939. https://doi.org/10.1021/nn403428m.
263. Xi, Y.; Song, J.; Xu, S.; Yang, R.; Gao, Z.; Hu, C.; Wang, Z. L. Growth of ZnO Nanotube Arrays and Nanotube Based Piezoelectric Nanogenerators. *J. Mater. Chem.*, **2009**, *19* (48), 9260. https://doi.org/10.1039/b917525c.
264. Manjula, Y.; Rakesh Kumar, R.; Swarup Raju, P. M.; Anil Kumar, G.; Venkatappa Rao, T.; Akshaykranth, A.; Supraja, P. Piezoelectric Flexible Nanogenerator Based on ZnO Nanosheet Networks for Mechanical Energy Harvesting. *Chem. Phys.* **2020**, *532*, 110699. https://doi.org/10.1016/j.chemphys.2020.110699.

21. [illegible] M.; [illegible] M.; [illegible] J.; Chapman, [illegible]; Sodano, [illegible] Enhanced Piezoelectric Energy Harvesting Using Arrays of Vertically Aligned [illegible] ZnO [illegible]. 2016, 19, [illegible]. https://doi.org/10.1016/[illegible]
22. [illegible] P.; [illegible], G.; Shen, Q.; Toyoda, T.; Yoshino, K.; [illegible] T.; [illegible] S. Enhanced Performance of ZnO [illegible] Solar Cells by [illegible]. 2015, 62, [illegible]. https://doi.org/10.1016/[illegible]
23. [illegible] Y.; [illegible] Wang, S.; Zhang, H.; [illegible] Y.; [illegible] 2015, [illegible] https://doi.org/10.[illegible]
24. [illegible] A.; [illegible] L.; [illegible] Forro, L.; [illegible] J.; [illegible] Hagfeldt, A.; [illegible] P.; [illegible] Dye-Sensitized Solar Cells [illegible] Angew. Chem. Int. Ed. 2009, 48, [illegible]. https://doi.org/10.1002/anie.200[illegible]
25. Chen, J.; Li, [illegible]; Zhang, [illegible]; [illegible] Nanotubes in [illegible] TiO2-Based Solar Cells. [illegible] 2012, [illegible]. https://doi.org/10.[illegible]
26. [illegible] W.; [illegible]; [illegible] S.; [illegible] P.; [illegible] S. [illegible] ZnO [illegible] Microstructures [illegible] Surfactant-Assisted [illegible] Dye-Sensitized Solar Cell Applications. ACS Appl. [illegible] 2016, [illegible]. https://doi.org/10.1021/acsami.[illegible]
27. [illegible], Zhang, [illegible]; [illegible] S.; Zhang, [illegible] Qu, [illegible] An [illegible] Growth [illegible] TiO2 Nanorod Arrays and Their Application [illegible] Solar Cells [illegible] 2016, [illegible]. https://doi.org/10.1039/[illegible]
28. [illegible] Wang, [illegible]; [illegible] Fabrication of Rutile TiO2 [illegible] Array for Enhanced [illegible] Perovskite Solar Cells. [illegible] 2018, [illegible] https://doi.org/10.[illegible]
29. [illegible] Photocatalytic [illegible] Processes and [illegible] 2012, [illegible]. https://doi.org/10.[illegible]
30. Wang, Z. L.; [illegible] Piezoelectric Nanogenerators Based on Zinc Oxide Nanowire Arrays. Science 2006, 312, [illegible]. https://doi.org/10.1126/science.1124005
31. [illegible]
32. [illegible] https://doi.org/10.[illegible]
33. [illegible] Zinc Oxide [illegible] https://doi.org/10.[illegible]
34. [illegible] Jaroniec, M.; [illegible] Synthesis of [illegible] Tetrahedral [illegible] 2020, 32, [illegible] https://doi.org/10.1016/[illegible].2020.[illegible]

2 Transition Metal-Oxide-Based Electrodes for Na/Li Ion Batteries

Paresh H. Salame

CONTENTS

2.1 INTRODUCTION

The generation of renewable energy and its storage is of paramount importance in this era of global warming due to increasing greenhouse gas effect caused by burning of traditional fossil fuel, in the form of coal, gasoline, oil, etc., to satisfy our energy needs. Moreover, the increasing consumption and limited fossil fuel resources are accelerating depletion of this precious resource in addition to increasing their prices. In view of this, renewable energy storage and conversion have been a propelling this topic amongst researchers all over the world. For energy storage, rechargeable lithium ion batteries (LIBs) have been explored and are being used at a great level. The emergence of high-energy density cathode materials have made them viable to use from portable devices to electrical vehicles (EVs). Advancement in the electrode materials is the key aspect of developing batteries for various commercial applications. Transition metal oxides (TMOs) have been used and studied by many groups in last three to four decades for making and developing high-capacity LIBs. The surge in this technology will be seeing a major boost in coming decades because many of the developing countries are encouraging the use of renewable energy resources and EVs. The demand for secondary rechargeable batteries will only be growing in such a scenario.

However, this boom in utilization of the rechargeable LIB for powering our gadgets to transport modes will lead to an increasing burden on lithium-rich minerals and salts, which are geographically concentrated only in few parts of the world (South America, China, etc.), thereby skyrocketing their prices. To counter the economical aspect and dependency on geographically limited materials, new materials capable of replacing lithium-based batteries have been explored in great detail. In the quest of exploring a new alternative, sodium (Na)-based cathode materials have found a renewed interest as an effective alternative. Na, being cheaper and more abundant (sea salt, rock salt, etc.) in addition to having similar intercalation chemistry to lithium, has potential to replace LIBs. However, being heavier when compared to lithium (Li), there will be limiting applications to Na-ion batteries (NIBs), more likely toward grid energy storage, where batteries will be stationary and weight will be a non-issue. Table 2.1 gives the information regarding the current electrode materials that are being used for the energy storage and conversion at various levels.

TMOs (CoO, NiO, MnO, FeO, etc.) have played a major role in enhancing the performance (energy density, cycling ability, and power density) of both the LIBs and NIBs.[1] In fact, lithium cobalt oxide ($LiCoO_2$), discovered by Goodenough et al.[2] was possibly the first positive electrode material that started the revolution in the field of intercalated rechargeable LIBs, which was later commercialized by Sony in 1991. In this chapter, development of TMO-based Li and Na batteries are discussed.

2.2 STRUCTURAL CLASSIFICATION OF TRANSITION METAL OXIDES USED IN LIBs AND NIBs

$LiCoO_2$ is isostructural to α-$NaFeO_2$ structure, as shown in Figure 2.1, with rhombohedral symmetry and *space group* belonging to $\mathbf{R\bar{3}m}$. This structure consists of alternating CoO_2 layers and Li layers at the (111) plane of a rock salt structure. Here, the Wykoff positions of various elements are: Co^{3+} occupies the 3a (0,0,0) site, O in 6c (0,0,1 4) site, while Li can be found in the 3b (0, 0, 1/2) site in a cubic closed-packed O^{2-} lattice. In this structure, cobalt (Co) as well as Li are octahedrally coordinated. Li and Co atoms are alternately located on octahedral sites between adjacent close-packed planes of oxygen. Various transition metals (Ni, Cr, V), when replaced at the Co site in $LiCoO_2$, result in similar structure as shown in Figure 2.1. However, when substituted by the Mn^{3+} ion, the cubic close-packing of the oxygen array is disturbed, primarily due to the Jahn-Teller distortion of Mn^{3+} ion. This was found to happen, at least when conventional high-temperature procedures are adopted for their synthesis. It is reported that for the Mn system (Li-Mn-O), rather than forming a metastable $LiMnO_2$, it is transformed into a ***spinel structured*** $LiMn_2O_4$ system.

TABLE 2.1
Electrode Materials Currently in Use and Commercialized for Various Applications

Cathode/Anode Material	Cell Voltage (V)	Specific Capacity (mAh/g)	Specific Energy (mWh/g)	Advantages	Disadvantages	Applications
Layered cathode structure $LiCoO_2$ (LCO)	~4	155–180	560	Good cycle life and energy density	Expensive, Co is toxic, thermal stability issue, restricted to 50% of theoretical capacity right now	Portable electronics
Spinel structure $LiMn_2O_4$ (LMO)	~4	100–120	480	Thermally stable, inexpensive, environmentally safe, high electronic-Li^+ ion conductivity, high power density	Moderate cycling life, low energy	High-power applications (tools)
$LiNi_{0.80}Co_{0.15}Al_{0.05}$ (NCA)	~3.7	180–190	470–530	Very good energy and power density, good cycle life	Thermal stability, moisture sensitive	Motive power and high-end electronic devices
$LiNi_xMn_yCo_{1-x-y}O_2$	~3.8	150–160	480–550	Very good energy density, power density and cyclability, thermally stable	Patented material	Portable as well as high-power applications and electrical vehicles
$LiFePO_4$	~3.5	160–180	560	Environmentally safe, inexpensive, extremely safe	Low electronic and Li^+ conductivity, require small particle size and carbon coating, which adds to its processing cost	Portable electronic devices
Graphite (as anode)	~0.1	370	—	Inexpensive, environmentally safe, low operating potential	Promotes SEI layer formation and lithium plating, could lead to safety concerns	Portable electronics devices

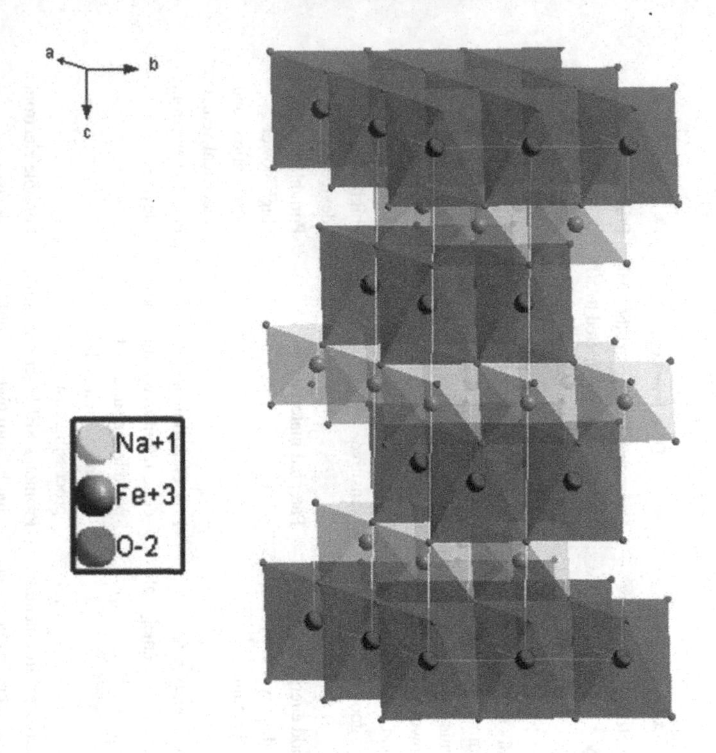

FIGURE 2.1 Crystal structure of α-$NaFeO_2$, to which $LiCoO_2$ belongs. (Permitted from Ellis, B.L. et al., *Chem. Mater.*, 22, 691–714, 2010. With permission.)

In this spinel system, $LiMn_2O_4$ oxide has a three-dimensional structure wherein Li^+ ions occupy the 8 tetrahedral sites, while Mn^{3+} ions occupy the 16 octahedral sites of the cubic close-packed oxygen.

In NIBs, TMOs Na_xTMO_2 primarily exist in two types of the structures viz. O3-type and P2-type structures, here O represents "octahedral," while P represents "prismatic," numbers "2" and "3" represent the stacking arrangement per unit of O ions.[4] In the O3-type structure, oxygen forms a sublattice in a cubic closed-pack arrangement, while transition metals (Mn, Fe, Co, Cr, Ni, etc.) form a octahedra coordinated by six oxygen atoms along the *c*-axis direction (see Figure 2.2). The octahedron formed by the transition metal with oxygen forms a network by sharing their octahedral edges, resulting into TMO_2 layers. Na^+ ions can occupy the octahedral site or prismatic site between these TMO_2 layers.

For the P2-type structure, the value of x is nearly 2/3, and here Na^+ ions can occupy two different prismatic sites; one shares **faces** between TMO_6 octahedra and the other shares **edges** between TMO_6 octahedra. In this P2-type structure, the transition metal ions are surrounded by oxygen frameworks with a stacking mode of ABBA.

For the O3 structure, value of $x = 1$ in Na_xTMO_2; here all Na^+ ions share one edge and one face with TMO_6 octahedra, and the oxygen frameworks are arranged in the ABCABC pattern.

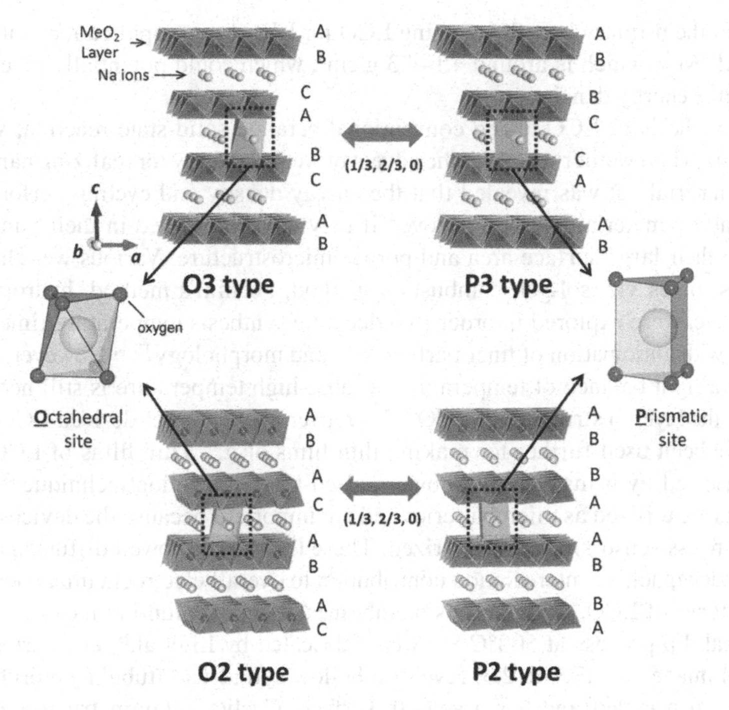

FIGURE 2.2 O3-type and P2-type structure of Na_xTMO_2-layered materials. (Reproduced with permission from Yabuuchi, N. et al., *Chem. Rev.*, 114, 11636–11682, 2014.)

2.3 TRANSITION METAL-OXIDE-BASED CATHODE MATERIALS FOR DEVELOPMENT OF Li-ION BATTERIES

2.3.1 $LiCoO_2$

$LiCoO_2$ (LCO) was the first material to exhibit the energy density, stability, and cyclability required to successfully commercialize the LIB technology. The onslaught of this modern battery material (along with $LiNiO_2$) started back in 1980s by Goodenough et al. at the University of Oxford.[2,6] This material was then later used by Akira Yoshino to develop the first lithium-ion rechargeable battery for the Asahi Kasei Corporation.[7,8] He fabricated the full cell by using $LiCoO_2$ as the cathode and graphitic-carbon as the anode. This technology was later assimilated by Sony Corporation for powering their portable devices. Over the years, many modifications in transition metals that the group used were made, and many more new modifications (Ni, Mn, Fe, etc.) were used in the molecular formula; in addition, many combinations of the mixed transition metals, bearing the general formula $LiTM_{1-x}TM_xO_2$ were also explored for improved performance of the LIBs. These layered structured materials were stable at ambient conditions and offered significantly higher Li-ion insertion voltage (3.5–4 V vs. Li/Li^+) as compared to earlier attempted batteries on Li/TiS_2 or Li/MoS_2 batteries.[9,10]

One of the primary benefits of using LCO for LIB storage application is its compaction density, which is around 4.1–4.3 g/cm^3, which could potentially give excellent volume energy density.

The synthesis of LCO, using conventional ceramic solid-state reaction, yielded particle sizes in few microns. With the advent of wet-chemistry for realizing nanopowders of materials, it was revealed that the energy density and cycling performance of the battery material can be improved if they are synthesized in their nanoform, owing to their large surface area and porous microstructure. Various wet-chemical synthesis routes viz. sol-gel, combustion method, Pechini's method, hydrothermal method, etc. were explored in order to reduce the synthesis temperature/time of the reaction with anticipation of finer particle size and morphology.[11–14] However, it need to strike a right balance of temperature because high temperature is still needed to stabilize the layered structure of LCO.[15] Moreover, these sol-gel-derived LCO materials have been used further for making thin films of it.[16] Thin films of LCO were also fabricated by using the high-power pulsed-laser deposition technique.[17] These thin films are utilized as microbatteries and are important because the devices (especially wireless sensors) are miniaturized. These thin films showed diffusion control and pseudocapacitive intercalation contribution to overall electrochemical activity.

Nanotubes of LCO, using porous membranes of anodic aluminum oxides via the dip-anneal-dip process at 500°C/8h, were fabricated by Li et al.[18] The corresponding SEM image (see Figure 2.3) reveals a hollow cylindrical (tubular) morphology, which is open ended and has a smooth surface. Cyclic voltammetry was carried over these nanotubes of LCO to evaluate its reversibility over the potential range of 2.5–4.3 V. Three peaks are clearly visible for the LCO system; the first peak can be

FIGURE 2.3 SEM images of (a) an overall view of $LiCoO_2$ nanotubes, (b) $LiCoO_2$ tubes with open tips, (c) an overall view of $LiNi_{0.8}Co_{0.2}O_2$ nanotubes, (d) face-on view of $LiNi_{0.8}Co_{0.2}O_2$ nanotube arrays with open-end tips, (e) $LiMn_2O_4$ nanotubes stacking up in disarray, and (f) $LiMn_2O_4$ tubes with open-ended tips. (Reproduced with permission from Li, X. et al., *J. Phys. Chem. B*, 109, 14017–14024, 2005.)

ascribed to Li insertion and the deinsertion process. The second and third peaks are due to order-disorder phase transition. The capacity of these nanotubes was found to be ~185 mAh/g and is found to be better than nanotubes of $LiNi_{0.8}Co_{0.2}O_2$ and $LiMn_2O_4$. Also, LCO nanotubes exhibited better capacity retention at higher discharge current than the nanotubes of other mentioned materials. Continuous efforts by scientists all over the world has made it possible to increase the initial operating range of Li insertion of 3.5–4 V vs. Li/Li^+ 4.2–4.45 V and volume energy density from 200 to 700 Wh/L capacity in excess of 200 mAh/g.[19,20]

Challenges of using LCO:

1. Compatibility issues of using LCO as a host and other battery components at high voltages
2. Thermal runaway and decomposition due to side reaction at electrode electrolyte interface

Possible remedy:

1. *Doping*: Substitution/doping at the appropriate site in LCO has been thought as a suitable way of improving overall storage capacity. With the doping of transition metal ions such as Ni, Zn, and Cu, the capacity of LCO was found to be increased. Whereas, when doped with other aliovalent metal ions viz. Mg, Al, Ca, etc. it can increase the delithiation potential. Table 2.2 gives a short summary of materials doped in LCO and their electrochemical performances.
2. *Surface coating*: In addition to doping, which enhances the performance and thermal stability of the LCO cathode materials in LIBs, it is extremely important to reduce the side reaction happening over the interfaces of the electrode and electrolyte. Coating is an excellent way to protect the surface of the LCO cathode powders. Usually, the coating materials include oxides (SnO_2, Al_2O_3, etc.), fluorides (AlF_3, CeF_3, etc.), phosphates ($AlPO_4$, Co_3PO_4, YPO_4), etc.

TABLE 2.2
Comparison of Various Elements Doped in the Structure of LCO on the Electrochemical Performance

Doped LCO		Capacity (mAh/g)	Cyclability	Potential Range (V)	References
Mg doping	$LiMg_{0.05}Co_{0.95}O_2$	160	50	3.0–4.5	22
	$LiMg_{0.2}Co_{0.8}O_2$	191	30	2.9–4.3	23
	$LiMg_{0.2}Co_{0.8}O_2$	205	25	3.0–4.5	24
Ca doping	$LiCa_{0.2}Co_{0.8}O_2$	199	25	3.0–4.5	24
	$LiCa_{0.1}Co_{0.9}O_2$	173.5	25	3.0–4.5	25
	$LiCa_{0.2}Co_{0.8}O_2$	178.5	25	3.0–4.5	24
Al doping	$LiAl_{0.15}Co_{0.85}O_2$	160	10	2.5–4.5	26
	$LiAl_{0.25}Co_{0.75}O_2$	182	09	2.0–4.5	27

2.3.2 $LiNiO_2$

$LiNiO_2$ (LNO) is isostructural to $LiCoO_2$; however, it is much more economical and has a much lower oxidation potential, which makes it less susceptible for corrosion and electrolyte decomposition, thus making it an attractive alternative to $LiCoO_2$.[28] However, due to oxidation of Ni^{2+} to Ni^{3+}, it renders it difficult to produce it on large scale. Moreover, high-temperature synthesis makes it Li^+ deficient (Li_xNiO_{2-x}), and these impurity phases are not ideal for Li^+ intercalation and deintercalation. This can be overcome if the compound can be synthesized via the wet-chemical "sol-gel" approach, which takes less time and temperature to obtain a phase pure $LiNiO_2$ material with minuscule chances of impurity phases. With the advantage of starting materials containing cations of Li and Ni being mixed at the atomic level giving a homogenous mixture, it can be synthesized at lower temperatures for a shorter time, thus minimizing the chances of the sample being Li^+ deficient and introducing corresponding impurity phases.[29] Choi et al.[29] reported the synthesis of LNO via the conventional solid-state reaction and mixing route as well as the sol-gel process. They were able to get the LNO sample via a conventional method while reacting it at 700°C for 12 hours, while for sol-gel synthesis they treated the precursors at 600°C for only 6 hours. They found a substantial capacity loss, a larger internal resistance drop during the first intermittent discharge for a conventionally prepared sample than the sol-gel-derived LNO electrodes. These sample exhibited capacity of 150 mAh/g and specific density of 500 Wh/g. Further, the electrodes of LNO nanoparticles derived from these sol-gel techniques were able to retain almost 90% of the capacity after 350 cycles of discharge.

To further improve the performance during cycling of LNO electrode material, doping with heteroatoms viz. Mn, Co, Al, etc. has been suggested in order to restrict the Ni^{4+} from going under reduction while cycling and stabilizing the LNO phase.

During the past few years, researchers have explored LNO-based layered structured material with doping with manganese and Co simultaneously, a ternary transition metal oxide. This lithium-nickel-manganese-cobalt oxide (often called NMC) has a layered structure and is nickel rich, having composition in the range $LiNi_{1-x-y}Co_xMn_yO_2$ ($0 \leq x \leq 0.5, 0 \leq y \leq 0.3$).[21,30,31] This compound was first proposed by Liu et al.[21] and prepared by heating $Ni_{1-x-y}Co_xMn_y(OH)_2$ and $LiNO_3$ in flowing oxygen for 10 hrs at 550°C followed by further heating at 750°C. Compared to LCO and LNO, due to presence of Mn^{4+}, which contributes in stabilizing the structure, the NMC can be charged to higher cut-off potentials. Furthermore, the practical capacity (>160 mAh/g) of NMC is comparable (or even better) than LCO, but comes at far reduced costs than LCO. Thus, NMC is considered a promising candidate to increase the energy density of LIBs. However, NMC with its higher energy density has a lot to desire in terms of cycle life and thermal stability vis-à-vis safety.[21,32] In this NMC composition, nickel excess could give rise to increase in energy density. Figure 2.4 shows the results over NMC samples published by Liu et al.[21] Figure 2.4 shows the cycle performance of $LiNiO_2$ and when LNO is doped with Co viz. $LiNi_{0.8}Co_{0.2}O_2$, $LiNi_{0.5}Co_{0.5}O_2$, and when doped with Co and Mn, making it a ternary TMO of $LiNi_{0.7}Co_{0.2}Mn_{0.1}O_2$ and $LiNi_{0.5}Co_{0.2}Mn_{0.3}O_2$. It is revealed from the data that cycling characteristics in the composition range with $x = 0.2$ and $x = 0.5$, are good; however,

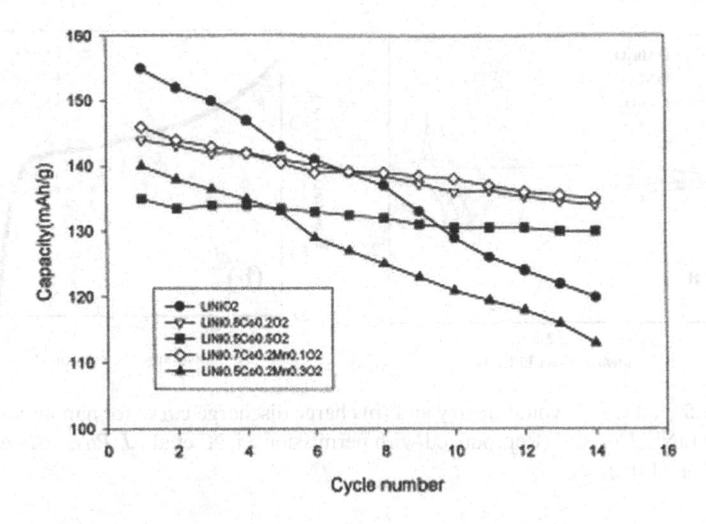

FIGURE 2.4 Charge and discharge cycle performance of $LiNi_{1-x}Co_xO_2$ ($x = 0, 0.2, 0.5$) and $LiNi_{1-y}Co_{0.2}Mn_yO_2$ ($y = 0.1, 0.3$) (current density: 0.2 mA/cm², voltage: 4.2–2.75 V vs. Li). (Reproduced with permission Liu, Z. et al., *J. Power Sources*, 81–82, 416–419, 1999.)

TABLE 2.3
Effect of Various Dopant on the Electrochemical Performance of LNO

Doped LNO		Capacity (mAh/g)	Cyclability	Potential Range (V)	References
Co doped	$LiNi_{0.8}Co_{0.2}O_2$	138	15	2.75–4.25	21
Ga	2%Ga doped LNO	190	100	3–4.3	33
		>200	100	3–4.5	
Zr Doped	$Li_{1.06}Ni_{0.937}Zr_{0.04}O_2$	247	100	2.7–4.3	34

when $x = 0.5$, the initial capacity is just marginally lower. When LNO is formed in ternary TMO form with Mn and Co doping, $LiNi_{0.7}Co_{0.2}Mn_{0.1}O_2$ shows good capacity and has very low loss of capacity as that of initial discharge capacity after cycling. Initially, the capacity for higher Mn doped $LiNi_{0.5}Co_{0.2}Mn_{0.3}O_2$ is quite as high as 150 mAh/g; however, after cycling, it could lead to large capacity loss, which is not favorable from an application point of view. A comparative chart showing effect of various dopant used for enhancing the electrochemical performance of LNO is shown in Table 2.3[21,33,34].

Li et al.[18] as demonstrated for LCO, synthesized nanotubes of Co and Ni co-doped system, namely, $LiNi_{0.8}Co_{0.2}O_2$, which displayed a similar behavior to the nanotubes of LCO. However, the three peaks in cyclic voltammetry are seen to appear at low lower potential value as compared to LCO (see Figure 2.5), which indicates that the $LiNi_{0.8}Co_{0.2}O_2$ cathode materials are less prone to corrosion and electrolyte oxidation.[18]

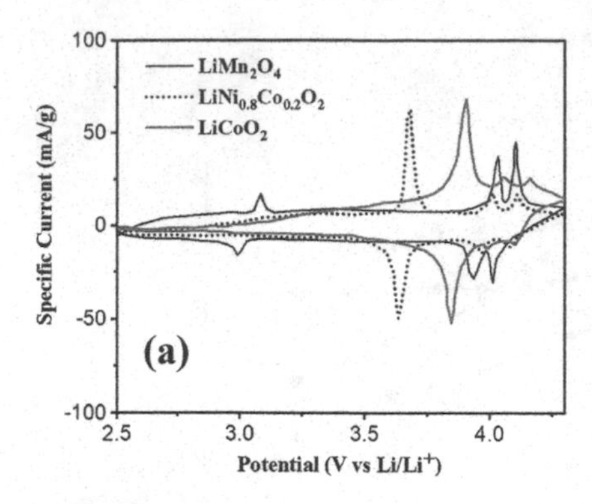

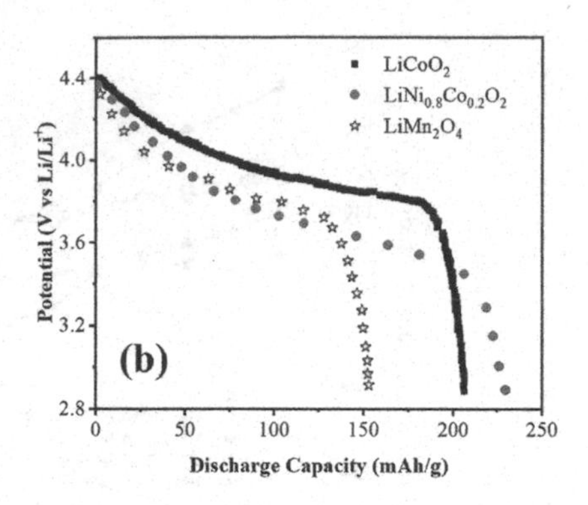

FIGURE 2.5 (a) Cyclic voltammetry and (b) charge discharge curve for nanotubes of LCO, LMO, and $LiNi_{0.8}Co_{0.2}O_2$. (Reproduced with permission Li, X. et al., *J. Phys. Chem. B*, 109, 14017–14024, 2005.)

2.3.3 $LiMn_2O_4$

Soon after the commercialization of LCO-based LIBs by Sony Corporation, J. C. Hunter (Eveready Laboratories) synthesized a spinel structured $LiMn_2O_4$ (LMO) cathode material, which could work at higher potentials akin to LCO-based batteries.[35,36] LMO was envisaged as potential cathode material for LIBs, owing to its high reduction potential, low cost, and low environmental hazard. These properties led to its commercialization for various portable electronic device applications using nonaqueous electrolytes. Structurally it is entirely different from LCO and other materials, as when the TMO in $LiTMO_2$ general formula is replaced with Mn, it acquires a more stable spinel LMO phase, which is different from Co and nickel-based $LiTMO_2$ systems. Moreover, compared to these nickel and Co systems, LMO is cheaper, abundant, and nontoxic. Also, this spinel (Mn) system has exhibited better thermal stability and power density when compared to its Co and nickel counterparts; however, they have limited cyclability and energy density. With these limitations and advantages, the LMO system has been widely used wherever high power is needed viz. power tools, electric motive power, etc.[37,38]

Kovacheva et al.[37] grew the nanocrystals of LMO, using a combustion synthesis technique and sucrose as a fuel, which after annealing at 700°C revealed single-phase cubic spinel structure, which was ~100°C lower than earlier reported[39] synthesis of LMO material employing a soft chemical synthesis route. Transition electron microscopy (TEM) images over the sample, with different sucrose contents, reveal the particle sizes to be in the range of 22–50 nm; the TEM images are shown in Figure 2.6a. The authors infer that the with increase in sucrose content, the average particle size became smaller with narrower particle size distribution.

The charge discharge profile of the cathode material synthesized by Kovacheva et al.[37] at various current densities is shown in Figure 2.6b. Two main plateaus can be seen around 4 V. In addition to this, two shorter plateaus are seen at 4.3 V in charge and ~3.3 V of discharge curve, when recorded at low current density of 0.20 mA/cm^2. As the sample is having excess lithium these two plateaus are typical of such

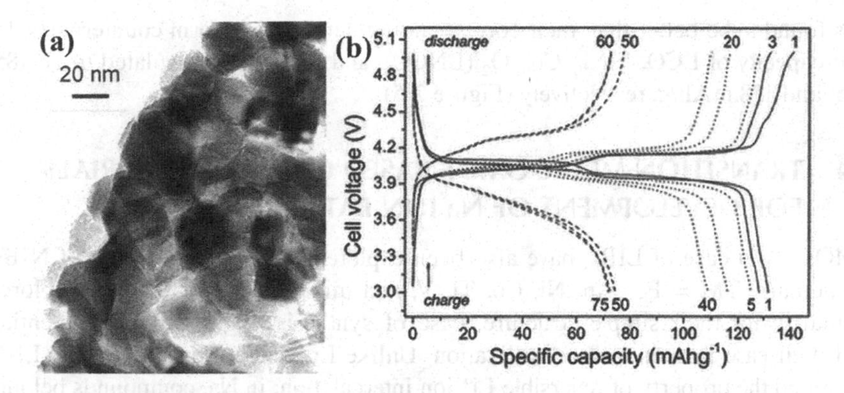

FIGURE 2.6 (a) TEM image of $Li_{1.09}Mn_{1.91}O_4$ sample and (b) charge discharge profile of the cathode materials at various current densities. (Reproduced with permission Kovacheva, D. et al., *J. Mater. Chem.*, 12, 1184–1188, 2002.)

compounds. At this current density, the authors found the discharge capacity of 131 mAh/g. This capacity was found to be decreased with an increase in the cycle number, and at around the 40th cycle, the capacity was reduced to ~85% from the original discharge capacity. However, this capacity is still higher by about 17% than the conventionally ceramic solid-state synthesized LMO particles with average particle size of ~1 μm.

Hirose et al.[40] synthesized the LMO powders with slight excess of Li, bearing the final composition $Li_{1.09}$ $Mn_{1.91}O_4$ using the spray pyrolysis method. After calcination at 800°C, $Li_{1.09}Mn_{1.91}O_4$ powders were realized into the spinel phase. When discharged at 1 C rate, the material displayed discharge capacity around 107 mAh/g. The cathode material also showed 99% retention at 1 C of the initial discharge even after 100 cycles and 90 retention at 10 C of the initial discharge capacity.

The nanotubes of LMO grown by Li et al.[18] were arranged in disordered fashion having irregular tips (see Figure 2.4), thus indicating the brittle nature of the tubes when compared to the nanotubes of LCO. Also, the walls of these nanotubes were thicker compared to LCO, probably due to undissolved trace remaining alumina in NaOH solution. Also, the TEM images revealed lower crystallinity of these nanotubes compared to LCO, thus suggesting the difficulty level is more in growing nanotubes of LMO than LCO. However, all the nanotubes grown by the templated method by Li et al. were hollow with the outer diameter measuring ~300 nm and wall thickness measuring 20–30 nm. The cyclic voltammetry studies carried over in these nanotubes reveal quite a different behavior than LCO or LNO. The reversible peak at ~3 V is ascribed to the lattice distortion due to the Jahn-Teller effect of Mn^{3+} ion, which transforms cubic crystal symmetry to tetragonal symmetry from original cubic crystal symmetry.

Figure 2.5 shows the discharge curves for the nanotubes of LCO, $LiMn_{0.8}Co_{0.2}O_2$, and LMO measured at constant current density of 10 mA/g. Out of these three materials, LMO can be seen to have a nearly constant voltage during 95% of the full discharge, thereby making this system well suited to be applied in electronic memory device backup applications. The capacity of these nanotube electrodes was found to be better than their corresponding nanoparticle form counterparts.[41,42] The capacity of LCO, $LiNi_{0.8}Co_{0.2}O_2$ (LNCO), and LMO was calculated to be 185, 205, and 138 mAh/g, respectively (Figure 2.5).

2.4 TRANSITION METAL-OXIDE-BASED CATHODE MATERIALS FOR DEVELOPMENT OF Na ION BATTERIES

TMOs, as in case of LIBs, have also been explored for the development of NIBs. Particularly TM = Fe, Mn, Ni, Co, Ti, V, and mixed TMOs have been explored primarily for their simple structure, ease of synthesis, high operating potential, and their ease for commercial utilization. Unlike Li, wherein only LCO and LNO exhibited the property of reversible Li^+ ion intercalation; in Na, compounds belonging to Na_xTMO_2 can all intercalate N^+ ion reversibly.[43,44] With replacing various transition metals on the transition metal site, we can get desired properties from Na_xTMO_2. The structural information of these compounds is already discussed in the previous section. Structural stability is one issue that is common to both LIBs and NIBs. Both the P2- as well as O3-layered phases suffered from the phase transition, which is induced by the extraction/insertion of Na^+ ions during the cycling process. This phase transition can be avoided by limiting the cutoff voltage because high cutoff voltage may result in bringing in the phase transition in the material at high voltages. However, limiting the cutoff voltage could result in lower specific capacity. Another way to inhibit the phase transition in these Na_xTMO_2 structures is to introduce various transition metals, Ni, Co, Fe, etc., into the structure to improve the cycling ability. Also, electrochemically inert elements viz. Mg, Al, Zn, etc. could be introduced to weaken the phase transition process at higher voltages.

2.4.1 Na_xCoO_2

Most of the cathode materials for NIBs were mimicked from their LIB counterparts because Na and Li had similar intercalation chemistry. In this regard, Na_xCoO_2 was explored widely being analogues to first commercial cathode material of LIBs, that is, $LiCoO_2$.[45,46] Lei et al.[46] investigated that pure O3, O′3, and P′3 phases can only be formed when the concentration of x in Na_xCoO_2 is fixed at 1.00, 0.83, and 0.67, respectively. In the composition range of $0.68 \leq x \leq 0.76$, the P2 phase exists. The charge-discharge curve over these composition ranges of Na concentration reveal various patterns (see Figure 2.7). The figure reveals potential drops that are reversible during charge and discharge, as can be seen from top right graph, indicating the reversibility of the intercalation process.[45]

Na_xCoO_2 was also synthesized in nanocrystalline form by D'Arienzo et al.[47] utilizing the hydrothermal synthesis route and solid-state reaction route. They first synthesized Co_3O_4 using the hydrothermal route, which was then the solid state reacted with either NaOH or Na_2CO_3. Both the products gave the same P2-$Na_{0.71}CoO_2$ phase, but morphology was drastically different. The material obtained from reacting Na_2Co_3 had narrow particle length and 2D platelet-like morphology; however, the product obtained from NaOH displayed large microcrystals, which were irregular in shape and had broad particle size distribution. The 2D microplatelets of $Na_{0.71}CoO_2$

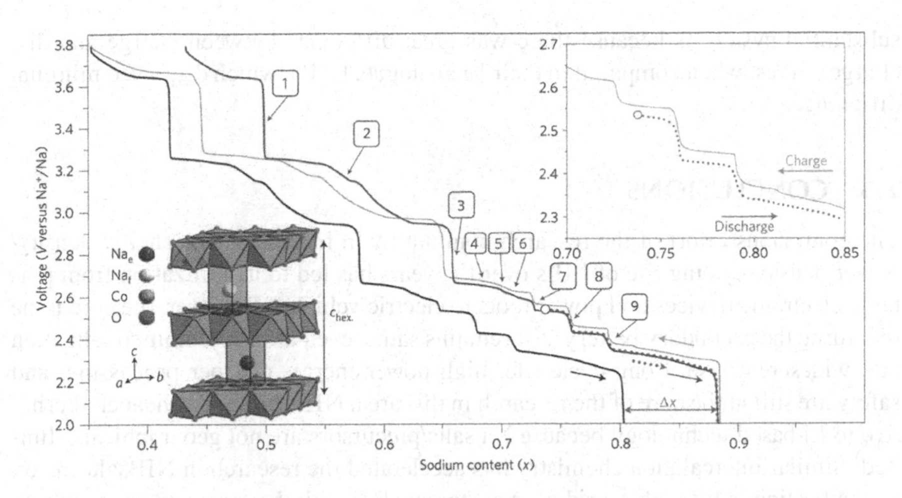

FIGURE 2.7 Cyclic voltammetry of Na_xCoO_2 cathode material, top right corner shows reversible intercalation of Na^+ ion in the structure. (Reproduced with permission Berthelot, R. et al., *Nat. Mater.*, 10, 74–80, 2011.)

exhibited superior behavior amongst them, with stable discharge specific capacity of 120, 105, and 80 mAh/g at 5, 20, and 40 mA/g, respectively, in the potential range of 2.0–3.9 V.

2.4.2 Na_xCrO_2

Investigation of Na_xCrO_2 for their electrochemical performance occurred in 1982, and at that time, only 0.15 Na was successfully cycled.[4] However, reinvestigation over these compositions now revealed as much as 0.45 of Na can be reversibly cycled giving discharge capacity of ~120 mAh/g within the voltage range of 2.0–3.6 V at current density of 25 mA/g. This is much higher than $LiCrO_2$, which is a structurally analogues polymorph and has the same transition metal element but showed capacity of only 50 mAh/g in the potential range of 3–4.5 V.[48]

2.4.3 Na_xMnO_2

When Na_xTMO_2 is substituted by manganese transition metal element, it can be realized in to 3D structures (where $x = 0.4$ and 0.44) as well as 2D structures ($x = 1.0$ and 0.7). In 2D structures, various structures can be possible viz. P2-structured $Na_{0.7}MnO_{2+y}$, α-$NaMnO_2$, and β-$NaMnO_2$, which depends upon synthesis condition and stoichiometry.

Ma et al.[49] reinvestigated the electrochemical performance over the monoclinic α-$NaMnO_2$. For this structure, nearly 0.85 Na^+ was deintercalated from $NaMnO_2$, and 0.8 Na^+ was intercalated back during intermittent charge and discharge. Galvanostatic charge-discharge, for the first cycle and C/10 rate, exhibited capacity of 185 mAh/g between a potential range of 2.0 and 3.8 V and gives 185. However, the capacity retention after 20 cycles was 132 mAh/g. This compound displayed substantial hysteresis because there was great difference between charge and discharge curves, when compared to their Li analogue, LMO, which displayed minimal difference.

2.5 CONCLUSIONS

The continuous effort of the research community in increasing the energy density/power density cycling life of LIBs over the years has led to its utilization from portable electronic devices and power tools to electric vehicles. However, the core issue regarding the secondary battery still remains same, even after its commercialization and widespread use. Long cycle life, high power/energy, cheaper processing, and safety are still at the core of the research in this area. NIBs present a cheaper alternative to Li-based technology because Na salts/precursors are not geographically limited. Similar intercalation chemistry has accelerated the research in NIBs; however, its application is limited to grid energy storage. For both these secondary rechargeable battery technologies, TMOs have played a key role in the development of the LIBs as well as NIBs. Many efforts have been carried out to stabilize the crystal structure and slow down the phase transition process. This has been made possible by incorporation of varying transition metals in the crystal structure. The stability of the structure can be maintained by varying the transition metal in the structure. Careful fabrication of electrode material in their nanoform and various nanostructures have resulted in improvement of energy and power density of these TMO-based cathode materials. Doping and surface coating/treatment has further improved the efficiency of these batteries.

REFERENCES

1. P. He, H. Yu, D. Li, and H. Zhou, "Layered lithium transition metal oxide cathodes towards high energy lithium-ion batteries," *Journal of Materials Chemistry*, 22(9) 3680–3695 (2012).
2. J. B. Goodenough, K. Mizushima, and T. Takeda, "Solid-solution oxides for storage-battery electrodes," *Japanese Journal of Applied Physics*, 19(S3) 305 (1980).
3. B. L. Ellis, K. T. Lee, and L. F. Nazar, "Positive electrode materials for Li-ion and Li-batteries," *Chemistry of Materials*, 22(3) 691–714 (2010).
4. C. Delmas, J.-J. Braconnier, C. Fouassier, and P. Hagenmuller, "Electrochemical intercalation of sodium in $NaxCoO_2$ bronzes," *Solid State Ionics*, 3–4, 165–169 (1981).
5. N. Yabuuchi, K. Kubota, M. Dahbi, and S. Komaba, "Research development on sodium-ion batteries," *Chemical Reviews*, 114(23) 11636–11682 (2014).
6. K. Mizushima, P. C. Jones, P. J. Wiseman, and J. B. Goodenough, "$LixCoO_2$ ($0 < x < -1$): A new cathode material for batteries of high energy density," *Materials Research Bulletin*, 15(6) 783–789 (1980).
7. A. Yoshino, K. Sanechika, and T. Nakajima, "Secondary Battery," in Edited by U. S. patent. *Asahi Kasei Kogyo Kabushiki Kaisha*, Osaka, Japan, United States, 1987.
8. A. Yoshini, "Polyacetylene Composite," in Edited by U. S. Patents. *Asahi Kasei Kogyo Kabushiki Kaisha*, United States, 1987.
9. M. S. Whittingham, "Electrical energy storage and intercalation chemistry," *Science*, 192 1126–1127 (1976).

10. M. S. Whittingham and R. R. Chianelli, "Layered compounds and intercalation chemistry: An example of chemistry and diffusion in solids," *Journal of Chemical Education*, 57(8) 569 (1980).
11. Y. K. Sun, I. H. Oh, and S. A. Hong, "Synthesis of ultrafine $LiCoO_2$ powders by the sol-gel method," *Journal of Materials Science*, 31(14) 3617–3621 (1996).
12. N. Ding, X. W. Ge, and C. H. Chen, "A new gel route to synthesize $LiCoO_2$ for lithium-ion batteries," *Materials Research Bulletin*, 40(9) 1451–1459 (2005).
13. N. A. Abdul Aziz, M. De Cunha, T. K. Abdullah, and A. A. Mohamad, "Degradation of $LiCoO_2$ in aqueous lithium–air batteries," *International Journal of Energy Research*, 41(2) 289–296 (2017).
14. H. Liang, X. Qiu, H. Chen, Z. He, W. Zhu, and L. Chen, "Analysis of high rate performance of nanoparticled lithium cobalt oxides prepared in molten KNO_3 for rechargeable lithium-ion batteries," *Electrochemistry Communications*, 6(8) 789–794 (2004).
15. W.-S. Yoon and K.-B. Kim, "Synthesis of $LiCoO_2$ using acrylic acid and its electrochemical properties for Li secondary batteries," *Journal of Power Sources*, 81–82 517–523 (1999).
16. H. Porthault, F. Le Cras, and S. Franger, "Synthesis of $LiCoO_2$ thin films by sol/gel process," *Journal of Power Sources*, 195(19) 6262–6267 (2010).
17. Y. Matsuda, N. Kuwata, and J. Kawamura, "Thin-film lithium batteries with 0.3–30 μm thick $LiCoO_2$ films fabricated by high-rate pulsed laser deposition," *Solid State Ionics*, 320 38–44 (2018).
18. X. Li, F. Cheng, B. Guo, and J. Chen, "Template-synthesized $LiCoO_2$, $LiMn_2O_4$, and $LiNi0.8Co0.2O_2$ nanotubes as the cathode materials of lithium ion batteries," *The Journal of Physical Chemistry B*, 109(29) 14017–14024 (2005).
19. O. Besenhard Jürgen and M. Winter, "Insertion reactions in advanced electrochemical energy storage," *Pure and Applied Chemistry*, 70 603 (1998).
20. S. Krueger, R. Kloepsch, J. Li, S. Nowak, S. Passerini, and M. Winter, "How do reactions at the anode/electrolyte interface determine the cathode performance in lithium-ion batteries?" *Journal of the Electrochemical Society*, 160(4) A542–A548 (2013).
21. Z. Liu, A. Yu, and J. Y. Lee, "Synthesis and characterization of $LiNi_{1-x-y}CoxMnyO_2$ as the cathode materials of secondary lithium batteries," *Journal of Power Sources*, 81–82 416–419 (1999).
22. R. Sathiyamoorthi, P. Shakkthivel, R. Gangadharan, and T. Vasudevan, "Layered $LiCo_{1-x}Mg_xO_2$ (x = 0.0, 0.1, 0.2, 0.3 and 0.5) cathode materials for lithium-ion rechargeable batteries," *Ionics*, 13(1) 25–33 (2007).
23. M. Mladenov, R. Stoyanova, E. Zhecheva, and S. Vassilev, "Effect of Mg doping and MgO-surface modification on the cycling stability of $LiCoO_2$ electrodes," *Electrochemistry Communications*, 3(8) 410–416 (2001).
24. R. Sathiyamoorthi, R. Chandrasekaran, A. Gopalan, and T. Vasudevan, "Synthesis and electrochemical performance of high voltage cycling $LiCo0.8M0.2O_2$ (M=Mg, Ca, Ba) as cathode material," *Materials Research Bulletin*, 43(6) 1401–1411 (2008).
25. R. Sathiyamoorthi, P. Shakkthivel, and T. Vasudevan, "New solid-state synthesis routine and electrochemical properties of calcium based ceramic oxide battery materials for lithium battery applications," *Materials Letters*, 61(17) 3746–3750 (2007).
26. H. Huang, G. V. S. Rao, and B. V. R. Chowdari, "$LiAlxCo_{1-x}O_2$ as 4 V cathodes for lithium ion batteries," *Journal of Power Sources*, 81–82 690–695 (1999).
27. Y.-I. Jang, B. Huang, H. Wang, G. R. Maskaly, G. Ceder, D. R. Sadoway, Y.-M. Chiang, H. Liu, and H. Tamura, "Synthesis and characterization of $LiAlyCo_{1-y}O_2$ and $LiAlyNi_{1-y}O_2$," *Journal of Power Sources*, 81–82 589–593 (1999).
28. J. R. Dahn, U. von Sacken, M. W. Juzkow, and H. Al-Janaby, "Rechargeable $LiNiO_2$/carbon cells," *Journal of the Electrochemical Society*, 138(8) 2207–2211 (1991).

29. Y.-M. Choi, S.-I. Pyun, S.-I. Moon, and Y.-E. Hyung, "A study of the electrochemical lithium intercalation behavior of porous $LiNiO_2$ electrodes prepared by solid-state reaction and sol-gel methods," *Journal of Power Sources*, 72(1) 83–90 (1998).
30. T. Ohzuku and Y. Makimura, "Layered lithium insertion material of $LiCo1/3Ni1/3Mn1/3O_2$ for lithium-ion batteries," *Chemistry Letters*, 30(7) 642–643 (2001).
31. J. Kasnatscheew, M. Evertz, B. Streipert, R. Wagner, R. Klöpsch, B. Vortmann, H. Hahn, S. Nowak, M. Amereller, A. C. Gentschev, P. Lamp, and M. Winter, "The truth about the 1st cycle Coulombic efficiency of $LiNi1/3Co1/3Mn1/3O_2$ (NCM) cathodes," *Physical Chemistry Chemical Physics*, 18(5) 3956–3965 (2016).
32. A. Manthiram, J. Choi, and W. Choi, "Factors limiting the electrochemical performance of oxide cathodes," *Solid State Ionics*, 177(26) 2629–2634 (2006).
33. Y. Nishida, K. Nakane, and T. Satoh, "Synthesis and properties of gallium-doped $LiNiO_2$ as the cathode material for lithium secondary batteries," *Journal of Power Sources*, 68(2) 561–564 (1997).
34. C. S. Yoon, M.-J. Choi, D.-W. Jun, Q. Zhang, P. Kaghazchi, K.-H. Kim, and Y.-K. Sun, "Cation ordering of Zr-doped $LiNiO_2$ cathode for lithium-ion batteries," *Chemistry of Materials*, 30(5) 1808–1814 (2018).
35. J. C. Hunter, "Preparation of a new crystal form of manganese dioxide: λ-MnO_2," *Journal of Solid State Chemistry*, 39(2) 142–147 (1981).
36. J. C. Hunter, "MnO_2 derived from $LiMn_2O_4$," in Edited by U. Patent. Union Carbide Corporation, New York, United States, 1981.
37. D. Kovacheva, H. Gadjov, K. Petrov, S. Mandal, M. G. Lazarraga, L. Pascual, J. M. Amarilla, R. M. Rojas, P. Herrero, and J. M. Rojo, "Synthesizing nanocrystalline $LiMn_2O_4$ by a combustion route," *Journal of Materials Chemistry*, 12(4) 1184–1188 (2002).
38. H.-S. Park, S.-J. Hwang, and J.-H. Choy, "Relationship between chemical bonding character and electrochemical performance in nickel-substituted lithium manganese oxides," *The Journal of Physical Chemistry B*, 105(21) 4860–4866 (2001).
39. V. Massarotti, D. Capsoni, M. Bini, G. Chiodelli, C. B. Azzoni, M. C. Mozzati, and A. Paleari, "Characterization of sol-gel $LiMn_2O_4$ spinel phase," *Journal of Solid State Chemistry*, 147(2) 509–515 (1999).
40. S. Hirose, T. Kodera, and T. Ogihara, "Synthesis and electrochemical properties of Li-rich spinel type $LiMn_2O_4$ powders by spray pyrolysis using aqueous solution of manganese carbonate," *Journal of Alloys and Compounds*, 506(2) 883–887 (2010).
41. Y. M. Hon, K. Z. Fung, S. P. Lin, and M. H. Hon, "Effects of metal ion sources on synthesis and electrochemical performance of spinel $LiMn_2O_4$ using tartaric acid gel process," *Journal of Solid State Chemistry*, 163(1) 231–238 (2002).
42. H. Chen, X. Qiu, W. Zhu, and P. Hagenmuller, "Synthesis and high rate properties of nanoparticled lithium cobalt oxides as the cathode material for lithium-ion battery," *Electrochemistry Communications*, 4(6) 488–491 (2002).
43. J. M. Paulsen, C. L. Thomas, and J. R. Dahn, "Layered Li-Mn-Oxide with the O_2 structure: A cathode material for Li-ion cells which does not convert to spinel," *Journal of The Electrochemical Society*, 146(10) 3560–3565 (1999).
44. Q. Liu, Z. Hu, M. Chen, C. Zou, H. Jin, S. Wang, S.-L. Chou, and S.-X. Dou, "Recent progress of layered transition metal oxide cathodes for sodium-ion batteries," *Small*, 15(32) 1805381 (2019).
45. R. Berthelot, D. Carlier, and C. Delmas, "Electrochemical investigation of the P2–$NaxCoO_2$ phase diagram," *Nature Materials*, 10(1) 74–80 (2011).
46. Y. Lei, X. Li, L. Liu, and G. Ceder, "Synthesis and stoichiometry of different layered sodium cobalt oxides," *Chemistry of Materials*, 26(18) 5288–5296 (2014).

47. M. D'Arienzo, R. Ruffo, R. Scotti, F. Morazzoni, C. M. Mari, and S. Polizzi, "Layered Na0.71CoO_2: A powerful candidate for viable and high performance Na-batteries," *Physical Chemistry Chemical Physics*, 14(17) 5945–5952 (2012).
48. S. Komaba, C. Takei, T. Nakayama, A. Ogata, and N. Yabuuchi, "Electrochemical intercalation activity of layered $NaCrO_2$ vs. $LiCrO_2$," *Electrochemistry Communications*, 12(3) 355–358 (2010).
49. X. Ma, H. Chen, and G. Ceder, "Electrochemical properties of monoclinic $NaMnO_2$," *Journal of the Electrochemical Society*, 158(12) A1307–A1312 (2011).

3 Layered Metal Oxides/Hydroxides for High-Performance Supercapacitor

Deepa B. Bailmare, Chandrasekar M. Subramaniyam, and Abhay D. Deshmukh

CONTENTS

3.1 INTRODUCTION

The supercapacitor (SC) is electrochemical energy storage device that can provide high impact on cyclic stability and power-delivering capability with fast charge-discharge rate. Nanomaterials are widely used to design the SC electrode that consists of faradic as well as non-faradic reactions, accordingly classified as electrochemical double layer capacitors (EDLCs) and pseudocapacitors. The EDLC-type electrodes provide great cyclic stability and specific surface area but lack in terms of energy-storing capability and specific capacitance [1]. However, pseudocapacitors provide high specific capacitance but have lower cycling stability. The cycle life of the device is a very important factor to be considered. Apart from this, the device's energy density and power density also play an important role in applications. Ultimately, the Ragone plot expresses the relationship between the energy density and power density shown in Figure 3.1a. In the Ragone plot, SCs are compared with the dielectric capacitors, batteries, and fuel cells. It can be clearly seen that batteries have high-energy density but lack power density. Similarly, dielectric

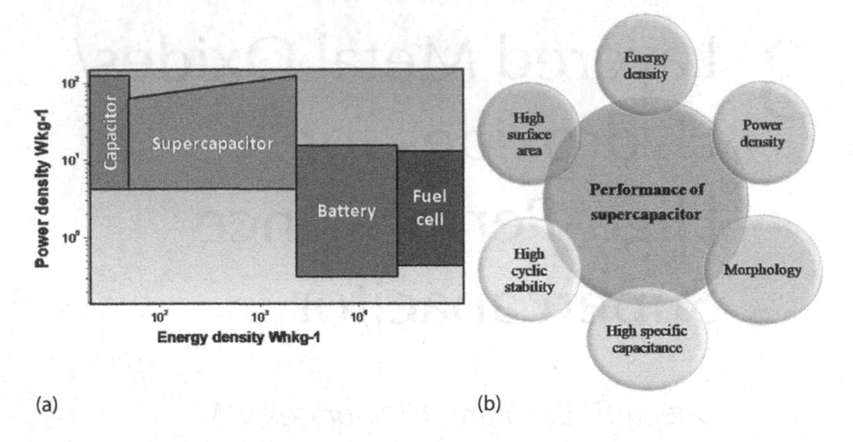

FIGURE 3.1 (a) Comparative Ragone plot showing energy density and power density of SCs with conventional capacitors, batteries, and fuel cells and (b) factors that affect the performance of supercapacitors.

capacitors or conventional capacitors provide high-power delivering capacity but limits in terms of energy storage. Apart from these, two SCs deliberately give high-energy density at the expense of their intrinsic power density. This makes SCs useful in current challenging applications when accessing the renewable sources of energy. SCs are also useful in various practical applications where high-power backups are essential, such as electric vehicles, aircraft, hybrid devices, and many power applications [1,2].

The major challenge concerning SCs is its energy density. SCs serve energy densities lower than batteries and fuel cells. The energy density of EDLCs is around 10–20 Whkg^{-1}, and in the case of hybrids or pseudocapacitors it reaches 40–50 Whkg^{-1}. Remarkably, experimental efforts were carried out to improve their energy-storing capabilities. Energy density depends upon two factors: specific capacitance and operating potential window, that is, $E = \frac{1}{2}CV^2$, where C = specific capacitance and V = operating potential window. Therefore, the performance of SCs can be improved by increasing either specific capacitance or the potential window of electrodes and electrolytes. The improvement in electrodes, electrolytes, and structure can be obtained by standardizing the correlation between them or optimizing the synergistic effect. To improve the performance of the SCs, various important factors are shown in Figure 3.1b.

As compared to the properties of electrode material, the interaction of electrodes with electrolytes is essential for the fabrication of SCs. The pore size of electrode material and the ion size of electrolytes should be idealized to access the ions at the electrode/electrolyte interface [3]. The EDLC and pseudocapacitance of the electrode material depends upon the type of electrolyte used. Specifically, dealing with SCs, high-energy density and cyclic stability are major issues; hence, to remove these barriers, the pseudocapacitors based on metal oxides and metal hydroxides are an important attraction. The transition metal improves high conductivity to the electrode material, thereby delivering high specific capacitance and high-energy density. Metal oxides and hydroxides are a special class of materials that are studied in various applications. These materials have excellent structures such that they can be used for many applications, such as oxygen evaluation reaction, absorbents, catalysts, and energy storage. The various methodologies are implemented to prepare metal oxides/hydroxides or mixed metal hydroxides and mixed metal oxides, for example, hydrothermal, sol-gel, chemical vapor deposition, and electrodeposition.

3.2 THE CHARGE STORAGE MECHANISM OF METAL OXIDES/HYDROXIDES

The charge storage mechanism of metal oxides/hydroxides is based on the pseudocapacitors or battery-type electrode materials. From Figure 3.2, it is speculated that, as compared to carbon-based electrode materials, pseudocapacitive materials follow the charge storage through both the faradic redox reaction and the electroabsorption/desorption at the electrode/electrolyte interfaces. The simultaneous processes were found to be non-electrostatic in nature, since the pseudocapacitors followed the redox reaction. The behavior of these materials is similar to the battery-type electrodes [4]. The metal oxides/hydroxide electrodes, along with EDLCs (carbon electrodes), form SC devices [5], and the metal oxides/hydroxides undergo a pseudocapacitive reaction with the faradic behavior of the materials. The following equation is used to estimate the theoretical pseudocapacitance of the material:

$$C = \frac{nF}{MV}$$

where, n = number of electrons that take part in the electrochemical process, F represents Faraday's constant, M represents mass of the electro active species, and V is the potential window [6]. The capacitance arises from pseudocapacitor electrode

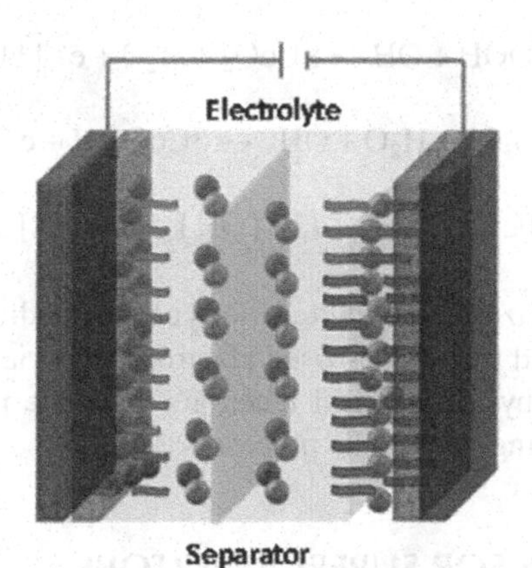

FIGURE 3.2 Charge storage mechanism of redox/pseudocapacitive materials for SCs.

material that depends on the operating potential, which induces the faradic current, which is further responsible for the redox reaction of the electroactive materials. The process of electroabsorption takes place due to the chemi absorption of the electron donating ions.

The charge storage mechanism of transition metal oxides and hydroxides basically follow a reversible redox reaction in the aqueous media, depending upon the valance electron present in the outer-most shell of preferred metal cation. The storage follows the intercalation/deintercalation of absorption/desorption at the surface of metal-based electrode material and electrolyte interfaces [7]. The general equation representing the typical redox mechanism of pseudocapacitive materials is given as follows:

$$MO + OH^- + H_2O \leftrightarrow MOOH + e^- \quad (3.1)$$

Here, M indicates metal cations.

There are many oxides and hydroxides prepared that follows the typical redox reactions as follows:

$$RuO_2 + xH^+ + xe^{-1} \rightarrow RuO_2 - x\left(OH\right)_x, 0 \le x \le 2\,[8] \quad (3.2)$$

where x denotes the average number of electron transfers in the electrochemical process.

$$NiO + zOH^- \rightarrow zNiOOH + \left(1-z\right)NiO + ze^-\,[9] \quad (3.3)$$

$$CuO + OH^- \rightarrow CuOOH$$

$$CuO + H_2O + 2OH^- \rightarrow Cu\left(OH\right)_4^- + e^-\ [10] \quad (3.4)$$

$$CoOOH + OH^- \leftrightarrow CoO_2 + H_2O + e^-\ [11]$$

$$Co_3O_4 + H_2O + OH^- \leftrightarrow 3CoOOH + e^- \quad (3.5)$$

$$FeO + 2OH^- \leftrightarrow Fe\left(OH\right)_2 + 2e^-\,[12] \quad (3.6)$$

All these are the generalized chemical reactions corresponding to the redox mechanism of metal oxides and hydroxide-based electrode material; although these reactions lack in detailed physicochemical reactions that take place in the electrode/electrolyte interface during the faradic processes.

3.3 METAL OXIDES FOR SUPERCAPACITORS

Besides, the EDLC-based carbon materials, metal oxides/hydroxides interestingly give high performances due to their nanostructures. The metal oxides/hydroxides give the advantage of highly conducting nanostructured architecture based on transition metals and provide varieties of nanostructures, such as nanorods, nanowires, nanoflakes, nanoflowers, nanopetals, and nanoplatelets. It is evidence that the transition metal-based oxides/hydroxides provide high electrical conductivity for fast ion transport and better electrochemical properties than other carbon-based electrode materials. It is a well-established fact that the metal oxide/hydroxide-based electrode materials follow the mechanism as it conducts a charge storage through both the faradic redox reaction and adsorption/desorption of ions at electrode/electrolyte interfaces [13]; hence, the metal oxides generally have high specific capacitance as compared to other carbon-based electrode materials.

For past few years, various metal oxides/hydroxides such as NiO, MnO_2, Fe_2O_3, Co_2O_4, and V_2O_5 and bimetallic hydroxides were also studied for their structural diversity. Recently, it is reported that the layered double hydroxides (LDHs) (bimetallic hydroxides) exhibit high electrical conductivity due to the synergistic effect of both the metal ions [12,13]. In recent years, the nickel (Ni) oxide-based electrode materials have attracted the attention of researchers because Ni possesses high electrocatalytic activity in aqueous electrolytes. Basically, in redox SCs, NiO plays a vital role as promising material that it provides high theoretical capacitance, high conductivity, easy synthesis process, and exhibits great reversibility. There were two contradictory theories regarding the charge storage mechanism of NiO-based materials. First, during the charge storage process, NiO possess the reversible redox reaction with NiOOH, and second, the NiO first gets changed to $Ni(OH)_2$ in a high aqueous electrolyte. The resultant consensus of these theories is that the Ni^{2+} ions basically gets oxidized to NiOOH during the chemical reaction with the alkaline electrolyte. The first theory is the most preferable with the researchers.

Despite the fact, there were some issues regarding the practical application of NiO-based electrode materials for SCs and low cyclic stability of the electrode material. Hence, researchers have developed various strategies to build nanostructured electrode materials using NiO. Vijaykumar et al. discussed the synthesis of NiO nanoflakes using facile microwave synthesis technique. This technique reduces the reaction time and accordingly increases the reproducibility. The microwave-synthesized NiO nanoflakes exhibited high specific capacitance of 401 Fg^{-1} at 0.5 $mAcm^{-2}$. The NiO nanoflake structure acts as a reservoir to avoid ion buffering and reduce the diffusion length [14]. Cao et al. investigated the microwave synthesis method of the NiO nanoflower. The NiO nanostructure delivered a high specific capacitance of 585 Fg^{-1} at the current density of 5 Ag^{-1} with excellent cyclic performance [15]. Zheng et al. showed the growth of the hexagonal NiO microrod array over a Ni substrate. The Ni surface was first etched with a hydrofluoric solution accompanied by a continuous flow of O_2 gas. The precursors give excellent strain and ductility and hence emerged as a superior flexible electrode material for the SC. The NiO free-standing electrode exhibits high volumetric capacitance of 785 Fcm^{-3} at 0.5 Acm^{-3} with excellent cyclic stability of 6,000 cycles with 92% capacitance retention [16]. Apart from NiO, MnO_2 also emerged as a supercapacitive electrode material due to its tremendous high theoretical capacitance. Based on the one electron redox reaction, MnO_2 exhibits specific capacitance of 1370 Fg^{-1} [17]; however, its inherently low electrical conductivity limits the practical applications. This is because the theoretical specific capacitance is realized only by the nanoscale ultrathin sheet or nanoparticles at very low current densities and scan rates, which are not suitable for devices. Hence, MnO_2 is mostly studied as a nanostructure electrode material supported by a conducting carbon substrate [18]. A $KMnO_4$ solution was used for the preparation of MnO_2 nanosheets over a carbon paper. The reaction taken place during the formation of MnO_2 in the precursors of $KMnO_4$ is given below:

$$4MnO_2 + 3C + H_2O \rightarrow 4MnO_2 + Co_3^{2-} + 2HCO_3 \tag{3.7}$$

The MnO_2 nanosheet exhibits wrinkle surface structure with few nanometer thickness, also the nanosheet exhibits high a specific capacitance of 306 Fg^{-1} with a good rate and cyclic performances. The paper discusses a methodology to prepare the MnO_2 nanosheet over a carbon fiber cloth as a substrate directly by carbonizing flex textile in order to explore a novel low cost, 3D structure, and scalable SC electrode material, such 3D carbon based architecture emerges promising applications in energy storage devices. The electrode material shows great advantages in terms of low cost, abundant sources, and high output [18]. Liu et al. reported a unique bowl-like MnO_2 nanosheet grown over a carbon cloth by using the hydrothermal technique. Such a novel nanostructure effectively reduces the empty space and provides additional electrochemically rich sites for the redox reaction, reduces the pathway for diffusion, and utilizes all the MnO_2 nanomaterial. The nanostructure exhibits high specific capacitance of 379 Fg^{-1} at 0.5 Ag^{-1} with long-term cyclic stability [19]. Nagaraju et al. reported that the core shell hexagonal nanoplate array over a carbon-cloth substrate with unique construction of two-electrode electrodeposition without a reference electrode, the MnO_2 hexagonal nanoplate arrays, and MnO_2 at nanostructured hierarchical structure exhibited a high specific capacitance of 244.54 Fg^{-1} at 0.5 A g^{-1} with good cyclic stability in the 1 M Na_2SO_4 electrolyte [20].

Hematite (Fe_2O_3), is another most frequently studied metal oxide for SCs. The various iron oxides were studied as anodic electrode material due to their variable oxidation state, natural abundance, and high theoretical capacitance [21]. But the less electronic conductivity limits its practical usage; hence, there has been a tremendous effort to improve their performances. Pai et al. reported the porous binder free-standing Fe_2O_3 electrode material fabricated through the scalable electrospinning technique. The iron precursor and the polymeric solution was electrospun to form a fiber material and then pyrolyzed in an inert environment. The iron oxide and iron carbide nanoparticles were embedded with the carbon and then converted into iron oxides. The in-situ iron-oxide-incorporated carbon fibers show specific capacitance as high as 460 Fg^{-1} at 1 Ag^{-1}, with excellent cyclic stability over 5,000 cycles and 82% capacitance retention [22]. Pal et al. reported a hydrothermal route to synthesize the Fe_2O_3 nanoparticle and Fe_2O_3/RGO hybrids. The prepared Fe_2O_3 electrode material when tested in Na_2SO_4 electrolyte results in the redox reaction between Fe^{2+} and Fe^{3+} and intercalation of SO_4^{2-} ions, and the redox reaction is given as:

$$2Fe^{II}O + SO_4^{2-} \rightarrow 4\left(Fe^{III}O\right) + SO_4^{2-}\left(Fe^{III}O\right) + 2e^- \tag{3.8}$$

The prepared electrode is tested with and without the presence of a magnetic field. In the presence of an external magnetic field, Fe_2O_3/RGO exhibited high specific capacitance of 868.89 Fg-1 almost two times vs. the absence of the magnetic field (450.61 Fg^{-1}) at a scan rate of 5 mVs^{-1}.The results may attribute to the fast electron transfer during the presence of the magnetic field, which significantly improves the charge density [23]. Zhao et al. herein, reported nanostructured Fe_2O_3 with 3D microporous nanosheets uniformly supported by the carbon fiber through the facile anodic deposition technique followed by thermal annealing. The Fe_2O_3 at carbon foam demonstrates high specific capacitance of 394.2 Fg^{-1} with excellent flexibility and rate performance. The iron oxides usually exhibit a negative potential range of −1 to 0 V in the aqueous electrolyte and hence can be used as a negative electrode. The low toxicity and environmental friendliness make them good candidates for the high-performance SC applications. The growth of iron oxides over a flexible carbon or graphene substrate provides high electronic conductivity to the electrode material. Therefore, there has been various reported strategies to implant iron oxide nanostructures over a conducting carbon or layered graphene substrate [24]. Achudan et al. reported the directly grown iron oxide nanoparticles over conducting multiwalled carbon nanotube utilized as a source of carbon material. The hybrid material was frequently characterized by using numerous physicochemical techniques. The hybrid design of iron oxides delivered a specific capacitance of 231 Fg^{-1} at the current density of 4 $Ag^{-1}$and good cyclic stability with an excellent capacitance retention of 77% for over 5,000 cycles [25]. Iron oxide-based electrodes are not effective pseudocapacitive electrode materials due to their poor capacitance and less cyclic stability. Another approach for the progress of metal oxides in the field of SCs is preparing oxide-based electrode material by using a cobalt metal ion because it is well-known for its high electric conductivity and highly valance electrons in the outer-most shell, making it chemical stable in the aqueous electrolyte. Numan et al. synthesized ultra-fine nanocrystals of Co_3O_4 using a novel process involving laser ablation in liquid. The electrochemical performance of Co_3O_4 was found to be as high as carbon-based electrode materials exhibiting an excellent cyclic stability of 100% capacitance retention for over 20,000 cycles. This advanced electrode material has emerged as a new approach to the transition metal oxides and opens the door of practical application of transition metal oxides as flexible supercapacitive devices [26]. The preparation of Co_3O_4 with the desirable morphology, valance state, and controlled structure is key to improving the efficiency of the electrode material [27]. Jang et al. reported a hydrothermal route to synthesize unique porous Co_3O_4 nanocubes. The unique porous Co_3O_4 nanocubes delivered a high specific capacity of 430.6 Fg^{-1} at a scan rate of 10 mVs^{-1} with excellent stability, and it was suggested

TABLE 3.1
Electrochemical Performance of Metal Oxide-Based Electrode Materials for Supercapacitors

Electrode Material	Structure and Morphology	Specific Capacitance (Fg^{-1}) at Current Density (Ag^{-1})	Cyclic Stability	References
NiO thin film	Nanoflakes sponge-like clusters	674 at 5 mVs^{-1}	72.5% at 2,000 cycles	[30]
$ZnCo_2O_4$/NiO	Nanoflakes	2797 at 1Ag^{-1}	90% at 3,000 cycles	[31]
Co_3O_4	Nanowire arrays	754 at 2 Ag^{-1}	100% at 4,000 cycles	[32]
Ni-Co oxides	Nanoflakes	506 at 1 Ag^{-1}	6% capacitance loss at 2,000 cycles	[33]
Co_3O_4	Oriented self-assembled 2D microsheets	263 at 1 Ag^{-1}	10.6% capacitance loss at 1,000 cycles	[34]
$NiCo_2O_4$/MWCNT	Mesoporous 3D network	1010 at 0.1 Ag^{-1}	83.4% at 2,000 cycles	[35]
$MnCo_2O_4$	Nanorods	845 at 1 Ag^{-1}	90.2% at 2,000 cycles	[36]
Mn_2SnO_4	Cube-like structure	298 at 1 $mAcm^{-2}$	89% at 5,000 cycles	[37]
Co_3O_4 at $NiCo_2O_4$	Hollow tubular structure	2430 at 1 Ag^{-1}	93.7% at 5,000 cycles	[38]
Fe_2O_3/RGO	Circular-like nanostructured	621 at 5 mVs^{-1}	77% at 14,000 cycles	[39]

as a manageable and measurable approach to develop a pseudocapacitor [28] electrode material with intercalative pseudocapacitive properties [29]. There have been various metal oxides studied up to now, but still the researchers are working on improving the efficiency of the materials for their practical applications (Table 3.1).

3.4 METAL HYDROXIDES FOR SUPERCAPACITORS

Recently, metal hydroxides have been frequently studied to improve their performance for SC applications. Likewise, $Ni(OH)_2$ [40] and $Co(OH)_2$ [41] are given tremendous attention due to their high electrical conductivities, stable valance states, and effective electronic and magnetic properties. Besides these, $Co(OH)_2$ exhibits a high charge storage capacity. Despite the fact that the pseudocapacitive performance of $Co(OH)_2$-based material is still limited. There is a need to develop different approaches to improve the performance of $Co(OH)_2$ and $Ni(OH)_2$. The novel nano-architecture of $Co(OH)_2$ gives high specific capacitance reported in literature [42]. Herein, the nanostructured $Co(OH)_2$ was electrodeposited on Ni film substrate. The electrodeposition technique is an efficient procedure for constructing nanostructure electrode materials. In general, this technique is implanted to any conducting substrate using an electrochemical protocol. This unique architecture of $Co(OH)_2$ exhibits a high specific surface area with specific capacitance of 2,800 Fg^{-1} and outstanding rate and redox capability. Xue et al. reported the $Co(OH)_2$ mesoporous nanowire array electrode synthesized through the electrodeposition technique, which delivered a high specific capacitance of 993 Fg^{-1} at 1Ag−1,which makes it as a potential electrode for SC application [43]. Apart from the metal hydroxides, the bimetallic combinations or binary mixture of metal hydroxides or layered double hydroxides emerged as a novel and effective class of electrode material owing to the advantages of their nanostructures with the layered architecture.

3.5 BIMETALLIC/LAYERED DOUBLE HYDROXIDES

Bimetallic or LDH is a class of multi-metal clay consisting of a metal cations octahedrally surrounded by the hydroxyl group forming an octahedral. This leads to the consideration that the LDH-based electrode materials give exceptionally high redox activities for effective utility of homogeneous transition metal cations and have an environmentally friendly nature. Mostly, the tenability of cations and ion exchange capabilities of anions makes them good candidates for the SC. Also, the facile exfoliation of the monolayer nanosheet and the preparation of LDH make them an excellent replacement of metal hydroxides. The LDHs are grown up in three types of nanostructures: 1D, 2D, and 3D. The nanostructure gives exceptionally high porosity to the electrode material with the unique architecture grown on a conductive surface. The 1D, 2D, and 3D nanostructure is usually depending on the type of substrates used [44]. The LDHs contain a fraction of metal cations (Mg^{2+}, Ni^{2+}, Cu^{2+}, Zn^{2+}, Fe^{2+}, etc.) and coordinates octahedrally with the trivalent metal ions (Al^{3+}, Cr^{3+}, Mn^{3+}, and Fe^{3+}). The interlayer charge compensating anion is mostly CO_3^{2-} due to its high charge affinity to the layered structure, but apart from this Cl^-, SO_4^{2-}, NO_3^{2-} is also considered. The unique architecture and easy processing of metal cations to the form layered structure open up an excellent chance to create a variety of nanostructures in the molecular and nanometer scales [45]. Apart from the compositional properties of LDHs, the high ion exchange properties of LDHs also play an important role by virtue of which various anions can be introduced at the interfaces of LDHs. Again, the surface OH^- groups attached with the layered structure give rise to the hierarchically porous architecture when they interact with any other nanomaterial. There are various direct and indirect methods to synthesize LDHs of different compositions and morphologies. Basically, to sustain the nanostructure architecture with chemical, temperature, and another environmental factor is a slightly difficult task. Hence, to improve their sustainability, various nanostructure-based LDHs have been synthesized with the versatile impact of morphology and anion exchange capability [46] wing to their high ion exchange process and flexible nanostructures as

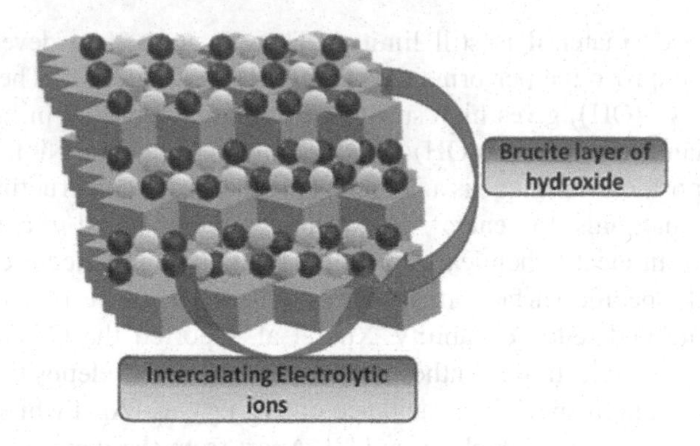

FIGURE 3.3 Schematic illustration of LDHs and their compositions.

shown in Figure 3.3. LDHs have been imparted in various fields of study, including energy storage devices, photochemistry, catalysis, and fire-retardant applications.

3.6 ELECTROCHEMICAL PERFORMANCES

Metal hydroxides, bimetallic/LDHs, are well-known for their nanostructure and unique architecture, and are extensively used in high-performance SCs. The $Ni(OH)_2$ hierarchal nanostructure consists of ultra-thin nanoflakes known to be exceptionally good pseudocapacitive electrode material with a thickness of ~7.4 nm. The ultra-thin nanoflakes of $Ni(OH)_2$ exhibit a high specific capacitance of 1,717 Fg^{-1} at the scan rate of 5 mVs^{-1} at a potential of 0.6 V with high-rate capability and excellent performance, making these nanostructured metal oxides an excellent alternative for high-performance SC applications [47]. The major challenge to electrode material is to fabricate in such a manner that it could have sufficient operating space with a constant mass loading at the desired area of the substrate to insure fast ion transportation [48]. Owing to the anisotropic structure, the 2D nanosheets provide the shortest diffusion path to elucidate transportation of electrons and ions, making them advantageous in various practical applications. Herein reported are the rationally translucent 2D $Ni(OH)_2$ nanosheets supported with Triethanol amine (TEA) as an alkali source and soft template. The prepared $Ni(OH)_2$ film was directly pressed into Ni foam and utilized as a 2D electrode material; it exhibited a high specific capacitance of 1,633 Fg^{-1} at 0.5 Ag^{-1} [49]. The pseudocapacitive materials though provide a high diffusion-controlled process with fast absorption-desorption, but limits in terms of cyclic and rate performance as compared to EDLC-based materials. The various metal oxide studies have been taking place in the respective fields but lack the cost effectiveness and performance. On the other hand, $Co(OH)_2$-based electrode material gives high performance with respect to their layered structure and interlayer spacing. Maile et al. reported the effect of pattern formation of $Co(OH)_2$ thin film with a different morphology. The electrode material dramatically exhibited a specific capacitance of 276 Fg^{-1} at 5 mVs^{-1}. Also, in the case of LDHs, the metal ions with the brucite layer form exceptionally good performances with 1D, 2D, and 3D nanostructures and hierarchically porous architecture. Basically, in terms of charge storage, the most preferable structure is 3D structures because they provide the self-assembly with great importance. The well-assembled 3D structures provide a highway for both the electron and electrolyte, which facilities high-energy storage [50]. The growth of nanosheets specially takes place vertically perpendicular to the surface of the substrate, and hence the structure will become dense and thickened. Such dense nanosheets provide easy access to the ions and act like ion reservoirs for the electrolyte. It should be noted that the LDH structures give comparatively the largest surface area and porosity to the electrodes, and hence emerged as an excellent alternative for high-performance SC applications. The various nanosheets have been studied as single-layer and double-layer electrode material for SC applications. Especially, Ni-Co as metal cations are studied frequently due to their exceptionally good electrochemical characteristics and electric and magnetic properties. Also, the theoretical capacitance of this material exceeds > 2300 Fg^{-1} with good rate of capability. The various devices has been prepared from metal hydroxides and bimettalic hydroxides/layered double hydroxides based elecrode materials. The recent study shows the remarkable performance of Ni and Co-based LDHs for asymmetric devices as shown in Table 3.2.

TABLE 3.2
Electrochemical Performance of Metal Hydroxides and Bimetalic Hydroxides/Layered Double Hydroxide Electrodes

Electrodes	Structure or Morphology	Specific Capacitance (Fg^{-1})	Cyclic Stability	References
$Ni(OH)_2$	Nanoboxes	2495 at 1 Ag^{-1}	1,200 cycles at 5Ag^{-1}	[51]
$Ni(OH)_2$/ graphene	Nanoflakes	2194 at 2 mVs^{-1}	2,000 cycles with 4.3% deterioration	[52]
Co-Mn LDH/ Ni foam	Nanoneedles	2422 at 1 Ag^{-1}	3,000 cycles with 86.5% retention	[53]
Mn-Co LDH	Hierarchically porous irregular nanosheets	511 at 2 Ag^{-1}	2,000 cycles with 91.6% retention	[54]
Co-Al LDH	2D nanosheets	1043 at 1 Ag^{-1}	—	[55]
Ni-Al-LDH	Urchin-like hollow microspheres	1578 at 1 Ag^{-1}	10,000 cycles with 97.3% capacitance retention	[56]
Ni-Mn LDH/ Carbon foam	Peony-like nanosheets	2128.53 at 0.5 Ag^{-1}	5,000 cycles with 94.3% retention	[57]
Ni-Mn LDH/ Ni foam	—	1511 at 2.5 Ag^{-1}	3,000 cycles 92.8% retention	[58]
Ni-Co LDH/Cu foam	3D nanowire network	2170 at 1 Ag^{-1}	2,000 cycles with 80.46% retention	[59]
Ni-Co LDH	3D nanowires	1563.1 at 1 Ag^{-1}	—	[60]
Zn-Co LDH	Uniform hexagonal platelets	3946.5 at 3 Ag^{-1}	—	[61]

3.7 APPLICATIONS OF METAL OXIDES/HYDROXIDES FOR SUPERCAPACITORS

The metal oxides and hydroxides are usually used as battery types or pseudocapacitive electrode materials for flexible devices or asymmetric SCs. Herein, we summarize the recent advancements of layered metal oxides and hydroxides with respects to their origin and nanomorphology because it is so obvious that no material is said to be an accurate or excellent material with knowing its practical applications. There were many devices being summed up to utilize the advancements of metal oxides/hydroxides. These devices are either rigid SCs or flexible SCs. Following, we are discussing the recent advancements of metal oxide hydroxide-based SC devices.

Basically, when taking this into account, the pseudocapacitive electrode materials takes the advantage of an excellent battery-type charge storage mechanism together with the excellent access of ions at electrolyte reservoirs. In most of the cases, the metal oxide/hydroxide-based electrode materials act as anodic and carbon materials and as cathodic electrode materials to assemble an asymmetric supercapacitive device. Mostly, aqueous alkaline electrolyte (commonly KOH) is used to facilitate the device for the electrochemical process and performance of metal oxide/hydroxide electrodes in a fast charging and discharging process. Moreover, the assembled device makes good use of potential windows, which results the increment in the specific energy. During the fabrication of a SC device, the mass balance between positive and negative electrode material improves the operating potential window of the overall device.

The charge carried by the material is calculated by the following equation:

$$Q = C \times \Delta E \times m \tag{3.9}$$

where C is the specific capacity (mAhg^{-1}), ΔE is the difference of the potential windows (V), and m is mass of the active materials (g). To balance the charges of the both electrodes, the mass balance equation is taken as

$$\frac{m_+}{m_-} = \frac{C_- X \, \Delta E_-}{C_+ X \, \Delta E_+} \tag{3.10}$$

Hence, based on the values of C and ΔE for the two electrodes of the cell, we can calculate the optimal mass ratios and finally apply them to the fabrication of asymmetric devices [6]. Whereas, the flexible devices are also developed for the development of wearable electronic devices, they come out with certain challenges regarding the flexible power sources, high energy storage, long-term cyclic stability, and rate capability. In practical applications, these devices have to be small, compressed, and flexible, and their performance should be robust under the bending, stretching, and compressing devices. Many flexible devices are studied by taking metal oxides/hydroxides or LDHs as positive electrode material and carbon-based materials as negative electrode material. Javed et al. reported the fabrication of a flexible asymmetric device composed of Zn-Co oxide at the carbon cloth and nano porous carbon (NPC) at the carbon cloth, respectively, as positive and negative electrode materials fabricated as an asymmetric device and have been tested in a (Polyvinyl alcohol) PVA/LiCl gel electrolyte. The device delivered exceptionally high specific capacitance of 210 Fg^{-1} at 1.5 Ag^{-1} current density, with good cyclic stability in a high potential window 0–2.0 V [62]. Bahareh et al. reported a flexible asymmetric device made up of Ni-Co LDH supported by 3D nitrogen-doped graphene as positive and negative electrodes, respectively, which delivered a specific capacitance of 109 Fg^{-1} at 0.5 Ag^{-1} current density with maximum energy density of 49 $Whkg^{-1}$. The flexible device provides excellent mechanical strength because it shows the high bending stability up to 450 bending cycles at 90°angle; hence, this device is said to be promising for wearable electronic devices [63]. There were various flexible substrates used such as carbon cloth, carbon fibers, and carbon textiles for the fabrication of flexible SC devices. Jeong et al. reported the binder-free Ni–Co LDH nanosheets on a Ni textile sheet as a positive and negative electrode material, respectively, which give rise to the high-bending mechanism for the flexible SC device as shown in Figure 3.4. The stacked device is made up of a face-to-face assembly, and the flexible device achieved volumetric energy density of 1.25 $mWhcm^{-3}$ at a power density of 47 $mWcm^{-3}$. It retained the maximum energy density and emerged as a promising flexible device for SCs [64]. Li et al. reported an ultra-thin Ni-Co LDH grown on a carbon fiber cloth with the simple electrodeposition technique. This technique creates a core shell-like structure with high electrical conductivity and cycle stability. The asymmetric device was made out of Ni-Co LDH as the positive and RGO/Ni foam as the negative electrode material exhibiting excellent performance in terms of high-energy density of 37 $Whkg^{-1}$ at high-power density 800 Wkg^{-1} and a long-term cyclic stability of 84.6% retention for over 10,000 cycles. This is expected to be a great performance in SC applications [65].

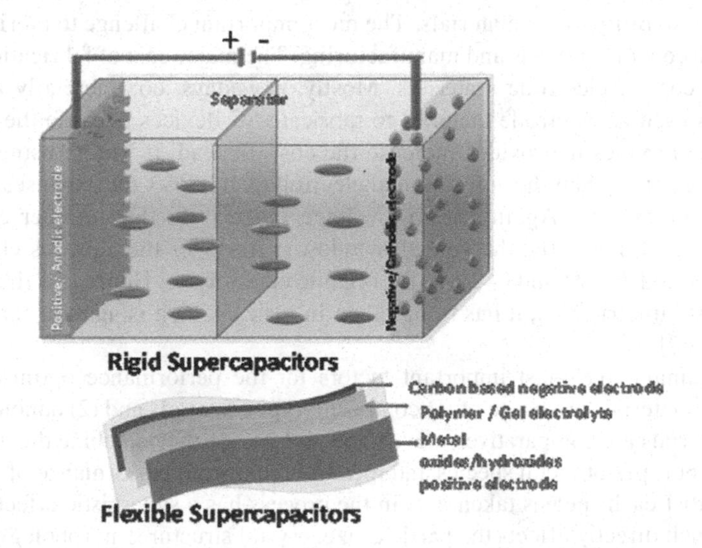

FIGURE 3.4 Schematic illustration of rigid SCs and flexible SCs for asymmetric SC configuration.

Although there has been a lot of research taking place with rigid SCs and flexible SCs, they are not up to the mark. In addition, the cyclic stability of these devices needs to be improved because they can be used in practical applications. During the charge/discharge process, the structural distortion, volume expansion, and phase transformation leads to the reduction of practical applications of these devices. In metal cations, the modifications in their host layer can help to prevent the phase transition and can increase its cyclic stability. Again, the formation of extra efficient materials with nanostructures, that is, the materials with highly amorphous structure, provides an excessive pathway for ion transportation and the use of solid-state electrolytes other than the conventional KOH electrolyte and provide better performance of flexible devices in SCs.

3.8 CONCLUSIONS

In this chapter, we discussed the metal oxides and hydroxides related work and their application in asymmetric devices for SCs. The metal oxides and hydroxide-based electrode materials are considered to be efficient electrode material in terms of their high specific capacitance, energy density, relatively easy synthesis technique, and unique nanostructure, tenability with metal ions, good ion reservoir, and tendency to achieve high theoretical capacitance. While still, the practical application of metal oxide/hydroxide-based electrode materials is restricted due to their low cycle life and high cost compared to the carbon-based electrodes. The challenge arises regarding the tunable synergistic effect between metal ions and host cations because it gives a direct impact on morphology of the material. Again, the mass balance between metal cations also plays a vital role that directly affects the phase structure and electrochemical performance of the materials. Hence, the modification in the host layer and interlayer spacing may cause an increase in rate capability and cyclic stability to the materials. The most important challenge to fabricate SCs is its high cost in synthesis and manufacturing. The major cost of fabrication arises with the cost of electrode materials. Mostly, nowadays, commercially activated carbon is used as electrode material to fabricate SC devices because the RuO_2 or other metal oxides/hydroxides increase the cost. Instead, if we are using organic electrolytes other than the conventional electrolyte, it boosts the expenses far from the expectations [5]. Again, the low voltage range is further another challenge to fabricate SCs. Mostly, the voltage window is fixed for the aqueous electrolyte between 0 and 1.5 V) and 0–2.7 V for organic electrolytes. Hence, for the fabrication of asymmetric SCs, it has to be taken in series, which is again a challenging task [66–67].

In summary, the most important factors for the performance optimization of electrode materials are usage of (1) composite-type materials and (2) nanomaterials. Nanomaterials are comparatively advantageous because they combine the properties of different types of metal species and increase the overall performance of SCs. It is worth it that each species taken part in the process has a synergistic effect on each other, which directly affects the particle size, crystal structure, morphology, surface area, and pore size distribution, facilitating fast ion transportation, preventing particles from agglomeration, inducing porosity, reducing the diffusion path, increasing active sites, protecting materials against mechanical degradation, and adding extra pseudo-capacitance. As explained above, the layered double hydroxide-based nanomaterials give excellent performance; this is the key aspect of nanostructured materials, which gives highly mesoporous microstructure that can accommodate fast electrolytic-ion transfer. The material's morphology is directly related to its surface area and the path for ion transportation; hence, dimensionally grown nanostructured electrode materials are said to be a great alternative for SC applications.

ABBREVIATIONS

EDLC	electrochemical double layer capacitor
LDH	layered double hydroxides
MWCNT	multiwalled carbon nanotube
RGO	reduced graphene oxide
SC	supercapacitor

REFERENCES

1. Ragupathy, P., Ranjusha, R., Jagan, R. Nanostructured metal oxides for supercapacitor applications. *Nanostructured Ceramic Oxides for Supercapacitor Applications*, 2014, 23, 119–152.
2. Jadhav, S., Chaturvedi, V., Shelke, M. Recent advances in flexible supercapacitors. *VKidney Organogenesis*, 2019, 3, 41–72.
3. Roldán, S., Granda, M., Menéndez, R., Santamaría, Blanco, C. Mechanisms of energy storage in carbon-based supercapacitors modified with a quinoid redox-active electrolyte. *The Journal of Physical Chemistry C*, 2011, 115(35), 17606–17611.
4. Xu, B., Wu, F., Chen, S., Zhang, C., Cao, G. Activated carbon fiber cloths as electrodes for high performance electric double layer capacitors. *Electrochemical Acta*, 2007, 4595–4598.
5. Muzaffara, A., Ahameda, M.B., Deshmukha, K., Thirumalaib, J. A review on recent advances in hybrid supercapacitors: Design, fabrication and applications. *Renewable and Sustainable Energy Reviews*, 2019, 101, 123–145.
6. Wang, Y., Song, Y., Xia, Y. Electrochemical capacitors: Mechanism, materials, systems, characterization and applications. *Chemical Society Reviews*, 2016, 45(21), 5925–5950.
7. Shi, F., Li, L., Wang, X., Gu, C., Tu, J. Metal oxide/hydroxide-based materials for supercapacitors. *RSC Advances*, (2014), 4(79). 41910–41921
8. Han, Z.J., Pineda, S., Murdock, A.T., Seo, D.H., Ostrikovba, K.K., Bendavid, A. RuO_2-coated vertical graphene hybrid electrodes for high-performance solid-state supercapacitors. *Journal of Materials Chemistry A*, 2017, 5, 17293–17301.
9. Xiao, H., Yao, S., Liu, H., Qu, F., Zhang, X., Wu, X. NiO nanosheet assembles for supercapacitor electrode materials. *Progress in Natural Science: Materials International*, 2016, 26(3), 271–275.
10. Zhang, W., Jia, G., Li, H., Liu, S., Yuan, C., Bai, Y. Fu, D. Morphology-modulated mesoporous CuO electrodes for efficient interfacial contact in nonenzymatic glucose sensors and high-performance supercapacitors. *Journal of the Electrochemical*, 2016, 164(2), B40–B47.
11. Tummala, R., Guduru, R.K., Mohanty, P.S. Nanostructured Co_3O_4 electrodes for supercapacitor applications from plasma spray technique. *Journal of Power Sources*, 2012, 209, 44–51.

12. Mishra, S.R., Adhikari, H., Kunwar, D.L., Ranaweera, C., Sapkota, B., Ghimire, M., Gupta, R., Alam, J. Facile hydrothermal synthesis of hollow Fe_3O_4 nanospheres effect of hydrolyzing agents and electrolytes on electrocapacitive performance of advanced electrodes. *Journal of Materials Physics*, 2017, 2, 1–15.
13. Xu, J., Wang, M., Liu, Y., Li, J., Cui, H. One-pot solvothermal synthesis of size-controlled NiO nanoparticles, *Advanced Powder Technology*, 2019, 30, 861–868.
14. Vijayakumar, S., Nagamuthu, S., Muralidharan, G. Supercapacitor studies on NiO nanoflakes synthesized through a microwave route. *ACS Applied Materials & Interfaces*, 2013, 5(6), 2188–2196.
15. Cao, C.-Y., Guo, W., Cui, Z.-M., Song, W.-G. Cai, W. Microwave-assisted gas/liquid interfacial synthesis of flowerlike NiO hollow nanosphere precursors and their application as supercapacitor electrodes. *Journal of Materials Chemistry*, 2011, 21(9), 3204.
16. Zheng, D., Zhao, F., Li, Y., Qin, C., Zhu, J., Hu, Q., Inoue, A. Flexible NiO micro-rods/nanoporous Ni/metallic glass electrode with sandwich structure for high performance supercapacitors. *Electrochimica Acta*, 2019, 297, 767–777.
17. Toupin, M., Brousse, T., Bélanger, D. Charge storage mechanism of MnO_2 electrode used in aqueous electrochemical capacitor. *Chemistry of Materials*, 2004, 16(16), 3184–3190.
18. He, S., Hu, C., Hou, H., Chen, W. Ultrathin MnO_2 nanosheets supported on cellulose based carbon papers for high-power supercapacitors. *Journal of Power Sources*, 2014, 246, 754–761.
19. Liu, P., Zhu, Y., Gao, X., Huang, Y., Wang, Y., Qin, S., Zhang, Y. Rational construction of bowl-like MnO_2 nanosheets with excellent electrochemical performance for supercapacitor electrodes. *Chemical Engineering Journal*, 2018, 350, 79–88.
20. Nagaraju, G., Ko, Y.H., Cha, S.M., Im, S.H., Yu, J.S. A facile one-step approach to hierarchically assembled core–shell-like MnO_2 at MnO_2 nanoarchitectures on carbon fibers: An efficient and flexible electrode material to enhance energy storage. *Nano Research*, 2016, 9(5), 1507–1522.
21. Yao, Y., Xu, N., Guan, D., Li, J., Zhuang, Z., Zhou, L., Mai, L. Facet-selective deposition of FeOx on α-MoO_3 nanobelts for lithium storage. *ACS Applied Materials & Interfaces*, 2017, 9(45), 39425–39431.
22. Pai, R., Singh, A., Simotwo, S., Kalra, V. In situ grown iron oxides on carbon nanofibers as freestanding anodes in aqueous supercapacitors. *Advanced Engineering Materials*, 2018, 20(6), 1701116.
23. Pal, S., Majumder, S., Dutta, S., Banerjee, S., Satpati, B., De, S. Magnetic field induced electrochemical performance enhancement in reduced graphene oxide anchored Fe_3O_4 nanoparticle hybrid based supercapacitor. *Journal of Physics D: Applied P*, 2018, 51(37), 375501.
24. Zhao, P., Wang, N., Hu, W., Komarneni, S. Anode electrodeposition of 3D mesoporous Fe_2O_3 nanosheets on carbon fabric for flexible solid-state asymmetric supercapacitor. *Ceramics International*, 2019, 45, 10420–10428.
25. Atchudan, R., Edison, T.N., Perumal, S., RanjithKumar, D., Lee, Y.R. Direct growth of iron oxide nanoparticles filled multi-walled carbon nanotube via chemical vapour deposition method as high-performance supercapacitors. *International Journal of Hydrogen Energy*, 2018, 44, 2349–2360.
26. Numan, A., Duraisamy, N., Saiha Omar, F., Mahipal, Y.K., Ramesh, K., Ramesh, S. Enhanced electrochemical performance of cobalt oxide nanocube intercalated reduced graphene oxide for supercapacitor application. *RSC Advances*, 2016, 6(41), 34894–34902.
27. Liu, X.Y., Gao, Y.Q., Yang, G.W. A flexible, transparent and super-long-life supercapacitor based on ultrafine Co_3O_4 nanocrystal electrodes. *Nanoscale*, 2016, 8(7), 4227–4235.

28. Jang, G.-S., Ameen, S., Akhtar, M.S., Shin, H.-S. Cobalt oxide nanocubes as electrode material for the performance evaluation of electrochemical supercapacitor. *Ceramics International*, 2018, 44(1), 588–595.
29. Kandalkar, S.G., Dhawale, D.S., Kim, C.-K., Lokhande, C.D. Chemical synthesis of cobalt oxide thin film electrode for supercapacitor application. *Synthetic Metals*, 2010, 160(11–12), 1299–1302.
30. Gund, G.S., Lokhande, C.D., Park, H.S. Controlled synthesis of hierarchical nanoflake structure of NiO thin film for supercapacitor application. *Journal of Alloys and Compounds*, 2018, 741, 549–556.
31. Yi, T.-F., Li, Y.-M., Wu, J.-Z., Xie, Y., Luo, S. Hierarchical mesoporous flower-like $ZnCo_2O_4$ at NiO nanoflakes grown on nickel foam as high-performance electrodes for supercapacitors. *Electrochimica Acta*, 2018, 284, 128–141.
32. Xia, X., Tu, J., Zhang, Y., Mai, Y., Wang, X., Gu, C., Zhao, X. Freestanding Co_3O_4 nanowire array for high performance supercapacitors. *RSC Advances*, 2012, 2(5), 1835.
33. Lu, X., Huang, X., Xie, S., Zhai, T., Wang, C., Zhang, P, Tong, Y. Controllable synthesis of porous nickel–cobalt oxide nanosheets for supercapacitors. *Journal of Materials Chemistry*, 2012, 22(26), 13357.
34. Xie, L., Li, K., Sun, G., Hu, Z., Lv, C., Wang, J., Zhang, C. Preparation and electrochemical performance of the layered cobalt oxide (Co_3O_4) as supercapacitor electrode material. *Journal of Solid State Electrochemistry*, 2012, 17(1), 55–61.
35. Jayaseelan, S.S., Radhakrishnan, S., Saravanakumar, B., Seo, M.-K., Khil, M.-S., Kim, H.-Y., Kim, B.-S. Mesoporous 3D $NiCo_2O_4$/MWCNT nanocomposite aerogels prepared by a supercritical CO_2 drying method for high performance hybrid supercapacitors. *Colloids and Surface A: Physicochemical and Engineering*, 2018, 538, 451–459.
36. Xu, J., Sun, Y., Lu, M., Wang, L., Zhang, J., Tao, E., Liu, X. Fabrication of the porous $MnCo_2O_4$ nanorod arrays on Ni foam as an advanced electrode for asymmetric supercapacitors. *Acta Materialia*, 2018, 152, 162–174.
37. Gnana Sundara Raj, B., Kim, H.-Y., Kim, B.-S. Ultrasound assisted formation of Mn_2SnO_4 nanocube as electrodes for high performance symmetrical hybrid supercapacitors. *Electrochimica Acta*, 2018, 278, 93–105.
38. Wang, Y., Zhang, M., Li, Y., Ma, T., Liu, H., Pan, D., Wang, A. Rational design 3D nitrogen doped graphene supported spatial crosslinked Co_3O_4 at $NiCo_2O_4$ on nickel foam for binder-free supercapacitor electrodes. *Electrochimica Acta*, 2018, 290, 12–20.
39. Zhu, S., Zou, X., Zhou, Y., Zeng, Y., Long, Y., Yuan, Z., Xiang, B. Hydrothermal synthesis of graphene-encapsulated 2D circular nanoplates of α-Fe_2O_3 towards enhanced electrochemical performance for supercapacitor. *Journal of Alloys and Compound*, 2018, 775, 63–71.
40. Xiong, X., Ding, D., Chen, D., Waller, G., Bu, Y., Wang, Z., Liu, M. Three-dimensional ultrathin $Ni(OH)_2$ nanosheets grown on nickel foam for high-performance supercapacitors. *Nano Energy*, 2015, 11, 154–161.
41. Gupta, V., Kusahara, T., Toyama, H. Gupta, S. Miura, N. Potentiostatically deposited nanostructured α-Co(OH)2: A high performance electrode material for redox-capacitors. *Electrochemistry Communications*, 2007, 9(9), 2315–2319.
42. Chang, J.K., Wu, C.M., Sun, I.W. Nano-architectured $Co(OH)_2$ electrodes constructed using an easily-manipulated electrochemical protocol for high-performance energy storage applications. *Journal of Materials Chemistry*, 2010, 20, 3729–3735.
43. Xue, T., Wang, X., Lee, J.M. Dual-template synthesis of $Co(OH)_2$ with mesoporous nanowire structure and its application in supercapacitors. *Journal of Power Sources*, 2012, 201, 382–386.
44. Li, X., Du, D., Zhang, Y., Xing, W., Xue, Q., Yan, Z. Layered double hydroxides toward high-performance supercapacitors. *Journal of Materials Chemistry A*, (2017), 5(30), 15460–15485.

45. Varadwaj, G.B.B., Nyamori, V.O. Layered double hydroxide- and graphene-based hierarchical nanocomposites: Synthetic strategies and promising applications in energy conversion and conservation. *Nano Research*, 2016, 9(12), 3598–3621.
46. Jiang, H., Zhao, T., Li, C., Ma, J. Hierarchical self-assembly of ultrathin nickel hydroxide nanoflakes for high-performance supercapacitors. *Journal of Materials Chemistry*, 2011, 21(11), 3818.
47. Li, X., Chen, R., Zhao, Y., Liu, Q., Liu, J., Yu, J., Wang, J. Layer-by-layer inkjet printing GO film anchored $Ni(OH)_2$ nanoflakes for high-performance supercapacitors. *Chemical Engineering Journal*, 2019, 375, 121988.
48. Tian, J., Xue, Y., Yu, X., Pei, Y., Zhang, H., Wang, J. 2D nanoporous $Ni(OH)_2$ film as an electrode material for high-performance energy storage devices. *RSC Advances*, 2019, 9(31), 17706–17716.
49. Maile, N.C., Shinde, S.K., Koli, R.R., Fulari, A.V., Kim, D.Y., Fulari, V. J. Effect of different electrolytes and deposition time on the supercapacitor properties of nanoflake-like $Co(OH)_2$ electrodes. *Ultrasonics Sonochemistry*, 2018, 51, 49–57.
50. Fu, Y., Song, J., Zhu, Y., Cao, C. High-performance supercapacitor electrode based on amorphous mesoporous $Ni(OH)_2$ nanoboxes. *Journal of Power Sources*, 2014, 262, 344–348.
51. Yan, J., Sun, W., Wei, T., Zhang, Q., Fan, Z., Wei, F. Fabrication and electrochemical performances of hierarchical porous $Ni(OH)_2$ nanoflakes anchored on graphene sheets. *Journal of Materials Chemistry*, 2012, 22(23), 11494.
52. Su, D., Tang, Z., Xie, J., Bian, Z., Zhang, J., Yang, D., Kong, Q. Co, Mn-LDH nanoneedle arrays grown on Ni foam for high performance supercapacitors. *Applied Surface Science*, 2018, 469, 487–494.
53. Wu, N., Low, J., Liu, T., Yu, J., Cao, S. Hierarchical hollow cages of Mn-Co layered double hydroxide as supercapacitor electrode materials. *Applied Surface Science*, 2017, 413, 35–40.
54. Wu, X., Jiang, L., Long, C., Wei, T., Fan, Z. Dual support system ensuring porous Co-Al hydroxide nanosheets with ultrahigh rate performance and high energy density for supercapacitors. *Advanced Functional Materials*, 2015, 25(11), 1648–1655.
55. Wang, W., Zhang, N., Shi, Z., Ye, Z., Gao, Q., Zhi, M., Hong, Z. Preparation of Ni-Al layered double hydroxide hollow microspheres for supercapacitor electrode. *Chemical Engineering Journal*, 2018, 338, 55–61.
56. Chen, D., Yan, S., Chen, H., Yao, L., Wei, W., Lin, H., Han, S. Hierarchical Ni–Mn layered double hydroxide grown on nitrogen-doped carbon foams as high-performance supercapacitor electrode. *Electrochimica Acta*, 2018, 292, 374–382.
57. Guo, X.L., Liu, X.Y., Hao, X.D., Zhu, S.J., Dong, F., Wen, Z.Q., Zhang, Y.X. Nickel-manganese layered double hydroxide nanosheets supported on nickel foam for high-performance supercapacitor electrode materials. *Electrochimica Acta*, 2016, 194, 179–186.
58. Liu, Y., Teng, X., Mi, Y., Chen, Z. A new architecture design of Ni–Co LDH-based pseudocapacitors. *Journal of Materials Chemistry A*, 2017, 5(46), 24407–24415.
59. Liang, H., Jia, H., Lin, T., Wang, Z., Li, C., Chen, S., Feng, J. Oxygen-vacancy-rich nickel-cobalt layered double hydroxide electrode for high-performance supercapacitors. *Journal of Colloid and Interface Science*, 2019, 554, 59–65.
60. Pan, Z., Jiang, Y., Yang, P., Wu, Z., Tian, W., Liu, L., Hu, L. In situ growth of layered bimetallic ZnCo hydroxide nanosheets for high-performance all-solid-state pseudocapacitor. *ACS Nano*, 2018, 12(3), 2968–2979.
61. Javed, M.S., Shaheen, N., Hussain, S., Li, J., Shah, S.S.A., Abbas, Y., Mai, W. Ultrahigh energy density flexible asymmetric supercapacitor based on hierarchical fabric decorated with 2D bimetallic oxide nanosheets and MOF derived porous carbon polyhydron. *Journal of Material Chemistry A*, 2018, 7, 946–957.

62. Mehrabimatin, B., Gilshteyn, E.P., Melandsø Buan, M.E., Sorsa, O., Jiang, H., Irajizad, A., Kallio, T. Flexible and mechanically dura flexible and mechanically durable asymmetric supercapacitor based on NiCo-layered double hydroxide and nitrogen-doped graphene using a simple fabrication method. *Energy Technology*, 2019, 7, 1801002.
63. Jeong, Y.-M., Son, I., Baek, S.-H. Binder–free of NiCo–layered double hydroxides on Ni–coated textile for wearable and flexible supercapacitors. *Applied Surface Science*, 2019, 467–468, 963–967.
64. Xu, J., Wang, M., Liu, Y., Li, J., Cui, H. One-pot solvothermal synthesis of size-controlled NiO nanoparticles. *Advanced Powder Technology*, 2019, 30, 861–868.
65. Li, Y., Shan, L., Sui, Y., Qi, J., Wei, F., He, Y., Liu, J. Ultrathin Ni–Co LDH nanosheets grown on carbon fiber cloth via electrodeposition for high-performance supercapacitors. *Journal of Materials Science: Materials in Electronics*, 2019, 30, 13360–13371.
66. He, S., Chen, W. Application of biomass-derived flexible carbon cloth coated with MnO_2 nanosheets in supercapacitors. *Journal of Power Sources*, 2015, 294, 150–158.
67. González, A., Goikolea, E., Barrena, J.A., Mysyk, R. Review on supercapacitors: Technologies and materials. *Renewable and Sustainable Energy Reviews*, 2016, 58, 1189–1206.

4 Transition Metal Oxide-Based Multiferroic Materials for Spintronics and Energy-Harvesting Applications

Rajan Singh and Jayant Kolte

CONTENTS

4.1 SPINTRONICS

The realization that the use of the electrical field can manipulate the electron spin gives rise to spintronics. Spin-based electronics is an emerging area and has many applications viz. magnetic random-access memory and quantum electronic devices. Conventional electronic devices use electron charge to store charge. In spintronics, the electron spin is used, which improves device efficiency. Spintronic-based memories store and process information based on the spin of the ferromagnetic materials [1]. Spintronic devices are made of trilayer with two ferromagnetic electrodes with a non-magnetic spacer. The spacer can be a metal or an insulator layer. These devices are called as multiferroic tunnel junctions (MFTJs). The resistance change depends upon the ferromagnetic electrode alignment (parallel or antiparallel). Here, the magnetic moment can be modified in two states viz. "0" and "1." These states are antiparallel and can be read by adjacent magnetic layers with fixed magnetization. We can classify the effect of magnetic field dependence properties of a material in many categories depending upon their functions. Magnetoresistance (MR) is a well-known phenomenon that is a change in the resistance state of any material by the application of a magnetic field. This can be positive or negative, depending on the difference in the resistance state. If the materials show a significant change in the resistance sate, it is called a giant magnetoresistance (GMR) effect [1]. In tunnel magnetoresistance (TMR), there is a magnetic tunnel junction. Here, the current conduction is due to spin-dependent tunneling, and the electron tunneling probability depends upon their spin state [1]. The following equation can define the TMR:

$$TMR = \left(R_{AP} - R_{P}\right) / R_{P} \tag{4.1}$$

where R_P and R_{AP} are the resistances of the parallel and antiparallel magnetic configurations, respectively. Some MFTJs also have the effect of ferroelectric (FE) polarization reversal on the resistance change. This phenomenon is known as tunneling electroresistance (TER) and can be defined as

$$TER = \left(R_{high} - R_{low}\right) / R_{low} \tag{4.2}$$

where R_{high} and R_{low} are the high and low resistances that occur during the ferroelectric polarization directions, respectively.

Anisotropic magnetic resonance (AMR) is the anisotropy in the direction of the spin axis. There is a change in the electrical resistance on the angle of applied the external magnetic field. The effect arises due to the spin-orbit interaction in the material [1].

To improve the power efficiency of the semiconductor devices, it is necessary to increase the data density as well as read and write behavior. By using the spintronics functionality, the flow of electrons can be used to control the magnetic state. As compared to the electric current to read the data, the change of spin by the current will consume less power and can avoid the heating effect in the sample. Therefore, the choice of a material, which gives rise to this phenomenon, is always a concern for scientists working in the field.

Multiferroic materials are the multifunctional materials that can possess more than one ferroic order [2]. There is a coupling in these ordering, and they are coupled. The magnetic state can be changed by the electric field [converse magnetoelectric (ME) effect]. Similarly, the electric state can be changed by the magnetic field (direct ME effect). This opens a new avenue to work on the rich functionalities of multiferroic materials, and a lot of research has focused upon the development of the use of these materials into various applications [2–8]. Since the magnetic field can control the electric (polarization) state of multiferroic materials, it requires less power, and the loss due to current flow will be minimum. Moreover, due to the insulating nature of the multiferroic materials, these are the best candidates to replace ferromagnetic (FM) materials in spintronics.

The multiferroic oxide materials have gained interest initially due to ease in synthesis and being single phase. The concept of ME random access memory has been proposed by Bibes in 2008 using $BiFeO_3$ (BFO) as an active material [9]. However, literature is limited to the oxides for use in spintronics devices. This is mainly due to the fact that they are not proper multiferroic materials, rather an antiferromagnetic one. Second is the multiferroic effect realization below room temperature. This does not make them a suitable candidate for spintronics. Due to this, a lot of research has focused upon a synthesis of artificial or composite multiferroic materials. In the composite system, one phase will be an oxide phase, and the coupling is interface mediated.

Here, we will mostly explore multiferroic materials that may not or may have direct ME coupling between them. The magnetic state of the system can be altered, and it can be characterized in many ways as depicted by Fusil et al. [10]. They can be classified into four types: TMR, GMR, AMR, and MR. Amongst these, TMR has gained interest recently due to low power consumption, high density, and multiple memory states that can be achieved. Also, this can be used for sensing applications. Here, we will first focus on the multiferroic materials used in the multiferroic tunnel junctions (called MFTJ). In multiferroic, single-phase multiferroic materials, as well as a layer of the ferroelectric and ferromagnetic layer, have been used.

4.1.1 Single-Phase Multiferroic Tunnel Junctions (MFTJ)

The magnetic tunnel junctions using single-phase multiferroic materials are scarce due to the non-insulating behavior of the materials and then ME coupling below room temperature. A single-phase multiferroic material should have empty d orbitals for magnetism, and this restriction is non-insulator with leaky behavior. $BiFeO_3$ and $BiMnO_3$ are the well-known multiferroic materials used in spintronics [11,12].

There is only one report on the $BiMnO_3$-based TMR devices from 2007. LSMO/LBMO (2 nm)/Au trilayers have been grown on (001) $SrTiO_3$ substrates using the pulsed-laser deposition technique [11]. Approximately 10% La has been substituted to the Bi site to minimize the volatility during deposition. Around 80% of the TMR change is observed in the prepared samples. Further, with the variation of voltage, the resistance state changes, and this proves a coupling between electric and magnetic ordering. But this phenomenon takes place at very low temperatures and not suitable for most of the practical applications. Therefore, alternate material such as BFO has been investigated for the feasibility study. Trilayers of $La_{0.6}Sr_{0.4}MnO_3/BiFeO_3/La_{0.6}Sr_{0.4}MnO_3$ have been used to study the TMR by Liu et al. [12]. The films have been deposited on (001) oriented $(LaAlO_3)_{0.3}(SrAl_{0.5}Ta_{0.5}O_3)_{0.7}$ (LSAT) substrates by using magnetron sputtering. The schematic of the device is represented in Figure 4.1. The TMR ratio found to be 3.4% due to the pinning of the FM layer by the antiferromagnetic ordering of BFO. The thickness of the BFO plays a major role here and leads to the scattering of conduction electrons. This can be verified using the polarization-dependent resistance change and the TER found to be 15%.

The TMR effect can be increased by decreasing the thickness of the ME material. This ME layer pins the magnetization of the LSMO layer [13]. Hambe and group have used $La_{0.67}Sr_{0.33}MnO_3$ (LSMO) as a top and bottom electrode with BFO as an ME layer in between. The thickness of the BFO film is ~3 nm and is grown on (001)-oriented $SrTiO_3$ substrate using pulsed-laser deposition. There is a significant 70% change in the MR with a small voltage pulse. This indicates the absence of LSMO magnetization pinning by the thin BFO layer.

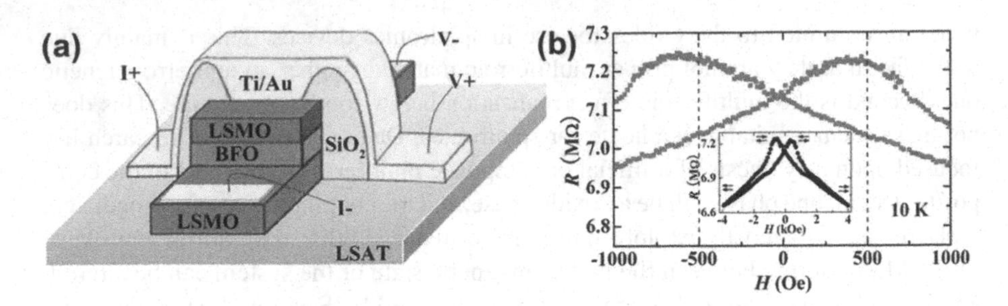

FIGURE 4.1 (a) The schematic spintronic device to study the TMR effect in LSMO/BFO/LSMO heterostructures and (b) the resistance change measured at 10 K without polarization. (Reprinted with permission from Liu, Y. et al., *Appl. Phys. Lett.*, 104, 043507, 2014.)

4.1.2 Multiferroic Tunnel Junctions with a Ferroelectric Layer

Ferroelectric and ferromagnetic layers have additional advantages in the spintronics devices. Ferroelectric materials are insulating and do not have leaky behavior as in the case of single-phase multiferroic materials. Much of the research in spintronics is devoted to the use of different ferroelectric and magnetic layers [14–19]. We will discuss here the ferroelectric materials, such as $BaTiO_3$, $PbTiO_3$, and $Pb_{1-x}Zr_xTiO_3$ (PZT) as ferroelectric layers, and Co and Fe as an FM layer.

Garcia et al. have fabricated a multiferroic heterostructure consisting of FM/FE/FM layers by using thin films of Fe/$BaTiO_3$ (BTO)/$La_{0.7}Sr_{0.3}MnO_3$ thin film on the $NdGaO_3$ substrate [14]. The thickness of the BTO film has been kept at ~1 nm. The authors have demonstrated the control of the magnetic state using the polarization of BTO. The typical resistance versus magnetic field curves indicates the negative TMR effect in the heterostructures. The TMR changes with the voltage pulse confirm that the change in TMR is due to the spin polarization of the Fe/BTO interface. The tunnel electromagnetoresistance (TEMR), that is, the change in the TMR with the electric field is ~450%, indicating the strong coupling in the junctions. The strong ME coupling can help to further change the magnitude and sign of the TMR as desired. This has been shown by Pantel et al. by fabricating Co/$PbZr_{0.2}Ti_{0.8}O_3$/$La_{0.7}Sr_{0.3}MnO_3$ MFTJs on the $SrTiO_3$ (STO) substrate [15]. The thickness of the PZT film is around 3 nm. It is interesting to note that the polarization state can strongly affect the TMR behavior. The resistance state changes with the polarization poled up and poled down by applying the voltage pulse. The TMR is low (10%) and possibly due to the defects at the barrier measured at 50 K. However, the TEMR value reaches to 230% due to the strong ME coupling in the PZT and Co interface. Quindeau et al. give the demonstration of four state memory levels. Authors have grown $La_{0.7}Sr_{0.3}MnO_3$ (LSMO), $PbTiO_3$ (PTO), and cobalt films with layer thicknesses of 20, 3.2, and 40 nm on the STO substrate [16]. In this report, four states of the TMR are achieved, and the exchange bias is clearly visible as shown in Figure 4.2a. The measurements have been carried out at 5 K and the main reason for the four-state resistance is attributed to the Co thin layer. The Co changes its density of states at the Fermi energy, and it is proven by experiment at the LSMO/FE interface. Therefore, the unidirectional anisotropy is only due to the FE/Co interface.

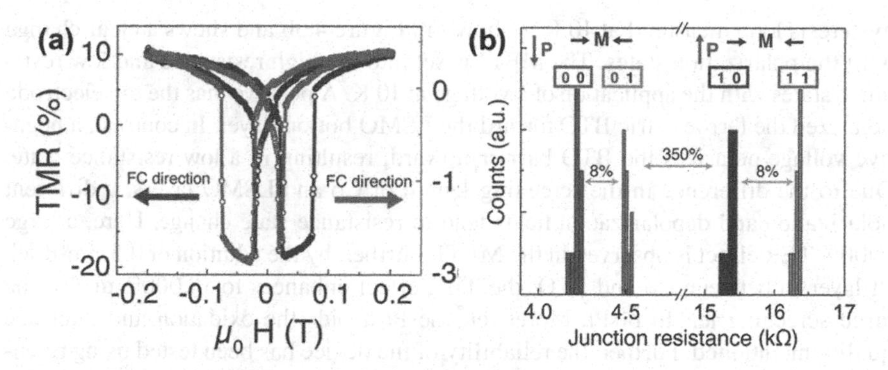

FIGURE 4.2 (a) Tunneling MR, that is, TMR versus magnetic field measured at 5 K of an LSMO/PTO/Co tunnel device. The curves on the left side and right side from the zero T field are showing the sample cooled down in a magnetic field of 800 mT and −800 mT respectively. The exchange bias is visible near the non-zero magnetic field and (b) clear separation between the four resistance states after 40 measurements. (Reprinted with permission from Quindeau, A. et al., *Sci. Rep.*, 5, 9749, 2015.)

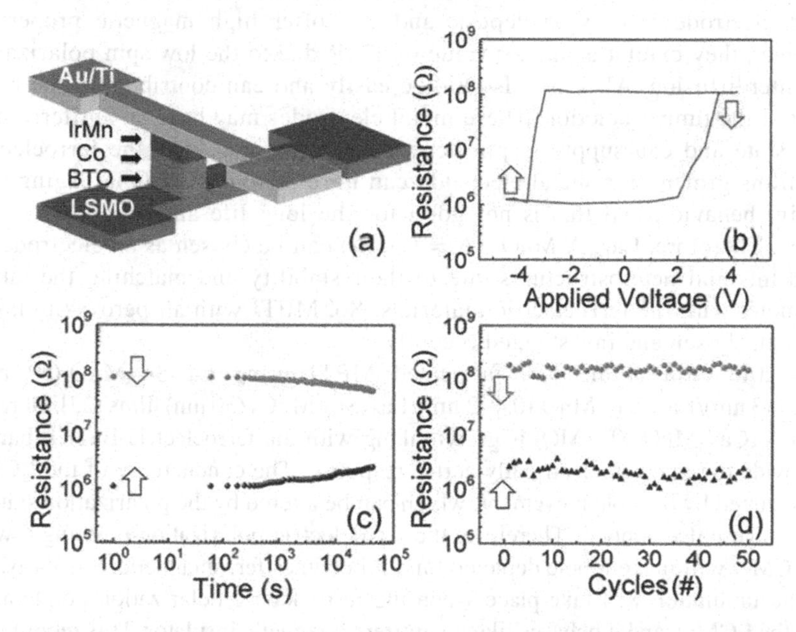

FIGURE 4.3 (a) Multiferroic tunnel junction made of LSMO/BTO/(Co/IrMn) layers, (b) Resistance versus applied voltage hysteresis loop of the junctions at 10 K, and (c,d) the resistance state with time (retention) and cycles (fatigue) represents the polarization orientation as given by the arrow. (Reprinted with permission from Mao, H. et al., *J. Appl. Phys.*, 116, 053703, 2014.)

The TMR and TER behavior can be manipulated by the addition of an interfacial layer in the MFTJ [20–22]. Mao et al. have grown LSMO/BTO/(Co/IrMn) thin films using pulsed-laser deposition [17]. The 6-nm thick IrMn layer has been chosen to increase the coercivity of the top ferromagnetic layer. The schematic of the MFTJ device to study the transport properties is depicted in Figure 4.3a. The resistance hysteresis loop measured at 10 K is shown in Figure 4.3b and shows a clear change with the polarization states. The MFTJ is set into the high resistance and low resistance states with the application of a voltage at 10 K. A positive bias the top electrode polarized the ferroelectric BTO toward the LSMO bottom layer. In contrast, a negative voltage polarizes the BTO barrier upward, resulting in a low resistance state. Due to the difference in the screening length of Co and LSMO layers, a different polarization and depolarization fields lead to resistance state change. Here, a large 1000% TER effect is observed in the MFTJ. Further, by the addition of 0.5-nm-thick Pt layer in between Co and BTO, the TER effect enhances to 10,000% due to the large screening length of Pt. Moreover, the Pt avoids the oxidation and interface quality maintained. Further, the reliability of the device has been tested using retention and fatigue test, and it remains in the same memory states even after repeated cycles.

4.1.3 MFTJ with All Perovskite Layers

Till now, we have seen the MFTJs made of one metal as an electrode. The 3D metal electrode is easy to deposit and can offer high magnetic properties. However, they exhibit a smaller value of TMR due to the low spin polarization and interdiffusion. Also, metals oxidize easily and can contribute to the resistance of the tunnel junction. These metal electrodes may have an antiferromagnetic state and can suppress spin-dependent tunneling. Also, the ferroelectric thin films grown on a metal electrode can have significant fatigue, aging, and imprint behavior, and this is not good for the long life and reliability of the device. Therefore, $La_{1-x}A_xMnO_3$ (A = Ca, Sr) can be chosen as an electrode for thin films and heterostructures due to their stability and matching the lattice parameter with the ferroelectric materials. So, MFTJ with all perovskite layers has been chosen and investigated recently.

In 2013, Yin et al. have fabricated MFTJ using $La_{0.7}Sr_{0.3}MnO_3$(50 nm)/ $BaTiO_3$(3 nm)/$La_{0.5}Ca_{0.5}MnO_3$(0.4–2 nm)/$La_{0.7}Sr_{0.3}MnO_3$ (30 nm) films [23]. Here the thin $La_{0.5}Ca_{0.5}MnO_3$ (LCMO) is grown along with the ferroelectric $BaTiO_3$ barrier to provide the magnetoelectrically active response. The conductivity of the LCMO can be tuned by the hole movement, which can be altered by the polarization state of the voltage pulse applied. Therefore, the ferroelectric polarization pointing toward the LCMO will make it hole depleted, and it becomes ferromagnetic. Similarly, the hole accumulation will take place when the ferroelectric polarization points away from the LCMO, and it behaves like an antiferromagnetic insulator. This modulation in the antiferromagnetic nature acts as a spin filter and allows only spin-dependent current [24]. Therefore, a large TER effect has been observed in the two polarization states only when the LCMO layer inserted. It reaches 5,000% at 40 K for a very thin layer about 2-unit cells as compared without the LCMO layer. In another study, there is an improvement by two orders of magnitude by insertion of a thin $SrTiO_3$ (STO) layer in the LSMRO/STO/BTO/LSMO heterostructure [20]. The TER ratio increases from 0.7% to 128% with the STO layer. The STO layer makes dielectric insulation between the ferroelectric BTO and the top electrode. This will not allow the charges at the top surface to move. Therefore, resistance states with low and high values are

TABLE 4.1
Summary of the TMR and TER in % for Different Multiferroic Tunnel Junctions

Material	TER (%)	TMR (%)	Temperature	Ref.
LSMO/BMO/Au	27	30	3 K	[11]
LSMO/BFO/LSMO	45	69	80 K	[13]
LSMO/BTO/Fe	16	45	4.2 K	[14]
LSMO/PZT/Co	1000	10	<200 K	[15]
LSMO/BTO/LCMO	8000	85	<100 K	[23]
LSMO/PTO/Co	350	8	<140 K	[16]
LSMO/BTO/Co/IrMn	10000	20	10 K	[17]
LSMO/BFO/LSMO	15	3	10 K	[12]
PMNPT/CoFeB/AlO/CoFeB	—	40	RT	[18]
PMNPT/Ta/CofeB/MgO/CoFeB/Ta	—	100	RT	[19]

created, and the asymmetry at the barrier profile with respect to two opposite polarization directions increases. The summary of the TMR and TER in MFTJs is given in Table 4.1.

4.2 MULTIFERROIC MATERIALS FOR ENERGY HARVESTING

4.2.1 Introduction

Multiferroics are the materials that possess ferroelectric properties as well as ferromagnetic properties simultaneously. Multiferroics are attractive due to their high electric polarization and high Curie temperature [4,25,26]. This property enables them to be used in miniature devices as a single component, as a current, and magnetic field sensor, tunable devices, actuators, transformers/gyrators, information storage devices, and energy harvesting [27].

A suitable combination of magnetostrictive and piezoelectric phases is required to achieve a good ME effect. There are many perovskite materials present that have both ferroelectricity and ferromagnetism viz. $BiFeO_3$, $BiMnO_3$, Boracites, $BaMF_4$ compounds (M, divalent transition metal ions), and hexagonal $RMnO_3$ (R, rare earth), but have rather weak ME coupling [3,28,29]. The ME coupling in $TbMnO_3$, $DyMnO_3$, and $TbMn_2O_5$ has low Curie temperature and weak ME coupling [30–33]. Single-phase multiferroic materials have weak ME coupling, so better alternatives are being investigated. ME composites with the best properties of the individual phases can be a good alternative to single-phase ME materials [34–37]. There is different connectivity of the artificial multiferroic materials such as 0–3 in which a particulate composite is used. A 2–2 composite is the bilayer composite, and a 1–3 composite is the vertical nanostructure of one phase and the matrix is another phase. In each of these connectivities, the polarization and magnetic field can be applied along different axes. One can choose the best material by looking at physical properties, such as piezoelectric voltage and strain coefficient, permittivity, permeability, sintering, and chemical reactivity. However, mostly the 2–2 bilayer system is largely investigated due to the insulating behavior and higher ME voltage coefficient. Many piezoelectric materials have been chosen for making ME composites. These include but are not limited to $BaTiO_3$ (BTO), $Pb(Zr, Ti)O_3$ (PZT) or $Pb(Mg_{1/3}Nb_{2/3})O_3$-$PbTiO_3$ (PMN-PT). The choice of magnetostrictive materials is amongst $Tb_{1-x}Dy_xFe_{2-y}$ (Permendur), $Ni_{1-x}Co_xFe_2O_4$ (NCFO), and $Co_{1-x}Zn_xFe_2O_4$ (CZFO) [38,39].

With the increase in the demand for Internet-of-Things-based devices, the power requirement has been rising. The semiconductor industry is making the device more sustainable by reducing power consumption. Their energy needs can be fulfilled in many ways, such as batteries, solar cells, and other non-renewable energy sources. Batteries have a limited lifetime, and solar cells work only with light, putting restrictions on the use of such devices. The energy harvesting using piezoelectric materials is the most common technique being used by researchers [40]. This phenomenon is known as vibration energy harvesting, and it can harvest power up to mW/cm^2. Energy harvesting-based devices are in demand and being researched to meet the power requirement of such devices. To enhance the efficiency of harvesting devices, new materials and new designs have been used that can harvest energy from multiple sources of energy. This section focuses on energy-harvesting studies using ME materials. These materials can harvest energy from both vibrational as well as magnetic fields and enhances the capability of the device.

4.2.2 Magnetoelectric Energy Harvesters

The harvesting characteristics of flexible low-cost Polyvinylidene fluoride (PVDF) and metglas laminated composites have been studied by Lashera et al. [41]. Metglas ($Fe_{64}Co_{17}Si_7B_{12}$) is in amorphous form glued to PVDF to form a flexible structure. The effect of the length of the harvester on the power characteristics has been studied. Three different lengths of around 0.5, 1, and 3 cm have been fabricated, and it is observed that the ME voltage coefficient increases with the length of the energy harvester. Authors have claimed that increasing the length of the harvesting device increases the ME voltage coefficient exponentially. Further, the optimized length of 3 cm has high power output ~1.5 mW/cm^3.

The multiferroic energy harvesting has been studied in a thin film by Onuta et al. [42] shows excellent power characteristics. The device consists of $Pb(Zr_{0.52}Ti_{0.48})O_3$ piezoelectric thin film and magnetostrictive $Fe_{0.7}Ga_{0.3}$ thin film heterostructures. The structure of the cantilever is represented in Figure 4.4. The peak power at 1 Oe is 0.7 mW/cm^3 at the resonant frequency of 3.8 kHz with a load of 12.5 kΩ. The outputs from such cantilever devices can be connected in series or parallel as per the requirement of voltage or current. Further, the power can be supplied to the sensor networks and ease of integration with peripheral circuits.

Most of the devices in energy harvesting rely on a single source of energy such as wind, solar, thermal, or vibrational. However, most of the forms of energy are in co-existence with other sources. The energy-harvesting system that can harvest energies simultaneously is highly desirable and can improve the efficiency of the device. The ME composite materials are beneficial of having two different forms of

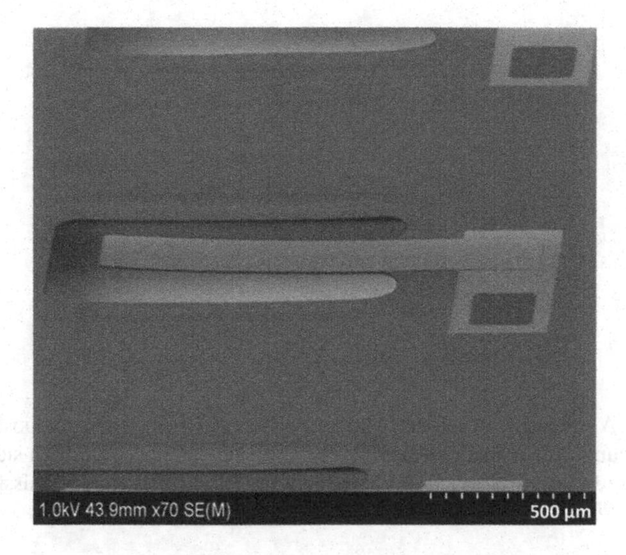

FIGURE 4.4 Microsocpic image of ~1 mm long multiferroic microcantilever energy harvester. (Reprinted with permission from Onuta, T.D. et al., *Appl. Phys. Lett.*, 99, 203506, 2011.)

materials that can harvest energy from different sources viz. vibrational and magnetic. This section deals with such materials and structures that can harvest energy from different sources, such as vibrational and magnetic as well.

Liu et al. [43] have investigated the bimorph cantilever-based magneto-mechano-electro (MME) composite harvester using Pb(Zr, Ti)O_3 (PZT-5H) and NdFeB as a proof mass. Bimorph cantilevers are chosen to enhance the output of the device. Here, the PZT-based cantilevers are the piezoelectric phase, and a tip mass is attached to the end to harvest magnetic energy. The size of the piezoelectric bimorph has been chosen as 35 mm (length) × 7 mm (width) × 0.5 mm (thickness) in size. The ambient AC magnetic field generates a magnetic force moment on the piezoelectric cantilever with a movement of the tip magnets. This results in the bending and vibration of the ME cantilever. Thus, during the motion, the residual magnetic energy can induce mechanical energy in the MME cantilever. The piezoelectric material converts this energy into the electrical power output. The harvester works on magnetic energy and the output around 11.73 $\mu W/Oe^2\ cm^3$ below 100 Hz (Figure 4.5) [42].

The output of the ME energy-harvesting device can be enhanced by using dual-phase energy harvesters. These harvest both mechanical and magnetic energies. Zhou et al. [44] in 2013 have reported a dual-phase energy harvester that harvests both magnetic and vibrational energy in the absence of the DC magnetic field. The synthesized piezoelectric unimorph structure uses a macrofiber of PZT and Ni plate (Figure 4.6) [43].

The quality and choice of the materials can have a strong effect on the ME behavior of the composite. To prove this, Annapureddy and the group have fabricated an MME energy harvester. They have used ME composite made of single-crystal fibers of $Pb(Mg_{1/3}Nb_{2/3})O_3$-Pb(Zr, Ti)O_3 (PMN PZT) and Nickel (Ni) [45]. Authors have studied the effect of loss factor on the ME response of the energy harvester.

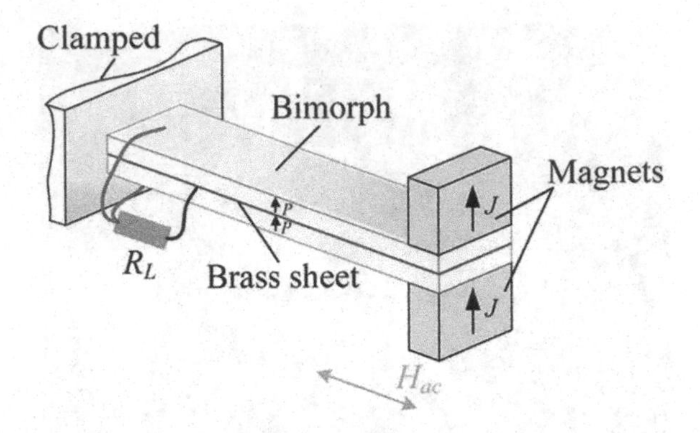

FIGURE 4.5 A schematic of the magnetoelectric mechano electrical composite cantilever. The bimorph cantilever is made of brass sheet and ME composite on both sides. The load resistance is used to draw power from the vibrations. (Reprinted with permission from Liu, G. et al., *Appl. Phys. Lett.*, 104, 032908, 2014.)

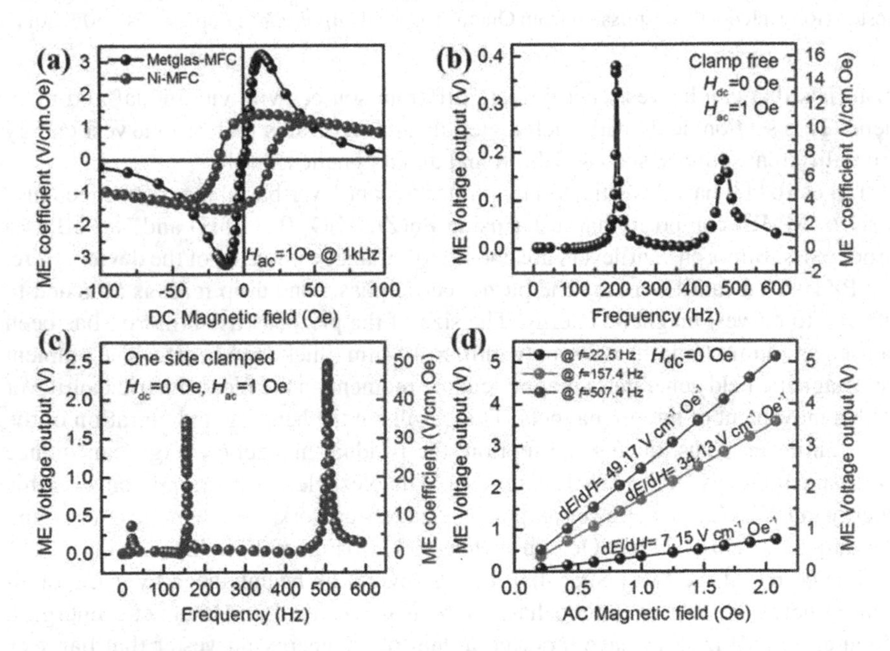

FIGURE 4.6 (a) Magnetoelectric voltage coefficient of the Metglas-MFC and Ni-MFC bilayer laminates with DC magnetic field. Here, no DC bias field is needed because the composite is self-biased in nature, (b) represents the ME voltage output of the free-standing composite with the application of frequency. The peak represents the resonance of the mechanical vibrations with the applied frequency, (c) ME voltage output of the clamped composite under DC-biased condition, and (d) relationship of the ME voltage coefficient of the composite at a fixed frequency and as a function of AC magnetic field. The linear response of the ME voltage coefficient with the AC magnetic field represents the intrinsic behavior of the device. (Reprinted with permission from Zhou, Y. et al., *Appl. Phys. Lett.*, 103, 192909, 2013.)

This is because the output of the piezoelectric material can be directly correlated to the dielectric loss and mechanical loss. This directly affects the performance of the device. They used piezoelectric materials with different loss factors to study and to improve the harvesting performance. Three different MME generators have been fabricated using an ME composite. The piezoelectric PMN-PZT single-crystalline materials in macrofibers form, and a magnetostrictive Ni plate have been chosen. The response of these ME composites directly indicates the performance in the terms of ME voltage coefficient response. The ME harvester with low loss has the highest ME voltage as compared to others. This clearly indicates that the quality of piezoelectric material is important while choosing ME energy harvester. Further, the sample with low loss is able to harvest power around 2.1 mW/cm^3 with a stray magnetic field in the surrounding.

Demonstration for harvesting stray magnetic energy is given by Ryu et al. using single crystal fibers of PMN-PZT and Ni plate as a bonding material [46]. Authors have built a harvester that can harvest stray magnetic field as low as 700 μT and 60 Hz with a power density of 46.3 mW cm^{-3} Oe^{-2}. Recently, Annapureddy and group have developed an enhanced power output in the milliwatt range for standalone applications [47]. Authors have used low loss piezoelectric Mn-doped $Pb(Mg_{1/3}Nb_{2/3})O_3$–$PbZrO_3$–$PbTiO_3$ (PMN–PZ–PT) single crystal fibers with textured Fe-Ga alloys. As depicted in Figure 4.7, they have demonstrated a strong ME coupling of around = 1330 V cm^{-1} without any magnetic bias and can harness weak magnetic energy. Due to the strong coupling and strong anisotropic properties of Fe-Ga alloy, the MME harvester is able to harvest stray magnetic energy down to 700 μT.

Recently, a broadband hybrid, that is, ME and vibrational energy harvester, is fabricated by Qiu and group. Here, the ME and electromagnetic energy can be harvested

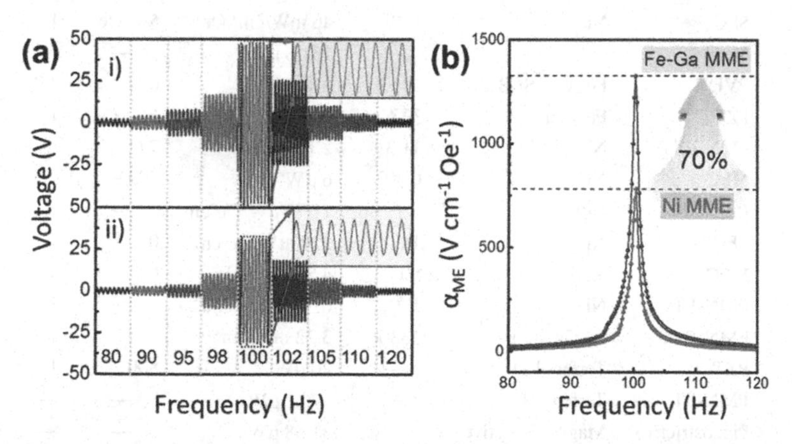

FIGURE 4.7 (a) Comparative open circuit voltage response of the (i) Fe-Ga based and (ii) Ni-based MME generator with frequency variation and (b) comparative ME response measured at 100 Hz. The Fe–Ga-based MME generator shows a maximum α_{ME} of 1330 Vcm^{-1} Oe^{-1} at 100 Hz, whereas the Ni MME generator exhibits a α_{ME} of only 781 Vcm^{-1} Oe^{-1}. The Fe-Ga-based MME harvester shows enhanced output as compared to Ni-based due to the textured behavior and anisotropy. (Reprinted with permission from Annapureddy, V. et al., *Energy Environ. Sci.*, 11, 818–829, 2018.)

simultaneously [48]. This system has a hybrid transducer and a double cantilever. A ME composite consisting of five-phase laminates of FeCuNbSiB/Terfenol-D/PZT/Terfenol-D/FeCuNbSiB (FMPMF) composites operated in the L-T mode. Here, L means the magnetostrictive layers (Terfenol-D and FeCuNbSiB) are magnetized along the longitudinal direction (L mode), while the piezoelectric layer (PZT) is polarized in the thickness direction (T mode). The Terfanol-D/PZT/Terfanol-D is the ME phase, and FeCuNbSiB will be used to harvest electromagnetic energy. This can be explained based on their relative permeability difference. The FeCuNbSiB has a permeability around 10,000, while for Terfanol-D is around 5. Therefore, the average magnetic induction strength of the harvester increases as compared to the ME harvester. The dual-energy harvesting can be verified from the output versus frequency graph, which shows more than one resonance peaks. The optimized power output is nearly equal to 36.8 mW, which is suitable for most of the wireless devices.

The summary of the results related to ME energy harvesting is summarized in Table 4.2. It is noted that almost all the ME composites are formed by using

TABLE 4.2
Summary of the Magnetoelectric Energy Harvester in Terms of the ME Voltage Coefficient and Power Output

Type	Piezo Strictive Phase	Magnetostrictive Phase	ME Voltage (V/cm.Oe)	Power/Power Density	DC Magnetic Field	AC Field Bias (Oe)	Ref.
ME	SFC	Ni	160	46 mW/cm^3·Oe^2	5–7 Oe	1	[46]
PE	MFC	—	—	151.6 μW	—	—	[49]
ME	PVDF	$Fe_{64}Co_{17}Si_7B_{12}$	145.6	6.4 μW	0.45 Oe	4.7	[41]
ME	PZT	$Fe_{0.7}Ga_{0.3}$	33.6	0.7 mW/cm^3	66.1 Oe	1	[42]
ME	PMN-PZT	Ni	11.3	2.1 mW/cm^3	7 Oe	—	[45]
ME	MFC	Ni	0.407	6 μW	50 Oe	0.5	[50]
ME	PZT	NdFeB	—	11.73 μW/Oe^{2}cm^3	—	1	[43]
ME	SFC	Ni	1.2	2.75 μW/Oe^{2}cm^3	0	1	[51]
ME	MFC	Ni	50	4.5 mW/cm^3	0	1	[44]
ME	PMN-PT	Ni	1.2	0.5 μW/Oe.cm^3		2	[52]
ME	PMN-PZT	Fe-Ga alloy	1330	3.22 mW/cm^3		1	[47]
ME	PZT	Terfenol-D	—	20 μW	800	1	[53]
ME	PMN-PT	Terfenol-D	—	970.2 μW	—	—	[54]
ME (theory)	Piezostrictive	Magnetostrictive	—	31.58 μW	—	—	[55]
ME	PZT	FeCuNbSiB/ Terfenol-D	—	12.6 mW	—	—	[48]

Abbreviations: ME, magnetoelectric; PE, Piezoelectric; MFC, Macro fiber composite; SFC, Single crystal fiber composite.

lead-based piezostrictive materials. Due to the environmental hazards, it is very necessary to find out an alternative to these lead-based materials.

4.3 CONCLUSIONS

We have presented an overview of the material and devices for spintronics and energy-harvesting applications. The multiferroic/ME materials are an excellent choice for these applications due to multiple functionalities. The MFTJs can produce multiple resistance states and can give rise to giant TMR and TER effects. These materials are excellent choices for multistate memory devices. Insulating ferroelectric oxide with ultrathin layers is a better candidate than the single-phase multiferroic materials. However, the low operating temperature of the device is a concern in multiferroic materials because it loses properties as the temperature increases. This room-temperature functionality is lacking in 3D metal-based as well as perovskite materials.

Energy harvesting by using ME materials has advantages because it can harvest energy from multiple sources. Most of the ME harvesters have used PZT-based materials as a piezoelectric phase due to excellent electromechanical properties. The emerging self-bias phenomena make the size of the device smaller and eliminate the use of DC bias for the device. Moreover, these devices can harvest the stray AC magnetic field and can supply power to portable devices such as a wireless sensor node. The bottleneck is to work on the processing of the materials and miniaturization of the device. Thus, energy harvesting using multiferroic materials has opened a new area, and it can reach its full potential with proper selection of materials and fine-tuning between the magnetostrictive and piezoelectric layers.

REFERENCES

1. Reig, C., S. Cardoso, and S.C. Mukhopadhyay (eds.). *Giant Magnetoresistance (GMR) Sensors From Basis to State-of-the-Art Applications.* Springer-Verlag, Berlin Heidelberg. 2013, p. 308.
2. Cheng, Y., et al., Recent development and status of magnetoelectric materials and devices. *Physics Letters A*, 2018. **382**(41): pp. 3018–3025.
3. Sun, N.X. and G. Srinivasan. Voltage control of magnetism in multiferroic heterostructures and devices. *SPIN*, 2012. **2**(3): p. 1240004.
4. Jayant Kolte, P.H.S., A.S. Daryapurkar, and P. Gopalan, Impedance and AC conductivity study of nano crystalline, fine grained multiferroic bismuth ferrite ($BiFeO_3$), synthesized by microwave sintering. *AIP Advances*, 2015. **5**(9): p. 097164.
5. Kolte, J., et al., Effect of substrate temperature on the structural and electrical properties of La and Mn co-doped $BiFeO_3$ thin films. *Thin Solid Films*, 2016. **619**: pp. 308–316.
6. Kolte, J., et al., Magnetoelectric properties of microwave sintered $BiFeO_3$ and $Bi_{0.90}La_{0.10}$ $Fe_{0.95}Mn_{0.05}O_3$ nanoceramics. *Materials Chemistry and Physics*, 2017. **193**: pp. 253–259.
7. Kolte, J., et al., Microwave sintered $Bi_{0.90}La_{0.10}Fe_{0.95}Mn_{0.05}O_3$ nanocrystalline ceramics: Impedance and modulus spectroscopy. *Ceramics International*, 2016. **42**(11): pp. 12914–12921.
8. Catalan, G. and J.F. Scott, Physics and applications of bismuth ferrite. *Advanced Materials*, 2009. **21**(24): pp. 2463–2485.

9. Bibes, M. and A. Barthélémy, Multiferroics: Towards a magnetoelectric memory. *Nature Materials*, 2008. **7**(6): p. 425.
10. Fusil, S., et al., Magnetoelectric devices for spintronics. *Annual Review of Materials Research*, 2014. **44**: pp. 91–116.
11. Gajek, M., et al., Tunnel junctions with multiferroic barriers. *Nature Materials*, 2007. **6**(4): p. 296.
12. Liu, Y., et al., Coexistence of four resistance states and exchange bias in $La_{0.6}Sr_{0.4}MnO_3$/$BiFeO_3$/$La_{0.6}Sr_{0.4}MnO_3$ multiferroic tunnel junction. *Applied Physics Letters*, 2014. **104**(4): p. 043507.
13. Hambe, M., et al., Crossing an interface: Ferroelectric control of tunnel currents in magnetic complex oxide heterostructures. *Advanced Functional Materials*, 2010. **20**(15): pp. 2436–2441.
14. Garcia, V., et al., Ferroelectric control of spin polarization. *Science*, 2010. **327**(5969): pp. 1106–1110.
15. Pantel, D., et al., Reversible electrical switching of spin polarization in multiferroic tunnel junctions. *Nature Materials*, 2012. **11**(4): p. 289.
16. Quindeau, A., et al., Four-state ferroelectric spin-valve. *Scientific Reports*, 2015. **5**: p. 9749.
17. Mao, H., et al., Interface-modification-enhanced tunnel electroresistance in multiferroic tunnel junctions. *Journal of Applied Physics*, 2014. **116**(5): p. 053703.
18. Li, P., et al., Electric field manipulation of magnetization rotation and tunneling magnetoresistance of magnetic tunnel junctions at room temperature. *Advanced Materials*, 2014. **26**(25): pp. 4320–4325.
19. Zhao, Z., et al., Giant voltage manipulation of MgO-based magnetic tunnel junctions via localized anisotropic strain: A potential pathway to ultra-energy-efficient memory technology. *Applied Physics Letters*, 2016. **109**(9): p. 092403.
20. Ruan, J., et al., Improved memory functions in multiferroic tunnel junctions with a dielectric/ferroelectric composite barrier. *Applied Physics Letters*, 2015. **107**(23): p. 232902.
21. Wang, L., et al., Overcoming the fundamental barrier thickness limits of ferroelectric tunnel junctions through $BaTiO_3$/$SrTiO_3$ composite barriers. *Nano Letters*, 2016. **16**(6): pp. 3911–3918.
22. Zhuravlev, M.Y., et al., Tunneling electroresistance in ferroelectric tunnel junctions with a composite barrier. *Applied Physics Letters*, 2009. **95**(5): p. 052902.
23. Yin, Y., et al., Enhanced tunnelling electroresistance effect due to a ferroelectrically induced phase transition at a magnetic complex oxide interface. *Nature Materials*, 2013. **12**(5): p. 397.
24. Jiang, L., et al., Tunneling electroresistance induced by interfacial phase transitions in ultrathin oxide heterostructures. *Nano Letters*, 2013. **13**(12): pp. 5837–5843.
25. Selbach, S.M., M.-A. Einarsrud, and T. Grande, On the thermodynamic stability of $BiFeO_3$. *Chemistry of Materials*, 2008. **21**(1): pp. 169–173.
26. Rojac, T., et al., $BiFeO_3$ ceramics: Processing, electrical, and electromechanical properties. *Journal of the American Ceramic Society*, 2014. **97**(7): pp. 1993–2011.
27. Chu, Z., M. PourhosseiniAsl, and S. Dong, Review of multi-layered magnetoelectric composite materials and devices applications. *Journal of Physics D: Applied Physics*, 2018. **51**(24): p. 243001.
28. Nan, C.W., et al., Multiferroic magnetoelectric composites: Historical perspective, status, and future directions. *Journal of Applied Physics*, 2008. **103**(3): p. 1.
29. Suryanarayana, S., Magnetoelectric interaction phenomena in materials. *Bulletin of Materials Science*, 1994. **17**(7): pp. 1259–1270.
30. Kimura, T., et al., Magnetic control of ferroelectric polarization. *Nature*, 2003. **426**(6962): p. 55.

31. Fiebig, M., et al., Observation of coupled magnetic and electric domains. *Nature*, 2002. **419**(6909): p. 818.
32. Hur, N., et al., Electric polarization reversal and memory in a multiferroic material induced by magnetic fields. *Nature*, 2004. **429**(6990): pp. 392–395.
33. Van Aken, B.B., et al., The origin of ferroelectricity in magnetoelectric $YMnO_3$. *Nature Materials*, 2004. **3**(3): p. 164.
34. Srinivasan, G., et al., Magnetoelectric interactions in hot-pressed nickel zinc ferrite and lead zirconante titanate composites. *Applied Physics Letters*, 2004. **85**(13): pp. 2550–2552.
35. Srinivasan, G., E. Rasmussen, and R. Hayes, Magnetoelectric effects in ferrite-lead zirconate titanate layered composites: The influence of zinc substitution in ferrites. *Physical Review B*, 2003. **67**(1): p. 014418.
36. Babu, S.N., T. Bhimasankaram, and S. Suryanarayana, Magnetoelectric effect in metal-PZT laminates. *Bulletin of Materials Science*, 2005. **28**(5): pp. 419–422.
37. Narita, F. and M. Fox, A review on piezoelectric, magnetostrictive, and magnetoelectric materials and device technologies for energy harvesting applications. *Advanced Engineering Materials*, 2018. **20**(5): p. 1700743.
38. Nan, C.-W., et al., A three-phase magnetoelectric composite of piezoelectric ceramics, rare-earth iron alloys, and polymer. *Applied Physics Letters*, 2002. **81**(20): pp. 3831–3833.
39. Laletin, V., et al., Frequency and field dependence of magnetoelectric interactions in layered ferromagnetic transition metal-piezoelectric lead zirconate titanate. *Applied Physics Letters*, 2005. **87**(22): p. 222507.
40. Wei, C. and X. Jing, A comprehensive review on vibration energy harvesting: Modelling and realization. *Renewable and Sustainable Energy Reviews*, 2017. **74**: pp. 1–18.
41. Lasheras, A., et al., Energy harvesting device based on a metallic glass/PVDF magnetoelectric laminated composite. *Smart Materials and Structures*, 2015. **24**(6): p. 065024.
42. Onuta, T.D., et al., Energy harvesting properties of all-thin-film multiferroic cantilevers. *Applied Physics Letters*, 2011. **99**(20): p. 203506.
43. Liu, G., P. Ci, and S. Dong, Energy harvesting from ambient low-frequency magnetic field using magneto-mechano-electric composite cantilever. *Applied Physics Letters*, 2014. **104**(3): p. 032908.
44. Zhou, Y., D.J. Apo, and S. Priya, Dual-phase self-biased magnetoelectric energy harvester. *Applied Physics Letters*, 2013. **103**(19): p. 192909.
45. Annapureddy, V., et al., Low-loss piezoelectric single-crystal fibers for enhanced magnetic energy harvesting with magnetoelectric composite. *Advanced Energy Materials*, 2016. **6**(24): p. 1601244.
46. Ryu, J., et al., Ubiquitous magneto-mechano-electric generator. *Energy & Environmental Science*, 2015. **8**(8): pp. 2402–2408.
47. Annapureddy, V., et al., Exceeding milli-watt powering magneto-mechano-electric generator for standalone-powered electronics. *Energy & Environmental Science*, 2018. **11**(4): pp. 818–829.
48. Qiu, J., et al., A tunable broadband magnetoelectric and electromagnetic hybrid vibration energy harvester based on nanocrystalline soft magnetic film. *Surface and Coatings Technology*, 2017. **320**: pp. 447–451.
49. Yang, Y., L. Tang, and H. Li, Vibration energy harvesting using macro-fiber composites. *Smart Materials and Structures*, 2009. **18**(11): p. 115025.
50. Kambale, R.C., et al., Magneto-mechano-electric (MME) energy harvesting properties of piezoelectric macro-fiber composite/Ni magnetoelectric generator. *Energy Harvesting and Systems*, 2014. **1**(1–2): pp. 3–11.
51. Patil, D.R., et al., Anisotropic self-biased dual-phase low frequency magneto-mechano-electric energy harvesters with giant power densities. *APL Materials*, 2014. **2**(4): p. 046102.

52. Chu, Z., et al., Dual-stimulus magnetoelectric energy harvesting. *Mrs Bulletin*, 2018. **43**(3): pp. 199–205.
53. Li, P., et al., A magnetoelectric energy harvester and management circuit for wireless sensor network. *Sensors and Actuators A: Physical*, 2010. **157**(1): pp. 100–106.
54. Dai, X., A vibration energy harvester with broadband and frequency-doubling characteristics based on rotary pendulums. *Sensors and Actuators A: Physical*, 2016. **241**: pp. 161–168.
55. Lin, Z., et al., Enhanced broadband vibration energy harvesting using a multimodal nonlinear magnetoelectric converter. *Journal of Electronic Materials*, 2016. **45**(7): pp. 3554–3561.

5 Advanced Graphene-Transition Metal-Oxide-Based Nanocomposite Photocatalysts for Efficient Degradation of Pollutants Present in Wastewater

S. Raut-Jadhav and M. V. Bagal

CONTENTS

5.1 INTRODUCTION

Water is the most important resource for every living organism because the existence of life on the earth is not possible without water. However, this valuable resource is getting polluted due to direct or indirect discharge of industrial, agriculture, and domestic waste in the surface or ground water resources. Water pollution has emerged as a major challenge that demands immediate and practical solutions (Dutta et al., 2019). Various conventional wastewater treatment methods such as physical, chemical, and biological processes are used widely for the treatment of wastewater (Singh et al., 2020). However, these methods have proved to be ineffective to remove toxic refractory organic compounds from wastewater. Hence, alternate effective treatment methods are imperative to treat these recalcitrant pollutants.

Photocatalysis has emerged as an effective advanced oxidation process (AOP) in the field of wastewater treatment. The basic mechanism of AOPs includes the generation and subsequent attack of highly reactive free hydroxyl radicals (oxidative potential –2.8 eV) on the pollutant to be treated (Santos et al., 2011). Similarly, mechanism of the photocatalytic process involves activation of photocatalyst in the presence of UV or visible light for the generation of hydroxyl radicals via simultaneous oxidation and reduction reactions. The transition metal oxide nanoparticles have attracted the attention of many researchers due to their wide range of applications in the field of wastewater treatment. Several transition metal oxides such as TiO_2, ZnO, CeO_2, Fe_3O_4, V_2O_5, MoO_3, and ZrO_2 having adequate band gaps can be applied for the photocatalytic degradation of organic pollutants (Devipriya and Yesodharan, 2005; Raizada et al., 2019). Titanium dioxide is the most promising photocatalyst for the degradation of large number of organic pollutants due to its stability, low cost, and non-toxicity (Khalid et al., 2013a). The widely used polymorphs of TiO_2 in photocatalytic applications are anatase (tetragonal) and rutile (tetragonal) with band gaps of 3.2 and 3.02 eV, respectively (Wunderlich et al., 2004). ZnO is also a low-cost photocatalyst material that is widely applied for the photocatalytic degradation of organic pollutants due to its adequate band gap 3.37 eV (Baruah and Dutta, 2009; Kim and Yong, 2012). TiO_2 and ZnO nanoparticles have extensively applied in photocatalytic applications since they can be tailored easily to absorb visible light by simple modifications.

Although bare transition metal oxides are applied widely for the photocatalytic degradation of organic pollutants, their applications are seriously limited by the following shortcomings (Gupta and Tripathi, 2011; Singh et al., 2020):

- Large band gaps and narrow light response range limiting their applications under ultraviolet light only
- Low separation probability of the photoinduced electron hole pairs
- Rapid recombination rates of photoinduced electron hole pairs
- Low activity in visible region of solar spectrum
- Poor adsorption of pollutants
- Tendency of agglomeration leading to lower surface areas

The photocatalytic activity of transition metal oxides can be improved by combining it with noble metal (Arabatzis et al., 2003; Yu et al., 2005), transition metal ions (or their oxides) (Yu et al., 2010a, 2010b) and other semiconductors (Li et al., 2001; Xu and Yu, 2011), or anchoring it on mesoporous or carbon-based materials having high surface areas. Mesoporous materials include zeolites, whereas carbon-based materials include activated carbon, carbon nanotubes, graphene, graphene oxide (GO), and reduced graphene oxide (rGO) (Mohamed, 2012; Banerjee et al., 2018).

The application of graphene for improving the performance of transition metal oxides has gained a lot of attention in wastewater treatment for the photocatalytic degradation of organic pollutants. Graphene is widely applied for synthesizing nanocomposite photocatalysts due to its two-dimensional structure, exceptional mobility of electrons, high flexibility, excellent thermal and mechanical properties, and large surface area (Wang et al., 2010, 2012; Xiang et al., 2012). Graphene is basically a single atomic layer of graphite, wherein sp^2 hybridized carbon atoms are arranged in a honeycomb network (Raizada et al., 2019). GO is an exfoliated carbonaceous material with a layered structure. It is actually an oxidized form of graphene that is decorated with various oxygen-bearing functional groups. Hydroxyl and epoxide functional groups are attached to sp^3-hybridized carbons on the top and bottom surfaces GO sheet, whereas carboxyl and carbonyl groups are attached to sp^2-hybridized carbons mostly on the edges (Stankovich et al., 2006). Due to the presence of oxygen functional groups on its surface, GO exhibits excellent dispersibility in various solvents for facile synthesis of various nanocomposites (Lawal, 2019). The functional groups on GO provides active sites for anchoring transition metal oxides on its surface. Generally, GO gets reduced to rGO while anchoring nanoparticles of transition metal oxides.

The present work has mainly focused on providing an overview on beneficial effects of anchoring transition metal oxides on graphene support, various methods of preparation of graphene-transition metal-oxide-based nanocomposite photocatalysts, and their applications in wastewater treatment. A detailed case study exhibiting the original work on UV-assisted synthesis of an rGO-TiO_2 nanocomposite for its application in photocatalytic degradation of methomyl has also been presented to acquire clear insight of the topic.

5.2 GENERAL ASPECTS OF GRAPHENE-BASED TRANSITION METAL OXIDE NANOCOMPOSITE

5.2.1 Beneficial Effects of Anchoring Transition Metal Oxide on Graphene

The addition of carbon-based material such as graphene has been observed to be one of the most effective methods to improve the photocatalytic activity of transition metal oxides. The anchoring of transition metal oxides on graphene leads to several beneficial effects. The important advantages are as mentioned below:

- Graphene is easy to produce by using inexpensive natural graphite through the formation of GO (Hummers and Offeman, 1958; Marcano et al., 2010).
- Its surface area is very large and has excellent mobility of charge carriers (Wang et al., 2010, 2012; Xiang et al., 2012).
- Oxygen-bearing functional groups present on GO and rGO facilitates the easy anchoring of transition metal oxides nanoparticles on its surface (Li et al., 2019; Smith et al., 2019).
- The photocatalytic activity of transition metal oxide enhances by combining it with graphene since graphene acts as an electron sink and reduces the charge carrier recombination (Kusiak-Nejman and Morawski, 2019). Further, it also acts as a photosensitizer that is capable of generating a high number of electron-hole pairs, leading to the higher generation of hydroxyl radicals and increased degradation rates of pollutants (Mohamed, 2012).

5.2.2 Mechanism of Photocatalysis Using Graphene-Based Transition Metal Oxide Nanocomposites

Photocatalytic process utilizes a transition metal oxide photocatalyst in presence of UV light for the generation of hydroxyl radicals. It is capable of degrading chemical pollutants by both oxidative and reductive pathways (McMurray et al., 2006). Schematic representation of the mechanism of photocatalysis is as shown in Figure 5.1.

When a graphene-transition metal oxide nanocomposite is irradiated with UV light, the transition metal oxide nanoparticles get activated and generate electron hole pairs. Valence band electrons (e^-) are promoted to the conduction band, leaving a hole (h^+) behind. These electron–hole pairs either recombine to produce heat or interact separately with other molecules, generating free radicals. Migration of these photogenerated electrons to the photocatalyst surface initiates simultaneous oxidation and reduction reactions on the surface of photocatalysts (Ahmed and Haider, 2018). In an aqueous suspension, h^+ reacts with H_2O to give $OH^\bullet$ radicals, whereas e^- reacts with adsorbed molecular O_2 to produce superoxide anion radical $^\bullet O_2^-$ which also contributes to the production of $OH^\bullet$ radicals (Mijin et al., 2009). In addition to this, electrons from the conduction band of photocatalyst also migrate

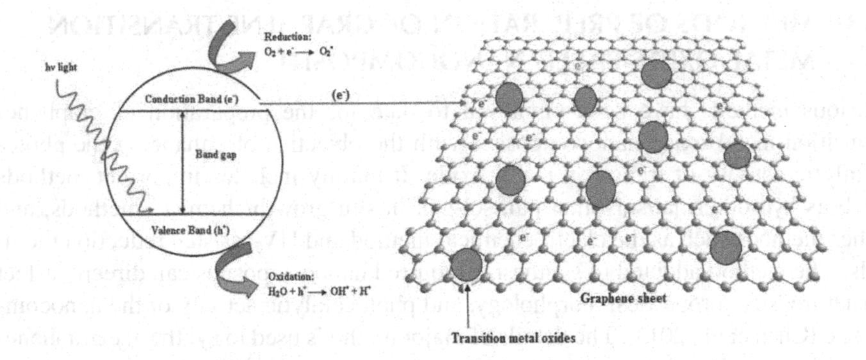

FIGURE 5.1 Schematic representation of the mechanism of heterogeneous photocatalysis.

to the graphene surface. Negatively charged graphene combines with molecular O_2 to produce the superoxide anion radical $^{\bullet}O_2^-$, which further reacts with hydrogen ions to produce hydroxyl radicals (Kusiak-Nejman and Morawski, 2019). Hence, graphene can act as an electron acceptor and can further reduce the recombination of electron hole pairs.

Generated $OH^{\bullet}$ radicals subsequently attack on the organic pollutants, which eventually leads to mineralization of these compounds. The series of oxidation and reduction reactions on transition metal oxide and graphene are summarized in the following equations (Dutta et al., 2019; Kusiak-Nejman and Morawski, 2019; Nadimi et al., 2019):

$$\text{Photocatalyst} + h\nu \rightarrow h^+ + e^-$$
$$h^+ + e^- \rightarrow \text{heat}$$

Oxidative reactions:

$$h^+ + H_2O \rightarrow H^+ + OH^{\bullet}$$
$$2\,h^+ + 2\,H_2O \rightarrow 2\,H^+ + H_2O_2$$

Reductive reaction:

$$e^- + O_2 \rightarrow {}^{\bullet}O_2^-$$
$${}^{\bullet}O_2^- + H^+ \rightarrow HO_2^{\bullet} \text{ or } OH^{\bullet}$$
$${}^{\bullet}O_2^- + HO_2^{\bullet} + H^+ \rightarrow H_2O_2 + O_2$$
$$H_2O_2 + e^- \rightarrow OH^- + OH^{\bullet}$$
$$e^- + GO \rightarrow \text{graphene}\,(e^-)$$
$$\text{graphene}\,(e^-) + O_2 \rightarrow \text{graphene} + {}^{\bullet}O_2^-$$
$$\text{graphene}\,(e^-) + H_2O_2 \rightarrow \text{graphene} + OH^- + OH^{\bullet}$$

Ultimately, both oxidative and reductive reactions lead to the generation of the highly reactive radicals. These highly oxidative radicals react with organic pollutants and oxidize them.

$$\text{Organic molecule} + OH^{\bullet} + {}^{\bullet}O_2^- + h^+ \rightarrow CO_2 + H_2O + \text{degradation products}$$

5.3 METHODS OF PREPARATION OF GRAPHENE-TRANSITION METAL-OXIDE-BASED NANOCOMPOSITE

Various methods have been employed to date for the preparation of graphene-transition metal oxide nanocomposites with the objective of enhancing the photocatalytic activity of transition metal oxide. It mainly includes important methods such as hydrothermal/solvothermal, sol-gel, in-situ growth/chemical methods, and other methods such as the electrochemical method and UV-assisted reduction methods. The method adopted to synthesize required nanocomposites can directly affect structure, size, properties, morphology, and photocatalytic activity of the nanocomposite (Chen et al., 2013). The details of major methods used to synthesize graphene-transition metal oxide nanocomposites are presented in the following sections.

5.3.1 HYDROTHERMAL/SOLVOTHERMAL METHOD

Hydrothermal and solvothermal methods are widely applied for the synthesis of graphene-transition metal oxide nanocomposites (Dutta et al., 2019). These methods are generally carried out in autoclaves operated at higher temperatures and pressures. A hydrothermal process is based on the aqueous solvent, whereas a solvothermal method is based on the non-aqueous solvent (Jianwei et al., 2013). The graphical illustration of the hydrothermal method for the synthesis of TiO_2-rGO nanocomposite is mentioned in Figure 5.2. It is clearly seen that GO sheet is laden with many oxygen-bearing functional groups that act as active sites for anchoring the transition metal oxides. The hydrothermal method leads to the reduction of GO by removing most of the functional groups and thereby anchoring transition metal oxides (Wang et al., 2013).

Wang and Zhang (2011) have synthesized rGO–TiO_2 (rGO–TiO_2) nanocomposites by using a hydrothermal route with commercial grade TiO_2 and GO as precursors.

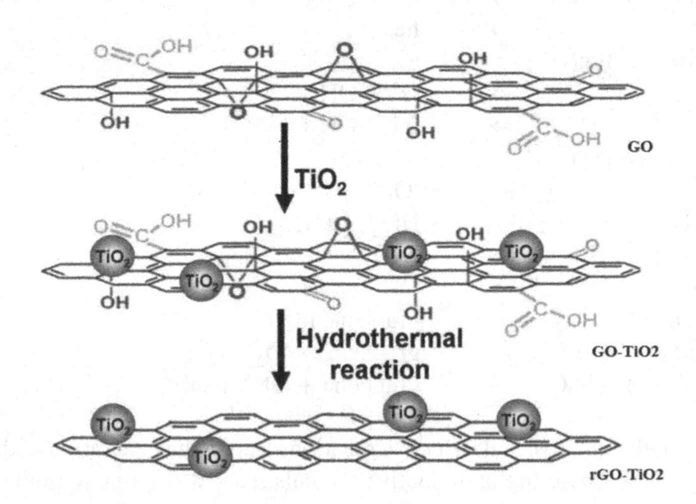

FIGURE 5.2 Graphical illustration for the synthesis of rGO–TiO_2 nanocomposite by the hydrothermal route. (Reprinted with permission from Wang, P. et al., *Appl. Catal. B*, 132–133, 452–459, 2013.)

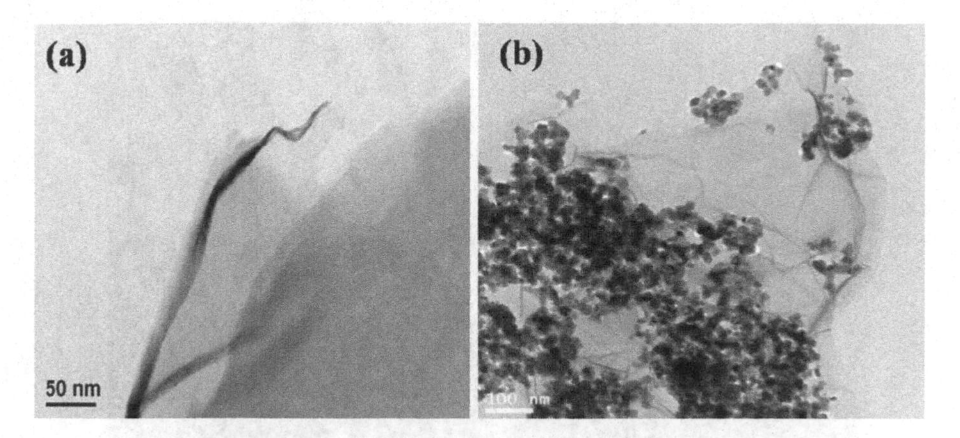

FIGURE 5.3 TEM images of (a) GO and (b) rGO–TiO_2 nanocomposite (1:20). (Reprinted with permission from Wang, F. and Zhang, K., *J. Mol. Catal. A*, 345, 101–107, 2011.)

Various rGO-TiO_2 nanocomposites were prepared by changing the weight ratio of GO to TiO_2 as 1:100, 1:40, 1:20, 1:10, and 1:3. The hydrothermal reaction was followed by calcination of dried nanocomposite at 400°C for the duration of 2 h. The transition electron microscopy (TEM) images of GO and rGO–TiO_2 nanocomposite with a weight ratio of 1:20 are shown in Figure 5.3a and 5.3b, respectively. TEM images of GO has clearly shown the layered structure of GO, whereas TEM images of (1:20) rGO–TiO_2 nanocomposite indicate the uniform dispersion of TiO_2 in the rGO matrix.

Wang et al. (2013) have produced rGO–TiO_2 nanocomposites by using the facile hydrothermal method. Before the hydrothermal reaction, the surface of TiO_2 was cleaned by pretreating it at the temperature of 550°C for the duration of 2 h. Pretreated TiO_2 was then added to the GO solution, stirred, and the resulting solution was transferred into a Teflon-sealed autoclave and maintained at the temperature of 150°C for 5 h. Fourier-transform infrared spectroscopy (FTIR) spectra of GO has shown many strong absorption peaks that correspond to various oxygen functional groups. After hydrothermal treatment of TiO_2-GO suspension, the resulting rGO–TiO_2 nanocomposite showed low-intensity absorption peaks corresponding to oxygen functional groups. It has confirmed the reduction of GO and the successful preparation of rGO–TiO_2 nanocomposite. Similarly, Khalid et al. (2013a) have also synthesized graphene TiO_2 nanocomposite by the hydrothermal method, which has resulted in the reduction of GO and loading of TiO_2. Further, Hayati et al. (2018) have successfully manufactured the ternary nanocomposite heterojunction of rGO/ZnO/TiO_2 through the facile hydrothermal method.

Cheng et al. (2016) have also synthesized ternary nanocomposites by anchoring P25TiO_2 and Fe_3O_4 nanoparticles on the rGO through a facile solvothermal reaction using ethylene glycol. TEM and high-resolution transmission electron microscopy (HRTEM) images of the rGO and ternary P25–graphene–Fe_3O_4 nanocomposite (PGF) depicted in Figure 5.4, has clearly indicated a layered structure of rGO

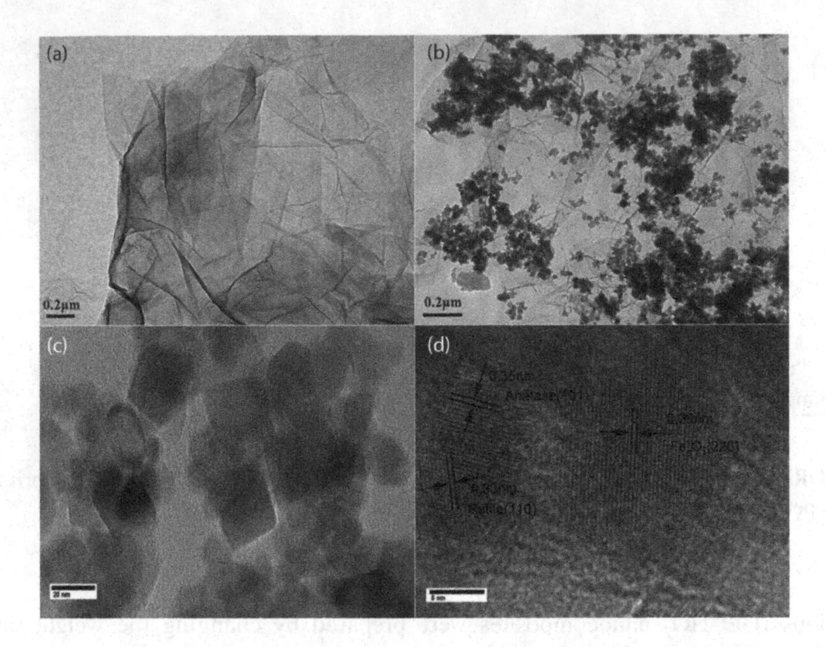

FIGURE 5.4 TEM images of pure rGO (a) and ternary P25–graphene–Fe_3O_4 nanocomposite, (b) 200 nm HRTEM images of PGF-3, (c) 20 nm, and (d) 5 nm. (Reprinted with permission from Cheng, L. et al., *Mater. Res. Bull.*, 73, 77–83, 2016.)

sheets with wrinkles and successful anchoring of spherical or rectangular Fe_3O_4 and P25 TiO_2 nanoparticles (20–40 nm) on rGO sheets.

Apart from these nanocomposites, hydrothermal method is also employed for synthesizing other graphene-transition metal oxide nanocomposites, such as GO-ZnO (Wu et al., 2010; Liu et al., 2019), GO-V_2O_5 (Aawani et al., 2019), GO-CeO_2 (Huang et al., 2014; Kumar and Kumar, 2017; Priyadharsan et al., 2017; Murali et al., 2019; Du et al., 2020), GO-ZrO_2 (Anjaneyulu et al., 2019), GO-CuO (Nuengmatcha et al., 2019) etc.

5.3.2 Sol-Gel Method

The sol-gel route provides a facile and inexpensive method for the preparation of graphene-transition metal oxide nanocomposites. The chemistry of the sol-gel method is usually based on hydrolysis, condensation, and drying processes. The formation of graphene-transition metal oxide nanocomposites by the sol-gel route involves different consecutive steps. Initially, the transition metal precursor undergoes rapid hydrolysis to produce the metal hydroxide solution in presence of GO. Condensation of resulting solution leads to the formation of gels, which is further washed, dried, and calcined to obtain the desired nanocomposite.

Ong et al. (2017) have grown ZnO nanoparticles on GO using a low-temperature sol-gel method. rGO-ZnO nanocomposites with weight percentages of ZnO as 5, 25, 30, 35, and 40 wt% were marked as ZG-1, ZG-2, ZG-3, ZG-4, and ZG-5, respectively.

Initially, the GO solution and zinc nitrate solution prepared in ethyl alcohol were mixed well, and then oxalic acid was added slowly to this solution under vigorous stirring until a gel was formed. Washing, drying, and calcination of the gel at 400°C for 1 h resulted in the formation of an rGO-ZnO nanocomposite. TEM images and X-ray diffraction (XRD) patterns of GO, ZnO, and rGO-ZnO nanocomposites are represented in Figure 5.5. The TEM image of pure ZnO has shown a hexagonal structure, and pure GO has shown a two-dimensional structure with wrinkles. The TEM image of the rGO-ZnO nanocomposite has confirmed the coupling of ZnO nanoparticles on the GO sheet. XRD analysis of pure GO has exhibited a characteristic peak at 10.2°, indicating the presence of oxygen functional groups (Xiong et al., 2010). After anchoring the ZnO nanoparticles on GO, the diffraction peak at 10.2° disappeared, which has indicated that GO has been successfully reduced to rGO as a result of loss of oxygen functional groups. In addition to this, the rGO-ZnO nanocomposite has also exhibited sharp diffraction peaks corresponding to (100), (101), (102), (103), (110), (112), (200), (201), (202), (002), and (004) planes, which are similar to the standard characteristic peaks of ZnO nanoparticles.

Atout et al. (2017) have prepared nanocomposites of TiO_2 and rGO by using a modified sol-gel method in which the sol-gel method was followed by the hydrothermal

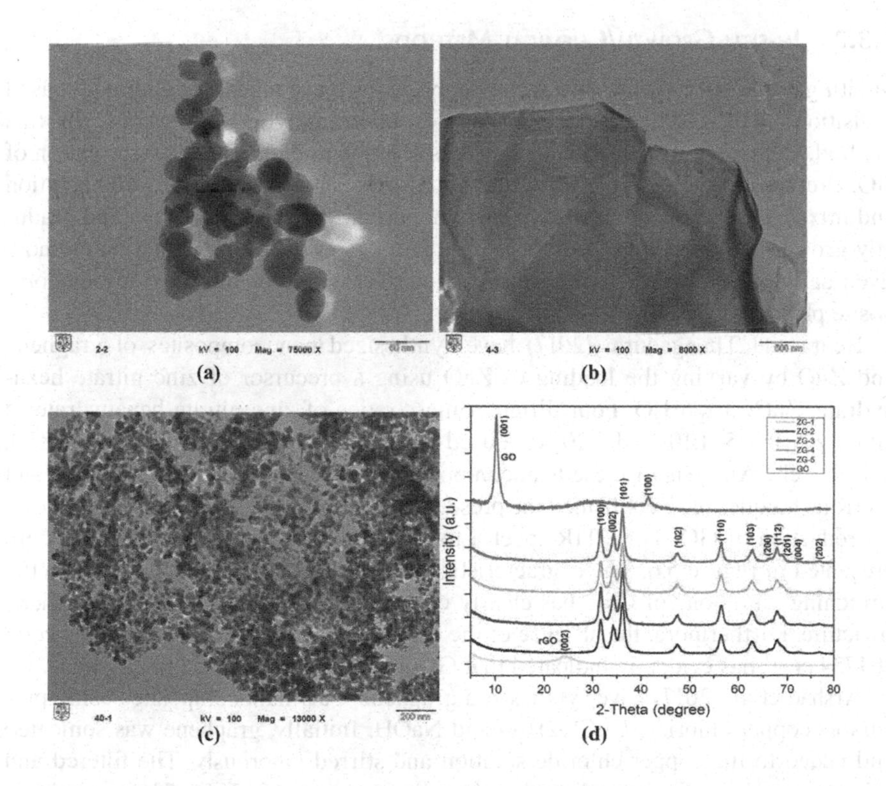

FIGURE 5.5 TEM images of (a) ZnO, (b) GO, (c) rGO–ZnO nanocomposite (ZG-5), and (d) XRD patterns of GO and rGO–ZnO nanocomposite. (Reprinted with permission from Ong, C.B. et al., *Process Saf. Environ. Prot.*, 112, 298–307, 2017.)

method. GO was first dispersed in water along with mechanical stirring followed by ultrasonication. Further, HNO_3 was added to the GO dispersion for adjusting the pH to 1.5. The required amount of GO dispersion was added dropwise to the solution of tetrabutyltitanate ($Ti(OC_4H_9)_4$) under ultrasonication. A hydrothermal reaction was carried out in Teflon-sealed autoclave, which was maintained at a temperature of 120°C for the duration of 5 h. Finally, the nanocomposite prepared was washed, dried, and calcined at 500°C for 5 h. TEM analysis has indicated that highly crystalline TiO_2 nanoparticles with sizes of about ~20 nm are well dispersed on the surface of rGO.

Banerjee et al. (2018) have synthesized ternary rGO-Fe_3O_4-TiO_2 nanocomposites using the simple sol-gel method by varying the weight ratios of rGO-Fe_3O_4-TiO_2 as (1:1:1), (2:1:1), (3:1:1), (1:2:1), (1:3:1), (1:1:2), and (1:1:3). Initially, GO and Fe_3O_4 were separately dispersed in ethanol and sonicated for 1 h. Further, both the solutions were mixed, followed by the addition of titanium tetra isobutoxide. Finally, the solution was transferred to a Teflon-lined stainless-steel autoclave that was maintained at a temperature of 180°C for the duration of 10 h. The resulted grayish precipitate was further washed and dried to obtain the rGO-Fe_3O_4-TiO_2 nanocomposite. XRD patterns, Raman spectra, and FTIR have confirmed the coupling of GO, TiO_2, and Fe_3O_4 in the ternary composite and also showed successful reduction of GO to rGO.

5.3.3 In-situ Growth/Chemical Methods

In-situ growth methods are also widely applied for the synthesis of graphene-based transition metal oxide nanocomposites for photocatalytic applications. In this method, GO is first dispersed in an aqueous solution to obtain a uniform solution of GO. Precursors to obtain the transition metal oxides are added to this GO solution and mixed well. Transition metal oxide nanoparticles are then nucleated and gradually grown on graphene by hydrolysis of precursors or the co-precipitation method. Eventually, GO is reduced to rGO, resulting in rGO-transition metal oxide nanocomposite photocatalysts.

Beura and Thangadurai (2017) have synthesized nanocomposites of graphene and ZnO by varying the loading of ZnO using a precursor of zinc nitrate hexahydrate, $Zn(NO_3)_2 \cdot 6H_2O$. Four different mass ratios of zinc nitrate hexahydrate to GO viz. 1:1, 1:5, 1:10, and 1:20 were used and marked as GZ1, GZ5, GZ10, GZ20, respectively. Analysis of these nanocomposites by XRD, Raman spectroscopy, and FTIR techniques have confirmed the presence of ZnO and graphene and the successful reduction of GO. The FTIR spectra for the graphene-ZnO nanocomposites are presented in Figure 5.6. The characteristic peak at 1573 cm^{-1} corresponding to the stretching vibrations of C=C has clearly confirmed the formation of the graphene structure. Furthermore, the absence of the characteristic peak of the carboxyl group at 1734 cm^{-1} has evidently indicated that GO is completely reduced to graphene.

Arshad et al. (2017) have synthesized graphene-CuO nanocomposites using precursors copper chloride, $CuCl_2 \cdot 2H_2O$ and NaOH. Initially, graphene was sonicated and added to the copper chloride solution and stirred vigorously. The filtered and dried material was further calcined in the tube furnace at 623 K for 5 h for synthesis of nanocomposite. The nanocomposite was characterized by X-ray powder diffraction (XRD), FE-SEM, FTIR, Raman, and UV–vis spectroscopy.

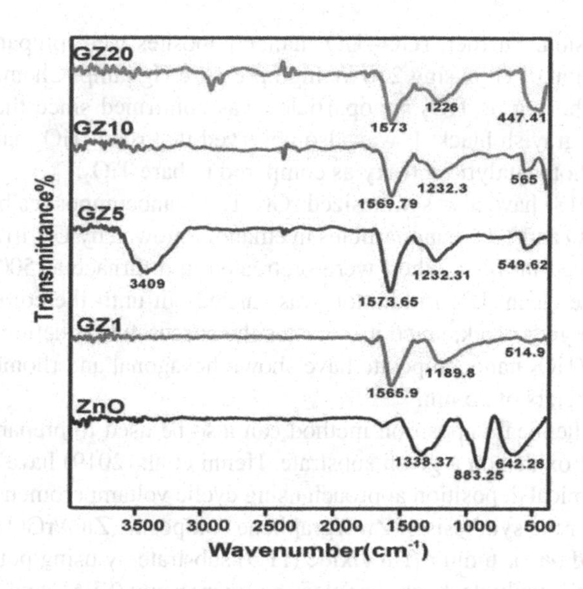

FIGURE 5.6 FTIR analysis of pure ZnO and graphene-ZnO nanocomposite. (Reprinted with permission from Beura, R. and Thangadurai, P., *J. Phys. Chem. Solids*, 102, 168–177, 2017.)

Dehghan et al. (2019) have also produced rGO/Fe_3O_4/ZnO ternary nanohybrid using the in-situ growth method. First of all, the rGO/Fe_3O_4 nanocomposite was prepared using the co-precipitation method in the presence of rGO with the precursors $FeCl_2$ and $FeCl_3$. The washed and dried rGO/Fe_3O_4 nanocomposite was further dispersed in water and sonicated to obtain its uniform solution. Finally, the rGO/Fe_3O_4/ZnO ternary nanohybrid was prepared using the precursor zinc nitrate, $Zn(NO_3)_2.6H_2O$. Characterization of the synthesized catalyst was carried out using FTIR, XRD, FE-SEM, Brunauer–Emmett–Teller (BET), photoluminescence (PL) spectra and UV–Vis diffuse reflection spectroscopy (DRS) techniques, and the successful preparation of rGO/Fe_3O_4/ZnO ternary composite was confirmed.

Nadimi et al. (2019) have also synthesized TiO_2/Fe_3O_4/GO nanocomposites by using hydrolysis of tetraisopropyl orthotitanat in the presence of an ultrasonicated suspension of Fe_3O_4 and GO. TEM images of TiO_2/Fe_3O_4/GO nanocomposites have indicated the two-dimensional structure of GO with wrinkles and has clearly shown that TiO_2 and Fe_3O_4 nanoparticles are uniformly dispersed on the surface of GO.

5.3.4 Other Methods

In addition to hydrothermal/solvothermal, sol-gel, and in-situ growth methods, some other methods such as the UV-assisted reduction method (Ali et al., 2018; Mohamed, 2012) and electrochemical method (Henni et al., 2019) are also used to synthesize the graphene-transition metal oxide nanocomposites.

Mohamed (2012) has synthesized rGO–TiO_2 nanocomposites using UV-assisted photocatalytic reduction of graphite oxide in the presence of TiO_2 nanoparticles. GO and TiO_2 nanoparticles were first dispersed in ethanol and sonicated to obtain the uniform dispersion. Further, rGO–TiO_2 nanocomposites were prepared by photocatalytic reduction of GO using 250 W high pressure Hg lamp. Chemical reduction of GO and anchoring of TiO_2 nanoparticles was confirmed since the color of the solution turned grayish black. It was also observed that rGO–TiO_2 nanocomposites exhibit better photocatalytic activity as compared to bare TiO_2.

Ali et al. (2018) have also synthesized rGO/TiO_2 nanocomposites by using ultrasonication of GO and TiO_2 nanoparticles in ethanol followed by UV irradiation. TiO_2 nanoparticles used in this method were pretreated in a furnace at 500°C for 2 h to keep the surface clean. UV irradiation was carried out until the color of the solution changed to grey-black, since it indicates the successful reduction of GO. TEM images of rGO/TiO_2 nanocomposite have shown hexagonal and rhombus structures with mean diameters of 25 nm.

The electrochemical deposition method can also be used to prepare the films of transition metal oxides on a given substrate. Henni et al. (2019) have used the one-step electrochemical deposition approach using cyclic voltamperometry and chronoamperometry for the synthesis of ZnO/graphene composite. ZnO/rGO coatings were electrodeposited on an Indium Tin Oxide (ITO) substrate by using potentiodynamic and potentiostatic methods. Scanning electron microscopy (SEM) and XRD analysis have confirmed that ZnO nanorods (wurtzite) and rGO nanosheets are successfully combined in the composite film.

5.4 APPLICATIONS OF GRAPHENE-TRANSITION METAL-OXIDE-BASED NANOCOMPOSITES IN WASTEWATER TREATMENT

Recently graphene-transition metal-oxide-based nanocomposites have emerged as an effective photocatalysts for wastewater treatment applications. In the present section, applications of various graphene-based nanocomposites such as graphene-TiO_2, graphene-ZnO, graphene-Fe_2O_3, graphene-CuO, graphene-V_2O_5, graphene-ZrO_2, graphene-Co_3O_4, and graphene-MoO_3 are presented.

5.4.1 GRAPHENE-TiO_2 NANOCOMPOSITES

Titanium dioxide (TiO_2) is a promising photocatalyst widely used in wastewater treatment applications due to its low cost, low toxicity, and high stability. The high rate of recombination of electron-hole pairs and large band gap are the main disadvantages of the TiO_2 photocatalyst, which limits its use in photocatalysis. The hybrid graphene-TiO_2 nanocomposite overcomes these limitations because it exhibits enhanced photocatalytic activity due to increased conductivity, higher transparency, and efficient charge separation of the system as compared to the TiO_2 nanoparticles and pristine graphene (Ng et al., 2010; Ramadoss and Kim, 2013; Cheng et al., 2017). TiO_2–graphene nanocomposites can be synthesized using various techniques such as hydrothermal, solvothermal, mechanical mixing with or without sonication, sol-gel techniques, deposition techniques of liquids, aerosol, chemical vapor, spin coating, and electrospinning (Tayel et al., 2018). Few studies reported in the literature are summarized below.

Wang and Zhang (2011) have synthesized the rGO–TiO_2 nanocomposite using a facile hydrothermal method followed by calcination. The different weight ratios of

GO to commercial grade TiO_2 (1:100, 1:40, 1:20, 1:10, and 1:3) were used to synthesize the nanocomposite. The photocatalytic performance of rGO–TiO_2 nanocomposites were examined by applying them for the degradation of Rhodamine B. The highest degradation efficiency was observed at the weight ratio of GO to TiO_2 as 1:20. The determination of optimum loading of GO has proven to be essential because an excessive GO loading may inhibit the electron hole recombination and obstruct the absorption of light, resulting in lowering the of rate of degradation of Rhodamine B. (Wang et al., 2005, 2008). In order to compare the photocatalytic performance of bare TiO_2 and graphene-based nanocomposite, experiments were conducted using commercial grade TiO_2 (P25), rGO–TiO_2 and a physical mixture of Gr–TiO_2 nanocomposites under UV and visible light. The results obtained are reported in Figure 5.7 (Wang and Zhang, 2011).

Bare commercial grade TiO_2 and the physical mixture of graphene and TiO_2 showed lower photocatalytic activity as compared to the rGO–TiO_2 nanocomposite under both visible light and UV light. In order to investigate the repeatability of as-prepared nanocomposites, the recycling experiments were also carried out using rGO–TiO_2—a 1:20 nanocomposite. It was observed that after the second cycle, there is a gradual decrease in the rate of degradation in subsequent cycles using UV light as a source of irradiation. This may be due to the reduction in the catalyst surface area for photon absorption and adsorption of Rhodamine B. Overall, the rGO–TiO_2 nanocomposite shows good repeatability.

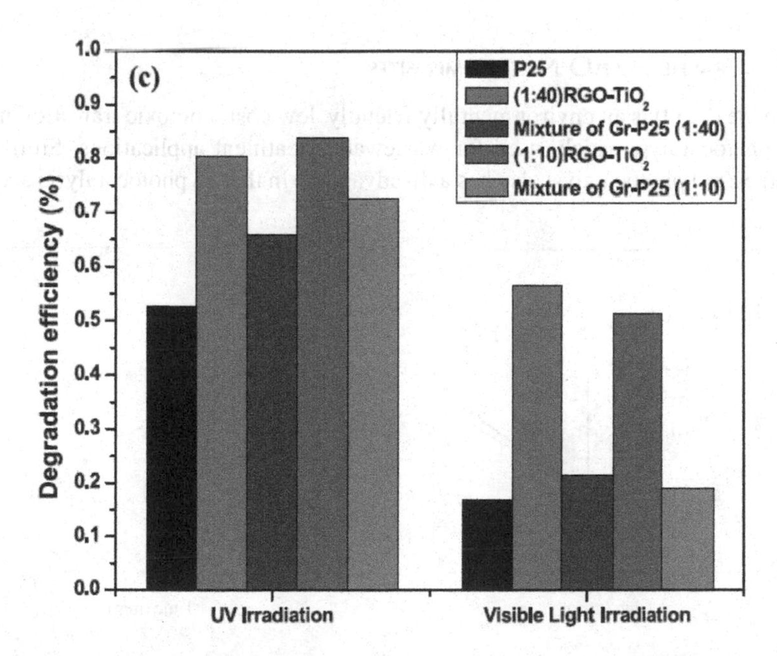

FIGURE 5.7 Comparison of photocatalytic degradation behaviors of Rh. B using various nanocomposites under UV and visible light. (Reprinted with permission from Wang, F., and Zhang, K., *J. Mol. Catal. A*, 345, 101–107, 2011.)

The photocatalytic degradation of Rhodamine B was also investigated by Zhang et al. (2013) using graphene-encapsulated TiO_2 nanospheres synthesized by using the hydrothermal method. Higher degradation efficiency (91%) was reported using TiO_2–graphene nanocomposites as compared to TiO_2 nanospheres (65%). Lee et al. (2012) have also synthesized low band gap TiO_2 nanoparticles wrapped by graphene using the one-step hydrothermal process. It was reported that this nanocomposite shows excellent photocatalytic activity under visible light when applied for the degradation of methylene blue (MB).

Zhang et al. (2014) have synthesized graphene TiO_2 nanocomposites using a hydrothermal process, and photocatalytic activity was evaluated by applying them for the degradation of sodium pentachlorophenol. It was reported that the extent of degradation of 31.1% obtained using TiO_2 has remarkably increased to 97% using graphene–TiO_2 nanocomposite. The kinetics and Total organic carbon (TOC) removal are depicted in Figure 5.8a and 5.8b, respectively. The photodegradation of sodium pentachlorophenol followed pseudo first-order degradation kinetics with an almost 18.36 times higher rate of degradation of sodium pentachlorophenol using a graphene–TiO_2 nanocomposite than that obtained using P25-TiO_2. The higher photocatalytic activity of graphene–TiO_2 nanocomposite than commercial P25-TiO_2 may be due to the fast-interfacial electron-transfer process and slow recombination of electrons and holes.

Thus, from various studies reported in the literature, it has been proved that graphene–TiO_2 nanocomposite is a promising photocatalyst for effective degradation of various pollutants in wastewater.

5.4.2 Graphene-ZnO Nanocomposites

Zinc oxide (ZnO) is an environmentally friendly, low cost, nontoxic transition metal oxide photocatalyst widely used for wastewater treatment applications. Similar to TiO_2, the ZnO photocatalyst also has a disadvantage in that its photocatalytic activity

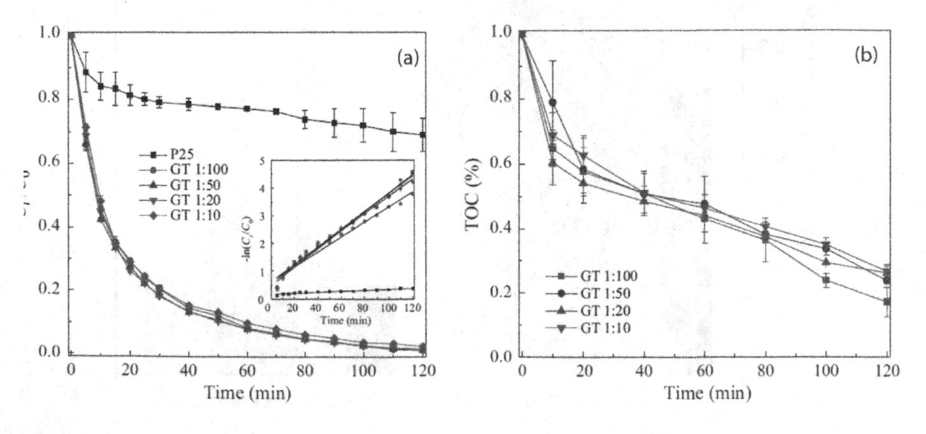

FIGURE 5.8 (a) Kinetics of photodegradation of PCP-Na using graphene–TiO_2 and P25 catalyst and (b) TOC variation. (Reprinted with permission from Zhang, Y. et al., *J. Environ. Sci.*, 26, 2114–2122, 2014.)

decreases due to a recombination reaction of photogenerated electron and hole pairs. In order to overcome this drawback, ZnO can be coupled with carbonaceous material like graphene, which reduces the charge carrier recombination, which in turn enhances the photocatalytic activity of the ZnO catalyst. Hence, research based on the ZnO-graphene nanocomposite for effective degradation of pollutant in wastewater is gaining the attention of a lot of researchers.

Beura and Thangadurai (2017) have investigated photocatalytic degradation of methyl orange (MO) using graphene-ZnO nanocomposites using UV light (8 W power) and direct sunlight as a source of irradiation and reported a maximum 97.1% and 98.6% degradation, respectively. Photocatalytic efficiency was reported 5.5 times higher using graphene–ZnO nanocomposite than that obtained using pure ZnO at optimized conditions. The rate of degradation is also dependent on surface area and PL lifetime. The loading of ZnO in graphene–ZnO nanocomposite also plays crucial role. With the increasing loading of ZnO, the surface area of graphene–ZnO nanocomposite decreases, resulting in a decrease in the exchange of electrons between the photocatalyst and the dye molecules. Also, a decrease in PL intensity of the nanocomposites as compared to that of pure ZnO was observed due to reduction of photoemission in the presence of graphene.

In another study, Qin et al. (2017) have synthesized various ZnO microsphere-rGO nanocomposites using simple solution method and applied it for the photodegradation of MB dye. The as-prepared ZnO microspheres–rGO composite with 20, 40, 80, and 120 mL of GO (0.5 mg/mL) were labeled as ZRGO20, ZRGO40, ZRGO80, and ZRGO120, respectively. In order to compare the photocatalytic efficiency, the experiments were carried out using pure ZnO microspheres, P25 TiO_2, and ZnO microsphere–rGO nanocomposite. Lower emission intensity was observed for all ZnO–rGO nanocomposites as compared to pure ZnO due to the inhibition of photogenerated electron hole recombination resulting in higher photocatalytic activity. Similarly, an increase in the surface area of the catalyst due to incorporation of rGO has also increased its photocatalytic activity. Stronger absorption intensity in UV and the visible region was reported with all prepared ZnO microspheres–rGO nanocomposites. The degradation efficiency in the order of ZRGO80 > ZRGO40 > ZRGO120 > ZRGO20 > TiO_2 > ZnO micro-spheres has indicated that the photocatalytic efficiency of ZnO microsphere can be enhanced by adding rGO at the optimum weight ratio. The highest photodegradation activity was reported using ZnO microsphere–rGO composites (with 4.06 wt% GO) for degradation of MB under UV light irradiation. This can be attributed to photogenerated electrons transfer between ZnO and rGO, which reduces the recombination of electron and hole.

In another study, the synthesis of ZnO/rGO nanocomposites by a facile one-step hydrothermal method have been reported by Liu et al. (2019), and photocatalytic activity of as-prepared nanocomposite have been investigated for photodegradation of MB in water using visible light. Various mass ratios of GO over the range of 2 wt% to 12 wt% (rGO-2 to rGO-12) have been studied and observed that ZnO/rGO-8 with 8 wt% of GO gives the highest photocatalytic activity. It was also reported that with an increase in the loading of GO, the size of the ZnO sheet decreases. The complete degradation of MB was observed using ZnO/rGO-8 nanocomposite since it leads to an effective separation of an electron hole pair due to an interfacial interaction between graphene and ZnO, resulting in higher photocatalytic activity.

Nguyen et al. (2019) have further reported synthesis and application of spherical-shaped, disk-shaped, and rod-shaped nanoparticles of ZnO loaded on rGO. The photocatalytic activity of all synthesized nanocomposites was investigated by performing degradation of MB (MB, C.I. 52015) and Rhodamine B (RhB, C.I. 45170) using a UV lamp (40 W) with 365 nm wavelength and solar simulator (150 W). Among all the prepared morphologies, the nanospherical ZnO/rGO (sZG) composite exhibited the highest removal efficiencies of selected dyes (MB and RhB). The spherical ZnO nanoparticle (sZG) catalyst exhibits rapid transport of excited electrons due to smallest particle size (15–35 nm), resulting in the largest specific area. For MB, the maximum degradation efficiency of 98.5% was obtained in 30 min using sZG catalyst (0.3 g/L) using UV illumination, whereas 99% of degradation was achieved using catalyst loading of 0.2 g/L under solar light irradiation in 100 min. It has also been reported that the morphology of the sZG nanocomposite remain unchanged even after 15 cycles and retained 96% of its photocatalytic activity. The sZG nanocomposite shows high photocatalytic activity due to the effective utilization of holes by the dye molecules that are adsorbed on the surface of rGO. Thus, low cost, effective recyclability, and long life of the sZG composite makes it a potential photocatalyst.

Similar investigations have been reported in literature by many researchers for the degradation of various pollutants present in wastewater. Wang et al. (2019) reported 100% degradation of MO using a porous graphene/ZnO nanocomposite due to the synergetic effect of photocatalysis and adsorption. Ong et al. (2017) reported effective degradation of perfluoro-octanoic acid and MO using ZnO/rGO nanocomposites. Henni et al. (2019) have also obtained better degradation efficiency of MB using a ZnO/graphene composite as compared to ZnO nanorods alone. Thus, the graphene-ZnO-based nanocomposite plays a crucial role in wastewater treatment application.

5.4.3 Graphene–CuO Nanocomposites

Nowadays, application of cupric oxide (CuO) and cuprous oxide (Cu_2O) nanoparticles in wastewater treatment is also attracting many researchers. The low cost, scalability, non-toxicity, ample availability, and easy preparation methods are some of the advantages of using CuO in wastewater treatment applications (Hu and Tang, 2013). The photocatalytic efficiency of CuO nanoparticles can be improved by coupling CuO with graphene because it slows down the rate of recombination of photogenerated electrons and holes and also increases the surface area for adsorption of organic compounds present in wastewater (Arshad et al., 2017). Many researchers have reported its application in wastewater treatment.

Arshad et al. (2017) have synthesized a CuO–graphene nanocomposite (GCuO) via a chemical route and evaluated its photocatalytic activity for the degradation of MB dye using a solar irradiation source. The extent of degradation of MB using CuO and CuO–graphene nanocomposite is depicted in Figure 5.9. The photocatalytic degradation of MB dye follows first-order kinetics with an enhanced reaction

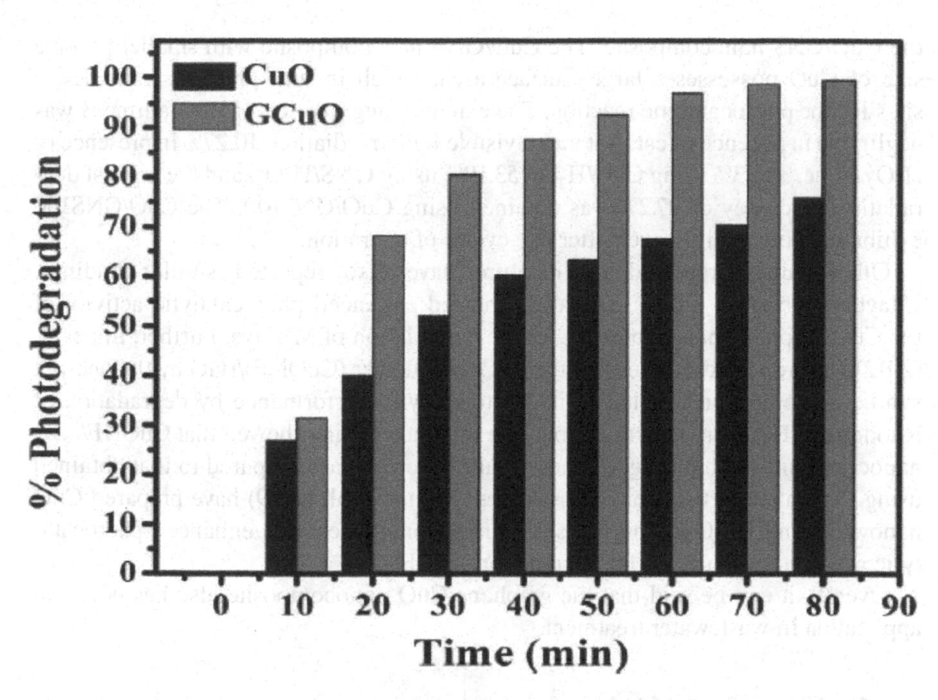

FIGURE 5.9 Photodegradation efficiency as a function of time for CuO and GCuO nanocomposite. (Reprinted with permission from Arshad, A. et al., *Ceram. Int.*, 43, 10654–10660, 2017.)

rate constant using CuO–graphene nanocomposite (0.06 min^{-1}) as compared to CuO alone (0.01 min^{-1}). Almost complete degradation of MB was obtained using the CuO–graphene nanocomposite, whereas degradation was 75% using CuO alone. The enhancement in the photocatalytic efficiency of the CuO–graphene nanocomposite may be due to availability of a larger surface area for the adsorption of dye molecules and a slower rate of recombination of photogenerated electron and hole pairs. The optimum loading of graphene on CuO nanoparticles is of utmost importance to obtain high photocatalytic activity (Hu and Tang, 2013).

Cheng et al. (2015) have synthesized a novel leaf-like CuO/graphene nanosheet (GNS) using a simple hydrothermal method. The photocatalytic activity of as-prepared CuO/GNS was investigated by applying it for photocatalytic degradation of Rhodamine B dye. Process intensification was also carried out by adding hydrogen peroxide (H_2O_2) under visible light irradiation. The CuO/GNS nanocomposite showed enhanced photocatalytic activity in the presence of H_2O_2 as compared to pure CuO for photocatalytic degradation of Rhodamine B. The improved photocatalytic performance was attributed to the higher rate of interelectron transfer at the interface due to very small particle size of the nanocomposites. The GNS immediately transfers the electrons generated on the surface of CuO, which further gets scavenged by H_2O_2 to produce a hydroxyl radical, thus reducing the rate of recombination, thereby increasing the utilization of the hole and enhances the rate of photocatalytic degradation of Rhodamine dye. In addition to this, the adsorption of dye molecules on the surface of GNS also enhances the photocatalytic activity of the CuO/GNS nanocomposite. The CuO/GNS nanocomposite with smaller particle size of CuO possesses a larger surface area, which in turn provides more active sites for the photocatalytic reaction. The extent of degradation of Rhodamine B was negligible in absence of catalyst under visible light irradiation, 10.27% in presence of H_2O_2 alone, 36.33% using CuO/H_2O_2, 53.19% using GNS/H_2O_2, and the highest degradation efficiency of 97.2% was obtained using CuO/GNS100. The CuO/GNS100 exhibited good stability even after five cycles of operation.

Other studies reported in literature have also reported similar findings. Udaybhaskar et al. (2019) have also reported enhanced photocatalytic activity of the CuO/graphene nanocomposite for the degradation of MB dye. Further, Liu et al. (2012) have reported synthesis of the CuO nanoflower (CuONF)/rGO by the one-pot synthesis method and evaluated its photocatalytic performance by degradation of Rhodamine B under UV irradiation. The obtained results showed that CuONF/rGO nanocomposites exhibit higher photocatalytic activity as compared to that obtained using CuO nanoparticles or rGO samples. Kumar et al. (2019) have prepared CuO nanowires and Cu_2O nanospheres/rGO nanocomposites, and enhanced photocatalytic activity was reported for degradation of MB.

Overall, it can be said that the graphene/CuO nanocomposite also has potential application in wastewater treatment.

5.4.4 Graphene–Fe_3O_4 Nanocomposites

Fe_3O_4 is another important transition metal oxide photocatalyst widely used in wastewater treatment due to its excellent magnetic properties. The photocatalytic activity of Fe_3O_4 can be enhanced by using graphene as a support material. Few photocatalytic applications of graphene–Fe_3O_4 nanocomposites are listed below.

Vinodhkumar et al. (2019) have synthesized rGO/magnetite (rGO/Fe_3O_4) nanocomposites by a solvothermal method. The varying mass ratios of GO to Fe_3O_4 [m GO: m $FeCl_3$ = 1.5:1 as (GF1.5:1), 2:1 as (GF2:1), 1:1.5 as (GF1:1.5), and 1:2 as (GF1:2)] were prepared. Synthesized rGO/Fe_3O_4 nanocomposites were further applied for the photocatalytic degradation of MB dye. The obtained results showed that nanocomposites with mass ratio 1:2 (GF1:2) led to highest extent of degradation of 74% in 60 min of operation. rGO/ Fe_3O_4 nanocomposites also showed good stability after four cycles of operation. The photocatalytic degradation of MB dye follows the order of, (GF1:1) < (GF1.5:1) < (GF2:1) < (GF1:1.5) < (GF1:2). The enhanced photocatalytic activity of rGO/Fe_3O_4 nanocomposite may be due to a strong electrostatic interaction between few negatively charged functional groups on the rGO surface and positively charged methyl nitride group $[(CH_3)_2N^+]$ of cationic MB dye. Also, the availability of a large surface area due to small particle size of Fe_3O_4 increases the number of reactive sites.

Arshad et al. (2019) have synthesized a graphene/Fe_3O_4 nanocomposite, and its performance was evaluated toward removal of MO. The nanocomposite with quantity of graphene as 0.050 and 0.075 g were marked as Fy and Fz nanocomposites, respectively. The results of concentration variation of MO and degradation kinetics are depicted in Figure 5.10a and 5.10b. The obtained results show an almost complete removal of dye using graphene/Fe_3O_4 nanocomposite using a solar light irradiation

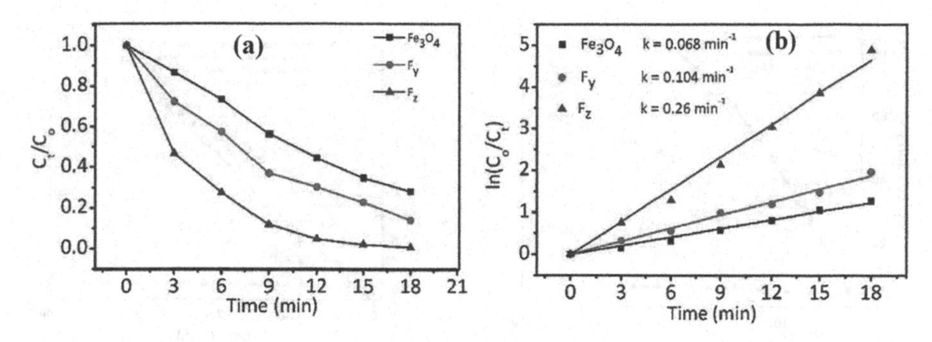

FIGURE 5.10 (a) Concentration variation of MO after addition of photocatalysts and (b) pseudo first-order reaction kinetics of photo-Fenton-type degradation of MO. (Reprinted with permission from Arshad, A. et al., *Appl. Surf. Sci.*, 474, 57–65, 2019.)

source, whereas only 72% degradation was obtained using Fe_3O_4 alone. Hence, the graphene /Fe_3O_4 nanocomposite proved to be effective catalyst with excellent reusability for effective degradation of MO.

Overall, Fe_3O_4 nanoparticles anchored on GO and rGO have been extensively researched to achieve effective degradation of several organic pollutants (Hua et al., 2014; Peik-See et al., 2014; Arshad et al., 2018).

5.4.5 Graphene–V_2O_5 Nanocomposites

Vanadium pentoxide (V_2O_5) synthesized with various morphologies can be used as a photocatalyst in wastewater treatment applications. However, it exhibits low photocatalytic activity as compared to other transition metal oxides, such as TiO_2 and ZnO. The addition of carbonaceous material such as graphene improves photocatalytic performance of V_2O_5 and leads to prolong cyclability (Boruah et al., 2018; Le et al., 2019).

Photocatalytic degradation of MB has been reported by Aawani et al. (2019) using rGO–V_2O_5 nanocomposites synthesized via the hydrothermal method. The photocatalytic performances of as-prepared nanocomposites were evaluated by its application for the degradation of MB dye using different irradiation sources, such as mercury-lamp, UV-light, and visible-light irradiations. The experiments were carried out with 50 ppm concentration of MB dye using V_2O_5 nanorods and rGO–V_2O_5 nanocomposite as photocatalysts, and the obtained results are depicted in Figure 5.11a. The higher extent of degradation of MB was observed (85% in 255 min runtime) using the rGO–V_2O_5 nanocomposite than that obtained using V_2O_5 nanorods using the mercury lamp as a source of irradiation. This may be due to the lower band gap of rGO–V_2O_5 nanocomposite as compared to V_2O_5 nanorods and PL quenching. The degradation of MB followed pseudo first-order kinetics. The obtained results are reported in Figure 5.11b. The highest rate constant value of 0.0078 min^{-1} was reported using mercury as a source of irradiation.

In another study, Shanmugam et al. (2015) have reported photocatalytic degradation of MB dye using UV, visible, and sunlight as sources of irradiations using

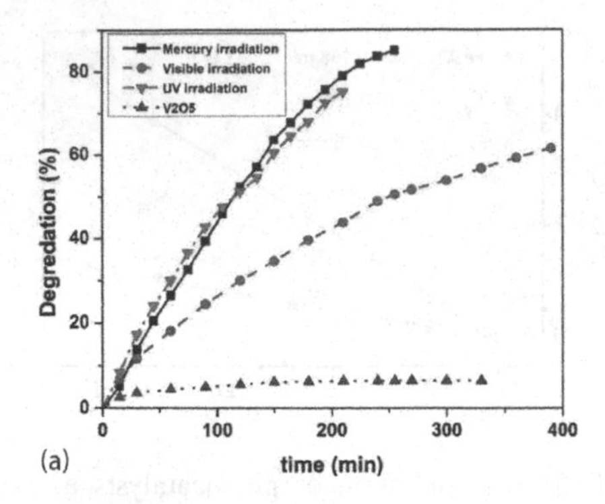

(a)

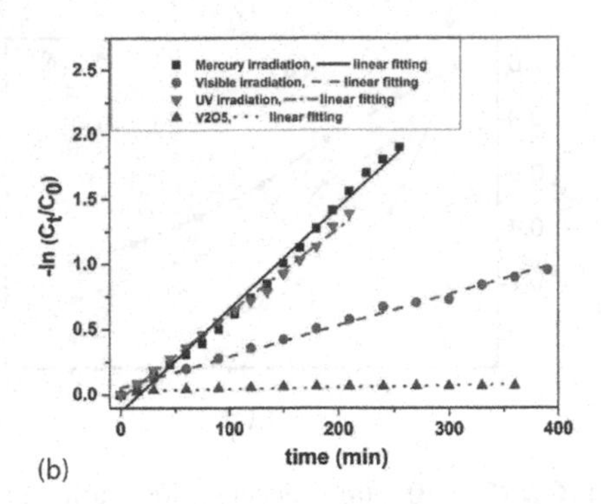

(b)

FIGURE 5.11 (a) Extent of degradation of MB and (b) MB degradation kinetics using various sources of irradiations. (Reprinted with permission from Aawani, E. et al., *J. Phys. Chem. Solids*, 125, 8–15, 2019.)

graphene–V_2O_5 nanocomposites. These nanocomposites were prepared by the solution mixing method in which V_2O_5 rods are decorated onto the graphene sheets. The prepared nanocomposites reported higher degradation efficiency of MB dye as compared to V_2O_5 nanorods, and also sunlight was observed as the most efficient source of irradiation as compared to other sources such as visible and UV light irradiations. It also has been reported that the degradation of MB follows first-order kinetics along with higher reaction rate constant and reduced half-life period using sunlight as a source of irradiation.

In order to overcome the separation issue of photocatalysts, many magnetic materials such as Fe_3O_4 have been widely used. Boruah et al. (2018) has reported synthesis of a novel Fe_3O_4/V_2O_5/graphene (FVG) photocatalyst and tested its photocatalytic efficiency by utilizing it in degradation of Bismarck brown and acid orange using direct sunlight as a source of irradiation. FVG has proven to be a highly stable, recyclable, and magnetically separable nanocomposite that holds potential value in wastewater treatment.

Therefore, the V_2O_5/rGO nanocomposite provides a new generation material with excellent photocatalytic activity.

5.4.6 Graphene–CoO Nanocomposites

Cobalt oxide (CoO or Co_3O_4) also serves as a potential photocatalyst and has been applied by few researchers for the photocatalytic degradation of various organic pollutants. Attempts have also been made to enhance the photocatalytic activity of CoO by making its composite with other transition metal oxides and anchoring them on a graphene surface. Sharma and Lee (2016) have reported synthesis of CoO loaded on TiO_2 nanoparticles and supported on rGO (TiO_2/CoO/rGO) using the sol-gel method. Further, its photocatalytic performance was evaluated by investigating the degradation efficiency of 2 chlorophenol (2-CP). It was reported that as-prepared nanocomposite (TiO_2/rGO/CoO) exhibits the highest degradation efficiency (98.2%) as compared to bare TiO_2 using visible light as a source of irradiation and follows pseudo first-order kinetics. The higher rate of photocatalytic degradation of 2-CP may be due to the enhanced charge transportation and generation of high quantum of hydroxyl radicals, which may further break the ring structure of 2-CP and convert it into carbon dioxide and water (Yang et al., 2014).

In another study, Ranjith et al. (2019) prepared hybrid rGO-TiO_2/Co_3O_4 nanocomposite by the co-precipitation method and utilized it for the decolorization of MB and crystal violet (CV) dye solution in visible light irradiation. The rGO-TiO_2/Co_3O_4 nanocomposite exhibits higher photocatalytic activity under visible light since the photogenerated electrons quickly get transferred to the rGO surface, thereby reducing the recombination of electron and hole pairs. The adsorption of MB and CV on the surface of rGO by π-π interaction has also enhanced their rate of degradation.

Hence, ternary composite of CoO with other transition metal oxide and graphene has proven to be an effective photocatalyst for the degradation of various organic pollutants.

5.4.7 Graphene–MoO_3 Nanocomposites

Transition metal oxides such as molybdenum trioxide (MoO_3) having wide band gap shows potential industrial applications in wastewater treatment due to their optical, electrical, and thermal properties. Because the dispersibility of MoO_3 in water is very poor, it can be combined with GO to improve its dispersibility. Few recent reports available in the literature have indicated successful applications of GO–MoO_3 nanocomposite in wastewater treatment. Kamalam et al. (2018) have synthesized GO–MoO_3 nanocomposites by the acidified sonochemical method, and photocatalytic activity was examined by applying it for the degradation of Victoria blue in the visible range. In order to investigate the effect of MoO_3 loading in the GO–MoO_3 nanocomposite, two different loadings of GO viz. 5 wt% (GM1) and 10 wt% (GM2) were used. The highest extent of degradation of dye (89%) was obtained using GM2 as compared to that of GM1 (59%) and MoO_3 (27%), which may be due to quenching of photoluminescence intensity resulting in a decreased rate of recombination of photogenerated electron and hole pairs. The photocatalytic degradation of Victoria blue followed pseudo first-order kinetics.

In another study, Ashraf et al. (2020) synthesized the NiS–MoO_3–GO nanocomposite using the hydrothermal method for extending its application in a wide area of light. The photocatalytic performance of the as-prepared nanocomposite was examined by the degradation of MO using UV and visible light as the sources of irradiation. The NiS–MoO_3–GO nanocomposite indicated a higher extent of degradation of MO (97%) as compared to MoO_3 and NiS-MoO_3 nanocomposite under UV light as a source of irradiation. Similar enhancement in the extent of degradation was also observed under visible light irradiation.

Hence, graphene-MoO_3 nanocomposite can successfully degrade various organic pollutants in wastewater.

5.4.8 Graphene–ZrO_2 Nanocomposites

ZrO_2 is another important transition metal oxide that can be combined with other semiconductors and graphenes for enhancing its photocatalytic activity.

The photocatalytic activity of Fe_2O_3-GO photocatalyst can be further improved by using ZrO_2. Anjaneyulu et al. (2019) reported synthesis of ZrO_2/Fe_2O_3/rGO nanocomposite using the in-situ hydrothermal method. The photocatalytic activity of ZrO_2/Fe_2O_3/rGO nanocomposite was evaluated by utilizing it for the degradation of Congo red (CR) and acetophenone. The degradation of CR showed highest photocatalytic activity using ZrO_2/Fe_2O_3/rGO, and also it follows pseudo first-order kinetics using visible light irradiation. The extent of degradation of CR was 45.5, 51.4, 67.1, and 98.4% using ZrO_2, Fe_2O_3, ZrO_2/Fe_2O_3, and ZrO_2/Fe_2O_3/rGO nanocomposites, respectively.

Overall, it can be said that graphene-transition metal-oxide-based nanocomposites have potential applications in wastewater treatment for the removal organic pollutants.

5.5 CASE STUDY: UV-ASSISTED SYNTHESIS OF rGO–TiO_2 NANOCOMPOSITE FOR PHOTOCATALYTIC AND SONO-PHOTOCATALYTIC DEGRADATION OF METHOMYL

A detailed cased study based on the synthesis of the rGO–TiO_2 nanocomposite and its application in photocatalytic and sono-photocatalytic degradation of methomyl has been presented for better understanding of application of graphene-transition metal-oxide-based nanocomposite photocatalysts for wastewater treatment. The research work presented in this case study is mainly divided into two parts for the sake of effortless understanding. The first part includes synthesis of the rGO–TiO_2 nanocomposite with a mass ratio of rGO:TiO_2 as 1:5, 1:10, 1:20, and 1:40 and its characterization. The second part presents application of the rGO–TiO_2 nanocomposite for the photocatalytic degradation of methomyl (carbamate group pesticide). The objective of this research work is to harness the beneficial effects of anchoring anatase TiO_2 nanoparticles on the sheets of rGO for the effective photocatalytic degradation of methomyl.

5.5.1 Synthesis and Characterization of rGO–TiO_2 Nanocomposite

GO has been synthesized via the modified Hummers method (Hummers and Offeman 1958; Marcano et al., 2010), and anatase TiO_2 has been synthesized via the simple chemical precipitation route using a precursor of $TiCl_4$ (Li et al., 2002; Li and Zeng, 2011). Further, rGO–TiO_2 was synthesized using UV-assisted photocatalytic reduction of GO (Williams et al., 2008; Mohamed, 2012) in presence of TiO_2 under ultrasound cavitation. A known amount of GO was first dispersed in 100 mL of and ethanol/water mixture (80:20 mL) and sonicated for 30 min at low amplitude. Later, a certain amount of anatase TiO_2 was added to the GO solution, and the mixture was further sonicated for 30 min to obtain the uniform dispersion. Four different samples of rGO–TiO_2 were prepared by using the various weight ratios of GO: TiO_2 as 1:05, 1:10, 1:20, and 1:40. UV-assisted photocatalytic reduction of GO was carried out in

FIGURE 5.12 Photographs of various samples (a) GO solution, (b) anatase TiO_2 suspension, (c) GO-TiO_2 suspension before UV treatment, and (d) rGO-TiO_2 suspension after UV treatment.

a 200 mL quartz reactor using two UV lamps (8 W each) for about 3 h. The suspension was sonicated during the UV treatment for uniform exposure of the particles to the UV light. Reduction of GO to rGO due to UV treatment was confirmed by change in the color of the suspension to grayish black. The photographs of GO solution, anatase TiO_2 suspension, and GO-TiO_2 suspension before a UV treatment and rGO-TiO_2 suspension after UV treatment are presented in Figure 5.12, which clearly indicates the successful preparation of the rGO–TiO_2 nanocomposite. The synthesized samples of rGO-TiO_2 (1:5), rGO-TiO_2 (1:10), rGO-TiO_2 (1:20), and rGO-TiO_2 (1:40) were then filtered, washed, and dried in the oven at 60°C. Dried samples were further calcined at a temperature of 400°C for the duration of 4 h.

The crystal structures of the catalyst samples were obtained by XRD by using the X-ray diffractometer (Bruker B8 advance). XRD patterns were recorded with CuKα as a radiation source (λ = 1.5406 nm) for 2θ in the range of 5°–70°. The JSM-6360 LV microscope was used to obtain scanning electron microscope (SEM) images. The CM 200-Philips electron microscope (operating voltage range of 20–200 kV and resolution of 2.4 A) was used to obtain the TEM images. The catalyst samples were dispersed in ethanol, and one drop of this solution was placed on a carbon–copper grid for TEM analysis.

XRD analysis of graphite and GO are depicted in Figure 5.13. It can be seen that the characteristic peaks of graphite at 26.28° and 54.74° corresponds to (002) and (004) planes, and the characteristic peak of GO at 11.16° corresponds to the (002) plane (Chen and Yan, 2010). The interlayer d-spacing of graphite and GO has been calculated using Bragg's equation and found that intercalation or oxidation of graphite has enhanced the d-spacing from 3.35 A to 7.98 A. The results obtained can be attributed to the introduction of oxygen-bearing functional groups to the carbon lattice of graphite during the oxidation.

XRD patterns of anatase TiO_2 and rGO-TiO_2 are depicted in Figure 5.14. The XRD pattern of TiO_2 clearly shows the characteristic peaks at 25.37°, 37.83°,

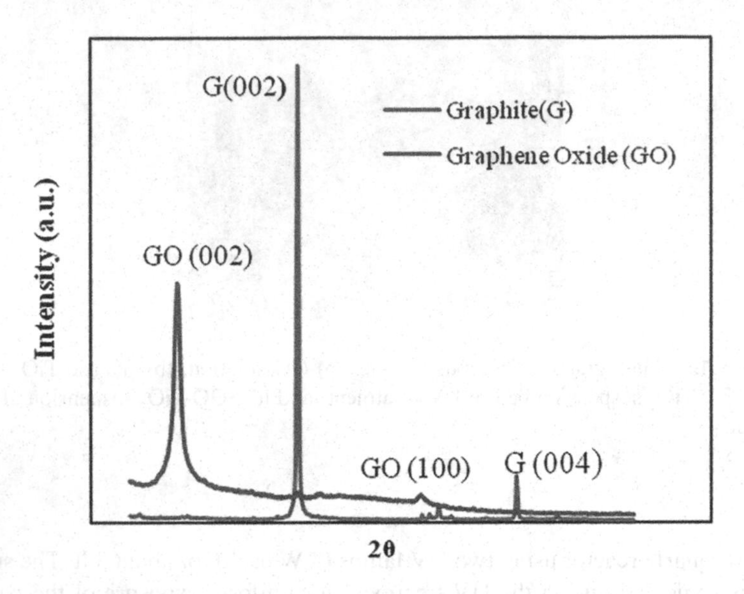

FIGURE 5.13 XRD pattern of graphite and GO.

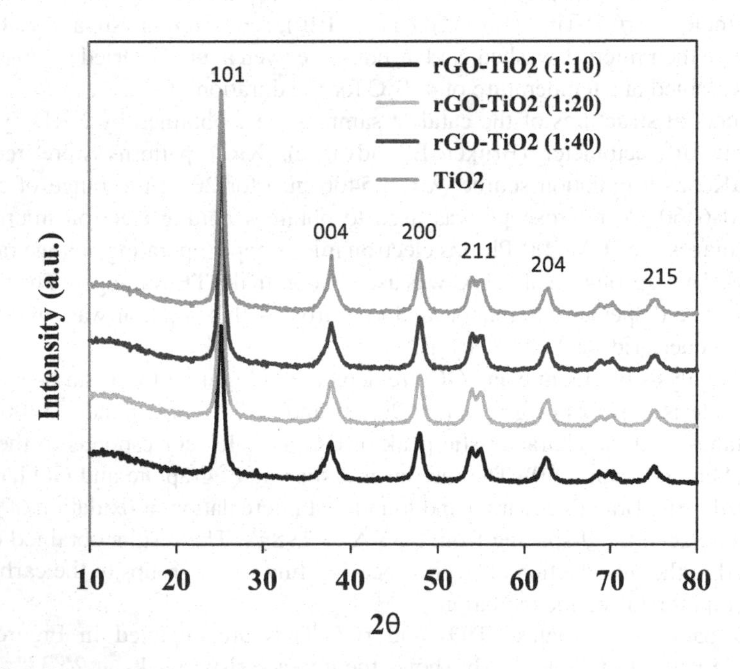

FIGURE 5.14 XRD pattern of anatase TiO_2 and rGO-TiO_2.

48.15°, 54.13°, 62.89°, 69.9°, and 74.9° corresponding to 101, 004, 200, 211, 204, 220, and 215 planes of anatase TiO_2, respectively. All the peaks obtained are analogous to the standard peaks of polycrystalline anatase phase of TiO_2 reported in (JCPDS-21-1272) (Khalid et al., 2013a). In the XRD patterns of rGO-TiO_2 (1:05, 1:10, 1:20, 1:30), the characteristic peak of GO was not detected. It may be due to the lower loading of GO as compared to the loading of anatase TiO_2. The identical peaks obtained in case of anatase TiO_2 and rGO-TiO_2 have clearly shown that the addition of GO does not change the structure of TiO_2 (Khalid et al., 2013b).

The surface morphology and shape of the GO have been obtained by using the SEM and TEM images, and the representative images are depicted in Figure 5.15. The SEM images show the layered or stacked structure of GO, and the TEM images clearly show that GO has the tendency to scroll and wrinkle (Wojtoniszak et al., 2012).

Further characterization of anatase TiO_2 and rGO-TiO_2 (1:20) has been carried out using TEM in order to obtain the size and shape of the particles. The representative TEM images are shown in Figure 5.15. It can be seen from the TEM images of anatase TiO_2 that the majority of the particles indicate nearly spherical morphology with a diameter <20 nm. Representative TEM images rGO-TiO_2 (1:20) clearly show the successful loading of TiO_2 on the surface of GO (Figure 5.16).

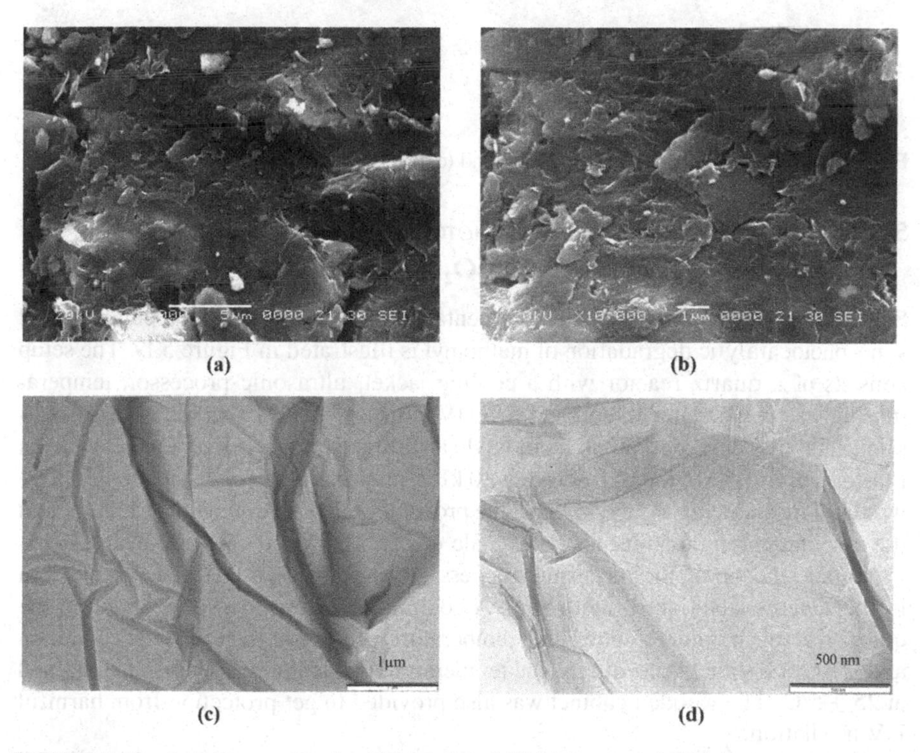

FIGURE 5.15 (a,b) Representative SEM and (c,d) TEM images of GO.

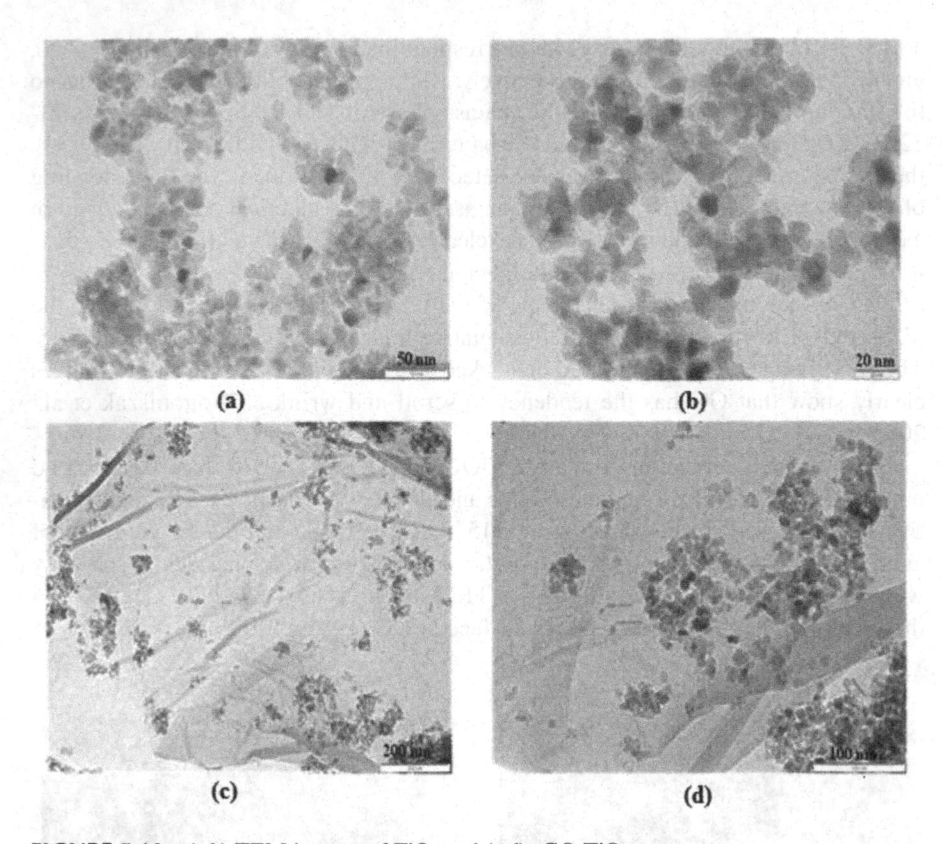

FIGURE 5.16 (a,b) TEM images of TiO_2 and (c,d) rGO-TiO_2.

5.5.2 Photocatalytic and Sono-photocatalytic Degradation of Methomyl Using rGO–TiO_2 Nanocomposite

Schematic representation of an experimental setup used for the photocatalytic and sono-photocatalytic degradation of methomyl is illustrated in Figure 5.17. The setup consists of a quartz reactor with a cooling jacket, ultrasonic processor, temperature indicator, wooden cabinet, and two UV lamps (8 W each). In the case of sono-photocatalytic degradation of methomyl, an ultrasonic processor (make-Johnson Plastosonic Pvt. Ltd., India; frequency-20 kHz, maximum power dissipation-500 W) was used for inducing the cavitation. The processor has a tip diameter of 13 mm, and the amplitude can be adjusted to the scale of 25, 50, 75 and 100%. During experimentation, the tip of the ultrasonic processor was immersed to the depth of 1 cm in the aqueous solution of methomyl. A cooling arrangement was provided to the quartz reactor in order to control the temperature rise due to the effects of cavitation and UV irradiation. Accordingly, the temperature inside the reactor was maintained at 25 ± 5°C. The wooden cabinet was also provided to get protection from harmful UV irradiation.

Change in the concentration of methomyl during experimentation was determined by using high-pressure liquid chromatography (make- JASCO, Japan) equipped with

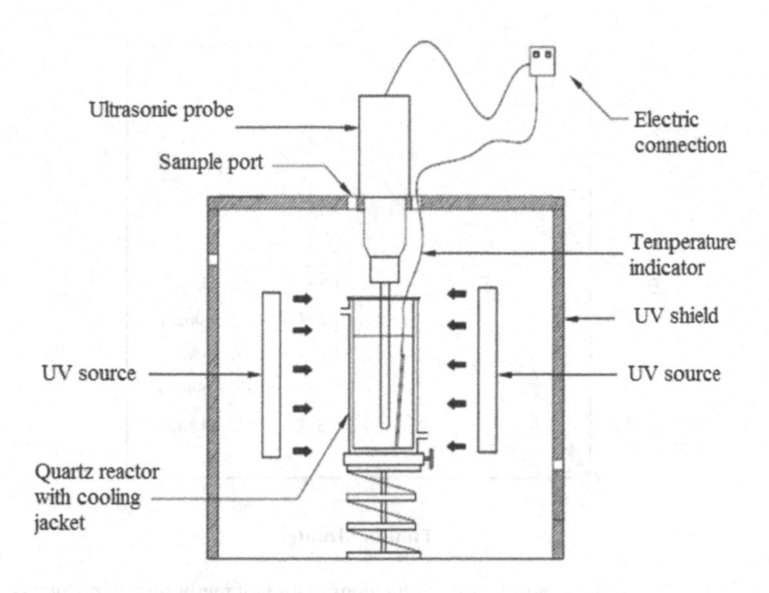

FIGURE 5.17 Schematic representation of experimental setup.

C18 column (dimensions- 4.6 × 250 mm) and a UV detector. A mixture of acetonitrile and water (30:70 volume ratio) was used as a mobile phase. The flow rate of the mobile phase was maintained at 1 mL/min, and the UV detector was adjusted to the wavelength of 234 nm. All the experiments were performed by using an initial concentration of methomyl as 25 ppm and volume as 250 mL. The optimal pH of 2.5 and optimal amplitude of ultrasonic processor as 50% with energy input of 250 W were used in the present work based on the findings of previous studies (Raut-Jadhav et al., 2016).

The optimization of loading of the photocatalyst plays an important role in maximizing the rate and extent of the degradation that can be achieved using photocatalytic or sono-photocatalytic degradation of organic pollutants. Initially, the sono-photocatalytic degradation of methomyl was carried out using anatase TiO_2 as a photo catalyst. The optimal amount of anatase TiO_2 has been evaluated by varying the loading of anatase TiO_2 viz. 50, 100, 200, 300, and 400 mg/L. The results obtained are depicted in Figure 5.18 and Table 5.1.

It has clearly indicated that the rate of sono-photocatalytic degradation of methomyl enhances with an increase in the amount of anatase TiO_2 up to 100 mg/L, owing to the increased number of reactive adsorption sites and effective absorption of UV light, which may lead to higher generation of hydroxyl radicals and their effective interaction (Anju et al., 2012; Cheng et al., 2012).

However, any further enhancement in the amount of anatase TiO_2 has rather decreased the rate of degradation of methomyl. It may be attributed to the fact that at higher loading of the catalyst, the penetration of UV light is reduced (due to scattering effect and increased opacity); subsequently, the photoactive volume of suspension is also reduced (Neppolian et al., 2011). Another reason may be the agglomeration of catalyst particles at higher loading causing the reduction in the

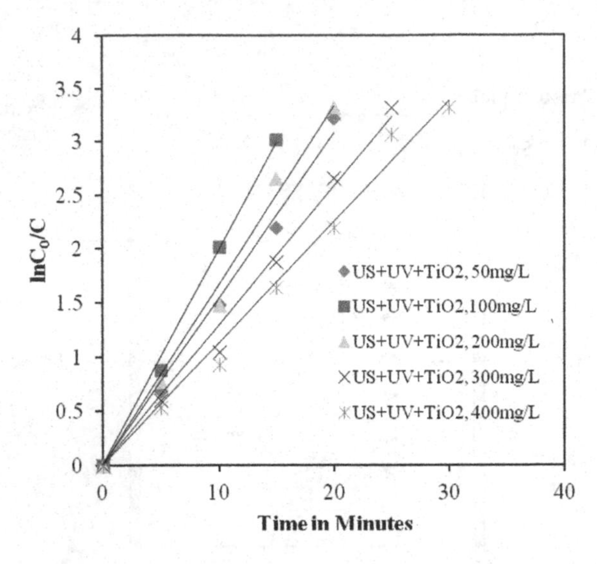

FIGURE 5.18 First-order sono-photocatalytic degradation of methomyl using anatase TiO_2.

TABLE 5.1
Sono-Photocatalytic Degradation of Methomyl at Various Loadings of Anatase TiO_2

Amount of Catalyst in US + UV + TiO_2 Process	Percent Degradation after 5 min	$k \times 10^3$ min^{-1}
50 mg/L	88.85	154.1
100 mg/L	95.09	199.568
200 mg/L	93.01	166.645
300 mg/L	84.72	129.344
400 mg/L	80.58	112.867

effective surface area available for the reaction (Anju et al., 2012). The results have clearly demonstrated that the loading of 100 mg/L of the catalyst is the optimal one with a maximum rate of degradation of 199.568×10^{-3} min^{-1} and extent of degradation of 95.09% in just 5.0 min. Prior studies of sono-photocatalytic degradation of organic pollutants have also indicated that the rate of degradation of pollutant reaches a maximum value at the optimal loading of catalyst, and any further increase beyond the optimal value leads to the detrimental effects (Neppolian et al., 2011; Anju et al., 2012).

It is well-reported in the literature that the photocatalytic efficiency of TiO_2 is seriously limited by the narrow light response range and low separation probability of the photoinduced electron hole pairs (Gupta and Tripathi, 2011). Therefore, it is essential to increase the efficiency of bare anatase TiO_2. The photocatalytic activity of anatase TiO_2 can be enhanced by anchoring it on the carbon materials such as graphene, since graphene possesses a huge surface area and also has an excellent mobility of charge carriers (Wang et al., 2010, 2012; Xiang et al., 2012). The presence of oxygen-bearing functional groups in GO and rGO makes them an excellent support material to anchor TiO_2 nanocrystals (Wang and Zhang, 2011). With this background, the rGO–TiO_2 nanocomposite with varied graphene content (mass ratio of rGO: TiO_2 such as 1:5, 1:10, 1:20, and 1:40) were synthesized and employed for the photocatalytic and sono-photocatalytic degradation of methomyl. The results of photocatalytic degradation of methomyl using anatase TiO_2 and rGO-TiO_2 are depicted in Figure 5.19.

It has been observed that the photocatalytic degradation of methomyl using anatase TiO_2 and rGO-TiO_2 (1:5, 1:10, 1:20, and 1:40) have fitted to first-order kinetics. It has been successfully demonstrated that the photocatalytic activity of anatase TiO_2 can be enhanced by anchoring the anatase TiO_2 on graphene material, since the reaction rate constant of 96.239 × 10^{-3} min^{-1} obtained in the case of anatase TiO_2 has significantly increased to 156.004 × $10^{-3}min^{-1}$ in the case of rGO-TiO_2 (1:40). A further increase in the graphene content of rGO–TiO_2 nanocomposite leads to even higher rate of degradation of methomyl with a reaction rate constant of 175.314 × 10^{-3} min^{-1} at rGO-TiO_2 (1:20). However, increasing the mass ratio of graphene to TiO_2 beyond 1:20 has shown negative impact on the rate of degradation of methomyl, since the reaction rate constant has appreciably reduced to 148.235 × $10^{-3}min^{-1}$ and 126.003 × $10^{-3}min^{-1}$ using rGO-TiO_2(1:10) and rGO-TiO_2(1:05), respectively.

The results of sono-photocatalytic degradation of methomyl by utilizing anatase TiO_2 and rGO–TiO_2 nanocomposites is depicted in Figure 5.20. The trend of the result obtained in the case of sono-photocatalytic and photocatalytic degradation of methomyl are found to be similar. The rate of sono-photocatalytic degradation

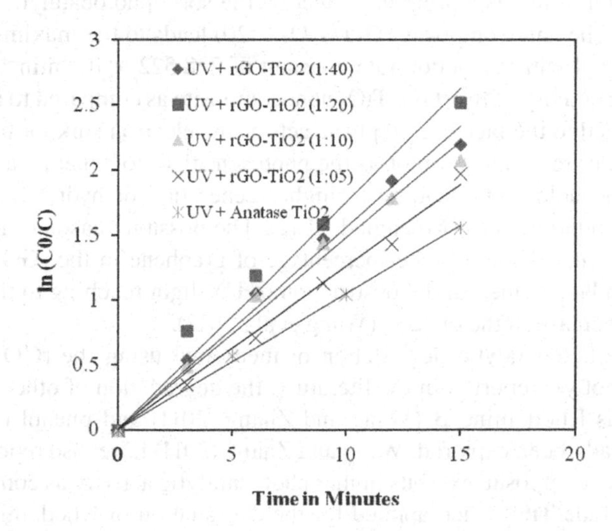

FIGURE 5.19 First-order photocatalytic degradation of methomyl by utilizing anatase TiO_2 and rGO–TiO_2 nanocomposite.

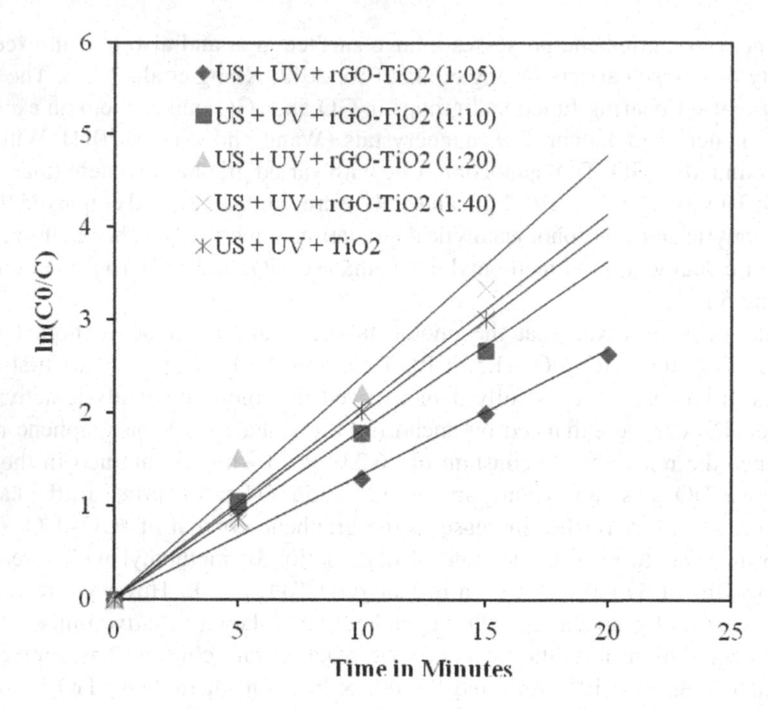

FIGURE 5.20 First-order sono-photocatalytic degradation of methomyl using rGO–TiO_2 nanocomposite.

of methomyl obtained in the case of bare TiO_2 has enhanced appreciably by using the composite of rGO–TiO_2. It was also observed that the graphene content in the rGO–TiO_2 composite has a profound effect on the sono-photocatalytic degradation of methomyl. The nanocomposite rGO–TiO_2 (1:20) leads to the maximum degradation of methomyl with a reaction rate constant of $238.522 \times 10^{-3} min^{-1}$. The higher photocatalytic activity of the rGO–TiO_2 nanocomposite as compared to anatase TiO_2 can be attributed to the fact that graphene acts as the electron sink for the hindrance of charge carrier recombination or as the photosensitizer to generate a greater density of electron hole pairs, leading to higher generation of hydroxyl radicals and increased degradation rates (Mohamed, 2012). The possible reason of low degradation rates of methomyl at a higher percentage of graphene in the rGO–TiO_2 nanocomposite can be ascribed to the obstruction of UV light reaching to the surface of TiO_2 and an increase in the opacity (Wang et al., 2013).

Although, photocatalytic degradation of methomyl using the rGO–TiO_2 nanocomposite is not yet reported in the literature, the degradation of other organic pollutants such as Rhodamine B (Wang and Zhang, 2011) and phenol (Wang et al., 2013) has already been explored. Wang and Zhang (2011) have also reported that the rGO–TiO_2 nanocomposite exhibits higher photocatalytic activity as compared to the commercial grade TiO_2 when applied for the degradation of Rhodamine B. It was also found that high graphene content in the rGO–TiO_2 nanocomposite leads to the detrimental effects due to shielding of UV light in the presence of excess graphene.

Similarly, Khalid et al. (2013a) have also observed higher photocatalytic activity of GR-TiO_2 as compared to bare TiO_2 when applied for the degradation of methyl orange.

The comparative study of efficacy of photocatalytic and sono-photocatalytic degradation of methomyl has also been carried out in order to obtain the synergistic effect of coupling ultrasound cavitation with the photocatalytic process. For any hybrid process to show the synergistic effect, it is obviously expected that the efficiency of the hybrid process (sono-photocatalytic) should be higher than the sum of the efficiencies of the two single processes (ultrasound cavitation and photocatalytic process) (Madhavan et al., 2010). The rate of degradation of methomyl obtained by using photocatalytic and sono-photocatalytic processes and the synergistic index have been reported in Table 5.2.

It has been observed that sono-photocatalytic process utilizing anatase TiO_2 leads to the synergistic effect, since the reaction rate constant of 96.239×10^{-3} min^{-1} and 4.861×10^{-3} min^{-1} obtained in case of photocatalytic and sonolytic degradation of methomyl have enhanced appreciably to 199.568×10^{-3} min^{-1} by using the sono-photocatalytic process. The synergistic index has been calculated as

$$\text{Synergistic index} = \frac{k_{(US+UV+TiO_2)}}{k_{US} + k_{(UV+TiO_2)}}$$

Similarly, the synergistic index has also been calculated by comparing the reaction rate constant obtained using photocatalytic and sono-photocatalytic degradation of methomyl using rGO-TiO_2. The sono-photocatalytic degradation of methomyl using the rGO-TiO_2-1:40, rGO-TiO_2-1:20, and rGO-TiO_2-1:10 leads to the synergistic index of 1.286, 1.324, and 1.183, respectively.

TABLE 5.2
Comparison of Reaction Rate Constant for the Photocatalytic and Sono-Photocatalytic Degradation of Methomyl and Synergistic Index

Process	Reaction Rate Constant ($k \times 10^3$ min^{-1})	US Based Hybrid Process	Reaction Rate Constant, ($k \times 10^3$ min^{-1})	Synergistic Index
UV + TiO_2, 100 mg/L	96.239	US + UV + TiO_2, 100 mg/L	199.568	1.97
UV + rGO-TiO_2 (1:40)	156.004	US + UV + rGO -TiO_2 (1:40)	206.792	1.286
UV + rGO-TiO_2 (1:20)	175.314	US + UV + rGO -TiO_2 (1:20)	238.522	1.324
UV + rGO-TiO_2 (1:10)	148.235	US + UV + rGO -TiO_2 (1:10)	181.067	1.183
UV + rGO-TiO_2 (1:05)	126.003	US + UV + rGO -TiO_2 (1:05)	134.162	1.025

The synergistic effect obtained may be attributed to the beneficial chemical and mechanical effects of ultrasound cavitation indicated below:

- Presence of photocatalyst particles enhances the ultrasound cavitation phenomenon since they provide additional nuclei for the formation of cavities, resulting in higher chemical effects of cavitation (Gogate, 2008). In addition to this, it also helps to break the microbubbles created by ultrasound cavitation into smaller ones, which subsequently helps in the formation of higher number of cavities, leading to higher generation of hydroxyl radicals (Anju et al., 2012).
- The mechanical effect of ultrasound wave reduces the agglomeration of catalyst particles, increases the mass transfer, and keeps the catalyst surface clean (Cheng et al., 2012).
- Higher local turbulence due to ultrasound cavitation improves the utilization of hydroxyl radicals (Gogate, 2008).

Anju et al. (2012) have also obtained similar results of the synergistic effect of the combination of ultrasound cavitation and photocatalytic processes when applied for the degradation of phenol using ZnO. Ahmedchekkat et al. (2011) have also achieved the synergistic effect of the combination of photocatalysis with sonolysis when applied for the degradation of Rhodamine B.

5.5.3 Key Findings and Conclusion

Key findings of photocatalytic and sono-photocatalytic degradation of methomyl have been summarized below:

- The rate of sono-photocatalytic degradation of methomyl was found to be dependent on the loading of anatase TiO_2, with a maximum rate of degradation of 199.568×10^{-3} min^{-1} and extent of degradation of 95.09% at the optimal loading of anatase TiO_2 of 100 mg/L.
- Photocatalytic efficiency of anatase TiO_2 has improved significantly by anchoring the anatase TiO_2 on graphene sheets. The rate of photocatalytic degradation of methomyl of 96.239×10^{-3} min^{-1} obtained by using anatase TiO_2 has significantly increased to $175.314 \times 10^{-3} min^{-1}$ when rGO-TiO_2 (1:20) was used with the same loading of 100 mg/L. Similarly, the first-order reaction rate constant of 199.568×10^{-3} min^{-1} obtained in the case of sono-photocatalytic degradation of methomyl (anatase TiO_2) has also significantly increased to $238.522 \times 10^{-3} min^{-1}$ using rGO-TiO_2 (1:20).
- The nanocomposite rGO-TiO_2 (1:20) demonstrated the highest photocatalytic activity as compared to rGO-TiO_2 with the ratio of graphene to TiO_2 as 1:5, 1:10, and 1:40. The sono-photocatalytic degradation of methomyl using anatase TiO_2 and rGO-TiO_2 (1:20) leads to the synergistic effect with the index of 1.97 and 1.324, respectively.

Overall, the efficacy of anatase TiO_2 that can be enhanced significantly by anchoring it onto GO and sono-photocatalytic process has proven to be more efficient as compared to individual processes for the degradation of methomyl.

5.6 CONCLUSION AND A WAY FORWARD

An extensive overview and detailed investigations of synthesis and applications of graphene-based transition metal oxide in the field of wastewater treatment has been carried out in the present study.

Although bare transition metal oxides such as TiO_2, ZnO, and Fe_3O_4 possess high photocatalytic properties, their use in the area of wastewater treatment are limited due to various drawbacks. The major drawback of transition metal oxides is the large band gap because they can absorb only a small fraction of photons in the visible region. Other drawbacks include poor stability, low surface area, fast recombination of electron-hole pairs, etc. Several strategies have been employed to improve photocatalytic efficiency of transition metal oxides. Graphene plays a crucial role to overcome the drawbacks associated with the applications of bare transition metal oxides.

Two main categories of graphene viz. GO and rGO are widely employed in the field of photocatalysis. The incorporation of graphene in transition metal oxide nanocomposite can improve adsorption capacity of pollutants on photocatalyst surfaces, extend light absorption in the visible range, and suppress the recombination of photogenerated electron–hole pairs. These beneficial effects further enhance the overall photocatalytic performance of transition metal oxides.

The efficiency of graphene-transition metal oxide nanocomposite photocatalysts depends on several parameters, such as the method of synthesis of nanocomposite, loading of graphene in graphene-transition metal oxide nanocomposite, source of irradiation, nature of pollutant, and reusability and stability of photocatalyst. Many researchers have observed that determination of optimum loading of graphene is very essential because excessive loading of graphene may further hinder the photocatalytic performance of the nanocomposite. The method of synthesis of graphene-transition metal-oxide-based nanocomposite also plays a vital role in enhancing photocatalytic activity of nanocomposites since it decides the morphology and surface area of the nanocomposite. Recent developments in the synthesis of nanocomposites have indicated that the hydrothermal/solvothermal method, sol-gel method, in-situ growth/chemical method, electrochemical method, UV-assisted reduction methods, etc. are widely applied for the synthesis of graphene-transition metal oxides.

Although significant progress has been made in the synthesis and applications of graphene-based transition metal oxide nanocomposites, many challenges need to be addressed while employing them on the industrial scale. The major challenge for graphene-based nanocomposites is the cost-effective production and their application on a large scale. Synthesis of nanocomposite with uniform properties is another major challenge in the development of practical applications of graphene-based transition metal oxide nanocomposites since reproducible results are highly imperative on the industrial scale.

Hence, extensive potential exists in the area of synthesis and application of graphene-based-transition metal oxide nanocomposites as a photocatalytic material on the industrial scale.

REFERENCES

Aawani, E., Memarian, N., Dizaji, H. R. (2019). Synthesis and characterization of reduced graphene oxide-V_2O_5 nanocomposite for enhanced photocatalytic activity under different types of irradiation. *Journal of Physics and Chemistry of Solids, 125(September 2018)*, 8–15.

Ahmed, S. N., Haider, W. (2018). Heterogeneous photocatalysis and its potential applications in water and wastewater treatment: A review. *Nanotechnology, 29(34)*, 342001.

Ahmedchekkat, F., Medjram, M. S., Chiha, M., Al-bsoul, A. M. A. (2011). Sonophotocatalytic degradation of Rhodamine B using a novel reactor geometry: Effect of operating conditions. *Chemical Engineering Journal, 178*, 244–251.

Ali, M. H. H., Al-Afify, A. D., Goher, M. E. (2018). Preparation and characterization of graphene—TiO_2 nanocomposite for enhanced photodegradation of Rhodamine-B dye. *Egyptian Journal of Aquatic Research, 44(4)*, 263–270.

Anjaneyulu, R. B., Mohan, B. S., Naidu, G. P., Muralikrishna, R. (2019). ZrO_2/Fe_2O_3/RGO nanocomposite: Good photocatalyst for dyes degradation. *Physica E: Low-Dimensional Systems and Nanostructures, 108(October 2018)*, 105–111.

Anju, S. G., Yesodharan, S., Yesodharan, E. P. (2012). Zinc oxide mediated sonophotocatalytic degradation of phenol in water. *Chemical Engineering Journal, 189–190*, 84–93.

Arabatzis, I. M., Stergiopoulos, T., Bernard, M. C., Labou, D., Neophytides, S. G., Falaras, P. (2003). Silver-modified titanium dioxide thin films for efficient photodegradation of methyl orange. *Applied Catalysis B: Environmental, 42*, 187–201.

Arshad, A., Iqbal, J., Ahmad, I., Israr, M. (2018). Graphene/Fe_3O_4 nanocomposite: Interplay between photo-Fenton type reaction, and carbon purity for the removal of methyl orange. *Ceramics International, 44(3)*, 2643–2648.

Arshad, A., Iqbal, J., Mansoor, Q. (2019). Graphene/Fe_3O_4 nanocomposite: Solar light driven Fenton like reaction for decontamination of water and inhibition of bacterial growth. *Applied Surface Science, 474*, 57–65.

Arshad, A., Iqbal, J., Siddiq, M., Ali, M. U., Ali, A., Shabbir, H., Saleem, M. S. (2017). Solar light triggered catalytic performance of graphene-CuO nanocomposite for waste water treatment. *Ceramics International, 43(14)*, 10654–10660.

Ashraf, M. A., Yang, Y., Fakhri, A. (2020). Synthesis of NiS–MoO_3 nanocomposites and decorated on graphene oxides for heterogeneous photocatalysis, antibacterial and antioxidant activities. *Ceramics International 46*(6), 8379–8384.

Atout, H., Álvarez, M. G., Chebli, D., Bouguettoucha, A., Tichit, D., Llorca, J., Medina, F. (2017). Enhanced photocatalytic degradation of methylene blue: Preparation of TiO_2/reduced graphene oxide nanocomposites by direct sol-gel and hydrothermal methods. *Materials Research Bulletin, 95*, 578–587.

Banerjee, S., Benjwal, P., Singh, M., Kar, K. K. (2018). Graphene oxide (rGO)-metal oxide (TiO_2/Fe_3O_4) based nanocomposites for the removal of methylene blue. *Applied Surface Science, 439*, 560–568.

Baruah, S., Dutta, J. (2009). Hydrothermal growth of ZnO nanostructures. *Science and Technology of Advanced Materials, 10*, 013001.

Beura, R., Thangadurai, P. (2017). Structural, optical and photocatalytic properties of graphene-ZnO nanocomposites for varied compositions. *Journal of Physics and Chemistry of Solids, 102*, 168–177.

Boruah, P. K., Szunerits, S., Boukherroub, R., Das, M. R. (2018). Magnetic Fe_3O_4@V_2O_5/rGO nanocomposite as a recyclable photocatalyst for dye molecules degradation under direct sunlight irradiation. *Chemosphere, 191*, 503–513.

Chen, J., Shi, J., Wang, X., Cui, H., Fu, M. (2013). Recent progress in the preparation and application of semiconductor/graphene composite photocatalysts. *Chinese Journal of Catalysis, 34*(4), 621–640.

Chen, W., Yan, L. (2010). Preparation of graphene by a low-temperature thermal reduction at atmosphere pressure. *Nanoscale, 2*, 559–563.

Cheng, G., Xu, F., Xiong, J., Wei, Y., Stadler, F. J., Chen, R. (2017). A novel protocol to design TiO-Fe_2O_3 hybrids with effective charge separation efficiency for improved photocatalysis. *Advanced Powder Technology, 28(2)*, 665–670.

Cheng, L., Wang, Y., Huang, D., Nguyen, T., Jiang, Y., Yu, H., Jiao, Z. (2015). Facile synthesis of size-tunable CuO/graphene composites and their high photocatalytic performance. *Materials Research Bulletin, 61*, 409–414.

Cheng, L., Zhang, S., Wang, Y., Ding, G., Jiao, Z. (2016). Ternary P25-graphene-Fe_3O_4 nanocomposite as a magnetically recyclable hybrid for photodegradation of dyes. *Materials Research Bulletin, 73*, 77–83.

Cheng, Z., Quan, X., Xiong, Y., Yang. L., Huang, Y. (2012). Synergistic degradation of methyl orange in an ultrasound intensified photocatalytic reactor. *Ultrasonics Sonochemistry, 19*, 1027–1032.

Dehghan, S., Jonidi, A., Farzadkia, M., Esra, A. (2019). Visible-light-driven photocatalytic degradation of Metalaxyl by reduced graphene oxide/Fe_3O_4/ZnO ternary nanohybrid: Influential factors, mechanism and toxicity bioassay. *Journal of Photochemistry & Photobiology A: Chemistry, 375*, 280–292.

Devipriya, S., Yesodharan, S. (2005). Photocatalytic degradation of pesticide contaminants in water. *Solar Energy Materials and Solar Cells, 86*, 309–348.

Du, X., Zhang, Z., Chen, H., Liang, P. (2020). Preparation of CeO_2 nanorods-reduced graphene oxide hybrid nanostructure with highly enhanced decolorization performance. *Applied Surface Science, 499*, 143939.

Dutta, V., Singh, P., Shandilya, P., Sharma, S., Raizada, P., Saini, A. K., Rahmani-Sani, A. (2019). Review on advances in photocatalytic water disinfection utilizing graphene and graphene derivatives-based nanocomposites. *Journal of Environmental Chemical Engineering, 7(3)*, 103–132.

Gogate, P. R. (2008). Treatment of wastewater streams containing phenolic compounds using hybrid techniques based on cavitation: A review of the current status and the way forward. *Ultrasonics Sonochemistry, 15*, 1–15.

Gupta, S. M., Tripathi, M. (2011). A review of TiO_2 nanoparticles. *Chinese Science Bulletin, 56*, 1639–1657.

Hayati, F., Isari, A. A., Fattahi, M., Anvaripour, B., Jorfi, S. (2018). Photocatalytic decontamination of phenol and petrochemical wastewater through ZnO/TiO_2 decorated on reduced graphene oxide nanocomposite: Influential operating factors, mechanism, and electrical energy consumption. *RSC Advances, 8(70)*, 40035–40053.

Henni, A., Harfouche, N., Karar, A., Zerrouki, D., Perrin, F. X., Rosei, F. (2019). Synthesis of graphene–ZnO nanocomposites by a one-step electrochemical deposition for efficient photocatalytic degradation of organic pollutant. *Solid State Sciences, 98*, 106039.

Hu, G., Tang, B. (2013). Photocatalytic mechanism of graphene/titanate nanotubes photocatalyst under visible-light irradiation. *Materials Chemistry and Physics, 138*(2–3), 608–614.

Hua, Z., Ma, W., Bai, X., Feng, R., Yu, L., Zhang, X., Dai, Z. (2014). Heterogeneous Fenton degradation of bisphenol A catalyzed by efficient adsorptive Fe_3O_4/GO nanocomposites. *Environmental Science and Pollution Research, 21*, 7737–7745.

Huang, K., Li, Y. H., Lin, S., Liang, C., Xu, X., Zhou, Y. F., Lei, M. (2014). One-step synthesis of reduced graphene oxide-CeO_2 nanocubes composites with enhanced photocatalytic activity. *Materials Letters, 124*, 223–226.

Hummers, W. S., Offeman, R. E. (1958). Preparation of graphitic oxide. *Journal of American Chemical Society, 80*(6), 1339–1342.

Kamalam, M. B. R., Inbanathan, S. S. R., Sethuraman, K. (2018). Enhanced photo catalytic activity of graphene oxide/MoO_3 nanocomposites in the degradation of Victoria Blue Dye under visible light irradiation. *Applied Surface Science, 449*, 685–696.

Khalid, N. R., Ahmed, E., Hong, Z., Ahmad, M., Zhang, Y., Khalid, S. (2013b). Cu-doped TiO_2 nanoparticles/graphene composites for efficient visible-light photocatalysis. *Ceramic International, 39*, 7107–7113.

Khalid, N. R., Ahmed, E., Hong, Z., Sana, L., Ahmed, M. (2013a). Enhanced photocatalytic activity of graphene-TiO_2 composite under visible light irradiation. *Current Applied Physics, 13(4)*, 659–663.

Kim, J., Yong, K. (2012). A facile, coverage controlled deposition of Au nanoparticles on ZnO nanorods by sonochemical reaction for enhancement of photocatalytic activity. *Journal of Nanoparticle Research, 14(8)*, 1–10.

Kumar, S., Kumar, A. (2017). Enhanced photocatalytic activity of rGO-CeO_2 nanocomposites driven by sunlight. *Materials Science and Engineering B: Solid-State Materials for Advanced Technology, 223*, 98–108.

Kumar, S., Ojha, A. K., Bhorolua, D., Das, J., Kumar, A., Hazarika, A. (2019). Facile synthesis of CuO nanowires and Cu_2O nanospheres grown on rGO surface and exploiting its photocatalytic, antibacterial and supercapacitive properties. *Physica B: Condensed Matter, 558*, 74–81.

Kuo, C. Y. (2009). Prevenient dye-degradation mechanisms using UV/TiO_2/carbon nanotubes process. *Journal of Hazardous Materials, 163(1)*, 239–244.

Kusiak-Nejman, E., Morawski, A. W. (2019). TiO_2/graphene-based nanocomposites for water treatment: A brief overview of charge carrier transfer, antimicrobial and photocatalytic performance. *Applied Catalysis B: Environmental, 253*, 179–186.

Lawal, A. T. (2019). Graphene-based nano composites and their applications: A review. *Biosensors and Bioelectronics, 141*, 111384.

Le, T. K., Kang, M., Tran, V. T., Kim, S. W. (2019). Relation of photoluminescence and sunlight photocatalytic activities of pure V_2O_5 nanohollows and V_2O_5/RGO nanocomposites. *Materials Science in Semiconductor Processing, 100(March)*, 159–166.

Lee, J. S., You, K. H., Park, C. B. (2012). Highly photoactive, low bandgap TiO_2 nanoparticles wrapped by graphene. *Advanced Materials, 24(8)*, 1084–1088.

Li, M-Fang, Liu, Y-Guo, Zeng, G-Ming, Liu, N., Liu, S-Bo. (2019). Graphene and graphene-based nanocomposites used for antibiotics removal in water treatment: A review. *Chemosphere, 226*, 360–380.

Li, W., Zeng, T. (2011). Preparation of TiO_2 anatase nanocrystals by $TiCl_4$ hydrolysis with additive H_2SO_4. *PloS One, 6(6)*, e21082.

Li, X. Z., Li, F. B., Yang, C. L., Ge, W. K. (2001). Photocatalytic activity of WO_x Photocatalytic activity of WO_x-TiO_2 under visible light irradiation. *Journal of Photochemistry and Photobiology A: Chemistry, 141*, 209–217.

Li, Y., Fan, Y., Chen, Y. (2002). A novel method for preparation of nanocrystalline rutile TiO_2 powders by liquid hydrolysis of $TiCl_4$. *Journal of Material Chemistry, 12*, 1387–1390.

Liu, S., Tian, J., Wang, L., Luo, Y., Sun, X. (2012). One-pot synthesis of CuO nanoflower-decorated reduced graphene oxide and its application to photocatalytic degradation of dyes. *Catalysis Science and Technology, 2*(2), 339–344.

Liu, W. M., Li, J., Zhang, H. Y. (2019). Reduced graphene oxide modified zinc oxide composites synergistic photocatalytic activity under visible light irradiation. *Optik, (August)*, 163778.

Madhavan, J., Grieser, F., Ashokkumar, M. (2010). Degradation of orange-G by advanced oxidation processes. *Ultrasonics Sonochemistry, 17*, 338–343.

Marcano, D. C., Kosynkin, D. V., Berlin, J. M., Sinitskii, A., Sun, Z. Z., Slesarev, A., Alemany, L. B., Lu, W., Tour, J. M. (2010). Improved synthesis of graphene oxide. *ACS Nano, 4*, 4806–4814.

McMurray, T. A., Dunlop, P. S. M., Byrne, J. A. (2006). The photocatalytic degradation of atrazine on nanoparticulate TiO_2 films. *Journal of Photochemistry and Photobiology A: Chemistry, 182*, 43–51.

Mijin, D., Savić, M., Snežana, P., Smiljanić, A., Glavaški, O., Jovanović, M., Petrović, S. (2009). A study of the photocatalytic degradation of metamitron in ZnO water suspensions. *Desalination, 249*, 286–292.

Mohamed, R. M. (2012). UV-assisted photocatalytic synthesis of TiO_2-reduced graphene oxide with enhanced photocatalytic activity in decomposition of Sarin in gas phase. *Desalination and Water Treatment, 50(1–3)*, 147–156.

Murali, A., Lan, Y. P., Sarswat, P. K., Free, M. L. (2019). Synthesis of CeO_2/reduced graphene oxide nanocomposite for electrochemical determination of ascorbic acid and dopamine and for photocatalytic applications. *Materials Today Chemistry, 12*, 222–232.

Nadimi, M., Ziarati Saravani, A., Aroon, M. A., Ebrahimian Pirbazari, A. (2019). Photodegradation of methylene blue by a ternary magnetic TiO_2/Fe_3O_4/graphene oxide nanocomposite under visible light. *Materials Chemistry and Physics, 225*, 464–474.

Neppolian, B., Ciceri, L., Bianchi, C. L., Grieser, F. (2011). Sonophotocatalytic degradation of 4-chlorophenol using Bi_2O_3/$TiZrO_4$ as a visible light responsive photocatalyst. *Ultrasonics Sonochemistry, 18*, 135–139.

Ng, Y. H., Lightcap, I. V., Goodwin, K., Matsumura, M., Kamat, P. V. (2010). To what extent do graphene scaffolds improve the photovoltaic and photocatalytic response of TiO_2 nanostructured films? *Journal of Physical Chemistry Letters, 1(15)*, 2222–2227.

Nguyen, V. Q., Baynosa, M. L., Nguyen, V. H., Tuma, D., Lee, Y. R., Shim, J. J. (2019). Solvent-driven morphology-controlled synthesis of highly efficient long-life ZnO/graphene nanocomposite photocatalysts for the practical degradation of organic wastewater under solar light. *Applied Surface Science, 486*, 37–51.

Nuengmatcha, P., Porrawatkul, P., Chanthai, S., Sricharoen, P., Limchoowong, N. (2019). Enhanced photocatalytic degradation of methylene blue using Fe_2O_3/graphene/CuO nanocomposites under visible light. *Journal of Environmental Chemical Engineering, 7(6)*, 103438.

Ong, C. B., Mohammad, A. W., Ng, L. Y., Mahmoudi, E., Azizkhani, S., Hayati Hairom, N. H. (2017). Solar photocatalytic and surface enhancement of ZnO/rGO nanocomposite: Degradation of perfluoro octanoic acid and dye. *Process Safety and Environmental Protection, 112*, 298–307.

Peik-See, T., Pandikumar, A., Ngee, L., Ming, H., Hua, C. C. (2014). Magnetically separable reduced graphene oxide/iron oxide nanocomposite materials for environmental remediation. *Catalysis Science and Technology, 4*, 4396–4405.

Priyadharsan, A., Vasanthakumar, V., Karthikeyan, S., Raj, V., Shanavas, S., Anbarasan, P. M. (2017). Multi-functional properties of ternary CeO_2/SnO_2/rGO nanocomposites: Visible light driven photocatalyst and heavy metal removal. *Journal of Photochemistry and Photobiology A: Chemistry, 346*, 32–45.

Qin, J., Zhang, X., Yang, C., Cao, M., Ma, M., Liu, R. (2017). ZnO microspheres-reduced graphene oxide nanocomposite for photocatalytic degradation of methylene blue dye. *Applied Surface Science, 392*, 196–203.

Raizada, P., Sudhaik, A., Singh, P. (2019). Photocatalytic water decontamination using graphene and ZnO coupled photocatalysts: A review. *Materials Science for Energy Technologies, 2(3)*, 509–525.

Ramadoss, A., Kim, S. J. (2013). Improved activity of a graphene-TiO_2 hybrid electrode in an electrochemical supercapacitor. *Carbon, 63*, 434–445.

Ranjith, R., Renganathan, V., Chen, S. M., Selvan, N. S., Rajam, P. S. (2019). Green synthesis of reduced graphene oxide supported TiO_2/Co_3O_4 nanocomposite for photocatalytic degradation of methylene blue and crystal violet. *Ceramics International, 45(10)*, 12926–12933.

Raut-Jadhav, S., Pinjari, D. V., Saini, D. R., Sonawane, S. H., Pandit, A. B. (2016). Ultrasonics sonochemistry intensification of degradation of methomyl (carbamate group pesticide) by using the combination of ultrasonic cavitation and process intensifying additives. *Ultrasonics Sonochemistry, 31*, 135–142.

Santos, M. S. F., Alves, A., Madeira, L. M. (2011). Paraquat removal from water by oxidation with Fenton's reagent. *Chemical Engineering Journal, 175*, 279–290.

Shanmugam, M., Alsalme, A., Alghamdi, A., Jayavel, R. (2015). Enhanced photocatalytic performance of the graphene-V_2O_5 nanocomposite in the degradation of methylene blue dye under direct sunlight. *ACS Applied Materials and Interfaces, 7(27)*, 14905–14911.

Sharma, A., Lee, B. K. (2016). Rapid photo-degradation of 2-chlorophenol under visible light irradiation using cobalt oxide-loaded TiO_2/reduced graphene oxide nanocomposite from aqueous media. *Journal of Environmental Management, 165*, 1–10.

Singh, P., Shandilya, P., Raizada, P., Sudhaik, A., Rahmani-Sani, A., Hosseini-Bandegharaei, A. (2020). Review on various strategies for enhancing photocatalytic activity of graphene based nanocomposites for water purification. *Arabian Journal of Chemistry 13*(1), 3498–3520.

Smith, A. T., LaChance, A. M., Zeng, S., Liu, B., Sun, L. (2019). Synthesis, properties, and applications of graphene oxide/reduced graphene oxide and their nanocomposites. *Nano Materials Science, 1(1)*, 31–47.

Stankovich, S., Piner, R. D., Nguyen, S. B. T., Ruoff, R. S. (2006). Synthesis and exfoliation of isocyanate-treated graphene oxide nanoplatelets. *Carbon, 44(15)*, 3342–3347.

Tayel, A., Ramadan, A. R., El Seoud, O. A. (2018). Titanium dioxide/graphene and titanium dioxide/graphene oxide nanocomposites: Synthesis, characterization and photocatalytic applications for water decontamination. *Catalysts, 8(11)*, 491.

Udayabhaskar, R., Suresh, R., Mangalaraja, R. V., Yáñez, J., Karthikeyan, B., Contreras, D. (2019). Unraveling the synergistic influences of graphene and CuO on the structural, photon and phonon properties of graphene: CuO nanocomposites. *Carbon, 152*, 766–776.

Vinodhkumar, G., Wilson, J., Inbanathan, S. S. R., Potheher, I. V., Ashokkumar, M., Peter, A. C. (2019). Solvothermal synthesis of magnetically separable reduced graphene oxide/ Fe_3O_4 hybrid nanocomposites with enhanced photocatalytic properties. *Physica B: Condensed Matter, 580*, 411752.

Wang, F., Zhang, K. (2011). Reduced graphene oxide-TiO_2 nanocomposite with high photocatalystic activity for the degradation of rhodamine B. *Journal of Molecular Catalysis A: Chemical, 345(1–2)*, 101–107.

Wang, L., Li, Z., Chen, J., Huang, Y., Zhang, H., Qiu, H. (2019). Enhanced photocatalytic degradation of methyl orange by porous graphene/ZnO nanocomposite. *Environmental Pollution, 249*, 801–811.

Wang, P., Wang, J., Wang, X., Yu, H., Yu, J., Lei, M., Wang, Y. (2013). One-step synthesis of easy-recycling TiO_2-rGO nanocomposite photocatalysts with enhanced photocatalytic activity. *Applied Catalysis B: Environmental, 132–133*, 452–459.

Wang, S., Shi, X., Shao, G., Duan, X., Yang, H., Wang, T. (2008). Preparation, characterization and photocatalytic activity of multi-walled carbon nanotube-supported tungsten trioxide composites. *Journal of Physics and Chemistry of Solids, 69(10)*, 2396–2400.

Wang, W. G., Yu, J. G., Xiang, Q. J., Cheng, B. (2012). Enhanced photocatalytic activity of hierarchical macro/mesoporous TiO_2–graphene composites for photodegradation of acetone in air. *Applied Catalysis B: Environmental, 119*, 109–116.

Wang, W., Serp, P., Kalck, P., Faria, J. L. (2005). Photocatalytic degradation of phenol on MWNT and titania composite catalysts prepared by a modified sol-gel method. *Applied Catalysis B: Environmental, 56(4)*, 305–312.

Wang, Y., Shi, R., Lin, J., Zhu, Y. (2010). Significant photocatalytic enhancement in methylene blue degradation of TiO_2 photocatalysts via graphene-like carbon in situ hybridization. *Applied Catalysis B: Environmental, 100*, 179–183.

Williams, G., Seger, B., Kamat, P. V. (2008). TiO_2-Graphene Nanocomposites: UV-assisted photocatalytic reduction of graphene oxide. *ACS Nano, 2*, 1487–1491.

Wojtoniszak, M., Chen, X., Kalenczua, R. J., Wajda, A., Łapczuk, J., Kurzewski, M., Drozdzik, M., Chu, P. K., Borowiak-Palen, E. (2012). Synthesis, dispersion and cytocompatibility of graphene oxide and reduced graphene oxide. *Colloid Surface B, 89*, 79–85.

Wu, J., Shen, X., Jiang, L., Wang, K., Chen, K. (2010). Solvothermal synthesis and characterization of sandwich-like graphene/ZnO nanocomposites. *Applied Surface Science, 256(9)*, 2826–2830.

Wunderlich, W., Oekermann, T., Miao, L. (2004). Electronic properties of nano-porous TiO_2-and ZnO-thin films-comparison of simulations and experiments. *The Journal of Ceramic Processing Research, 5*, 343–354.

Xiang, Q. J., Yu, J. G., Jaroniec, M. (2012). Graphene-based semiconductor photocatalysts. *Chemical Society Reviews, 41*, 782–796.

Xiong, Z., Zhang, L. L., Ma, J., Zhao, X. S. (2010). Photocatalytic degradation of dyes over graphene-gold nanocomposites under visible light irradiation. *Chemical Communications, 46*(33), 6099–6101.

Xu, Z. H., Yu, J. G. (2011). Visible-light-induced photoelectrochemical behaviors of Fe-modified TiO_2 nanotube arrays. *Nanoscale, 3*, 3138–3144.

Yang, J., Cui, S., Qiao, J. Q., Lian, H. Z. (2014). The photocatalytic dehalogenation of chlorophenols and bromophenols by cobalt doped nano TiO_2. *Journal of Molecular Catalysis A: Chemical, 395*, 42–51.

Yu, H., Irie, H., Hashimoto, K. (2010b). Conduction Band energy level control of titanium dioxide: Toward an efficient visible-light-sensitive photocatalyst. *Journal of American Chemical Society, 132*, 6898–6899.

Yu, H., Irie, H., Shimodaira, Y., Hosogi, Y., Kuroda, Y., Miyauchi, M., Hashimoto, K. (2010a). An efficient visible-light-sensitive Fe (III)-grafted TiO_2 photocatalyst. *Journal of Physical Chemistry, 114*, 16481–16487.

Yu, J. G., Xiong, J. F., Cheng, B., Liu, S. W. (2005). Fabrication and characterization of Ag-TiO_2 multiphase nanocomposite thin films with enhanced photocatalytic activity. *Applied Catalysis B: Environmental, 60*, 211–221.

Zhang, J., Zhu, Z., Tang, Y., Feng, X. (2013). Graphene encapsulated hollow TiO_2 nanospheres: Efficient synthesis and enhanced photocatalytic activity. *Journal of Materials Chemistry A, 1(11)*, 3752–3756.

Zhang, Y., Zhou, Z., Chen, T., Wang, H., Lu, W. (2014). Graphene TiO_2 nanocomposites with high photocatalytic activity for the degradation of sodium pentachlorophenol. *Journal of Environmental Sciences (China), 26(10)*, 2114–2122.

6 Metal Oxides for High-Performance Hydrogen Generation by Water Splitting

Sagar D. Balgude, Shrikant S. Barkade, and Satish P. Mardikar

CONTENTS

6.1 INTRODUCTION

Day-to-day increase in globalization has obscured that the exhaustion of fossil sources and environmental degradation have become severe obstacles to further social and economic growth. The interest for vitality requires the quest for sustainable and clean elective assets to enhance and in the end supplant our reliance on petroleum products [1]. Sun-powered vitality is a free, copious, and inexhaustible clean vitality and is a standout amongst the most encouraging choices to facilitate the vitality and natural strain. Subsequently, the reap and change of sun-powered vitality into a usable vitality by utilizing photocatalytic procedures is exceedingly attractive [2,3]. Meanwhile, in 1972, Fujishima and Honda reported TiO_2-mediated photoelectrochemical (PEC) and solar H_2 production. This pioneering report has turned into a promising strategy for solar-based clean and carbon-neutral substance vitality [4]. It has been recommended that the photocatalytic H_2 generation innovation procedure could be utilized for commercial application when the sun-based vitality transformation productivity is above 12% [5].

Key variables affecting solar fuel synthesis effectiveness include: (1) electron-hole separation, (2) light harvesting, and (3) the molecular diffusion of chemical reaction. In this way, the structure of proficient photocatalysts for creating subatomic H_2 from water part by gathering sun-based vitality is one of the principal challenges in the improvement of a sunlight-based hydrogen economy. In the previous 40 years, incredible endeavors have been made to incorporate new materials and improve the photocatalytic hypothesis [6–8]. In the previous 10 years, the subsidizing for clean vitality research has been expanding in China and from different governments around the globe, which has enormously advanced the improvement of photocatalytic hydrogen generation. Several semiconductor materials have been explored and employed as photocatalytic hydrogen generation materials. This chapter provides a quick insight into recent development of metal oxide photocatalysts for hydrogen production by means of H_2O splitting followed by a detailed discussion on a case study on Sn_3O_4-based nanophotocatalysts for hydrogen production.

6.2 MECHANISTIC ASPECTS OF WATER SPLITTING

Photocatalyst-mediated H_2O splitting for hydrogen production can be realized as a counterfeit photosynthesis process, since it resembles sunlight-mediated natural photosynthesis process in plants [9]. On the similar streak, the hydrogen generation from water could be accomplished by utilizing natural sunlight [10]. During this procedure, while chemical energy is obtained from sunlight, water part response elevates to develop free energy [11]. H_2 generation can be achieved by two techniques, which can be: (1) transformation of organics in the presence of a photocatalyst and (2) photocatalytic water production. In the first method, organic ingredients provide e^- and get oxidized to produce protons; however, in second method, water experiences a redox response with e^-/h^+. The proton particles are at last changed over to hydrogen by the connotation of electrons [12]. In general, the photocatalytic reaction framework for hydrogen production requires photocatalyst material, sunlight, a reactant, and a photoreactor [13]. Water itself or an ethanol/water mixture can be utilized as a reactant. For a photocatalyst, it ought to work with light, either UV or visible light.

For capable H_2 generation, a productive synergy between reactants, catalysts, and light is necessary [14]. The photocatalysis process commences with the absorption of

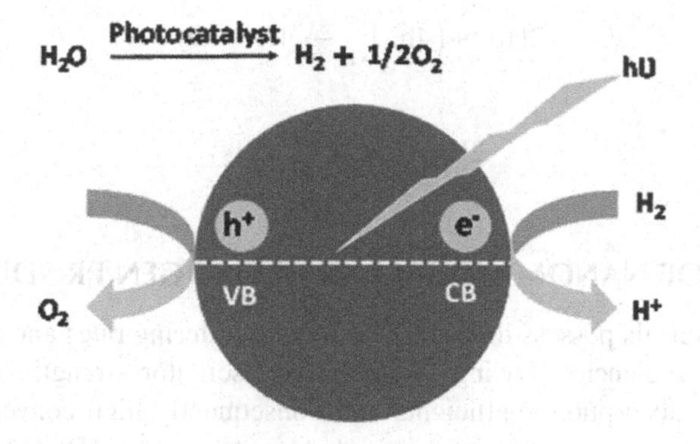

FIGURE 6.1 Schematic diagram of the mechanism for water splitting.

light energy greater than or equivalent to band energy of a semiconductor, separating from the occupied valence band (VB) and vacant conduction band (CB), energizing an e^- in VB legitimately into CB to the generation of an e^-/h^+ pair. The photoinduced electron and holes are involved in various redox process [15]. The mechanism of semiconductor-based hydrogen generation is shown in Figure 6.1. The photocatalysis include several reactions, which are as follows: (1) light absorption by semiconductor; (2) electron-hole pair generation; (3) recombination of generated electron-hole; (4) charge separation; (5) transition of photogenerated charges; (6) entrap of charges carriers; and (7) transport of charges to surface of adsorbed jots [16]. Initially, photocatalysis begin with illumination with light energy more than or equivalent to the band gap energy of a semiconductor. Generally, the semiconductor comprises a VB and a CB, isolated from each other by a specific energy. Photocatalyst by absorbing the light energy and generates the e^-/h^+ pairs [Eq. (6.1)]. In next step, these photoinduced electron/hole pairs get separated, and the electron gets transferred in the CB, generating holes in the VB. The photogenerated charge carriers (e^-/h^+ pairs) bring about a variety of redox reactions with water. For an oxidation reaction, the hole present in the VB oxidizes a water molecule to H^+ and generates an O_2 molecule [Eq. (6.2)], while the H^+ reacts with the electron present in conduction to form an H_2 molecule [Eq. (6.3)]. The overall process relies on band positions of the photocatalyst and levels of VBs and CBs. For generation of hydrogen, the CB with added negative potential and VB with more positive potential is obligatory requirement [17]. The suitable band gap (1.23 to 3.00 eV) of the photocatalyst is required to initiate the water-splitting process. However, the improper (lesser/grater) photon energy compared to the photocatalyst band gap acts as an activation barrier for the charge transfer process and thus hampers the overall process to significantly reduced rates. Additionally, the rate of backward reaction (i. e., materialization of hydrogen and oxygen to procedure water) must be strictly inhibited, and the photocatalyst material should also be sufficiently stable throughout the photocatalysis.

$$\text{Photocatalyst} + h\vartheta \rightarrow \left(h^+\right)_{VB} + \left(e^-\right)_{CB} \tag{6.1}$$

$$2H_2O + \left(4h^+\right)_{VB} \rightarrow 4H^+ + O_2 \quad (6.2)$$

$$2H^+ + \left(2e^-\right)_{CB} \rightarrow H_2 \quad (6.3)$$

6.3 ROLE OF NANOMATERIALS IN HYDROGEN PRODUCTION

Nanoscale materials possess drastically reduced scattering rates and increased carrier trapping efficiencies. Owing to increased oscillator strength, nanomaterials possess strong absorption coefficients and consequently high conversion efficiencies. The variation in size of nanomaterials Quantum dots (QD) can be monitored to fine-tune the band gap and thus the absorption in particular region of the spectrum. The requisite modifications can be brought about simply by the doping process. The facile bottom-up growth approach, comprising of utilization of smaller constituents to construct multifaceted structures, offers accessible creation of single crystalline nanomorphs under trivial reaction parameters. Nanoparticles (NPs) are the basic building components for fabricating photocatalysts since their corpus and charge transfer is fast. The NPs can be employed in the form of coating, suspensions, or dispersions in aqueous/solvent media to visualize amended photoactivities.

6.4 METAL OXIDES FOR HYDROGEN GENERATION VIA WATER SPLITTING

6.4.1 TITANIUM OXIDE (TiO_2)-MEDIATED H_2 GENERATION

TiO_2 is considered to be a benchmark for solar water splitting among the metal-oxide semiconductors. It exhibits three different stable crystal phases: (1) anatase phase, (2) Rutile form, and (3) Brookite phase. Anatase phase (with bad gap ~ 3.2 eV) is reflected to be the paramount active catalyst for photocatalysis. Rutile form with a narrow band gap (~3.0 eV) displays photoactivities while inferior to that of anatase. The Brookite phase with high purity has been developed recently and is yet to be inspected in detailed for its photocatalytic characteristics [18].

Presently, numerous research efforts have been devoted to synthesize various hierarchical nanomorphs of TiO_2, including nanowires, -rods, -fibers, -belts, and -tubes. Hierarchically nanostructured TiO_2 materials are expected to exhibit practical stability, high quantum efficiency, and effective light reaping, which are necessary for effective water splitting. Several synthesis methods adopted for fabricating TiO_2-based hybrid photocatalytic materials include soft/hard templated reactions, microwave acceleration, emulsion method (micro, mini, and reverse), hydrothermal processes, gamma irradiation, chemical precipitation, electrospinning, atomic layer deposition, and laser ablation. Wang et al. have reported Ni NP-anchored TiO_2 nanowire for PEC water splitting [19]. The investigation reveals that pristine and TiO_2 nanowire arrays both exhibited good absorption in the visible region upon application of Ni Atomic layer deposition (ALD). It was speculated that interparticle interaction of the Ni NPs resulted in an abundant concentration of charge carriers and consequently a higher water-splitting activity. Peng et al. reported Ag-decorated TiO_2 nanowires [20]. The water-splitting performance of Ag-decorated TiO_2 NWs was credited to the surface Plasmon resonance (SPR)-enthused transfer of electrons.

A simple hydrothermal method for preparation of a (Sn)–anion (C) co-doped TiO_2 nanowires for solar water splitting was reported by Aragaw et al. [21]. These TiO_2-based photoanodes demonstrated photocurrent density ~ 2.8 mA/cm^2 with an energy conversion rate of 1.32%. These improved results were regarded to reduced recombination rates due to cation-anion co-doping and prolonged recombination of photogenerated charge carriers. Amino group functionalized TiO_2 nanotubes synthesized by the diethylenetriamine (DETA)-mediated hydrothermal route have been reported by Hejazi et al. for water splitting [22]. Luo et al. reported Au/TiO_2 nanotube array-based photoanodes for visible-light-driven water splitting [23]. The synthesized plasmonic Au/TiO_2 nanotube arrays exhibited photocurrent density ~ 202 mA/cm^2, which was maximum for Au-TiO_2 systems. Higher PEC water-splitting performance was due to the meticulous interaction between Au and TiO_2 nanotubes.

In general, hydrogen-generation performance of TiO_2 is governed by charge separation rates from the CB of one polymorph to the CB of the other polymorph [24,25]. In this context, Mahadik et al. reported rutile TiO_2 nanorods modified with anatase nanograins on Fluorine-doped Tin Oxide (FTO) via the hydrothermal and dip-coating methods for dual application comprising hydrogen generation and dye degradation [26].

Recently, compared to a bulk semiconductor, synthesis of metal-semiconductor-carbon multicomponent heterojunctions have proven to be most successful schemes for enhancing the overall performance of the catalyst material. In this context, El-bery et al. have reported the assembly of a M/TiO_2/reduced graphene oxide (rGO) (M = Au or Pt) composite by the facile hydrothermal method [27]. The improved activity of the composite was ascribed to several factors including: (1) facile transport of e^-s from TiO_2 to graphene, (2) availability of extended photocatalytic reaction sites (TiO_2 as well as rGO), (3) formation of p-n heterojunction between rGO and TiO_2,(4) rGO-driven increased surface area, and (5) lastly, the presence of platinum serving as a basin for electrons, which prevented photogenerated charge recombination. The overall mechanism can be schematically represented as shown in Figure 6.2.

Visible light active and metal-free 2D g-C_3N_4 have also arose as potential photocatalysts. Alcudia-Ramos et al. have reported the hydrothermal method for fabrication of the g-C_3N_4-coupled TiO_2 heterojunction photocatalyst [28].

Besides fabrication of composites with carbon, employment of co-catalysts has also been comprehensively used for photocatalytic H_2 generation. In particular, copper oxide (CuO) has been broadly cast off for the hydrogen generation reaction. CuO can improve the catalytic performance by efficient separation of charge carriers. Guerrero-Araque et al. reported the electrochemical method for fabrication of CuO-decorated ZrO_2-TiO_2 (ZT) photocatalyst material for photocatalytic H_2 evolution [29]. Varied amount of CuO were employed for decorating the ZrO_2-TiO_2 photocatalyst. Among the synthesized samples, 1 wt. % CuO exhibited superior performance for H_2 evolution. Experimental results revealed that hydrogen-generation performance increased in parallel with CuO amount up to 1 wt%; besides which the decreased performance was observed. The observed performance was attributed to the extended

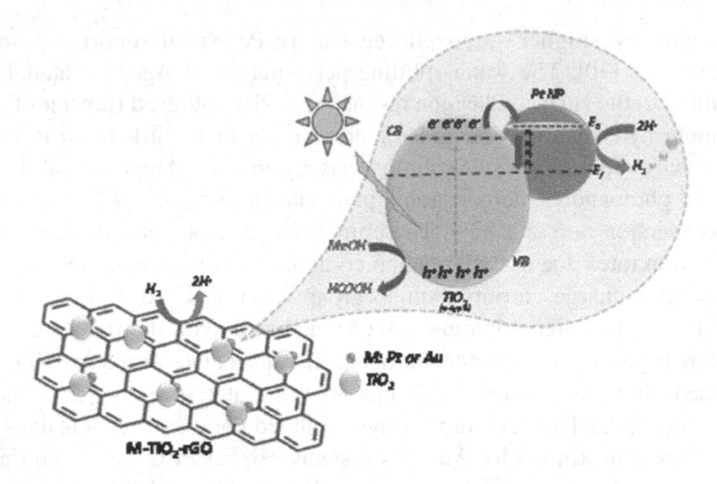

FIGURE 6.2 Representative schematic diagram showing H_2-splitting mechanism over M-TiO_2-rGO catalyst (M = Pt/Au).

lifetime of photogenerated charge carriers. The maximum H_2 production rate exhibited by sample was estimated to be 1877 mmol$h^{-1}g^{-1}$ compared to the pristine sample ZT (698 mmol$h^{-1}g^{-1}$). Polystyrene colloidal microspheres templated sol-gel synthesis of Pt/TiO_2-ZrO_2 composites were reported by An et al. [30]. The experimental results revealed that the Pt loading significantly affected the absorption of composite in visible region. The enhanced photocatalytic properties were attributed to prolonged carrier lifetime due to a larger work function and plasma resonance effect. The H_2 evolution rate of Pt loaded TiO_2-ZrO_2 composite was (7082.8 μmolg^{-1}), which was a 600-fold excess than that of TiO_2 (11.3 μmolg^{-1}) within 8 hr of the experiment. Thus, exhibiting various synthetic strategies for enhancing the hydrogen production rate of TiO_2 photocatalyst.

6.4.2 Zinc Oxide (ZnO)-Mediated H_2 Generation

ZnO-based nanostructures synthesized by different synthetic methods including various solution/vapor phase methods are utilized for many applications. However, the reports pertaining to the application of ZnO nanostructures for hydrogen production are comparatively less than that of TiO_2-based materials. Luévano-Hipólito et al. [31] reported sonochemical preparation of ZnO particles used as photocatalysts for hydrogen generation. The H_2 production rate was influenced by the physical properties of ZnO, including surface area and particle size. Nsib et al. [32] fabricated hybrid photocatalysts based on PANI-Ni-ZnO utilized for H_2 generation. The hydrogen generation activity of the Ni-ZnO/PANI photocatalyst was dependent on Ni content in the composite. Additionally, the extent of polyaniline (PANI) during hybridization of PANI-ZnO also affected the photocatalytic activity. The enhanced hydrogen generation activity was attributed to the chemical interaction between ZnO and PANI and increased electron density in ZnO due to the coupling of PANI.

Some reports suggest that efforts have also been devoted to upgrade the utilization of sunlight by extending the absorption of ZnO in the visible region [33,34]. These include metal doping, decoration by noble metals, and coupling with narrow band-gap semiconductors. Coupling with co-catalysts can lead to operative contact on the interface and consequently increased charge transfer/separation. For example, Chouhan et al. [35] reported cetyl trimethyl ammonium bromide (CTAB)-mediated hydrothermal synthesis of co-doped ZnO nanorods. Modification of pristine ZnO lead to the emergence of deformities in CdS/Co-ZnO nanorods compared to the ZnO and Co-ZnO nanorods. The 1.5% Pt/CdS/Co-ZnO nanorods demonstrated superior hydrogen-generation performance (~67.20 mmol/H_2 g).

Xi et al. [36] reported Pt/TiO_2–ZnO for water splitting. And 0.5 wt% Pt/TiO_2–ZnO catalyst (Ti/Zn = 10) demonstrated best hydrogen production rate (2150 $\mu molh^{-1}g^{-1}$). Incorporation of Pt increased the hydrogen yield about 10 times than that for pristine TiO_2-ZnO. The increased performance was attributed to Pt for facilitating electron-hole separation.

Despite the considerable hydrogen generation capacity of noble-metal-decorated composite photocatalysts, the rare abundance and costlier nature severely restricts their large-scale applications. To recoup these problems, noble metals were replaced with low-cost earth-abundant co-catalysts [37,38]. Cobalt (Co), nickel (Ni), and copper (Cu) are the most common co-catalysts reported in the literature. Incorporation of these metal co-catalysts can lead to formation of the Schottky barrier at the interface. Such kinds of junctions could stimulate charge separation. Moreover, transition metals are capable of reducing protons. Thus, increased H_2 generation activity could be expected upon employment of metal co-catalyst-loaded semiconductor photocatalysts. The schematic diagram showing the relationship between the co-catalyst loading amount and the photoactivity of the photocatalyst is depicted in Figure 6.3.

Recently, Madhusudan et al. [39] reported copper-hydroxide-adorned ZnO/ZnS hierarchical composite photocatalysts for H_2 generation. The charge migration distance can be shortened by attaching the $Cu(OH)_2$ on ZnO/ZnS composite, which further provided active sites and extended absorption in visible light. The composite

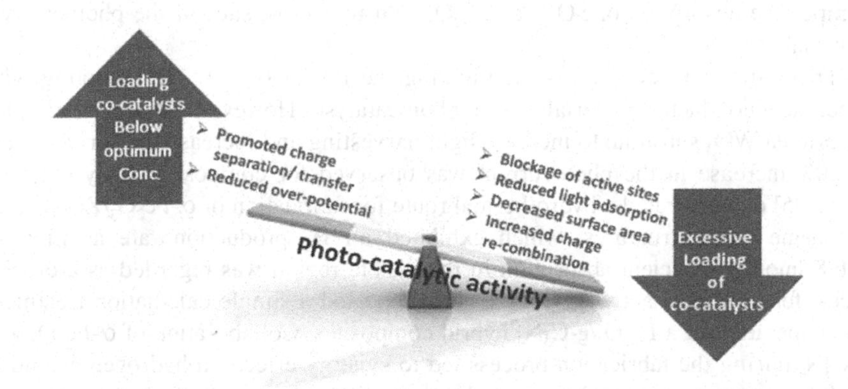

FIGURE 6.3 Schematic illustration presenting the relationship between the co-catalyst loading amount and the photocatalytic activity.

photocatalyst with optimum weight of $Cu(OH)_2$ along with ZnO/ZnS improved the H_2 generation rate more than pure $Zn_5(CO_3)_2(OH)_6$ and ZnO/ZnS by 46 and 13 times higher, respectively.

6.4.3 Iron Oxide (Fe_2O_3)-Mediated H_2 Generation

Hematite (band gap ~ 2.1 eV) possesses many characteristics, such as low cost, abundant availability, non-toxic nature, and ability to absorb in the visible region, which makes it a potential candidate for photoanode material for PEC water splitting [40]. The estimated theoretical water oxidation current density of α-Fe_2O_3 is 12.6 $mAcm^{-2}$ under AM 1.5G illumination. Still, moderately high peripheral bias is essential for water oxidation because of some drawbacks, such as high charge recombination, poor conductance, and deprived kinetics for water oxidation.

Guan's group described enhanced photocatalytic recital of flower-like Fe_2O_3 hierarchical structures than that of bulk Fe_2O_3 counterparts [41]. Utilization of biological materials as the template derived a great inspiration to obtain hierarchical structure. In this context, Zhu et al. [42] demonstrated fabrication of *Papilioparis butterfly* wings' templated Fe_2O_3/TiO_2 hierarchical quasi-honeycomb photocatalysts for H_2 production. H-Fe_2O_3/TiO_2 composites established a distinctly enhanced visible light-absorbing property due to distinct morphology, which enabled repeated reflection and absorption of light. Remarkably, the H_2 production rate of Fe_2O_3/TiO_2 hierarchical structures was 217.6 $lmolh^{-1}$; it was a 1.8-fold excess to that of Fe_2O_3/TiO_2. These results imply a positive synergistic effect of the composition and morphology for photocatalytic activity enhancement.

A stable photocatalyst with good efficiency comprehends its practical applications. Preethi et al. [43] reported magnetically separable Fe_2O_3 coupled with ZnS and CdS for H_2 production from industrial sulfide effluent. (CdS+ZnS)/Fe_2O_3 composite exhibited a relatively good H_2 production rate, which was a 12-fold excess than that of pristine Fe_2O_3 NPs. SO_3^{2-} and S^{2-} ions were employed in an aqueous solution as sacrificial reagents. The H_2 production rate was decreased parallel with increase in the sulfite ion concentration. The decreased H_2 production rate was attributed to competitive absorption of SO_3^{2-} and $S_2O_3^{2-}$ on the active sites of the photocatalyst material.

There are very few reports highlighting the use of α-Fe_2O_3 for coupling with other semiconductor materials as a photocatalyst. However, Sivula et al. [44] reported a WO_3 substrate to increase light harvesting and increase the surface area. A 20% increase in the photocurrent was observed for composite catalyst. Imran et al. [45] demonstrated a hydrothermal route for fabrication of α-$Fe_2O_3/Zn_{0.4}Cd_{0.6}S$ Z-scheme heterostructures, which exhibited an H_2 production rate as high as 536.8 $lmolh^{-1}$. Efficient absorption in the visible region was regarded as a crucial factor for improved activity. Li et al. [46] reported a simple calcination treatment for preparation of a-Fe_2O_3/g-C_3N_4 hybrid composites. Collaboration of α-Fe_2O_3 and g-C_3N_4 during the fabrication process led to synergy effects in hydrogen evolution performance compared to that of g-C_3N_4 alone. The enhanced light absorbance competence increased the surface area, and (110) faceted structures of α-Fe_2O_3 were responsible for the enhanced H_2 evolution performance. Shen et al. [47] reported

RGO-Cu_2O/Fe_2O_3 composites for H_2 production. The rGO played an important role for efficient transport of Fe_2O_3 conduction band electrons to Cu_2O valence band. Additionally, tetracycline was used to devour holes during H_2 production. As a result, the H_2 production capacity of G50-7/3 composite from the tetracycline (TC) solution was about an eightfold excess than that in pure water and five times advanced than that from the methanol solution.

6.4.4 Tungsten Oxide (WO_3)-Mediated H_2 Generation

Several parameters associated with WO_3 include a band gap ~2.5–2.7 eV, highly crystalline and porous nature, good solar-light absorption ability (~12%), temperate hole diffusion length, and virtuous chemical stability with cost-effective preparation method that offers novel opportunities for a development of photocatalyst material for H_2 generation [48–50].

Ye et al. [51] reported an interfacial seeding growth strategy for fabrication of a core-shell WO_3@$ZnIn_2S_4$ photocatalyst. The WO_3@$ZnIn_2S_4$ hybrid photocatalyst exhibited superior H_2 generation performance of 3900 mmolg^{-1}h^{-1} due to strong electronic interaction between two semiconductors. Spanu et al. [52] investigated Pt-WO_3-TiO_2-stacked structure as a specific platform for photocatalytic H_2 generation. The emphasis was given toward site-specific and successive decoration of W and Pt onto TiO_2 nanotubes. These highly ordered structures demonstrated muscularly amended photocatalytic hydrogen-generation efficiency than other countercatalyst system.

Hu et al. [53] reported a WO_3/CdS-diethylenetriamine (CdS-DETA) Z-scheme catalyst for hydrogen production. The CdS-DETA-based inorganic–organic offered abundant active sites and a large surface area. Among synthesized samples, 5% WO_3/CdS-DETA offered the highest H_2 evolution rate (15522 lmolg^{-1}h^{-1}), which was a twofold excess than that of pristine CdS-DETA. The observed improvement in photocatalytic H_2 evolution rate was attributed to improved separation of photogenerated e^-/h^+, restricted intrinsic photocorrosion associated with CdS and appropriate band gap between CdS-DETA and WO_3. Zhang et al. [54] utilized CdS@WO_3 photocatalyst. In a lactic acid aqueous solution as a sacrificial agent, the observed H_2 production rate was 736.89 mmol.

Recently, Song et al. [55] reported modification of ZnCdS with WO_3 for improved hydrogen evolution efficiency. A 35 wt % WO_3-loaded ZnCdS exhibited the hydrogen production activity of 98.68 mmol/mg, which was ~10 times higher than that of pure ZnCdS (10.28 mmol/mg).

He et al. [56] reported hydrothermal method for fabrication of a graphene-modified WO_3/TiO_2 heterojunction. The S-scheme heterojunction was fabricated by local adherence of TiO_2 and WO_3 NPs to rGO. Higher photocatalytic H_2 generation activities of the S-scheme heterostructure was attributed to TiO_2-WO_3 strong interactions at the surface and to rGO with abundant surface-active sites, which provided facile electron transfer through a unique photothermal effect in the composite.

Tahir et al. investigated a sol-gel photodeposition method for designing Au-decorated WO_3/TiO_2 for photocatalytic H_2 evolution from a glycerol/water mixture [57]. And 0.5% Au-WO_3/TiO_2 exhibited the highest hydrogen evolution efficiency of 17200 ppmh^{-1}gcat^{-1}, which was approximately 5 and 6 times higher than WO_3/

TiO_2 and TiO_2, respectively. The synergy between the WO_3/TiO_2 heterojunction attuned the required redox potential and embedded Au-NPs-enabled efficient charge separation via SPR which resulted in enhanced H_2 evolution performance of the composite.

6.4.5 Bismuth Vanadate ($BiVO_4$)-Mediated H_2 Generation

$BiVO_4$ (band gap ~ 2.4 eV) has attracted extensive attention as a visible-light-driven photocatalyst for water oxidation [58]. Abdi et al. have recently reported a solar-to-hydrogen efficiency of the W:$BiVO_4$ photoanode for water splitting to be ~4.9%. [59]. However, several research groups have endeavored to increase the efficiency of $BiVO_4$, particularly, through other semiconductors coupling. Zhou and co-workers [60] have recently reported a CdS NP-anchored $BiVO_4$ nanowire (CdS/$BiVO_4$ NWs) Z-scheme composite with broader light absorption for photocatalytic H_2 generation under visible light. CdS/$BiVO_4$ NWs in 1:2 wt. ratio demonstrated a twofold increase in water-splitting performance compared to bare CdS.

Veldurthi and coworkers [61] have reported a solution combustion method for fabrication of a co-catalyst-free Ag/$LaVO_4$/$BiVO_4$-based Z-schematic photocatalytic system for hydrogen production. Experiments were performed in the presence of Na_2S and Na_2SO_3 as hole foragers. The Ag/$LaVO_4$/$BiVO_4$ composite demonstrated a photocatalytic performance accounting to be 1.17 μmolh^{-1} compared to bare $LaVO_4$ and $BiVO_4$ both.

Zhu et al. [62] reported a $ZnIn_2S_4$-$BiVO_4$ Z scheme photocatalyst system for hydrogen generation. The synthesized composite demonstrated H_2 evolution performance in absence of a sacrificial agent, yet the performance was drastically improved in collaboration with graphene. 1.0La-$ZnIn_2S_4$/1.5RGO/1.0RuO_2/$BiVO_4$ (1:5) demonstrated the best H_2 evolution activity of 4.1 mmolg^{-1}h^{-1}.

6.4.6 Nanocarbon-Mediated H_2 Generation

Transition metal oxides or their NPs represent conventional photocatalytic materials. However, metal-free carbon nanomaterials could be a vital alternative to these catalysts due to their exclusive properties, such as abundant surface area, vide variety of morphology, nontoxic nature, and high stability. Nanocarbon-based materials offer excellent support as conductive materials in composite photocatalysts for photocatalytic H_2 production. One-dimensional carbon nanotubes (CNTs) as one of the allotropes of nanocarbon exhibit large specific surface area and good mechanical/electronic properties [63]. In particular, nanocylindrical morphology associated with multiwalled CNTs (MWCNTs) facilitates free movement of electrons due to the conjugated π-electron network throughout their structure [64]. Thus, a major role of photogenerated electron transfers to active sites during H_2-evolution can be anticipated by CNTs.

Reddy and coworkers [65] reported the hydrothermal method for fabrication of metal-free MWCNT-TiO_2 composites for photocatalytic hydrogen production. The calcination temperature during synthesis and MWCNT loading amount was monitored for optimizing the best H_2 production rate in a 5 vol% glycerol aqueous solution. The 0.1 wt% CNT-loaded TiO_2 (450°C/2 h) demonstrated an H_2 production rate of 8.8 mmolg^{-1}h^{-1}.

Yu and coworkers [66] reported hydrothermal synthesis of a MWCNT-modified $Cd_{0.1}Zn_{0.9}S$ composite visible-light photocatalyst. Modification of $Cd_{0.1}Zn_{0.9}S$ with a quantified amount of CNTs drastically improved the photocatalytic H_2-production performance. According to the study, 0.25 wt% CNT-$Cd_{0.1}Zn_{0.9}S$ showed the maximum rate of 1563.2 $mmolh^{-1}g^{-1}$ surpassing 3.3 times more than that of pristine $Cd_{0.1}Zn_{0.9}S$ without any noble metal co-catalyst. The leading influence of CNTs in CNT–$Cd_{0.1}Zn_{0.9}S$ composites is ascribed to their role as electron acceptors and transfer networks, which effect enhanced charge separation efficiency.

The effective electron-hole pair separation electron density contributing to the reduction reaction determines the overall photocatalytic performance. In this context, She and co-workers [67] reported a lucid synthetic route for fabrication of *n*-type semiconducting tenery composites based on carbon nanotubes, g-C_3N_4 nanosheets, and plasmonic Ag nanocubes. The improved photocatalytic H_2 evolution performance of the Ag-C_3N_4-CNTs composite was ascribed to increased electron density and unidirectional flow from plasmonic Ag nanocubes to g-C_3N_4 conduction to reactants.

Gopannagari and coworkers [68] have investigated the influence of surface functionalized MWCNT on H_2 evolution performance of CdS nanohybrids. The surface-active sites and electrochemical properties of CNTs were modified by acid functionalization. Binary composites comprising amine, sulfonic, and ascorbic acids functionalized CNTs and CdS nanorods and demonstrated significant development in photocatalytic H_2 production. However, Pt-decorated-Af-CNTs-CdS nanohybrids exhibited an H_2 production rate as high as 120.1 $mmolh^{-1}g^{-1}$, which was a 48-fold excess comparative to pure CdS.

Unlike CNTs, 2D graphene can also act as a provision for the growth/modification of other H_2-generation photocatalysts. Huo and coworkers [69] demonstrated in-situ photodeposition of CuO NPs onto hydrothermally synthesized 3D mesoporous graphene. In this study, Eosin Y was employed as a sensitizer. The H_2 evolution rate of the fabricated composite was 5.85 $mmolg^{-1}h^{-1}$ under optimal conditions, which was a 2.3-fold excess than that of CuO-rGO. The porous structure of the 3D graphene assembly functioned in a key role by efficiently improving the electron transfer.

Lv et al. [70] reported synthesis of novel 3D nest-like $LaCO_3OH$ and flower-like $Ni(OH)_2$@graphene (rGO) hierarchical composites for H_2 generation. During $Ni(OH)_2$@rGO nanostructure synthesis, the concentration of Ni^{2+} ions drifting from perovskite to graphene oxide (GO) was regulated using lattice-confined effect. GO served the role of both sacrificial carbon source and morphology regulator. The H_2 evolution rate of $LaCO_3OH$-$Ni(OH)_2$@rGO was found to 1.3807 $mmolh^{-1}g^{-1}$, which was 13 times higher than pure $LaCO_3OH$. The structurally improved light-absorbing ability of nest and flower-like morphology of $LaCO_3OH$-$Ni(OH)_2$@RGO promoted a rapid transfer of electrons and consequently improved H_2 evolution rates.

6.5 CHALLENGES FOR SEMICONDUCTOR PHOTOCATALYSTS

In recent decades, many researchers have studied semiconductor photocatalysis using a variety of nanostructured materials. However, practical applications are still limited. The major research focus is the development of a semiconductor that is capable of absorbing and converting the sunlight into chemical energy efficiently [71]. Basically, photocatalysis involves three processes: (1) surface transfer of photoinduced charge carriers, (2) excitation, and (3) bulk diffusion. Thus, a good photocatalyst must fulfil the requirement related to the band gap, decreased electron-hole recombination, crystallinity, and surface properties. However, it is difficult to address all these issues by means of a single photocatalytic material.

Various semiconductors such as silicones or sulfides of metals can absorb the sunlight effectively, but they are unable to drive the redox reaction and have less stability. In contrast, several metal oxides possess photochemical stability and exhibit strong redox potential; however, they cannot absorb sunlight competently. Thus, substantial challenges remain in devising highly efficient and stable photocatalysts [72].

Presently, oxides of tin (SnO, SnO_2, and Sn_3O_4) are the most widely explored photocatalytic materials owing to their high catalytic efficiencies, good stability, and relatively lower costs. However, the wide band gap of these oxides makes it inefficient for utilization of solar light. Many strategies have been employed for the improvement of the photocatalytic performance of tin oxide, which includes: (1) increasing the surface-to-volume ratio, (2) enhanced light harvesting, and (3) improving electron-hole separation. These can be achieved by controlling morphology, porosity, doping with metal and non-metals, and coupling tin oxides with other metal oxides. Even though the photocatalytical performance of tin oxide was enhanced by these strategies, they are usually achieved separately. Very few methods are used for integrating these strategies to obtain multifunctional photocatalysts with improved performance.

The emphasis of our investigation is to design tin oxide photocatalysts with multiple functionalities and capabilities to harvest maximum solar energy, which can provide the solution to hydrogen production. Our approach is to study the effects of various precursors and preparation routes on structural and morphological properties of tin oxides and their photocatalytical performances.

6.6 MIXED VALENCE Sn_3O_4 FOR HYDROGEN GENERATION UNDER VISIBLE LIGHT—CASE STUDY

6.6.1 Materials and Methods

1. **Synthesis of heterovalent Sn_3O_4:** Heterovalent Sn_3O_4 nanostructures were synthesized by facile hydrothermally using substrate stannous oxalate (SnC_2O_4) and malic acid as capping agent. Stannous oxalate (SnC_2O_4), malic acid, and sodium hydroxide were purchased from Merck. In a typical synthesis procedure, aqueous solutions of stannous oxalate (0.02M) and malic acid (0.023M) were admixed using a magnetic stirrer. To this solution, aq. NaOH solution (0.2M) was added within 30 min. After complete addition, the white suspension that was formed was stirred for 30 min and then hydrothermally treated at 180°C for 16 h. After treatment, the autoclave was cooled down naturally to room temperature. The yellow precipitate so formed was centrifuged and washed several times with deionized

water and ethanol and dried at 50°C in oven for 12 h. Finally, these dried heterovalent Sn_3O_4 nanostructures were then subjected to various physicochemical characterizations prior to hydrogen generation.

2. **Characterization:** The structural analysis and phase purity of the heterovalent Sn_3O_4 sample was examined by powder X-ray diffraction (XRD) analysis (model Rigaku Miniflex X-ray diffractometer) with Cu Kα irradiation at $\lambda = 1.5406$ Å. FESEM model JEOL-JSM 6700F was used for analyzing the morphology of the as-prepared sample. Raman spectra was recorded on a Jobin Yvon T64000 triple grating spectrometer equipped with a liquid-nitrogen-cooled charge-coupled device. UV-visible absorption spectra of the aqueous suspension were recorded on a JASCOV-570 spectrophotometer. The particle size distribution was done by using a PSS-NICOMP particle sizing system (Santa Barbara, California, USA).
3. **Photocatalytic hydrogen production**
   a. **From H_2O splitting:** The as-synthesized photocatalysts were tested for H_2 evaluation from water splitting. Typical experiments were performed in 250 mL round-bottom flasks containing 100 mL of double-distilled water and 25 mL of methanol as the sacrificial reagent. The dissolved oxygen was removed from the reaction mixture by purging argon. During experiments, 100 mg of catalyst was added to this solution. The photocatalytic H_2 evolution was studied under solar light irradiation. The reaction progress was monitored by gauging the evolved gas with the help of a eudiometer. The amount of gas evolved as a function of time was noted, and data was used for further calculations. The quantification of hydrogen gas evolved was analyzed by using gas chromatography operated using a 5 Å capillary column and a thermal conductivity detector (Model Shimadzu GC-2014, Porapak-Q packed column, TCD, N2-UHP carrier).
   b. **From H_2S splitting:** The cylindrical quartz photochemical thermostatic reactor was filled with 700 mL of 0.5 M aqueous KOH and purged with Ar for 1 h. H_2S was bubbled through the solution at a rate of 2.5 mLmin^{-1} at 298 K. H_2S was continuously fed into the system during the photoreduction. The 0.5 gm of as-prepared sample was introduced into the reactor and irradiated with normal solar light with constant stirring. The excess H_2S was trapped in the NaOH solution. The amount of hydrogen evolved was measured using a graduated glass burette.

6.6.2 Result and Discussion

XRD analysis was used to investigate the crystallinity of Sn_3O_4 nanostructures. The XRD pattern of as-prepared of Sn_3O_4 nanostructures is shown in Figure 6.4. The diffraction peaks observed at 2θ-values of 24.11°, 27.05°, 31.75°, 33.75°, 37.27°, 50.68°, and 51.72° can be attributed to lattice planes (101), (111), (–210), (–122), (130), (–311), and (–132), respectively. The broad and intense diffraction peaks indicate the

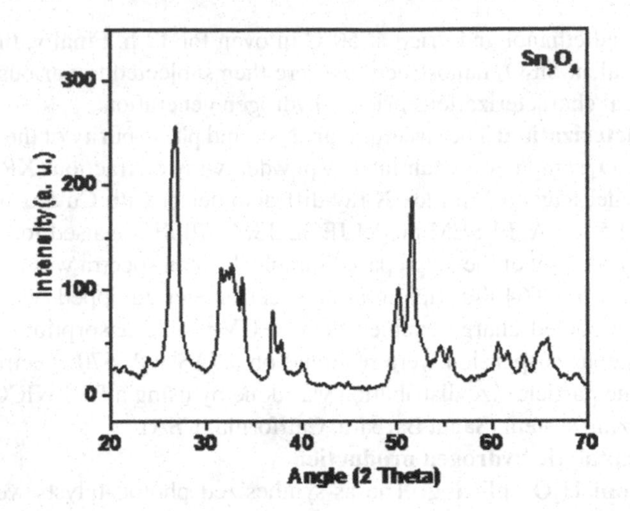

FIGURE 6.4 XRD spectra of as-synthesized Sn_3O_4 nanostructures.

formation of highly crystalline nanostructures having a triclinic phase. The experimental XRD pattern matches well with JCPDS 16-0737, revealing the successful formation of Sn_3O_4 [73], and the obtained XRD confirmed that there were no diffraction peaks obtained from impurities like SnO and SnO_2, indicating high purities of Sn_3O_4. The crystallite sizes of Sn_3O_4 nanostructures were estimated using Debye-Schrerrer's equation and were found to be approximately 40 nm [74].

Further, Raman scattering was performed for Sn_3O_4 nanostructures to confirm the formation and impurity detection. The Raman spectrum results are shown in Figure 6.5. From the figure it can be observed that there are two typical Raman bands located at 134 and 168 cm^{-1}, which correspond to the characteristic bands of

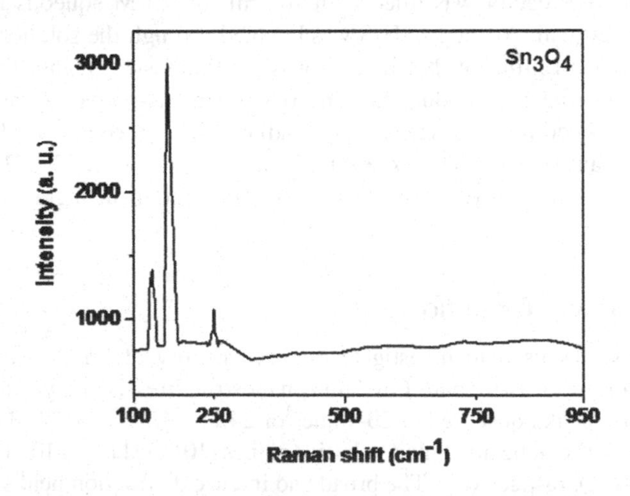

FIGURE 6.5 Raman spectra of as-synthesized Sn_3O_4 nanostructures.

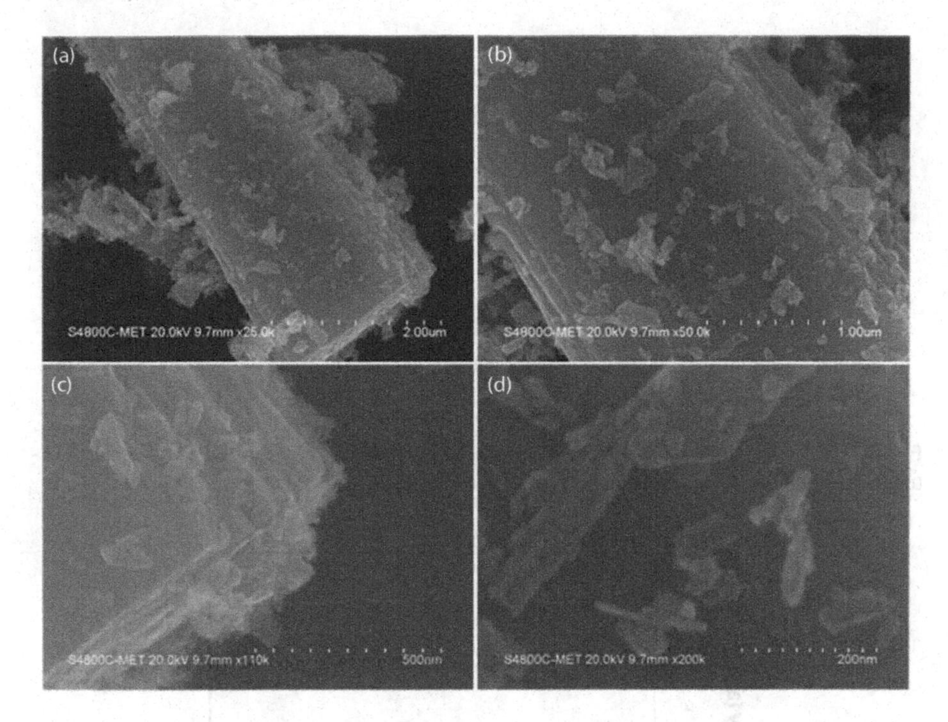

FIGURE 6.6 (a–d) FE-SEM images of as-synthesized Sn_3O_4 nanostructures at different Magnifications.

the triclinic phase Sn_3O_4 [75]. The peak located at 246 cm^{-1} may be due to oxidation of the sample and is in well agreement with what the literature reports [76]. No impurity bands of SnO_2 were centered at 460 and 769 cm^{-1} in recorded spectra Sn_3O_4, which clearly indicates good crystallinity of as-synthesized Sn_3O_4.

The surface morphology of the as-synthesized Sn_3O_4 nanostructures is shown in Figure 6.6. Figure 6.6a and b show a rod-like morphology of Sn_3O_4 nanostructures. The morphology is characterized by an assembly of small nanosheets of nearly uniform size, forming a rod-like morphology (Figure 6.6c). The low magnification image (Figure 6.6d) indicates that these rod-like nanostructures are decorated with some flat NPs with sizes 20–200 nm.

In malic-acid-assisted hydrothermal synthesis, the capping agent plays an important role for growth of rod-like Sn_3O_4. The growth of the rod-like nanostructure proceeds via. agglomeration of irregular counter-like Sn_3O_4 NPs and their consequent growth into a rod-shaped nanostructure. Initially, NPs are self-aligned under hydrothermal conditions and exhibit zero-dimensional growth. Therefore, we observe a rod-like structure of Sn_3O_4.

Figure 6.7a shows the UV-visible diffuse reflectance spectra of Sn_3O_4 nanostructures. The figure reveals that Sn_3O_4 exhibits a strong absorption in the UV-visible region. The absorption onset of Sn_3O_4 is observed at 462 nm, which can be credited to the triclinic phase of Sn_3O_4. The corresponding Tauc's plot of as-synthesized $Sn_3O_4$was plotted for clarity in the band gap (Figure 6.7b). The extrapolation of the

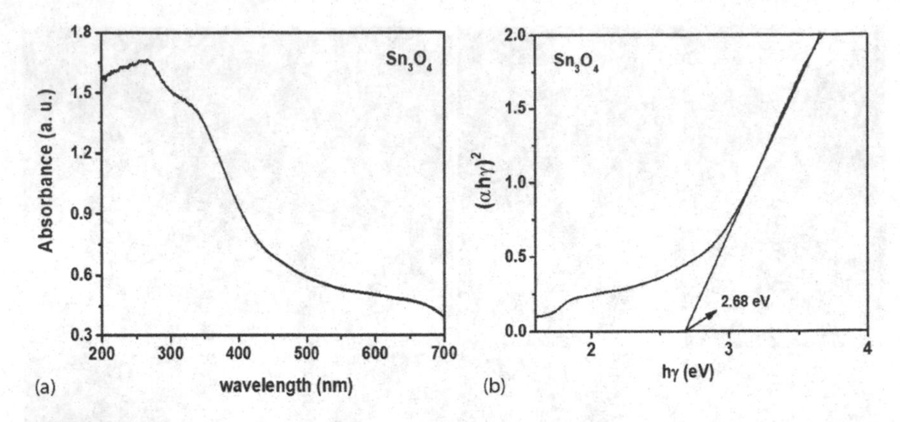

FIGURE 6.7 (a) UV-visible diffuse reflectance spectra of as-synthesized Sn_3O_4 nanostructures, and (b) corresponding Tauch's plot.

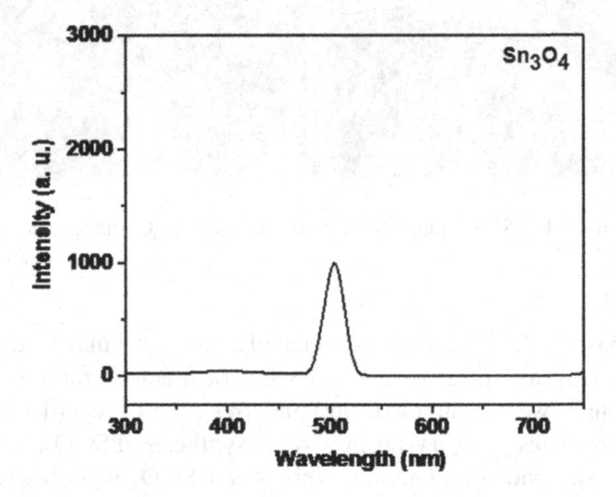

FIGURE 6.8 Photoluminescence (PL) spectra of as-synthesized Sn_3O_4 nanostructures.

fitting curve to the photon energy axis gives a good approximation of the energy band gap for the Sn_3O_4 as 2.68 eV. From the energy gap, it can be seen that the Sn_3O_4 shows good response to visible light. These results are in good agreement with reported literature [77].

Photoluminescence was performed to study electron-hole (e^- & h^+) pair recombination of the photocatalyst. Figure 6.8 shows photoluminescence (PL) spectra of as-synthesized Sn_3O_4. Generally, a greater PL emission peak intensity results from a greater recombination of photogenerated electron-hole pairs, whereas low-emission intensities show reduced recombination rates and hence the increased involvement of photoexcited charging carriers in the photoreaction [78]. The Sn_3O_4 nanostructures show the band-edge emission peak at 462 nm. The PL spectrum also shows a strong and dominating peak at ~508 nm for Sn_3O_4, which is attributed to the recombination of h^+ and e^- in the VB and CB. The lowering of PL intensity for Sn_3O_4 indicates that more excited electrons are trapped and transferred stably through the interface.

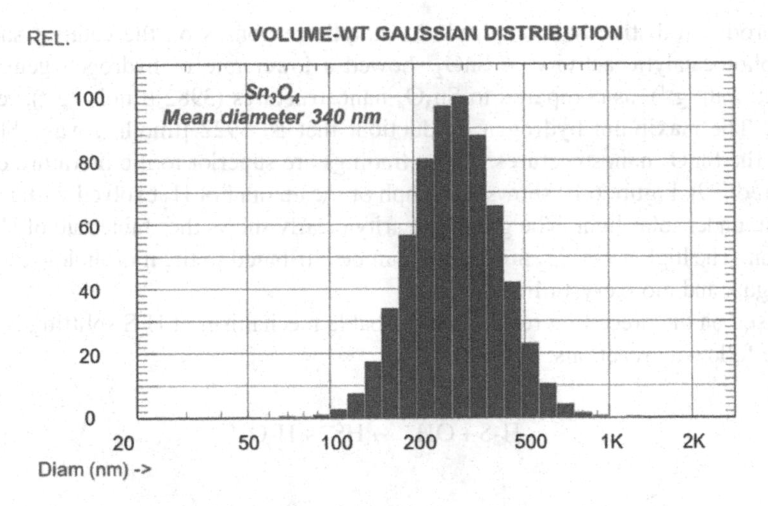

FIGURE 6.9 Particle-size distribution of as-synthesized Sn_3O_4 nanostructures.

The particle size distribution results of as-synthesized Sn_3O_4 is shown in Figure 6.9. From the figure, it can be observed that the calculated mean Gaussian diameter for Sn_3O_4 was 340 nm.

6.6.3 Photocatalytic Activity

As previously discussed, Sn_3O_4 nanostructures possess narrow band gaps in the visible region. Photocatalytic activity of Sn_3O_4 nanostructures were investigated because of their phenomenal response to solar light. Here we report the photocatalytic evolution of H_2 from H_2S and water splitting under solar light. Additionally, the water-splitting activity was also compared to SnO_2.

6.6.4 Photocatalytic H_2 Production via H_2S Splitting

The photocatalytic activity of the SnO_2 and Sn_3O_4 nanostructures for hydrogen evolution via H_2S splitting was performed. Table 6.1 shows the rate of photocatalytic hydrogen generation via H_2S splitting using SnO_2 and Sn_3O_4 under solar light. Control investigations demonstrated that no considerable hydrogen production was identified without the light of the photocatalyst, recommending that hydrogen

TABLE 6.1
Photocatalytic Hydrogen Generation via. H_2S Splitting

Catalyst	Hydrogen Evolution Rate (μmol/h/0.5 g)
SnO_2	891
Sn_3O_4	3922

was produced distinctly through photocatalytic responses on the catalyst surface. The photocatalytic activity of SnO_2 showed a lower rate of hydrogen generation (891 μmolh^{-1}g^{-1}), as compared to Sn_3O_4 nanostructures (3982 μmolh^{-1}g^{-1}), respectively. The maximum hydrogen production, that is, 3922 μmolh^{-1}, was obtained using the Sn_3O_4 nanostructures. These findings are superior to the literature earlier reported [79]. Figure 6.10 shows the graph of the amount of H_2 evolved with respect to time under solar light. The graph linearity clearly shows the stable rate of H_2 production. The higher rate H_2 production can be attributed to the morphology, narrow band gap, and more crystalline nature.

Based on the preceding results, the probable mechanism of H_2S splitting is given by the following reactions:

$$H_2S + OH^- \rightarrow HS^- + H_2O \tag{6.4}$$

$$Sn_3O_4\left(\text{photocatalyst}\right)\underline{h\vartheta}\left(h^+\right)_{VB} + \left(e^-\right)_{CB} \tag{6.5}$$

$$2HS^- + \left(2h^+\right)_{VB} \rightarrow 2H^+ + 2S \tag{6.6}$$

$$2H^+ + \left(2e^-\right)_{CB} \rightarrow H_2 \tag{6.7}$$

During H_2S splitting, Sn_3O_4 nanostructures absorb sunlight and produce e^-/h^+ pairs. H_2S is a diprotic acid, which dissociates under basic conditions and maintains the equilibrium with hydrosulfide (HS^-). The photogenerated e^-/h^+ pairs drive the

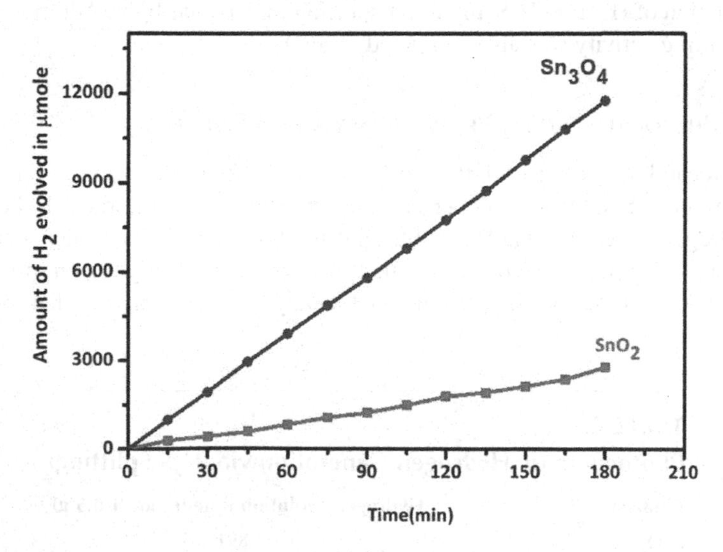

FIGURE 6.10 Graph of amount of H_2 evolved with respect to time under solar light in the presence of as-synthesized Sn_3O_4.

oxidation-reduction reaction. The e^- present in conduction band of Sn_3O_4 reduces protons to generate H_2, while the hole in the valence band of Sn_3O_4 oxidizes HS^- into elemental sulfur [80].

6.6.5 Photocatalytic H_2 Production via H_2O Splitting

Considering the good performance of the catalyst for hydrogen production under solar light, the photocatalytic activities of the Sn_3O_4 nanostructures for hydrogen evolution from water splitting under solar light irradiation have been investigated. Figure 6.11 represents the time-dependent graph of hydrogen evolution using SnO_2 and Sn_3O_4 nanostructures. The data for H_2 generation via water splitting is summarized in Table 6.2.

The hydrothermally synthesized rod-like Sn_3O_4 nanostructures showed utmost hydrogen generation, that is, 184.54 μmol/h/0.1 g, whereas SnO_2 showed 55.63 μmol/h/0.1 g, which is much lower than that of Sn_3O_4. A similar trend of hydrogen generation by means of H_2S splitting is seen on account of water splitting. The rate of evolution of H_2 due to water splitting is much lower than that of H_2S due to

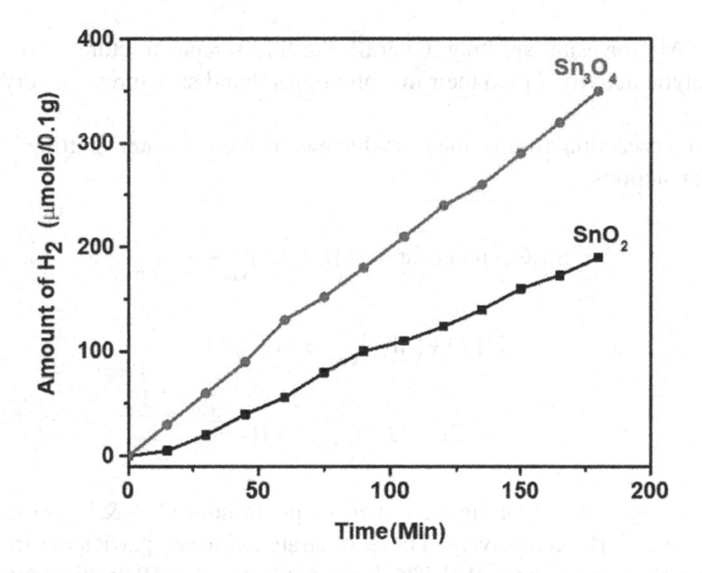

FIGURE 6.11 Graph of amount of H_2 evolved with respect to time under solar light in the presence of as-synthesized Sn_3O_4 and SnO_2.

TABLE 6.2
Photocatalytic Hydrogen Generation via H_2O Splitting

Catalyst	Hydrogen Evolution Rate (μmol/h/0.1 g)
SnO_2	55.63
Sn_3O_4	184.54

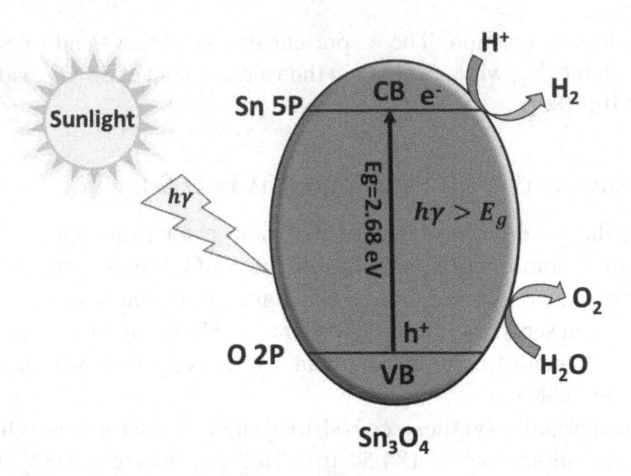

FIGURE 6.12 Schematic illustration of mechanism of water splitting over as-synthesized Sn_3O_4 nanostructures.

the grater ΔG for water splitting. Overall, the Sn_3O_4 nanostructures show superior photocatalytic activity due to their morphologies, band structures, and crystallinity of Sn_3O_4.

Based on preceding results, the probable mechanism of water splitting is given by following reactions:

$$Sn_3O_4\left(\text{photocatalyst}\right)\xrightarrow{h\vartheta}\left(h^+\right)_{VB}+\left(e^-\right)_{CB} \tag{6.8}$$

$$2H_2O+\left(4h^+\right)_{VB}\rightarrow 4H^+ + O_2 \tag{6.9}$$

$$2H^+ + \left(2e^-\right)_{CB}\rightarrow H_2 \tag{6.10}$$

During water splitting, upon light irradiation, photoinduced e^- & h^+ get transferred to the CB and VB, respectively. These separated charges participate in different oxidation-reduction reactions [81]. The holes present in the VB oxidize water to generate H^+ ions while electrons in the CB reduce H^+ to generate H_2. The schematic representation is shown in Figure 6.12.

6.7 CONCLUSIONS

In conclusion, the facile hydrothermal method has been demonstrated for synthesis of Sn_3O_4 nanostructures using stannous oxalate (SnC_2O_4) and malic acid. The synthesized samples have been characterized by several physicochemical characterization techniques. Further, the H_2 generation capabilities of synthesized Sn_3O_4-based photocatalysts were tested by water splitting and H_2S splitting. The excellent photocatalytic activity of the rod-like Sn_3O_4 can be ascribed to its photoabsorption ability due to the narrowing band gap, rod-like morphology, and good crystallinity.

6.8 FUTURE PERSPECTIVES

Although, a variety of nanostructured semiconductor metal-oxide photocatalysts have been utilized for solar water splitting, improper band gap, uncontrolled charge recombination, and poor absorption in the visible region strictly hinder their application. In this context, the coupled photocatalyst system based on TiO_2 have been exclusively studied. In order to meet the challenges, appropriate methodologies can be employed to develop the photocatalyst materials with respect to particle size morphology and chemical composition. Particularly, the water-splitting performance can be upgraded by fabricated heterostructured material assemblies with increased surface area, crystallinity, and specific faceted crystal structures. Doping as the most widely used methodology, along with decoration with metal/nonmetal NPs, metal oxides, metal sulfides, and different carbon allotropes, can be adopted for increasing the visible-light absorptivity of photocatalyst material. Recently, novel hierarchical photocatalyst materials have been explored for H_2 production. A comprehensive literature search on these methodologies for photocatalyst material development reveals that many inorganic oxide materials are yet to reconnoiter in relations of their photoactivity for H_2 generation. The efforts should be devoted to utilize these materials in combination with TiO_2 for anticipated high photocatalytic H_2 production efficiency. Particular emphasis should be given in manipulating the multifunctional characteristics of the resulting composite photocatalyst. It is therefore of utmost importance to employ multistep processes for designing photocatalyst material with high efficiencies. At the same time, it is also necessary to recognize the thermodynamic and kinetic aspects related to H_2 evolution mechanism. Profound understanding about the dynamic charge-transfer processes at the nanoscale photocatalyst surface during H_2 evolution is also highly anticipated. More research efforts should be focused on developing photocatalysts with advanced stabilities that can effectively absorb natural sunlight (>10%) for prolonged durations.

REFERENCES

1. Armaroli, N., Balzani, V. Towards an electricity-powered world. *Energy, Environ. Sci.*, 2011, 4, 3193–3222. Höök, M., Tang, X., 2013. Depletion of fossil fuels and anthropogenic climate change—A review. *Energy Policy*, 2013, 52, 797–809.
2. Acar, C., Dincer I. Comparative assessment of hydrogen production methods from renewable and non-renewable sources. *Int. J. Hydrogen Energy*, 2014, 39, 1–12.
3. Ni, M., Leung, M.K.H. Leung, D.Y.C., Sumathy, K. A review and recent developments in photocatalytic water-splitting Renew. *Sustain Energy Rev.*, 2007, 11, 401–425.
4. Fujishima, A., Honda, K. TiO_2 photoelectrochemistry and photocatalysis. *Nature*, 1972, 238, 37–38.
5. Chen, X.B., Shen, S.H., Guo, L.J., Mao, S.S. Semiconductor-based photocatalytic hydrogen generation. *Chem. Rev.*, 2010, 110, 6503–6570.
6. Hde, L., Rosales, B.S., Moreira, J., Valades-Pelayo, P. Efficiency factors in photocatalytic reactors: Quantum yield and photochemical thermodynamic efficiency factor. *Chem. Eng. Technol.*, 2015, 39, 51–65.

7. Zhao, Y., Hoivik, N., Wang, K. Recent advance on engineering titanium dioxide nanotubes for photochemical and photoelectrochemical water splitting. *Nano Energy*, 2016, 30, 728–744.
8. Goetzberger, A., Hebling, C., Schock, H.W. Photovoltaic materials. History, status and outlook. *Mater. Sci. Eng. R*, 2003, 40, 1–40.
9. Shi, N., Li, X., Fan, T., Zhou, H., Zhang, D., Zhu, H. Artificial chloroplast: Au/chloroplast-morph-TiO_2 with fast electron transfer and enhanced photocatalytic activity. *Int J. Hydrogen Energy*, 2014, 39, 5617–5624.
10. Clarizia, L., Spasiano, D., Di Somma, I., Marotta, R., Andreozzi, R., Dionysiou, D.D. Copper modified-TiO_2 catalysts for hydrogen generation through photoreforming of organics. A short review. *Int. J. Hydrogen Energy*, 2014, 39, 16812–16831.
11. Dincer, I., Acar, C. Review and evaluation of hydrogen production methods for better sustainability. *Int. J. Hydrogen, Energy*, 2015, 40, 11094–11111.
12. Kondarides, D.I., Daskalaki, V.M., Patsoura, A., Verykios, X.E. Hydrogen production by photo-induced reforming of biomass components and derivatives at ambient conditions. *Catal. Lett.*, 2007, 122, 26–32.
13. Tahir, M., Amin, N.S. Advances in visible light responsive titanium oxide-based photocatalysts for CO_2 conversion to hydrocarbon fuels. *Energy Convers. Manag.*, 2013, 76, 194–214.
14. Labidine Messaoudani, Z., Rigas, F., Hamid, M.D.B. Che Rosmani Che Hassan. Hazards, safety and knowledge gaps on hydrogen transmission via natural gas grid: A critical review. *Int. J. Hydrogen Energ.*, 41(39), 2016, 17511–17525.
15. Acar, C., Dincer, I., Zamfirescu, C. A review on selected heterogeneous photocatalysts for hydrogen production. *Int. J. Energy Res.*, 38, 2014, 1903–1920.
16. Linsebigler, A.L., Lu, G.Q. Photocatalysis on TiO_2 Surfaces: Principles, mechanisms, and selected results. *Chem. Rev.*, 95, 1995, 735–758; Kudo, A., Miseki, Y., Heterogeneous photocatalyst materials for water splitting. *Chem. Soc. Rev.*, 38, 2009, 253–278.
17. Balgude, S., Sethi, Y., Kale, B., Amalnerkar, D., Adhyapak, P. ZnO decorated Sn_3O_4 nanosheets nano-heterostructure: Stable photocatalyst for water splitting and dye degradation under natural sunlight. *RSC Adv.*, 9, 2019, 10289–10296.
18. Moss, B., Lim, K.K., Beltram, A., Moniz, S., Tang, J., Fornasiero, P., Barnes, P., Durrant, J., Kafizas, A. Comparing photoelectrochemical water oxidation, recombination kinetics and charge trapping in the three polymorphs of TiO_2. *Sci. Rep.*, 7, 2017, 2938.
19. Wang, C.C., Chou, P.-H., Yu, Y.-H., Kei, C.-C. Deposition of Ni nanoparticles on black TiO_2 nanowire arrays for photoelectrochemical water splitting by atomic layer deposition. *Electrochimica Acta*, 284, 2018, 211–219. doi:10.1016/j.electacta.2018.07.164.
20. Peng, C., Wang, W., Zhang, W., Liang, Y., Zhuo, L. Surface plasmon-driven photoelectrochemical water splitting of TiO_2 nanowires decorated with Ag nanoparticles under visible light illumination. *Appl. Surf. Sci.*, 420, 2017, 286–295. doi:10.1016/j.apsusc.2017.05.101.
21. Aragaw, B.A., Pan, C.-J., Su, W.-N., Chen, H.-M., Rick, J., Hwang, B.-J. Facile one-pot controlled synthesis of Sn and C codoped single crystal TiO_2 nanowire arrays for highly efficient photoelectrochemical water splitting. *Appl. Catal. B-Environ.*, 163, 2015, 478–486. doi:10.1016/j.apcatb.2014.08.027.
22. Hejazi, S., Nguyen, N.T., Mazare, A., Schmuki, P. Aminated TiO_2 nanotubes as a photoelectrochemical water splitting photoanode. *Catalysis Today*, 281, 2017, 189–197. doi:10.1016/j.cattod.2016.07.009.
23. Luo, J., Chen, J., Wang, H., Liu, H. Ligand-exchange assisted preparation of plasmonic Au/TiO_2 nanotube arrays photoanodes for visible-light-driven photoelectrochemical water splitting. *J. Power Sources*, 303, 2016, 287–293. doi:10.1016/j.jpowsour.2015.11.016.

24. Yao, H., Fu, W., Liu, L., Li, X., Ding, D., Su, P., Feng, S., Yang, H. Hierarchical photoanode of rutile TiO_2 nanorods coupled with anatase TiO_2 nanosheets array for photoelectrochemical application. *J. Alloys Compd.*, 680, 2016, 206–211.
25. Scanlon, D.O., Dunnill, C.W., Buckeridge, J., Shevlin, S.A., Logsdail, A.J., Woodley, S.M. et al., Band alignment of rutile and anatase. *Nat. Mater.*, 12, 2013, 798–801.
26. Mahadik, M.A., An, G.W., David, S., Choi, S.H., Cho, M., Jang, J.S. Fabrication of A/R-TiO_2 composite for enhanced photoelectrochemical performance: Solar hydrogen generation and dye degradation. *Appl. Surf. Sci.*, 426, 2017, 833–843. doi:10.1016/j.apsusc.2017.07.179.
27. El-Bery, H.M., Matsushita, Y., Abdel-moneim, A. Fabrication of efficient TiO_2–RGO heterojunction composites for hydrogen generation via water-splitting: Comparison between RGO, Au and Pt reduction sites. *Appl. Surf. Sci.*, 423, 2017, 185–196. doi:10.1016/j.apsusc.2017.06.130.
28. Alcudia-Ramos, M.A., Fuentez-Torres, M.O., Ortiz-Chi, F., Espinosa-González, C.G., Hernández-Como, N., García-Zaleta, D.S., Godavarthi, S. Fabrication of g-C_3N_4/TiO_2 heterojunction composite for enhanced photocatalytic hydrogen production. *Ceram. Int.*, 2019. doi:10.1016/j.ceramint.2019.08.228.
29. Guerrero-Araque, D., Acevedo-Pena, P., Ramırez-Ortega, D., Calderon, H.A., Gomez, R. Charge transfer processes involved in photocatalytic hydrogen production over CuO/ZrO_2-TiO_2 materials. *Int. J. Hydrogen Energy*, 42, 2017, 9744e97. 53
30. An, M., Li, L., Cao, Y., Ma, F., Liu, D., Gu, F. Coral reef-like Pt/TiO_2–ZrO_2 porous composites for enhanced photocatalytic hydrogen production performance. *Mol. Catal.*, 475, 2019, 110482. doi:10.1016/j.mcat.2019.110482.
31. Luévano-Hipólito, E., Torres-Martínez, L.M. Sonochemical synthesis of ZnO nanoparticles and its use as photocatalyst in H2 generation. *Mater. Sci. Eng. B*, 226, 2017, 223e33.
32. Nsib, M.F., Saafi, S., Rayes, A., Moussa, N., Houas, A. Enhanced photocatalytic performance of Ni–ZnO/Polyaniline composite for the visible-light driven hydrogen generation. *J. Energy Inst.*, 89(4), 2016, 694–703. doi:10.1016/j.joei.2015.05.001.
33. Hsu, M.H., Chang, C.J., Weng, H.T. Efficient H2 production using Ag_2S-coupled ZnO@ZnS core–shell nanorods decorated metal wire mesh as an immobilized hierarchical photocatalyst. *ACS Sustain. Chem. Eng.*, 4, 2016, 1381–1391.
34. Lonkar, S.P., Pillai, V.V., Alhassan, S.M. Facile and scalable production of heterostructured ZnS-ZnO/Graphene nano-photocatalysts for environmental remediation. *Sci. Rep.*, 8, 2018, 13401.
35. Chouhan, N., Ameta, R., Meena, R.K., Mandawat, N., Ghildiyal, R. Visible light harvesting Pt/CdS/Co-doped ZnO nanorods molecular device for hydrogen generation. *Int. J. Hydrogen Energ.*, 41(4), 2016, 2298–2306. doi:10.1016/j.ijhydene.2015.11.019.
36. Xie, M.-Y., Su, K.-Y., Peng, X.-Y., Wu, R.-J., Chavali, M., Chang, W.-C. Hydrogen production by photocatalytic water-splitting on Pt-doped TiO_2–ZnO under visible light. *J. Taiwan. Inst. Chem. E*, 70, 2017, 161–167. doi:10.1016/j.jtice.2016.10.034.
37. Ran, J.R., Zhang, J., Yu, J.G., Jaroniec, M., Qiao, S.Z. Earth-abundant cocatalysts for semiconductor based photocatalytic water splitting. *Chem. Soc. Rev.* 43, 2014, 7787–7812.
38. Ran, J.R., Zhang, J., Yu, J.G., Qiao, S.Z. Enhanced visible-light photocatalytic H2 production by ZnxCd1–xS modified with earth-abundant nickel-based cocatalysts. *ChemSusChem* 7, 2014, 3426–3434.
39. Madhusudan, P. Wang, Y., Chandrashekar, B.N., Wang, W., Wang, J., Miao, J., Shi, R., Liang, Y., Mi, G., Cheng, C. Nature inspired ZnO/ZnS nanobranch-like composites, decorated with Cu(OH)2 clusters for enhanced visible-light photocatalytic hydrogen evolution. *Appl. Catal. B Environ.* 253, 2019, 379–390. https://doi.org/10.1016/j.apcatb.2019.04.008.

40. Sivula, K., Le Formal, F., Grätzel, M. Solar water splitting: Progress using hematite (α-Fe_2O_3) photoelectrodes. *ChemSusChem*, 2011, 4, 432–439.
41. Guan, J., Xiao, Z., Tong, G., Mou, F., Fan, X., Flower-like porous hematite nanoarchitectures achieved by complexation–mediated oxidation–hydrolysis reaction, X. Huang. *Colloid Interface Sci.*, 357, 2011, 36–45.
42. Zhu, S., Yao, F., Yin, C., Li, Y., Peng, W., Ma, J., Zhang, D. Fe_2O_3/TiO_2 photocatalyst of hierarchical structure for H2 production from water under visible light irradiation. *Micropor. Mesopor. Mat.*, 190, 2014, 10–16. doi:10.1016/j.micromeso.2014.01.018.
43. Preethi, V., Kanmani, S. Photocatalytic hydrogen production using Fe_2O_3–based core shell nano particles with ZnS and CdS. *Int. J. Hydrogen Energ.*, 39(4), 2014, 1613–1622. doi:10.1016/j.ijhydene.2013.11.029.
44. Sivula, K., Formal, F.L., Grätzel, M. WO_3–Fe_2O_3 Photoanodes for water splitting: A host scaffold, guest absorber approach. *Chem. Mater.*, 21(13), 2009, 2862–2867. doi:10.1021/cm900565a.
45. Imran, M., Yousaf, A.B., Kasak, P., Zeb, A., Zaidi, S.J. Highly efficient sustainable photocatalytic Z-scheme hydrogen production from an α-Fe_2O_3 engineered ZnCdS heterostructure. *J. Catal.*, 353, 2017, 81–88. doi:10.1016/j.jcat.2017.06.019.
46. Li, Y., Li, F., Wang, X., Zhao, J., Wei, J., Hao, Y., Liu, Y. Z-scheme electronic transfer of quantum-sized α-Fe_2O_3 modified g-C3N4 hybrids for enhanced photocatalytic hydrogen production. *Int J Hydrogen Energ.*, 42(47), 2017, 28327–28336. doi:10.1016/j.ijhydene.2017.09.137.
47. Shen, H., Liu, G., Yan, X., Jiang, J., Hong, Y., Yan, M., et al. All-solid-state Z-scheme system of RGO-Cu_2O/Fe_2O_3 for simultaneous hydrogen production and tetracycline degradation. *Mater. Today Energy*, 5, 2017, 312–319. doi:10.1016/j.mtener.2017.07.008.
48. Yang, L., Jie, L., Wenzhang, L., Haizhou, H., Yahui, Y., Yaomin, L. et al. Electrochemical doping induced in situ Homo-species for enhanced photoelectrochemical performance on WO_3 nanoparticles film photoelectrodes. *Electrochim Acta*, 210, 2016, 251e60.
49. Wenzhang, L., Faqi, Z., Jie, L., Canjun, L., Yanghui, Y., Yaomin, L. et al. Enhancing photoelectrochemical water splitting by aluminum-doped plate-like WO_3 electrodes. *Electrochim. Acta*, 160, 2015, 57e63.
50. Fàbrega, C., Murcia-López, S., Monllor-Satoca, D., Prades, J.D., Hernández-Alonso, M.D., Penelas, G., et al. Efficient WO_3 photoanodes fabricated by pulsed laser deposition for photoelectrochemical water splitting with high faradaic efficiency. *Appl. Catal. B*, 189, 2016, 133e40.
51. Ye, L., Wen, Z. $ZnIn_2S_4$ nanosheets decorating WO_3 nanorods core-shell hybrids for boosting visible-light photocatalysis hydrogen generation. *Int. J. Hydrogen Energ.*, 2019. doi:10.1016/j.ijhydene.2018.12.093.
52. Spanu, D., Recchia, S., Mohajernia, S., Schmuki, P., Altomare, M. Site-selective Pt dewetting on WO_3–coated TiO_2 nanotube arrays: An electron transfer cascade-based H_2 evolution photocatalyst. *Appl. Catal. B-Environ.*, 237, 2018, 198–205. doi:10.1016/j.apcatb.2018.05.061.
53. Hu, T., Li, P., Zhang, J., Liang, C., Dai, K. Highly efficient direct Z-scheme WO3 / CdS-diethylenetriamine photocatalyst and its enhanced photocatalytic H_2 evolution under visible light irradiation. *Appl. Surf. Sci.*, 442, 2018, 20–29. doi:10.1016/j.apsusc.2018.02.146.
54. Zhang, Y., Hao, X., Ma, X., Liu, H., Jin, Z. Special Z-scheme CdS@WO_3 heterojunction modified with CoP for efficient hydrogen evolution. *Int. J. Hydrogen Energ.*, 2019. doi:10.1016/j.ijhydene.2019.03.168.

55. Song, L., liu, D., Zhang, S., Wei, J. WO_3 cocatalyst improves hydrogen evolution capacity of ZnCdS under visible light irradiation. *Int. J. Hydrogen Energ.*, 2019. doi:10.1016/j.ijhydene.2019.04.284.
56. He, F., Meng, A., Cheng, B., Ho, W., Yu, J. Enhanced photocatalytic H_2–production activity of WO_3/TiO_2 step-scheme heterojunction by graphene modification. *Chin. J. Catal.*, 41, 2020, 9–20. doi:10.1016/s1872-2067(19)63382-6.
57. Tahir, M., Siraj, M., Tahir, B., Umer, M., Alias, H., Othman, N. Au-NPs embedded Z–scheme WO_3/TiO_2 nanocomposite for plasmon-assisted photocatalytic glycerol-water reforming towards enhanced H_2 evolution. *Appl. Surf. Sci.*, 2019, 144344. doi:10.1016/j.apsusc.2019.144344
58. Kudo, A., Ueda, K., Kato, H., Mikami, I. Photocatalytic O_2 evolution under visible light irradiation on BiVO 4 in aqueous $AgNO_3$ solution. *Catal. Lett.*, 53, 1998, 229–230.
59. Abdi, F.F., Han, L., Smets, A.H.M., Zeman, M., Dam, B., van de Krol, R. Efficient solar water splitting by enhanced charge separation in a bismuth vanadate-silicon tandem photoelectrode. *Nat. Commun.*, 4(2195), 2013, 1–7.
60. Zhou, F.Q., Fan, J.C., Xu, Q.J., Min, Y.L. $BiVO_4$ nanowires decorated with CdS nanoparticles as Z-scheme photocatalyst with enhanced H_2 generation. *Appl. Catal. B-Environ.*, 201, 2017, 77–83. doi:10.1016/j.apcatb.2016.08.027.
61. Veldurthi, N.K., Eswar, N.K., Singh, S.A., Madras, G. Cocatalyst free Z-schematic enhanced H_2 evolution over LaVO4 /BiVO4 composite photocatalyst using Ag as an electron mediator. *Appl. Catal. B-Environ.*, 220, 2018, 512–523. doi:10.1016/j.apcatb.2017.08.082.
62. Zhu, R., Tian, F., Cao, G., Ouyang, F. Construction of Z scheme system of ZnIn2S4 / RGO/BiVO4 and its performance for hydrogen generation under visible light. *Int. J. Hydrogen Energy*, 42(27), 2017, 17350–17361. doi:10.1016/j.ijhydene.2017.02.091
63. Iijima, S. Helical microtubules of graphitic carbon. *Nature*, 354, 1991, 56.
64. Ma, L.L., Sun, H.Z., Zhang, Y.G., Lin, Y.L., Li, J.T., Wang, E., Yu, Y., Tan, M., Wang, J.B. Preparation, characterization and photocatalytic properties of CdS nanoparticles dotted on the surface of carbon nanotubes. *Nanotechnology*, 2008, 19, 115709; Liu, X.J., Zeng, P., Peng, T.Y., Zhang, X.H., Deng, K.J. Preparation of multiwalled carbon nanotubes/Cd0.8Zn0.2S nanocomposite and its photocatalytic hydrogen production under visible-light. *Int. J. Hydrogen Energy*, 2012, 37, 1375.
65. Reddy, N.R., Kumari, M.M., Cheralathan, K.K., Shankar, M.V. Enhanced photocatalytic hydrogen production activity of noble metal free MWCNT-TiO_2 nanocomposites. *Int. J. Hydrogen Energy*, 43(8), 2018, 4036–4043. doi:10.1016/j.ijhydene.2018.01.011.
66. Yu, J., Yang, B., Cheng, B. Noble-metal-free carbon nanotube-$Cd_{0.1}Zn_{0.9}S$ composites for high visible-light photocatalytic H_2–production performance. *Nanoscale*, 4, 2012, 2670–2677. doi:10.1039/C2NR30129F.
67. She, X., Wu, J., Xu, H., Mo, Z., Lian, J., Song, Y., Liu Daolin Du, L., Li, H., Enhancing charge density and steering charge unidirectional flow in 2D non-metallic semiconductor-CNTs-metal coupled photocatalyst for solar energy conversion. *Appl. Catal. B-Environ.*, 202, 2017, 112–117. doi:10.1016/j.apcatb.2016.09.013.
68. Gopannagari, M., Kumar, D.P., Park, H., Kim, E.H., Bhavani, P., Reddy, D.A., Kim, T.K. Influence of surface-functionalized multi-walled carbon nanotubes on CdS nanohybrids for effective photocatalytic hydrogen production. *Appl. Catal. B-Environ.*, 236, 2018, 294–303. doi:10.1016/j.apcatb.2018.05.009.
69. Huo, J., Liu, X., Li, X., Qin, L., Kang, S.-Z. An efficient photocatalytic system containing Eosin Y, 3D mesoporous graphene assembly and CuO for visible-light-driven H_2 evolution from water. *Int. J. Hydrogen Energ.*, 42(23), 2017, 15540–15550. doi:10.1016/j.ijhydene.2017.05.033.

70. Lv, T., Xu, Z., Hong, W., Li, G., Li, Y., Jia, L. Graphene oxide mediated self-sacrificial synthesis of $LaCO_3OH$-$Ni(OH)_2$@graphene hierarchical composite for photocatalytic H2 evolution and supercapacitor. *Chem. Eng. Trans.*, 123021, 2019. doi:10.1016/j.cej.2019.123021.
71. Li, C., Wang, H., Naghadeh, S.B., Zhang, J.Z., Fang, P. Visible light driven hydrogen evolution by photocatalytic reforming of lignin and lactic acid using one-dimensional NiS/CdS nanostructures. *Appl. Catal. B-Environ.*, 2018, 227, 229–239.
72. Zhou, N., Wang, Y., Zhou, Y., Shen, J., Zhou, Y., Yang, Y., Star-shaped multi-arm polymeric ionic liquid based on tetraalkylammonium cation as high performance gel electrolyte for lithium metal batteries. *Electrochimica Acta*, 301, 2019, 284–293.
73. Balgude, S.D., Sethi, Y.A., Kale, B.B., Munirathnam, N.R., Amalnerkar, D.P., Adhyapak, P.V., Nanostructured layered Sn_3O_4 for hydrogen production and dye degradation under sunlight. *RSC Adv.*, 6, 2016, 95663–95669.
74. Adhyapak, P.V., Meshram, S.P., Mulla, I.S., Pardeshi, S.K., Amalnerkar, D.P. Controlled synthesis of zinc oxide nanoflowers by succinate-assisted hydrothermal route and their morphology-dependent photocatalytic performance. *Mater. Sci. Semicond. Process*, 17, 2014, 197–206.
75. Berengue, O.M., Simon, R.A., Chiquito, A.J., Dalmaschio, C.J., Leite, E.R., Guerreiro, H.A., Guimaraes, F.E.G. Semiconducting Sn_3O_4 nanobelts: Growth and electronic structure. *J. Appl. Phys.*, 107, 2010, 033717.
76. Huang, M., Yu, J., Li, B., Deng, C., Wang, L., Wu, W., Dong, L., Zhang, F., Fan, M. Intergrowth and coexistence effects of TiO_2–SnO_2 nanocomposite with excellent photocatalytic activity. *J. Alloys Compd.*, 629, 2015, 55–61.
77. Manikandan, M., Tanabe, T., Li, P., Ueda, S., Ramesh, G.V., Kodiyath, R., Wang, J. Photocatalytic water splitting under visible light by mixed-valence Sn_3O_4. *ACS Appl. Mater. Interfaces*, 6, 2014, 3790–3793.
78. Yu, X., Wang, L., Zhang, J., Guo, W., Zhao, Z., Qin, Y., Mou, X., Li, A., Liu, H. Hierarchical hybrid nanostructures of Sn_3O_4 on N doped TiO_2 nanotubes with enhanced photocatalytic performances, *J. Mater. Chem. A*, 2015, 3, 19129–19136.
79. He, Y., Li, D., Chen, J., Shao, Y., Xian, J., Zheng, X., Wang, P. Sn_3O_4: A novel heterovalent-tin photocatalyst with hierarchical 3D nanostructures under visible light. *RSC Adv.*, 4, 2014, 1266.
80. Mahadadalkar, M.A., Kale, S.B., Kalubarme, R.S., Bhirud, A.P., Ambekar, J.D., Gosawi, S.W., Kulkarni, M.V., Park, C.J., Kale, B.B. Architecture of the $CdIn_2S_4$/graphene nano-heterostructure for solar hydrogen production and anode for lithium ion battery. *RSC Advances*, 6, 2016, 34724–34736.
81. Sethi, Y.A., Panmand, R.P., Kadam, S.R., Kulkarni, A.K., Apte, S.K., Naik, S.D., Munirathnam, N., Kulkarni, M.V., Kale, B.B. Nanostructured CdS sensitized $CdWO_4$ nanorods for hydrogen generation from hydrogen sulfide and dye degradation under sunlight. *J. Colloid Interface Sci.*, 487, 2017, 504–512.

7 Electrocatalysts in Solid Oxide Fuel Cells

A. P. Khandale and S. S. Bhoga

CONTENTS

7.1 INTRODUCTION

An increase in the carbon emission and demand for the electricity from non-conventional resources have influenced the growth of fuel cell (FC) technology. The evolution of electric and/or hybrid electric vehicles has further extended the adoption scope of FCs. The FC is an electrochemical device that converts the chemical energy of a fuel directly to electricity and heat via a redox reaction at the cathode and anode, giving a much higher efficiency than traditional energy conversion systems, without producing a significant amount of pollutants. Consequently, FCs are seen as ideal energy sources in transport, stationary, and distributed power generators.

In FCs, the anode and cathode are just charge-transfer media, and the active masses undergoing redox reaction are delivered from outside the cell, either from the environment (e.g., oxygen from air) or from a tank (e.g., fuels such as hydrogen and hydrocarbons) (Winter and Brodd, 2004). FCs are usually classified on the basis of an electrolyte used and its operating temperature. Different types of FCs that have been realized and are currently in use and under development are summarized in Table 7.1 (Carrette et al., 2001).

TABLE 7.1
Different Types of Fuel Cells

Type of Fuel Cell	Operating Temperature (°C)	Charge Carrier in Electrolyte	Anode Reaction	Cathode Reaction
AFC (Alkaline)	<100	OH^-	$H_2 + 2OH^- \rightarrow 2H_2O + 2e'$	$\frac{1}{2}O_2 + 2H_2O + 2e' \rightarrow 2OH^-$
PEMFC (Polymer Electrolyte Membrane)	60–120	H^+	$H_2 \rightarrow 2H^+ + 2e'$	$\frac{1}{2}O_2 + 2H^+ + 2e' \rightarrow H_2O$
DMFC (Direct Methanol)	60–120	H^+	$CH_3OH + H_2O \rightarrow CO_2 + 6H^+ + 6e'$	$\frac{3}{2}O_2 + 6H^+ + 6e' \rightarrow 3H_2O$
PAFC (Phosphoric Acid)	160–220	H^+	$H_2 \rightarrow 2H^+ + 2e'$	$\frac{1}{2}O_2 + 2H^+ + 2e' \rightarrow H_2O$
MCFC (Molten Carbonate)	600–800	CO_3^{2-}	$H_2 + CO_3^{2-} \rightarrow H_2O + CO_2 + 2e'$	$\frac{1}{2}O_2 + CO_2 + 2e' \rightarrow CO_3^{2-}$
SOFC (Solid Oxide)	800–1000 low temperature (500–600) possible	O^{2-}	$H_2 + O^{2-} \rightarrow H_2O + 2e'$	$\frac{1}{2}O_2 + 2e' \rightarrow O^{2-}$

Solid oxide fuel cells (SOFCs), based on oxide-ion conducting solid-electrolyte, have several advantages over other types of FCs, such as relatively less-expensive materials, relatively low sensitivity to impurities in the fuel, and very high efficiency (up to 85% energy efficiency, when combined with gas turbine). Moreover, SOFCs are capable of directly utilizing available hydrocarbon fuels (hydrogen to natural gas, coal gas, reformed gasoline or diesel, gasified carbonaceous solids, etc.) without reforming them to hydrogen, which can decrease the complexity and cost of the FC system (Faes et al., 2012; Boaro, 2017; Rokni, 2017; Yang et al., 2017). However, commercial SOFCs are expensive and have only a handful of high-profile clients, who have recently adopted SOFC technology (Behling, 2013). One of the major challenges the new generation of low cost SOFC encountered is the design of novel materials with unique compositions, structures, morphologies, and architectures that promote fast electrocatalytic activity and transport of ionic and electronic defects, facilitate rapid surface electrochemical reactions, and enhance the tolerance to the contaminants at low temperatures (Liu et al., 2011). In this chapter, we have addressed some of the recent progresses in electrocatalyst materials.

7.2 BASIC OPERATING PRINCIPLE OF SOFCs

The SOFC comprises an anode electrode (exposed to fuel), a cathode electrode (exposed to oxygen/air), and a solid electrolyte (SE). The SE separates the cathode and anode and facilitates the oxy-ion transport required for oxidation of the fuel (Singh and Minh, 2005). Figure 7.1 shows the schematic of the SOFC and electrode reactions. Typically, both the electrodes are porous for the gas (oxidant and fuel) permeation, whereas dense electrolytes ensure negligible gas crossover. The fuel (H_2, CO, or hydrocarbons) is oxidized at the anode, whereas the oxidant (oxygen or air) is reduced at the cathode according to the following electrochemical reactions (Eqs. 7.1 and 7.2).

$$\frac{1}{2}O_2 + 2e' \rightarrow O^{2-} \tag{7.1}$$

$$H_2 + O^{2-} \rightarrow H_2O + 2e' \tag{7.2}$$

Oxygen ions, thus formed at the cathode–electrolyte interface, are transported to the anode through an oxy-ion conducting electrolyte, where it reacts with fuel to form H_2O or CO_2. The overall cell reaction is given in Eq. (7.3).

$$H_2 + \frac{1}{2}O_2 \rightarrow H_2O \tag{7.3}$$

The open-circuit voltage corresponding to the reaction given in Eq. 7.3 is described by the following Nernst equation (Brett et al., 2008; Struchtrup, n.d.):

$$E = E^{\circ} + \frac{RT}{2F}\ln\left(\frac{p_{H_2} p_{O_2}^{0.5}}{p_{H_2O}}\right) \tag{7.4}$$

where E is the equilibrium potential, E° is the standard potential, R is the universal gas constant, F is the Faraday constant, p is the gas pressure, and T is the absolute temperature.

Equation 7.4 describes the relationship between ideal standard potential E°, for the FC at standard conditions to the reactions and ideal or open-circuit potential E at any other given temperature and pressure of the reactants and products. The ideal

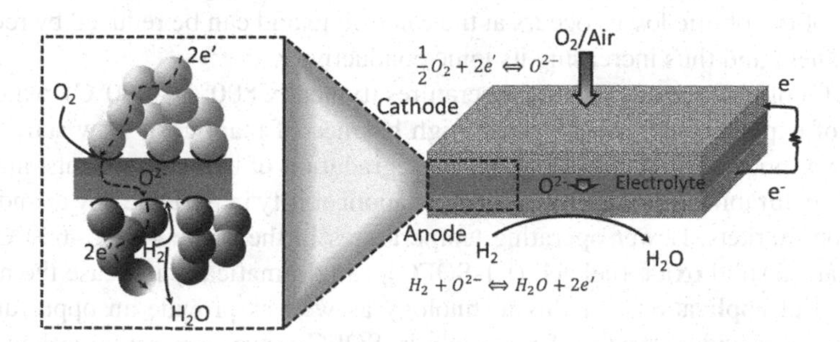

FIGURE 7.1 Schematic of operating principle of SOFC.

$E°$ for the H_2/O_2 FC depends on the state of H_2O. It is 1.229 V for the liquid phase and 1.18 V for the gaseous phase. The actual cell potential is reduced from its equilibrium potential when the current is drawn. This reduction in the potential is attributed to various losses or polarizations, namely, activation, ohmic, and concentration, depending on the electrode and the electrolyte materials, their microstructures, cell design, and operating temperature.

At lower currents, the activation barrier for the reactions is greater; thus, activation losses are predominant. The activation polarization is represented by the following Butler-Volmer equation (Srinivasan et al., 2006):

$$i = i_0 \left[\exp\left(\frac{\alpha_1 F \eta_a}{RT} \right) - \exp\left(\frac{-\alpha_2 F \eta_a}{RT} \right) \right] \tag{7.5}$$

where η_a is the activation polarization, T is the absolute temperature, R is the universal gas constant, α_1 and α_2 are the charge-transfer coefficients, F is the Faraday constant, i is the current density, and i_0 is the exchange current density.

The concentration losses are predominant at high current densities due to diffusion limitations in the electrodes at these currents. The concentration polarization can be represented as (Srinivasan et al., 2006):

$$\eta_{\text{conc.}} = \frac{RT}{nF} \ln\left(1 - \frac{i}{i_L} \right) \tag{7.6}$$

where, i_L is the limiting current density, and n is the number of electrons participating in the reaction.

The consumption of the reactant gas at the electrode–electrolyte interface leads to the formation of a concentration gradient in the electrodes, which results in reduced cell voltage.

The ohmic polarization is due to the resistance to the flow of electrons and ions and is present over the entire range of the currents. This resistance can be due to the ionic flow in the electrolyte and electronic flow through the electrodes, the interconnect plates, and the external electrical circuit. The ohmic polarization is generally given as $\eta_{\text{ohm}} = iR_{\text{ohm}}$, where, R_{ohm} is the cell resistance (Vijay et al., 2011). The major portion of the ohmic losses occurs at the electrolyte and can be reduced by reducing its thickness and thus increasing its ionic conductivity.

SOFCs often operate at high temperatures (typically 800°C–1000°C), which lead to use of expensive cell components, high balance of plant cost, slow start-up and shutdown, poor cycling capability, faster degradation of cell components, and poor long-term durability, eventually limiting its applicability in portable power and transportation markets. Lower operating temperatures in the range 400°C–650°C [low-temperature solid oxide fuel cell (LT-SOFC)] can dramatically increase the number of potential applications for this technology as well as provide an opportunity to incorporate a wider variety of materials in SOFC power-generation systems with greater reliability and lower cost. In fact, overall efficiency of SOFC depends on thermodynamics (attained voltage relative to theoretical open circuit potential and fuel used) and kinetics (polarization losses) during operation. Lowering SOFC operating temperature causes a significant loss in electrode polarization, eventually reducing overall cell performance.

Therefore, new materials must be studied and developed to improve performance and thus compensate for the envisioned temperature decrease.

Appropriate electrocatalysts selection governs the operating temperature of SOFC for high conversion efficiency.

7.3 ELECTROCATALYSTS IN SOFC

Electrodes are the crux of the entire electrochemical system. In SOFCs, electrodes accomplish three major activities: (1) promotion of catalytically activated surface reactions resulting in a charge transfer between electrode and gas molecules, (2) charge collection and transport through the electrode, and (3) charge and mass transfer across the electrode–electrolyte interface (Thampi et al., 1995). The entire electrochemical reaction occurs at the three-phase boundary (TPB), a site where the ion-conducting electrolyte, the electron-conducting phase, and the gas all meet. A schematic illustration of the electrode–electrolyte interface highlighting the TPB is show in Figure 7.2.

SOFCs are made of multilayered sandwich structures with abrupt interfaces. These abrupt interfaces, present between electrode and electrolyte as well as within electrode structures, can represent regions of high-energy barriers. Dissolution of such structures into the large matrix with few or no abrupt interfaces results in better material continuity and higher dispersion of active species, which facilitates a better electrocatalysis.

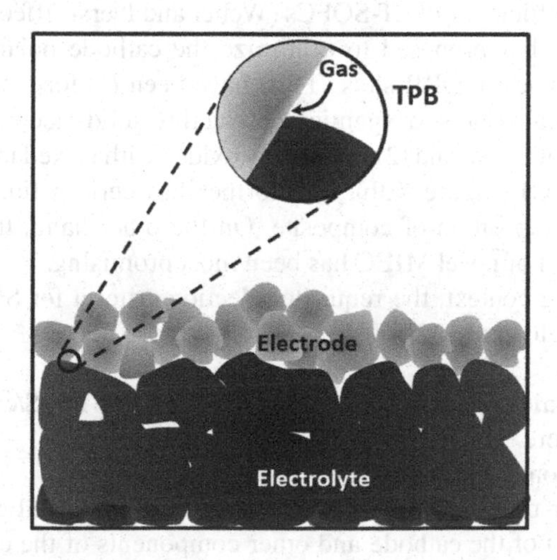

FIGURE 7.2 Schematic of electrode–electrolyte interface and TPB.

7.3.1 Cathode

The pioneering studies on kinetics of high-temperature electrochemical reactions have been focused on metal electrodes in contact with solid oxide electrolytes (e.g., YSZ). The electrochemical processes, in the simplest cases, have been often understood to occur at TPBs. The length of a TPB normally depends on the cell microstructure formed during fabrication. As a matter of fact, oxygen reduces at the cathode and the reduced oxygen ion O^{2-}_{ads} goes into the cathode lattice vacancy $V^{\cdot\cdot}_{O(cathode)}$; the overall reaction can be written using the Kröger–Vink notation as (Eq. 7.7):

$$\frac{1}{2}O_{2(gas)} + 2e^{'} + V^{\cdot\cdot}_{O} \rightarrow O^{x}_{O}, \tag{7.7}$$

where O^{x}_{o} is oxygen at the oxygen lattice site with the required charge. Details of sequential oxygen reduction reaction (ORR) steps at electrochemically active sites are given below (Eq. 7.8):

$$\left.\begin{array}{l} \text{i. } O_{2(gas)} \rightleftharpoons O_{2(gas\ adsorb)} \rightleftharpoons 2O_{ads(cathode)} \\ \text{ii. } O_{ads(cathode)} + 2e^{-}_{cathode} \rightleftharpoons O^{2-}_{ads(TPB)} \\ \text{iii. } O^{2-}_{ads(TPB)} + V^{\cdot\cdot}_{O(cathode)} \rightleftharpoons O^{''}_{O(cathode)} \\ \text{iv. Diffusion of } O^{''}_{O(cathode)} \text{ through bulk material} \end{array}\right\} \tag{7.8}$$

where $O^{2-}_{ads(TPB)}$, $V^{\cdot\cdot}_{O(cathode)}$ and $O''_{O(cathode)}$ are oxygen ions with two negative charges at the TPB, oxygen vacancy at oxygen lattice site with two positive charges, and an oxygen ion with two negative charges at its lattice site, respectively. High-activation energy of the O_2 electrode reaction (ORR) at the cathode has been the key factor for the highest loss of efficiency of LT-SOFCs (Weber and Ivers-Tiffée, 2004). Two complementary approaches proposed to minimize the cathode polarization resistance through increasing active ORR sites (TPB) have been (1) formation of composites (electron conducting oxide + oxygen ion conducting solid-electrolyte) as schematically shown in Figure 7.3a, and (2) use of new oxides with mixed ionic and electronic conductivity (MIEC) (Figure 7.3b). The former has certain limitations regarding optimization of composition of composite. On the other hand, the latter approach comprising a search of novel MIEC has been most promising.

In the preceding context, the requisite/selection criteria for SOFC cathodes are given below (Mahato et al., 2015):

- High electronic conductivity (preferably more than 100 S/cm in an oxidizing atmosphere
- High oxide ion conductivity
- Minimum or no mismatch between the values of thermal expansion coefficient (TEC) of the cathode and other components of the cell, such as the electrolyte and interconnect materials

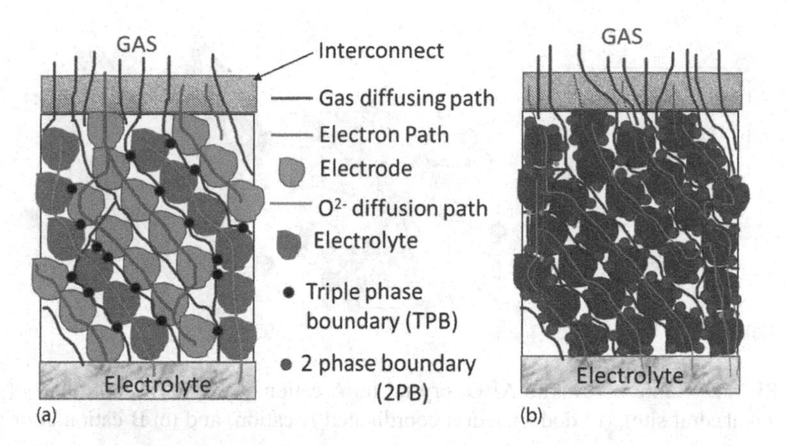

FIGURE 7.3 (a) Electron:electrolyte conducting composite, and (b) mixed-ionic electronic conductor. White lines = electron-conducting paths, black lines = O_2-conducting paths, and gray lines = gas-diffusing paths. White grains = electrolyte, black grains = electron conductor, and small dots = TPBs.

- Good chemical compatibility with the electrolyte and interconnect materials at operating temperatures
- Sufficient porosity to allow fast diffusion of O_2 gas from interconnect–cathode interface surface to cathode–electrolyte interface
- Good stability under oxidizing atmosphere during fabrication as well as operation
- High catalytic activity for ORR (desired value of polarization resistance less than 0.15Ω cm^2)
- Cost effective

In the recent past, most of available cathode materials for SOFCs have been oxides. A number of oxide phases with perovskite, double-perovskite, rutile, fluorite, pyrochlore, and tungsten bronze structures have been extensively explored as potential electrode materials. Among them, the most popular oxide phases have been perovskites and perovskite-related oxides (e.g., double-perovskites and Ruddlesden-Popper structures) (Su et al., 2015). Perovskites and related compounds have been much economical compared to old popular noble metal catalysts (e.g., platinum). Except being electronically low conducting (La, Sr)(Mn, Fe)O_3, $YCoO_3$, or (Y, Ca) FeO_3, almost all perovskites can be considered as cathode materials for SOFCs. Some of the details on perovskite and related oxides are discussed below.

7.3.1.1 Perovskite-Related Oxides

Perovskites and related oxides have been a very important class of functional materials. They exhibit a range of chemical stoichiometries and crystal structures. The phase composition, stoichiometry, and crystal structure have a great influence on electronic and ionic conductivities and in turn the electrochemical catalytic activity. Two different representations of crystal structures of cubic perovskites (chemical formula ABO_3)

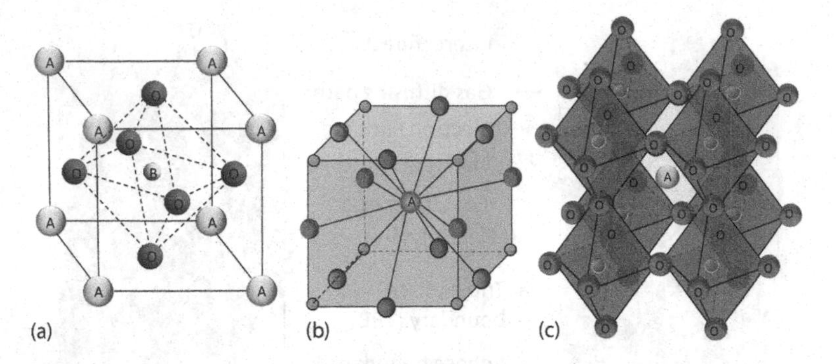

FIGURE 7.4 Cubic perovskite ABO_3 crystal (a) A cation at origin of coordinates (A cation at octahedral site), (b) dodecahedral coordinated A cation, and (c) B cation at origin of coordinates.

are shown in Figure 7.4a and c when the origin of coordinates is occupied by A-cation and B-cation, respectively. Essentially, in these perovskite structures, alkaline earth or rare earth metal cations occupy tetrahedral B-sites (six oxygen ions) with a strong covalent bond (Figure 7.4a) while transition metal ions at A-sites coordinate with 12 oxygen ions (Figure 7.4b with a strong ionic bond). The B-cation occupies the body center lattice site of a cubic unit cell (Figure 7.4a). The B-cations in an alternative structure (Figure 7.4c) occupy corners of a cubic unit cell, and oxygen anions occupy the cube edge centers (Figure 7.4b and c). The oxygen octahedra containing small B-cation (Figure 7.4c) form a cubic symmetry through edge sharing (Figure 7.4c), while the dodecahedra surrounding A-cation share faces. Oxygen anions shield A- and B-cations from one another. The anion vacancies at oxygen lattice sites are introduced (aliovalent doping at A- and/or B-site) whenever the sum of valances of A- and B-site cations (valance A + valance of B) is less than the required six.

A number of perovskites have distorted cubic symmetry. The cation displacements within the octahedra and tilting of the octahedra, which have been related to the properties of A and B substituted atoms, cause distortions. The degree of distortion in these perovskites can be determined using the Goldschmidt tolerance factor (t) given below:

$$t = \frac{r_A + r_O}{\sqrt{2}\left(r_B + r_O\right)} \tag{7.9}$$

where r_A, r_B, and r_O are effective radii of the A-site cation, B-site cation, and oxide ion, respectively. Shannon's ionic radii corresponding to respective coordination numbers has usually been considered for estimation of this tolerance factor (Shannon, 1976). The perovskite crystal structure has been the ideal cubic, when $t \approx 1$. Smaller A relative to B or bigger B relative to A cations result in a decrease in tolerance factor (0.71–0.9) causing cation displacement and tilting of corner-sharing BO_6 octahedra, leading to an orthorhombic symmetry instead of a cubic one. On the other hand, when $t > 1$, (larger A than B), a hexagonal structure like a $BaNiO_3$

type forms. A redox catalytic property has usually been corelated to properties of B-site cations in most cases. Further, octahedral symmetry around this transition metal promote metallic or semiconducting band structure, which induces high electronic conductivity in these oxides. Whereas, oxygen vacancies have been the key for high oxygen ion conductivity/diffusivity. The oxygen vacancies arise from either partial substitution of A^{3+} by A^{2+} or from the partial reduction of B^{3+} or B^{4+} to B^{2+} and B^{3+}, respectively. Obviously, the partial substitution of B′ at B-site of $ABB'O_3$ perovskites introduces oxygen non-stoichiometry (δ) while changing the average valence state (n) of the B-site cation for constancy of charge. The mixed valance of the B-cation has been crucial for improved catalytic activity. The oxygen non-stoichiometry (δ) at different temperatures can be obtained from the thermogravimetric data using Eq. (7.10).

$$\delta = \frac{(M \times m_o) - (M - 15.9994 \times \delta_o) \times m}{15.9994 \times m_o}, \tag{7.10}$$

where m_o is the original weight (before heating), m is the actual weight after heating of the sample, M is the molar mass of stoichiometric form of $ABB'O_3$, and δ_0 is the oxygen non-stoichiometry of the sample at the room temperature.

In the absence of an interaction among oxygen vacancies and ordering phenomena, ORR catalytic activity has been correlated with the oxygen non-stoichiometry (Zhou et al., 2008b). The bulk oxygen vacancies, which facilitate diffusion of oxygen ions, effectively extend the ORR sites from TPB to the whole cathode (two phase boundary-2PB). The surface oxygen vacancies, particularly, act as the active sites for oxygen adsorption, oxygen dissociation, and reduction [Steps (i) through (iii) of Eq. 7.8] at the cathode surface. Perovskite materials facilitate manipulation of nature and concentration of defects resulting into excellent electrocatalytic activity.

The undoped perovskite oxides mostly have been poor oxygen ion conductors in air. Partial substitution (doping) at the A-site (or B-site) with acceptor cations such as Sr^{2+}, Ca^{2+}, and Ba^{2+}, or the reduction of B^{3+} to B^{2+} (B = Mn and Co) has resulted in hypostochiometric perovskites such as $La_xSr_{1-x}CoO_{3-\delta}$ in which oxygen vacancies facilitate O^{2-} conductivity. The O^{2-} can be transported through these oxides via hopping of O^{2-} from its regular occupied site to nearby equivalent vacant sites by traversing the potential barrier (activation energy E_a). The magnitude of O^{2-} conductivity can be obtained through complex impedance spectroscopy (Chang and Park, 2010) or the isotope exchange depth profile method (Chater et al., 1992). On the other hand, electron conduction occurs by a small polaron hopping mechanism, where a lattice electron (polaron) hops between B-site cations. The electronic conduction in these oxides has been dominant over O^{2-} conduction. Conventional electrical conductivity measurement, hence, yields the magnitude of electronic conductivity. The chemical stability of the B-site cation against the reducing atmosphere ensures the phase stability of these oxides.

Among all perovskites, only $La_{1-x}Sr_xMnO_3$ [lanthanum strontium manganite (LSM)] has been widely attended and used as cathodes in SOFCs. Sr doping in LSM oxidizes Mn instead of introducing additional oxygen vacancies (as usually observed for most of other perovskites) to achieve charge neutrality as per defect chemistry expressed using the Koger-Vink notation given below:

$$Mn_{Mn}^{x} + SrO \rightarrow Sr_{La}^{'} + Mn_{Mn}^{\cdot} + O_{O}^{x} \quad (7.11)$$

where Mn_{Mn}^{x}, $Sr_{La}^{'}$ and $Mn_{Mn}^{\cdot}$ are Mn at the Mn lattice site (no additional charge), Sr at the La lattice site with negative charge (′), and Mn oxidizes (Mn^{3+} to Mn^{4+}) at the Mn lattice site (one positive charge represented by a dot (·). The Sr-doped lanthanum perovskites have shown large oxygen-excess (hyperstoichiometry) under oxidizing conditions, whereas a large oxygen deficiency under reducing gas atmospheres. Excess oxygen results in metal vacancies. LSM has good electrode properties for YSZ electrolyte-based SOFCs. Extremely low oxygen diffusivity at low temperatures (<800°C) restricts its use in LT-SOFCs). Therefore, perovskites exhibiting mixed oxygen ionic and electronic conductivity have been in demand for LT-SOFCs and oxidation catalysts (Shao and Halle, 2004).

7.3.1.2 Disordered MIEC Perovskites

The manganite perovskites containing different rare-earth cations at A-site instead of La such as $Ln_{0.6}Sr_{0.4}MnO_3$ (Ln = Nd, Sm, Gd, Yb, or Y) MIEC oxides have shown high catalytic ORR activity along with compatible TECs with established electrolytes (Sun et al., 2010). Amongst all these perovskites, Sr-doped $PrMnO_3$ have not only exhibited the best electrode performance but also thermochemical stability in a wide temperature range (Ishihara et al., 1995). Further, smaller lanthanides, especially $Ln_{1-x}Sr_xMnO_3$ (Ln = Pr, Nd and Sm), have been found potential oxides (Sakaki et al., 1999). An increase of the atomic number of lanthanides along with high Sr dopant concentration has resulted in being TEC compatible with the established electrolytes. Moreover, Ca-doped manganites have been more promising as compared to Sr-doped ones in terms of TEC and thermochemical stability (Afzal et al, 2016). An altogether different $Sr_{0.8}Ce_{0.2}Mn_{0.8}Co_{0.2}O_3$ cathode has high catalytic activity for ORR in the temperature range of 700°C–800°C (Gu et al., 2008).

In the near recent past, perovskite-type cobaltates such as $La_{1-x}Sr_xCoO_{3-\delta}$ (LSC) and related perovskites have shown higher ionic and electronic conductivities as compared to LSM as well as other above-mentioned oxides. Also, they have been extensively explored oxides as intermediate temperature solid oxide fuel cell (IT-SOFC) cathodes (Sun et al., 2010). Conventionally, thus, LSM has been, in general, considered as an electronic conductor, and $La_{0.75}Sr_{0.25}Cr_{0.5}Fe_{0.5}O_{3-\delta}$ (LSCF) has invariably been referred to as having MIEC. Also, the latter has been the most popular and representative perovskite MIEC cathode material. In this oxide, large La and Sr cations occupy A-sites; whereas, smaller Co- and Fe-cations preferably have been populated at the B-sites. Depending on chemical composition, temperature, and defect concentration, they may crystallize into orthorhombic or rhombohedral or cubic structures. Although use of these oxides as cathodes can effectively improve the cell performance, they should be selected carefully because of their higher TECs than the YSZ electrolyte. Additionally, although it has good stability and catalytic properties for oxidation of CH_4, the activity degrades under reducing (H_2) and oxidizing (O_2) atmospheres (Tao and Irvine, 2004). The operating temperature has a significant effect on the Sr segregation, Cr deposition, and formation of $SrCrO_4$ phase on the LSCF surface (Wang et al., 2014). Particularly, a decrease in the operating temperature and partial pressure of gaseous significantly reduced the segregated SrO and Cr species, respectively, which led to the significant reduction in Cr deposition. The absorption as well as desorption rates of oxygen gas has increased at higher dopant Sr content, whereas, substitution of Co at the Fe-site enhanced the ORR catalytic activity (Teraoka et al., 1985). In other words, despite their impressive electrochemical performance, these perovskites have become less popular because of their high chemical expansion coefficients, susceptibility to Cr poisoning, and long-term stability issues (Tucker et al., 2006). A thin layer of gadolinia-doped ceria (GDC) in between the LSCF and YSZ-electrolyte has been mandated to avoid unfavorable reactions and eventually stop deterioration of cell performance. Jiang has presented details highlighting scientific and technological importance in respect of developed strategies for optimization. Also, his focus has been on rational design and development of LSCF-based ORR active and stable phases for IT- or LT-SOFCs (Jiang, 2012).

Other ABO_3-type perovskites such as $Ba_{0.5}Sr_{0.5}Co_{0.8}Fe_{0.2}O_{3-\delta}$ (BSCF), $Sc_{0.2}Co_{0.8}O_{3-\delta}$ (SSC), and $SrNb_{0.1}Co_{0.9}O_{3-\delta}$ (SNC) cathodes have been projected as promising cathode for LT-SOFCs (Liu et al., 2006; Zhou et al., 2008a, 2008b). The susceptibility to CO_2 and high TEC has been the major constraints that restrict their commercialization. Among the cobaltates with perovskite structures, BSCF has been most attended and found utmost potential cathode material. In this perovskite, the Ba-cation (large ionic radius) has been the key component for expanding cubic lattice structure in addition to promoting oxygen vacancy formation and migration, and therefore low electrode polarization (Merkle et al., 2012). It also increases ORR kinetics owing to an improvement in O-surface exchange. In the recent past, within a perovskite framework, $SrNb_{0.1}Co_{0.9-x}Fe_xO_{3-\delta}$ (SNCF, $0.1 \leq x \leq 0.5$) has been found an exceptionally potential cathode for LT-SOFC applications (Zhu et al., 2014). The presence of Co has been understood as the main reason for deterioration and limiting the use of Co-rich perovskites. Therefore, heavy doping of ferrites and/or manganites have been preferred at the B-site to overcome above difficulties in addition to increased ORR activity, for example, $Pr_{0.3}Sr_{0.7}Co_{0.3}Fe_{0.7}O_3$ (Kim et al., 2008a). Also, Ln-doped cubic perovskites $Ln_xBa_{1-x}Co_{0.7}Fe_{0.3}O_{3-\delta}$ (Ln = La, Pr, Nd, $x = 0.1, 0.2$) exhibit enhanced electrocatalytic activity compared to undoped $BaCo_{0.7}Fe_{0.3}O_{3-\delta}$ with optimum/best performance for $Pr_{0.1}Ba_{0.9}Co_{0.7}Fe_{0.3}O_{3-\delta}$ (Zhang et al., 2020). Details on magnitude of ionic- and electron-conductivity along with TEC of various perovskites have been available in literature (Sun et al., 2010).

Increasing interest in the development of alternative cobalt-free MIECs has directed attention on the iron-based perovskites because they have been economic, readily available, thermochemically stable under cathodic environment, excellent CO_2 tolerance, and acceptable electrocatalytic ORR activity. In addition, $La_{1-x}Sr_xFeO_{3-\delta}$, $Ba_{0.5}Sr_{0.5}Zn_{0.2}Fe_{0.8}O_{3-\delta}$, $Ba_{0.5}Sr_{0.5}Cu_{0.2}Fe_{0.8}O_{3-\delta}$, and $Sm_{0.5}Sr_{0.5}Fe_{0.8}Cu_{0.2}O_{3-\delta}$ have been proposed as promising alternatives. $Bi_{1-x}Sr_xFeO_{3-\delta}$ (BSF), have been another most studied iron-based perovskite oxide, with very attractive electrocatalytic activity (Ling et al., 2011; Park et al., 2011; Zhao et al., 2010). Exceptional physicochemical performance of BSF has been correlated with intrinsic perovskite structure and specific composition. Bi^{3+} with the 6s lone pair at the A-site has been found to improve surface oxygen exchange ability and density of oxygen vacancy. The Nb, Co, Ta, Cu, and Mn have been successfully doped into $Bi_{0.5}Sr_{0.5}FeO_{3-\delta}$ so as to reduce TEC and improve thermochemical stability as well as electrochemical performance (Gao et al., 2019). The decomposition of these oxides under CO_2 gas has been understood to be due to alkali metals; however, low oxygen vacancy concentration can overcome this issue. The doping with high-valence metals at the B-site has been expected to reduce the amount of oxygen vacancy and in turn enhance tolerance to CO_2. Concurrently, Sb^{5+}-doped BSF and Ti-doped $SrFeO_{3-\delta}$ have been reported as promising oxygen reduction electrocatalysts with excellent resistance to CO_2. $Ba_{1-x}La_xFeO_3$ oxides have been successfully employed as ceramic membranes for oxygen separation from air. More recently, $Ba_{0.95}La_{0.05}FeO_3$ (BLF), used as an electrode of proton conducting SOFCs, has shown encouraging performance (Dong et al., 2012; Yan et al., 2011). Unconventional phosphorus, P, (nonmetal) doping in $Ba_{1-x}La_xFeO_{3-\delta}$ has been seen to reduce the energies required for the formation of oxygen vacancies, and their migration in turn increased diffusion and ORR activity. Optimum catalytic performance has been reported for $Ba_{0.95}La_{0.05}Fe_{0.95}P_{0.05}O_{3-\delta}$ (Liu et al., 2019). $Ba_{0.5}Sr_{0.5}Fe_{0.8}M_{0.2}O_{3v\delta}$ (M = Ni, Cu, Zn) perovskites have also exhibited competitive electronic and ionic conductivities in addition to good thermodynamic stability at high temperatures (600°C–700°C) (Basbus et al., 2014). Further, a slow degradation of these oxides while stored in air at room temperature, ascribed to adsorption of H_2O and CO_2 near grain boundaries, can be restored after a treatment at high temperatures. The value of tolerance factor $t = 1.4$ for $Ba_{0.5}Sr_{0.5}Fe_{0.8}Co_{0.2}O_{3-\delta}$ being slightly larger than optimum value, crystallized with hexagonal structure instead of a cubic one. The cobalt-free cathodes have been projected as potential materials for LT-SOFC. Similarly, $Y_{0.9}In_{0.1}BaCo_3ZnO_{7+\delta}$ oxide has also been an attractive cathode (West et al., 2014). Besides the compatible TEC with electrolyte, the above discussed Cobalt-free perovskites have acceptable polarization resistance. In the nearest recent past, $BaFe_{0.9}Zr_{0.1}O_{3-\delta}$ has been found to exhibit high thermochemical stability under reducing as well as oxidizing atmospheres in addition to acceptable electrical conductivity (Park et al., 2013). Literature revealed that $La_{0.6}Sr_{0.4}Fe_{0.9}Nb_{0.1}O_{3-\delta}$ may be one of the other choices of Co-free potential symmetric electrode material for hydrogen and carbon monoxide SOFCs (Bian et al., 2019). A short review on these perovskites has been reported in literature (Baharuddin et al., 2017). A systematic analysis and review highlighting the process parameters and their effect on electrical conductivity, electrochemical properties, TEC, and mechanical properties (toughness and hardness) of different cathode materials can be found in literature (Kaur and Singh, 2019). Also, the significance of selection of initial composition, dopants, and their valence, which decide properties, has been included. In fact, ordering partially or over the entire available crystallographic sites of ABO_3-perovskites results, when charges or the ionic radii of host- and dopant-cations differ considerably to reduce the lattice energy, and termed as ordered perovskites or double perovskites.

7.3.1.3 Ordered Perovskites or Double-Perovskites

The general formula for double-perovskites $A_{0.5}A'_{0.5}BO_{3-\delta}$ (or $AB_{0.5}B'_{0.5}O_{3-\delta}$) can alternatively be written as $AA'B_2O_{5+\delta}$ (or $A_2BB'O_{5+\delta}$) provided the size difference between the two cations is significant. Details of these oxides have been dealt extensively in literature (Vasala and Karppinen, 2015). Three different types of ordering have been envisaged for A- or B-site cations (King and Woodward, 2010). The rock-salt ordering has been the most symmetric owing to the alternate arrangement of B and B′ (or A and A′) cations similar to the rock-salt structure (Figure 7.5a). In remaining the two arrangements, cations can order into columns or layers as shown in Figures 7.5b and c, respectively.

The rock-salt structure is often found for $A_2BB'O_{6-\delta}$, whereas the layered one is for $AA'B_2O_{6-\delta}$ (King and Woodward, 2010). The partial aliovalent acceptor cation ($A = Sr^{2+}$) substitution for Ln^{3+} at A-site leads to charge imbalance, which can be compensated either electronically where B^{3+} oxidizes to B^{4+} or ionically by forming oxygen vacancies $\left(V_O^{\bullet\bullet}\right)$ or through both. The defect reaction for charge compensation in 2 mole% Sr-doped $La_{0.8}Sr_{0.2}BO_3$ (for simplicity, change in A′ and B′ are not considered) through both electronic and ionic compensations is given below (Eq. 7.12).

$$0.8La_{La}^{x} + (1.0)B_{B}^{x} + (3.0)O_{O}^{x} + (0.2)SrO \rightleftharpoons$$

$$(0.8)La_{La}^{x} + (0.2)Sr_{La}^{'}(0.8)B_{B}^{x} + (0.2-2\delta)B_{B}^{\bullet} + (3-\delta)O_{O}^{x} + \delta V_{O}^{\bullet\bullet} + \frac{\delta}{2}O_2, \tag{7.12}$$

where $B_B^{\bullet} = h^{\bullet}$ is the hole.

Recently, several researchers have focused their attention on $LnBaM'M''O_{5+\delta}$ layered/double-perovskites as promising MIECs for ORR catalysis. In these oxides, equal amounts of different 3D transition metal cations (M′, M′ = Mn, Fe, Co, Ni, Cu) have been preferred at the B-lattice sites. These double-perovskites have much higher chemical diffusion and surface exchange coefficients as compared to ABO_3-type perovskites. The alternating lanthanide and alkali-earth planes of these layered oxides anticipated to facilitate facile oxygen transport (Chroneos et al., 2011).

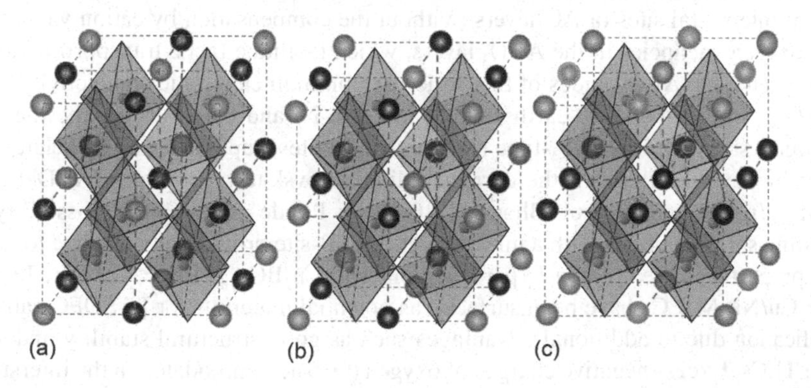

FIGURE 7.5 $AA'B_2O_{5+\delta}$ (or $A_2BB'O_{5+\delta}$)-type structures: (a) rock-salt, (b) columns, and (c) layers.

The $AA'B_2O_{5+\delta}$-type $LnBaCo_2O_{5+\delta}$ (LnBCO where Ln = Pr, Nd, Sm, Gd and Y) have been extensively investigated owing to high mixed ion-electron conductivity along with high oxygen surface exchange rates (Burriel et al., 2012; Chen et al., 2013). The $PrBaCo_2O_{5+\delta}$ (PBCO) has exhibited lowest electrode polarization resistance (R_p) among all LnBCO oxides (Liu et al., 2017) and (Zhang et al., 2008). Various approaches including A-site or B-site cationic doping have been followed to improve further electrochemical catalytic performance. The substitution of Sr for Ba in $LnBaCo_2O_{5+\delta}$ and $GdBaCo_2O_{5+\delta}$ improved the conductivity and catalytic activity along with chemical stability (Kim et al., 2008b, 2009). Similar to distorted perovskites, these Co-containing double-perovskites have high TEC, poor stability, and high cost, making them unsuitable. Partial substitution of different cations instead of Co has been considered to compensate without compromising superior electrochemical activity of cobalt-containing cathode materials, including $LnBaCo_{2-x}Cu_xO_{5+\delta}$ (Ln = Nd, Gd), $YBaCo_{2-x}Cu_xO_{5+\delta}$, $GdBaCo_{2/3}Fe_{2/3}Cu_{2/3}O_{5+\delta}$, $PrBaM_xCo_{2-x}O_{5+\delta}$ (M = Cu, Fe) and $SmBa_{0.5}Sr_{0.5}Co_{1.5}Cu_{0.5}O_{5+\delta}$, (Jun et al., 2013). The composition, electrical conductivity, ionic conductivity, oxygen diffusion coefficient D, oxygen surface exchange coefficient k, and TEC of these perovskites along with for Ruddlesden-Popper phases have been systematically presented in literature (Nikonov et al., 2018).

7.3.1.4 Ruddlesden-Popper Phases

Another interesting cathode material from LT-SOFC viewpoint and the phases related to perovskite structure have been Ruddlesden-Popper phases having general formula of $A_{n+1}\,B_nO_{3n+1}$ (n = 1, 2, 3). The A_2BO_4 (n = 1), a simplest and most widely studied member of the series, has a K_2NiF_4-type lattice structure. Many mixed oxides of the type A_2BO_4 (A = rare earth, alkaline earth; B = transition metal) crystallized with tetragonal K_2NiF_4-type crystal structure (space group I4/mmm). K_2NiF_4-type structure has an outer intergrowth of alternating perovskite (ABO_3) and rock-salt (AO) layers stacked along the tetragonal c-axis (Figure 7.6).

Alternatively, this structure has been consisting of an intergrowth of BO_2 infinite layer sheets and A_2O_2 fluorite-type layers along the c-axis, containing corner-shared BO_4 square planes and AO_8 cubes. Such structure allows incorporation of excess oxygen at interstitial sites of AO layers (without the compensation by cation vacancies) and oxygen vacancies in the ABO_3 layers, which facilitate facile transport/diffusion of oxygen ions. Also, oxides of this series exhibit high electronic conductivity (e.g., $A_2BO_{4\pm\delta}$ where A = rare earth cations and B = Ni and Cu). Further, the thermo-chemical stability of K_2NiF_4-type manganites, ferrites, cuperates, and nickelates has been higher than those of the corresponding perovskite-type oxides (Al Daroukh et al., 2003; Khandale and Bhoga, 2010, 2014; Punde et al., 2014). Possibility of forming solid-solutions with a mixed valence of B-site provides good scope to tailor the properties. The oxygen hyperstoichiometric $Ln_2BO_{4+\delta}$ (Ln = La, Sm, Pr and B = Cu/Ni) MIECs have been surfaced as potential materials for IT-SOFC cathode application due to additional advantages such as good structural stability and suitable TECs. Excess negative charges of oxygen (O^{2-}) accommodated in the interstitial sites have been compensated by the formation of holes $\left(B_B^{\cdot}\right)$ to achieve charge neutrality. The partial substitution of aliovalent (other than valance +3) cations such as

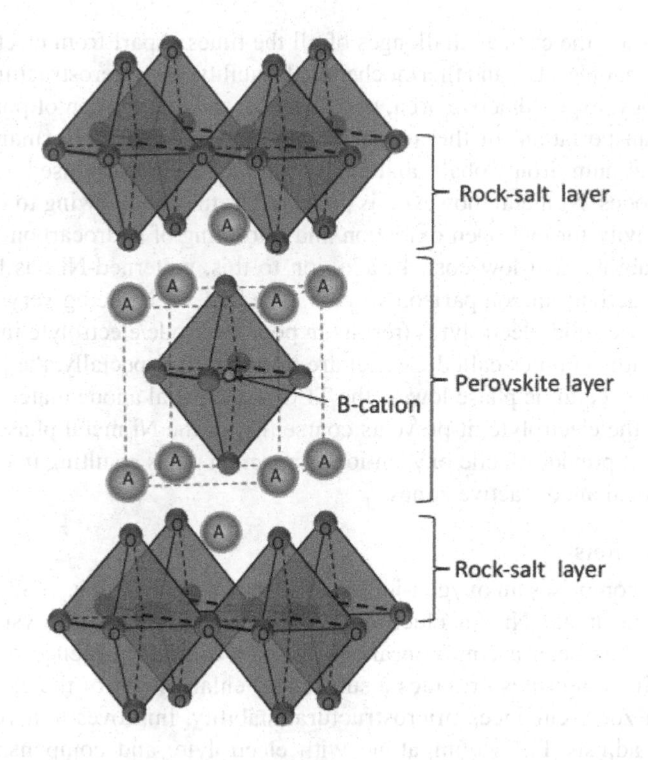

FIGURE 7.6 $AABO_4$ (or K_2NF_4)-type structures.

Sr/Ce at the Ln^{3+} regular lattice site brings changes in the valence of B-cations and the oxygen stoichiometry. The defect reaction can be represented as given in Eq. (7.13).

$$
\begin{aligned}
&(2-y)\mathrm{Ln}_{\mathrm{Ln}}^{x} + (y)\mathrm{SrO} + \mathrm{B}_{\mathrm{B}}^{x} \rightleftharpoons \\
&(2-y)\mathrm{Ln}_{\mathrm{Ln}}^{x} + (y)\mathrm{Sr}_{\mathrm{A}}^{'} + (1-y)\mathrm{B}_{\mathrm{B}}^{x} + (y-2\delta)\mathrm{B}_{\mathrm{B}}^{\cdot} + (\delta)\mathrm{V}_{\mathrm{O}}^{\cdot\cdot} + \frac{\delta}{2}\mathrm{O}_2
\end{aligned}
\tag{7.13}
$$

The changes in interstitial oxygen and hole concentrations have been correlated with control over oxygen migration and electrical conductivity, respectively (Al Daroukh et al., 2003).

7.3.2 Anode

Electrochemical oxidation of fuels takes place at the anode preferentially at the TPB. Thus, the anode material should be preferably able to handle fuel flexibility (and not chemically react) with commonly used fuel gases, for example, hydrogen, CO, natural gas, and other hydrocarbons in addition to tolerating carbon deposition, sulfur poisoning and reoxidation. Understanding the charge- and mass-transfer through the bulk, along the surfaces, and across interfaces and suitably develop convenient, efficient and economic processing methods along with engineering of fabrication technologies are the critical challenges of all the times. Apart from electrocatalytic activity, compatible TEC and thermochemical stability, and microstructure (in terms of morphology, exposed/active area, porosity, size and distribution of pores for efficient gas transportation) of the anode largely influences the performance. Metals including platinum, iron, cobalt, and nickel were conventionally used as anodes as well as cathodes. Ni metal, however, is preferred as the anode owing to its excellent catalytic activity for hydrogen oxidation and reforming of hydrocarbon fuels, good chemical stability, and low cost. In addition to this, patterned-Ni has better electrochemical activity in comparison with Pt. The TEC of Ni being very large compared to known solid-electrolytes (results in poor electrode/electrolyte interface), its composite with ceramics called Cermet are introduced. Especially, the polycrystalline electrolyte ceramic phase lowers the TEC of the metal anode material to match with that of the electrolyte, it prevents coarsening of the Ni metal phase, and most importantly, it provides facile oxygen-ion conduction paths resulting in extension of electrochemical anodic active zones.

7.3.2.1 Cermets

The Cermet comprises an oxygen-ion-conducting electrolyte (e.g., YSZ) and metal (e.g., Ni). The metal Ni, an electronic conductor, acts as a catalyst (oxidation of fuel) and has been a common anode for SOFCs. The presence of the solid-electrolyte in composites provides a substantial enlargement of the electrochemical reaction zone, enhances microstructural stability, improves adherence to the electrolyte, adjusts TECs compatible with electrolyte, and compensates strains related to chemical expansion due to overpotential variations. It, in turn, necessarily provides sufficient power density and durability to the cell. The composite anode essentially provides percolation paths for electrons, oxygen ions, and gas, thus, microstructure-related factors, including processing conditions and pre-history also decide its performance.

The electrochemical oxidation of H_2 at Ni/YSZ Cermet anode can be given in terms of the Kröger–Vink notation as in Eq. (7.14):

$$H_2 + O^{2-}_{YSZ} \rightleftharpoons H_2O + 2e^{'} + V^{\cdot\cdot}_{O,\,YSZ} \tag{7.14}$$

where O^{2-}_{YSZ} and $V^{\cdot\cdot}_{O,\,YSZ}$are the oxide ions at its site and the lattice oxygen vacancy in YSZ, respectively. Further, based on complex impedance spectroscopy, Mogensen and Lindegaard (1993) proposed the following steps for hydrogen oxidation at the Ni/YSZ Cermet anode:

$$\left.\begin{array}{l} H_2 \rightleftharpoons 2H_{ad,\,Ni} \\ 2H_{ad,\,Ni} \rightleftharpoons H^{+}_{ad,Ni} + e^{-} \\ \text{Diffusion of to } H^{+}_{ad,Ni} \text{ Ni-YSZ boundary} \\ H^{+}_{ad,Ni} + O^{2-}_{YSZ} \rightleftharpoons OH^{-}_{YSZ} \end{array}\right\} \tag{7.15}$$

Reformulation of this while taking into account incorporation of H^+ into bulk of Ni as well as interstitial lattice site "I" of YSZ along with at surface results in:

$$\left.\begin{aligned} & H_2 \rightleftharpoons 2H_{Ni}^{+} + 2e^{-} \\ & H_{Ni}^{+} \rightleftharpoons H_{i,YSZ}^{+} \\ & H_{i,YSZ}^{+} + O_O^{x} \rightleftharpoons H_2O_{ad,YSZ} + V_O^{\cdot\cdot} \\ & H_2O_{ad,YSZ} \rightleftharpoons H_2O(g) \end{aligned}\right\} \quad (7.16)$$

In a nutshell, different steps involved in the anodic H_2-oxidation process may strongly effluence the rate of reaction as summarized below (Mizusaki et al., 1994):

1. On the anode electrocatalyst surface: The first step is the mass transport of gaseous fuel to the electrochemically active region of porous anode material followed by adsorption of fuel species onto electrocatalyst surface. The surface diffusion of active adsorbate species to the TPB region occurs, which is followed by a desorption of any electrocatalyst-adsorbed oxidation products from the anode electrocatalyst surface.
2. On the electrolyte surface at the vicinity of the anode: The O^{2-} ions migrate from the bulk of electrolyte up to the electrolyte surface, ensued with the surface diffusion of ionic species to the TPB region and desorption (if necessary) of any electrolyte-adsorbed oxidation products from the electrolyte surface.
3. At the three-phase boundaries: Release of electrons formed during charge transfer reaction into the electrocatalyst bulk.

Minimizing the interfacial resistance via increasing TPB length has been one of the prime approaches for Ni-YSZ anodes. In order increase TPB length, submicron/nanocrystalline Ni particles with high YSZ-to-Ni particle ratio has been employed. Mechanically derived microstructures comprising uniform distributions of submicron Ni and YSZ along with micropores has improved anode performance significantly. Alternatively, fabrication of a number of graded functional layers has also been tried in order to enhance the performance (Schneider et al., 2007).

A considerable amount of attention has been focused on the theoretical modeling to form the guidelines for achieving the desired composition and microstructure (Barfod et al., 2007). The percolation threshold (maximum performing mixture composition) has been expected around 30 Vol% of metallic phase, and approximately observed experimentally. The fabrication of YSZ and anode processing have also been equally important for its optimum fraction and eventually, for electrochemical performance. However, coarsening of the metal particles while operating at high temperatures induces significant changes in microstructure and properties of Cermets (high Ni content), resulting in fast anode performance degradation. The rise of overpotentials has been another factor, may cause microstructural reconstruction, strains, and failure that make it necessary to decrease metal concentration in Cermets. The use of noble metals may obviate above said drawbacks, but they have not been preferred due to economic reasons. Since 1970, consistent efforts have been directed to use other transition metals to attend the durability issues of Ni-based Cermets related to carbon deposition and sulfur poisoning (Niakolas, 2014). Although, Cu based Cermets have been considered to develop optimized anode for hydrocarbon fuel based SOFCs but, failed to obtain desired electrochemical performance (Kim et al., 2002). Also, the use of Fe and Co posed serious drawbacks associated with metal oxidation under high current densities. In fact, with the sulfur-desorption process being rather slower than the adsorption process, the active site binds sulfur at a higher rate than its release. Two alternative approaches are proposed to overcome sulfur poising viz. (1) remove sulfur completely from the fuel or (2) develop sulfur-tolerant anode materials. The former approach resulted in costlier fuel processing, and thus the latter approach attracted researchers to develop novel materials based on ScSZ (Scandia stabilized zirconia) and GDC in place of YSZ in Ni–YSZ Cermets or using metals/alloys other than Ni.

Since the beginning (1960s), the use of CeO_2-based additives as the SOFC anode has been considered as one of the most promising approaches of anode developments (Gorte, 2005). Its major advantages include very high catalytic activity to the combustion reactions involving oxygen and is most useful particularly for hydrocarbons and biogas based SOFCs. In addition, reduced CeO_2 and its derivatives have good MIECs; the accommodation of acceptor-type dopant further improves its transport properties and reducibility, which ultimately improves electrode performance. Further, by doping with acceptor oxides (such as CaO, Y_2O_3, GdO_3, and Sm_2O_3) its ionic conductivity can be controlled. In addition, it acts as a catalyst for direct oxidation of hydrocarbon containing fuels, such as methane, ethane, 1-butene, *n*-butane, and toluene. Furthermore, it also has excellent compatibility with adjoining cell component materials. Particularly, the GDC exhibits the higher electronic conductivity ($\sigma_{8YSZ} = 1 \times 10^{-5}$ S/cm; $\sigma_{10GDC} = 2 \times 10^{-5}$ S/cm at 850°C and 97% H_2 + 3% H_2O) and adequate surface catalytic activity. Although no carbon deposition on the $Ce_{0.6}Gd_{0.4}O_2$ electrode while operated for 1,000 h at 1000°C (steam to carbon ratio of 0.3) was observed, the electrocatalytic activity of this ceramic for direct CH_4-oxidation has been found inadequate without extra additives (Marina et al., 1999). These ceramics, however, have an edge over other ceramics under LT-SOFC operation conditions. The needed improvement in the electrochemical performance of these ceramics can be achieved through infiltration of nanocrystalline Ni onto the cermet anode surface (enlarge TPB and electrode surface area). The better performance of Ni-GDC anodes among all has been attributed to good mixed ionic- and electronic-conductivity and high H_2 adsorption rate on GDC. Thus, in spite of blockage of Ni surface by the adsorption of S, GDC can continue to allow the required electrochemical reactions. In Ni-GDC, oxygen gets adsorbed into the anode material due to the release of electrons to Ni particles, and about eight-times higher power density can be drawn out of the cell when compared to that of Ni–YSZ (Zhao et al., 2008). However, with other fuels, viz., hydrocarbons, CO, etc., the poisoning mechanisms are different and still under investigation in the research community. It was demonstrated that the carbon deposit was greatly suppressed on the Cu–Ni alloy + GDC compared to that on pure Ni-GDC (Kim et al., 2002). Ceria upon loading with noble metals viz. Pt, Pd, and Rh acts as one of the best catalysts for total oxidation of hydrocarbons relative to ceria alone or any other oxide (Futamura et al., 2019).

7.3.2.2 Perovskite-Related Oxides

The quest for alternative anode materials cantered attention on perovskite-related structures owing to more tolerant to extensive cation substitution and possessing better transport properties to oxides of other families viz., spinels, fluorites, pyrochlores, C-type oxides, or garnets (Fergus, 2006; Hui and Petric, 2001; Tsipis and Kharton, 2008; Vernoux, 1997). However, cobaltite-, manganite-, and ferrite-based oxides cannot be stable thermodynamically under anodic conditions. However, rare-earth vanadates and molybdates have displayed attractive properties in reducing environments but change in phase results in low conductivity materials on oxidation (Kharton et al., 2001). Consequently, Cr- and Ti-containing oxides have been emerged as promising materials because of their stability in the reducing conditions. These oxides, such as (La, Sr)CrO_3 and (Sr, La)TiO_3 are stable and have acceptable electronic conductivity with appropriate doping under SOFC fuel conditions. However, electrode polarization resistance, R_p, of these oxides has been reported much higher than Ni-based Cermets, resulting in low cell-power density. Substitution of a reducible transition-metal cation on the B-site has considerably increased oxygen vacancy concentrations, and in turn, oxygen ionic conductivity resulted in improved cell performance. Reduction in R_p has been attributed to fuel oxidation occurring on the entire oxide surface. The cell with $SrTi_{0.3}Fe_{0.7}O_{3-\delta}$ (STF)-based anodes has shown $R_p = 0.13\ \Omega cm^2$ at 800°C in humidified hydrogen; however, R_p has been observed to increase rapidly with decreasing temperature and H_2 partial pressure (Cho et al., 2013; Fowler et al., 2015; Zhu et al., 2016). Rate limitation by H_2 dissociative adsorption has been found to be responsible for increased R_p because there is no metallic catalyst to promote hydrogen dissociation (Zhu et al., 2016).

Introduction of nanometal catalysts onto the oxide surface has been shown to reduce R_p. One of the conventional ways to produce metal nanoparticles at the surface is an infiltration technique (Nielsen et al., 2012; Tucker, 2010). This technique leads to formation of non-uniformed nanoparticle distribution on the oxide surface. On the other hand, exsolution of reducible cations from oxide anodes, in reducing conditions, has been shown to yield well-dispersed metallic nanoparticles embedded into the oxide surface, which prevents nanoparticle coarsening (Neagu et al., 2015). Reduction in R_p has been reported for Ru nanoparticles exsoluated on $La_{0.8}Sr_{0.2}Cr_{0.82}Ru_{0.18}O_{3-\delta}$ on exposure to humidified hydrogen (Madsen et al., 2007). Numerous doped chromite and titanate anodes have been reported to exsolve transition-metal cations, including Fe, Co, Ni, Pd, and Ru. Improved cell performance for Ru-exsolved $La_{0.8}Sr_{0.2}Cr_{0.82}Ru_{0.18}O_{3-\delta}$ anodes has been ascribed to an increased rate of dissociative H_2 adsorption process (Zhu et al., 2016). Moreover, these oxide anodes with exsolved metal nanoparticles have been reported to provide excellent and stable performances with H_2 as well as hydrocarbon fuels along with tolerance to H_2S contaminant (Neagu et al., 2015; Yang et al., 2012; Zhu et al., 2015). Exsolution of alloy metal-nanoparticles has been observed for oxides containing two reducible metal cations (Fe–Co or Fe–Ni alloy nanoparticle formation in $(Pr/La)_{0.4}Sr_{0.6}Co_{0.2}Fe_{0.7}Nb_{0.1}O_{3-\delta}$, $La_{0.7}Sr_{0.3}Cr_{0.85}Ni_{0.1125}Fe_{0.0375}O_{3-\delta}$, and $LaNi_{0.6}Fe_{0.4}O_{3-\delta}$) (Luo et al., 2015; Yang et al., 2012, 2015). Furthermore, exsolved nanometal catalyst oxides have shown promising performance in solid-oxide steam electrolysis cells (Myung et al., 2016).

7.4 CONCLUSIONS AND OUTLOOK

In this chapter, the state of the art SOFC electrocatalyst materials are discussed in order to lower the operating temperature (400°C–650°C) without compromising the cell performance. Various factors affecting the cell performance, such as electrocatalytic activity, compatible thermal expansion coefficient, thermochemical stability, and microstructure (in terms of morphology, exposed/active area, porosity, size, and distribution of pores for efficient gas transportation), are deliberated in selecting suitable electrocatalysts for hydrogen oxidation and oxygen reduction. Tailored nanostructured electrode materials with enhanced electrocatalytically active reaction sites are envisaged for improved cell performance at 400°C–650°C.

REFERENCES

Afzal, M., Xia, C., & Zhu, B. (2016). Lanthanum-doped calcium manganite ($La_{0.1}Ca_{0.9}MnO_3$) cathode for advanced solid oxide fuel cell (SOFC). In *Materials Today: Proceedings* (Vol. 3, pp. 2698–2706). https://doi.org/10.1016/j.matpr.2016.06.014

Al Daroukh, M., Vashook, V. V., Ullmann, H., Tietz, F., & Arual Raj, I. (2003). Oxides of the AMO_3 and A_2MO_4-type: Structural stability, electrical conductivity and thermal expansion. *Solid State Ionics, 158*(1–2), 141–150. https://doi.org/10.1016/S0167-2738(02)00773-7

Baharuddin, N. A., Muchtar, A., & Somalu, M. R. (2017). Short review on cobalt-free cathodes for solid oxide fuel cells. *International Journal of Hydrogen Energy, 42*(14), 9149–9155. https://doi.org/10.1016/j.ijhydene.2016.04.097

Barfod, R., Mogensen, M., Klemensø, T., Hagen, A., Liu, Y. L., & Vang Hendriksen, P. (2007). Detailed characterization of anode-supported SOFCs by impedance spectroscopy. *Journal of the Electrochemical Society, 154*(4), 371–378. https://doi.org/10.1149/1.2433311

Basbus, J. F., Prado, F. D., Caneiro, A., & Mogni, L. V. (2014). A comparative study of high temperature properties of cobalt-free perovskites. *Journal of Electroceramics, 32*(4), 311–318. https://doi.org/10.1007/s10832-014-9901-9

Behling, N. H. (2013). *Fuel cells: Current technology challenges and future research needs.* Amsterdam, the Netherlands: Elsevier.

Bian, L., Wang, L., Duan, C., Cai, C., Song, X., & An, S. (2019). Co-free $La_{0.6}Sr_{0.4}Fe_{0.9}Nb_{0.1}O_{3-\delta}$ symmetric electrode for hydrogen and carbon monoxide solid oxide fuel cell. *International Journal of Hydrogen Energy, 44*, 1–9. https://doi.org/10.1016/j.ijhydene.2019.10.090

Boaro, M. (2017). *Advances in Medium and High Temperature Solid Oxide Fuel Cell Technology* (Vol. 574). https://doi.org/10.1007/978-3-319-46146-5

Brett, D. J. L., Atkinson, A., Brandon, N. P., & Skinner, S. J. (2008). Intermediate temperature solid oxide fuel cells. *Chemical Society Reviews, 37*. https://doi.org/10.1039/b612060c

Burriel, M., Peña-Martínez, J., Chater, R. J., Fearn, S., Berenov, A. V., Skinner, S. J., & Kilner, J. A. (2012). Anisotropic oxygen ion diffusion in layered $PrBaCo_2O_{5+\delta}$. *Chemistry of Materials*, *24*(3), 613–621. https://doi.org/10.1021/cm203502s

Carrette, L., Friedrich, K. A., & Stimming, U. (2001). Fuel cells: Fundamentals and applications. *Fuel Cells*, *1*(1), 5–39.

Chang, B.-Y., & Park, S.-M. (2010). Electrochemical impedance spectroscopy of composite adhesive joints. *Annual Review of Analytical Chemistry (Palo Alto, Calif.)*, *3*(6), 207–229. https://doi.org/10.1146/annurev.anchem.012809.102211

Chater, R. J., Carter, S., Kilner, J. A., & Steele, B. C. H. (1992). Development of a novel SIMS technique for oxygen self-diffusion and surface exchange coefficient measurements in oxides of high diffusivity. *Solid State Ionics*, *53–56*(PART 2), 859–867. https://doi.org/10.1016/0167-2738(92)90266-R

Chen, Y. C., Yashima, M., Peña-Martínez, J., & Kilner, J. A. (2013). Experimental visualization of the diffusional pathway of oxide ions in a layered perovskite-type cobaltite $PrBaCo_2O_{5+\delta}$. *Chemistry of Materials*, *25*(13), 2638–2641. https://doi.org/10.1021/cm4010618

Cho, S., Fowler, D. E., Miller, E. C., Cronin, J. S., Poeppelmeier, K. R., & Barnett, S. A. (2013). Fe-substituted $SrTiO_{3-\delta}$–$Ce_{0.9}Gd_{0.1}O_2$ composite anodes for solid oxide fuel cells. *Energy & Environmental Science*, *6*(6), 1850. https://doi.org/10.1039/c3ee23791e

Chroneos, A., Yildiz, B., Tarancón, A., Parfitt, D., & Kilner, J. A. (2011). Oxygen diffusion in solid oxide fuel cell cathode and electrolyte materials: Mechanistic insights from atomistic simulations. *Energy and Environmental Science*, *4*(8), 2774–2789. https://doi.org/10.1039/c0ee00717j

Dong, F., Chen, D., Chen, Y., Zhao, Q., & Shao, Z. (2012). La-doped $BaFeO_{3-\delta}$ perovskite as a cobalt-free oxygen reduction electrode for solid oxide fuel cells with oxygen-ion conducting electrolyte. *Journal of Materials Chemistry*, *22*(30), 15071–15079. https://doi.org/10.1039/c2jm31711g

Faes, A., Hessler-Wyser, A., Zryd, A., & Van Herle, J. (2012). A review of RedOx cycling of solid oxide fuel cells anode. *Membranes*, *2*(3), 585–664. https://doi.org/10.3390/membranes2030585

Fergus, J. W. (2006). Oxide anode materials for solid oxide fuel cells. *Solid State Ionics*, *177*(17–18), 1529–1541. https://doi.org/10.1016/j.ssi.2006.07.012

Fowler, D. E., Messner, A. C., Miller, E. C., Slone, B. W., Barnett, S. A., & Poeppelmeier, K. R. (2015). Decreasing the polarization resistance of (La, Sr)$CrO_{3-\delta}$ solid oxide fuel cell anodes by combined Fe and Ru substitution. *Chemistry of Materials*, *27*(10), 3683–3693. https://doi.org/10.1021/acs.chemmater.5b00622

Futamura, S., Muramoto, A., Tachikawa, Y., Matsuda, J., Lyth, S. M., Shiratori, Y., … Sasaki, K. (2019). SOFC anodes impregnated with noble metal catalyst nanoparticles for high fuel utilization. *International Journal of Hydrogen Energy*, *44*(16), 8502–8518. https://doi.org/10.1016/j.ijhydene.2019.01.223

Gao, J., Li, Q., Xia, W., Sun, L., Huo, L. H., & Zhao, H. (2019). Advanced electrochemical performance and CO_2 tolerance of $Bi_{0.5}Sr_{0.5}Fe_{1-x}TixO_{3-\delta}$ perovskite materials as oxygen reduction cathodes for intermediate-temperature solid oxide fuel cells. *ACS Sustainable Chemistry and Engineering*, 0–30. https://doi.org/10.1021/acssuschemeng.9b05086

Gorte, R. J. (2005). Recent developments towards commercialization of solid oxide fuel cells. *AIChE Journal*, *51*(9), 2377–2381. https://doi.org/10.1002/aic.10621

Gu, H., Chen, H., Gao, L., Zheng, Y., Zhu, X., & Guo, L. (2008). Effect of Co doping on the properties of $Sr_{0.8}Ce_{0.2}MnO_{3-\delta}$ cathode for intermediate-temperature solid-oxide fuel cells. *International Journal of Hydrogen Energy*, *33*(17), 4681–4688. https://doi.org/10.1016/j.ijhydene.2008.06.025

Hui, S., & Petric, A. (2001). Conductivity and stability of $SrVO_3$ and mixed perovskites at low oxygen partial pressures. *Solid State Ionics*, *143*(3–4), 275–283. https://doi.org/10.1016/S0167-2738(01)00870-0

Ishihara, T., Kudc, T., Matsudc, H., & Takita, Y. (1995). Doped $PrMnO_3$ perovskite oxide as a new cathode of solid oxide fuel cells for low temperature operation. *Journal of the Electrochemical Society*, *142*(5), 1519–1524. https://doi.org/10.1149/1.2048606

Jiang, S. P. (2012). Nanoscale and nano-structured electrodes of solid oxide fuel cells by infiltration: Advances and challenges. *International Journal of Hydrogen Energy*, *37*(1), 449–470. https://doi.org/10.1016/j.ijhydene.2011.09.067

Jun, A., Shin, J., & Kim, G. (2013). High redox and performance stability of layered $SmBa_{0.5}Sr_{0.5}Co_{1.5}Cu_{0.5}O_{5+\delta}$ perovskite cathodes for intermediate-temperature solid oxide fuel cells. *Physical Chemistry Chemical Physics*, *15*(45), 19906–19912. https://doi.org/10.1039/c3cp53883d

Kaur, P., & Singh, K. (2019). Review of perovskite-structure related cathode materials for solid oxide fuel cells. *Ceramics International*, 0–1. https://doi.org/10.1016/j.ceramint.2019.11.066

Khandale, A. P., & Bhoga, S. S. (2010). Combustion synthesized $Nd_{2-x}CexCuO_4$ (x = 0–0.25) cathode materials for intermediate temperature solid oxide fuel cell applications. *Journal of Power Sources*, *195*(24), 7974–7982. https://doi.org/10.1016/j.jpowsour.2010.06.044

Khandale, A. P., & Bhoga, S. S. (2014). Effect of Sr doping on structural, electrical and electrochemical properties of Nd_2CuO_4 for IT-SOFC application. *Solid State Ionics*, *262*(5), 416–420. https://doi.org/10.1016/j.ssi.2014.01.048

Kharton, V. V., Naumovich, E. N., Yaremchenko, A. A., & Marques, F. M. B. (2001). Research on the electrochemistry of oxygen ion conductors in the former Soviet Union-IV. Bismuth oxide-based ceramics. *Journal of Solid State Electrochemistry*, *5*(3), 160–187. https://doi.org/10.1007/s100080000141

Kim, H., Lu, C., Worrell, W. L., Vohs, J. M., & Gorte, R. J. (2002). Cu-Ni cermet anodes for direct oxidation of methane in solid-oxide fuel cells. *Journal of the Electrochemical Society*, *149*(3), 8–12. https://doi.org/10.1149/1.1445170

Kim, J. H., Baek, S. W., Lee, C., Park, K., & Bae, J. (2008a). Performance analysis of cobalt-based cathode materials for solid oxide fuel cell. *Solid State Ionics*, *179*(27–32), 1490–1496. https://doi.org/10.1016/j.ssi.2008.01.086

Kim, J. H., Cassidy, M., Irvine, J. T. S., & Bae, J. (2009). Advanced electrochemical properties of Ln $Ba_{0.5}Sr_{0.5}CO_2O_{5+\delta}$ (Ln=Pr, Sm, and Gd) as cathode materials for IT-SOFC. *Journal of the Electrochemical Society*, *156*(6), 682–689. https://doi.org/10.1149/1.3110989

Kim, J. H., Prado, F., & Manthiram, A. (2008b). Characterization of $GdBa_{1-x}Sr_xCo_2O_{5+\delta}$ ($0 \leq x \leq 1.0$) Double Perovskites as Cathodes for Solid Oxide Fuel Cells. *Journal of the Electrochemical Society*, *155*(10), 4–9. https://doi.org/10.1149/1.2965792

King, G., & Woodward, P. M. (2010). Cation ordering in perovskites. *Journal of Materials Chemistry*, *20*(28), 5785–5796. https://doi.org/10.1039/b926757c

Ling, Y., Yu, J., Lin, B., Zhang, X., Zhao, L., & Liu, X. (2011). A cobalt-free $Sm_{0.5}Sr_{0.5}Fe_{0.8}Cu_{0.2}O_3$-δ-$Ce_{0.8}Sm_{0.2}O_{2-\delta}$ composite cathode for proton-conducting solid oxide fuel cells. *Journal of Power Sources*, *196*(5), 2631–2634. https://doi.org/10.1016/j.jpowsour.2010.11.017

Liu, J., Wang, J., Belotti, A., & Ciucci, F. (2019). P-substituted $Ba_{0.95}La_{0.05}FeO_{3-\delta}$ as a cathode material for SOFCs. *ACS Applied Energy Materials*, *2*(8), 5472–5480. https://doi.org/10.1021/acsaem.9b00624

Liu, M., Lynch, M. E., Blinn, K., Alamgir, F. M., & Choi, Y. (2011). Rational SOFC material design: New advances and tools. *Materials Today*. https://doi.org/10.1016/S1369-7021(11)70279-6

Liu, Q. L., Khor, K. A., & Chan, S. H. (2006). High-performance low-temperature solid oxide fuel cell with novel BSCF cathode. *Journal of Power Sources*, *161*(1), 123–128. https://doi.org/10.1016/j.jpowsour.2006.03.095

Liu, S., Zhang, W., Li, Y., & Yu, B. (2017). $REBaCo_2O_{5+\delta}$ (RE = Pr, Nd, and Gd) as promising oxygen electrodes for intermediate-temperature solid oxide electrolysis cells. *RSC Advances*, *7*(27), 16332–16340. https://doi.org/10.1039/c6ra28005f

Luo, T., Liu, X., Meng, X., Wu, H., Wang, S., & Zhan, Z. (2015). In situ formation of $LaNi_{0.6}Fe_{0.4}O_{3-\delta}$-carbon nanotube hybrids as anodes for direct-methane solid oxide fuel cells. *Journal of Power Sources*, *299*, 472–479. https://doi.org/10.1016/j.jpowsour.2015.09.035

Madsen, B. D., Kobsiriphat, W., Wang, Y., Marks, L. D., & Barnett, S. A. (2007). Nucleation of nanometer-scale electrocatalyst particles in solid oxide fuel cell anodes. *Journal of Power Sources*, *166*(1), 64–67. https://doi.org/10.1016/j.jpowsour.2006.12.080

Mahato, N., Banerjee, A., Gupta, A., Omar, S., & Balani, K. (2015). Progress in material selection for solid oxide fuel cell technology: A review. *Progress in Materials Science*, *72*, 141–337. https://doi.org/10.1016/j.pmatsci.2015.01.001

Marina, O. A., Bagger, C., Primdahl, S., & Mogensen, M. (1999). A solid oxide fuel cell with a gadolinia-doped ceria anode: Preparation and performance. *Solid State Ionics*, *123*(1–4), 199–208. https://doi.org/10.1016/s0167-2738(99)00111-3

Merkle, R., Mastrikov, Y. A., Kotomin, E. A., Kuklja, M. M., & Maier, J. (2012). First principles calculations of oxygen vacancy formation and migration in $Ba_{1-x}Sr_xCo_{1-y}FeyO_{3-\delta}$ perovskites. *Journal of the Electrochemical Society*, *159*(2). https://doi.org/10.1149/2.077202jes

Mizusaki, J., Tagawa, H., Saito, T., Yamamura, T., Kamitani, K., Hirano, K., … Hashimoto, K. (1994). Kinetic studies of the reaction at the nickel pattern electrode on YSZ in H_2/H_2O atmospheres. *Solid State Ionics*, *70–71*(PART 1), 52–58. https://doi.org/10.1016/0167-2738(94)90286-0

Mogensen, M., & Lindegaard, T. (1993). The kinetics of hydrogen oxidation on a Ni-YSZ SOFC electrode at 1000°C. *ECS Proceedings Volumes*, *1993–1994*, 484–493. https://doi.org/10.1149/199304.0484pv

Myung, J., Neagu, D., Miller, D. N., & Irvine, J. T. S. (2016). Switching on electrocatalytic activity in solid oxide cells. *Nature*, *537*(7621), 528–531. https://doi.org/10.1038/nature19090

Neagu, D., Oh, T.-S., Miller, D. N., Ménard, H., Bukhari, S. M., Gamble, S. R., … Irvine, J. T. S. (2015). Nano-socketed nickel particles with enhanced coking resistance grown in situ by redox exsolution. *Nature Communications*, *6*, 8120. https://doi.org/10.1038/ncomms9120

Niakolas, D. K. (2014). Sulfur poisoning of Ni-based anodes for solid oxide fuel cells in H/C-based fuels. *Applied Catalysis A: General*, *486*, 123–142. https://doi.org/10.1016/j.apcata.2014.08.015

Nielsen, J., Klemenso, T., & Blennow, P. (2012). Detailed impedance characterization of a well performing and durable Ni: CGO infiltrated cermet anode for metal-supported solid oxide fuel cells. *Journal of Power Sources*, *219*, 305–316. https://doi.org/10.1016/j.jpowsour.2012.07.031

Nikonov, A. V., Kuterbekov, K. A., Bekmyrza, K. Z., & Pavzderin, N. B. (2018). A brief review of conductivity and thermal expansion of perovskite-related oxides for SOFC cathode. *Eurasian Journal of Physics and Functional Materials*, *2*(3), 274–292. https://doi.org/10.29317/ejpfm.2018020309

Park, C. Y., Lee, T. H., Dorris, S. E., & Balachandran, U. (2013). A cobalt-free oxygen transport membrane, $BaFe_{0.9}Zr_{0.1}O_{3-\delta}$, and its application for producing hydrogen. *International Journal of Hydrogen Energy*, *38*(15), 6450–6459. https://doi.org/10.1016/j.ijhydene.2013.02.119

Park, J., Zou, J., Yoon, H., Kim, G., & Chung, J. S. (2011). Electrochemical behavior of $Ba_{0.5}Sr_{0.5}Co_{0.2}$-xZnx$Fe_{0.8}O_{3-\delta}$ (x = 0–0.2) perovskite oxides for the cathode of solid oxide fuel cells. *International Journal of Hydrogen Energy*, *36*(10), 6184–6193. https://doi.org/10.1016/j.ijhydene.2011.01.142

Punde, J. D., Khandale, A. P., & Bhoga, S. S. (2014). Influence of synthesis route on electrical and electrochemical properties of $Nd_{1.8}Sr_{0.2}NiO_{4+\delta}$. *Solid State Ionics*, *262*, 701–706. https://doi.org/10.1016/j.ssi.2014.01.021

Rokni, M. (2017). Addressing fuel recycling in solid oxide fuel cell systems fed by alternative fuels. *Energy*, *137*, 1013–1025. https://doi.org/10.1016/j.energy.2017.03.082

Sakaki, Y., Takeda, Y., Kato, A., Imanishi, N., Yamamoto, O., Hattori, M., ... Esaki, Y. (1999). $Ln_{1-x}Sr_xMnO_3$ (Ln=Pr, Nd, Sm and Gd) as the cathode material for solid oxide fuel cells. *Solid State Ionics*, *118*(3–4), 187–194. https://doi.org/10.1016/s0167-2738(98)00440-8

Schneider, L. C. R., Martin, C. L., Bultel, Y., Dessemond, L., & Bouvard, D. (2007). Percolation effects in functionally graded SOFC electrodes. *Electrochimica Acta*, *52*(9), 3190–3198. https://doi.org/10.1016/j.electacta.2006.09.071

Shannon, R. D. (1976). Revised effective ionic radii and systematic studies of interatomic distances in halides and chalcogenides. *Acta Crystallographica Section A*, *32*(5), 751–767. https://doi.org/10.1107/S0567739476001551

Shao, Z., & Halle, S. M. (2004). A high-performance cathode for the next generation of solid-oxide fuel cells. *Nature*, *431*(7005), 170–173. https://doi.org/10.1038/nature02863

Singh, P., & Minh, N. Q. (2005). Solid oxide fuel cells: Technology status. *International Journal of Applied Ceramic Technology*, *1*(1), 5–15. https://doi.org/10.1111/j.1744-7402.2004.tb00149.x

Srinivasan, S., Krishnan, L., & Marozzi, C. (2006). Fuel cell principles. *Fuel Cells*, 189–233. https://doi.org/10.1007/0-387-35402-6_4

Struchtrup, H. (n.d.). Henning Struchtrup 123.

Su, C., Wang, W., Liu, M., Tadé, M. O., & Shao, Z. (2015). Progress and prospects in symmetrical solid oxide fuel cells with two identical electrodes. *Advanced Energy Materials*, *5*(14), 1500188. https://doi.org/10.1002/aenm.201500188

Sun, C., Hui, R., & Roller, J. (2010). Cathode materials for solid oxide fuel cells: A review. *Journal of Solid State Electrochemistry*, *14*(7), 1125–1144. https://doi.org/10.1007/s10008-009-0932-0

Tao, S., & Irvine, J. T. S. (2004). Catalytic properties of the perovskite oxide $La_{0.75}Sr_{0.25}Cr_{0.5}Fe_{0.5}O_{3-\delta}$ in relation to its potential as a solid oxide fuel cell anode material. *Chemistry of Materials*, *16*(21), 4116–4121. https://doi.org/10.1021/cm049341s

Teraoka, Y., Zhang, H.-M., & Yamazoe, N. (1985). Oxygen-sorptive properties of defect perovskite-type La_{1-x} Sr_xCo_{1-y} $Fe_yO_{3-\delta}$. *Chemistry Letters*, *14*(9), 1367–1370. https://doi.org/10.1246/cl.1985.1367

Thampi, K. R., Mcevoy, A. J., & herle, J. Van. (1995). Electrocatalysis in solid oxide fuel cell electrode domains. *Journal of the Electrochemical Society*, *142*(2), 506–513. https://doi.org/10.1149/1.2044089

Tsipis, E. V., & Kharton, V. V. (2008). Electrode materials and reaction mechanisms in solid oxide fuel cells: A brief review: I Electrochemical behavior vs. materials science aspects. *Journal of Solid State Electrochemistry*, *12*(11), 1367–1391. https://doi.org/10.1007/s10008-008-0611-6

Tucker, M. C. (2010). Progress in metal-supported solid oxide fuel cells: A review. *Journal of Power Sources*, *195*(15), 4570–4582. https://doi.org/10.1016/j.jpowsour.2010.02.035

Tucker, M. C., Kurokawa, H., Jacobson, C. P., De Jonghe, L. C., & Visco, S. J. (2006). A fundamental study of chromium deposition on solid oxide fuel cell cathode materials. *Journal of Power Sources*, *160*(1), 130–138. https://doi.org/10.1016/j.jpowsour.2006.02.017

Vasala, S., & Karppinen, M. (2015). A2B′B″O$_6$ perovskites: A review. *Progress in Solid State Chemistry* (Vol. 43). https://doi.org/10.1016/j.progsolidstchem.2014.08.001

Vernoux, P. (1997). Lanthanum chromite as an anode material for solid oxide fuel cells. *Ionics, 3*(3–4), 270–276. https://doi.org/10.1007/BF02375628

Vijay, P., Samantaray, A. K., & Mukherjee, A. (2011). Bond Graph Modelling of Engineering Systems. *Bond Graph Modelling of Engineering Systems*. New York, NY: Springer. https://doi.org/10.1007/978-1-4419-9368-7

Wang, C. C., Becker, T., Chen, K., Zhao, L., Wei, B., & Jiang, S. P. (2014). Effect of temperature on the chromium deposition and poisoning of $La_{0.6}Sr_{0.4}Co_{0.2}Fe_{0.8}O_{3-\delta}$ cathodes of solid oxide fuel cells. *Electrochimica Acta, 139*, 173–179. https://doi.org/10.1016/j.electacta.2014.07.028

Weber, A., & Ivers-Tiffée, E. (2004). Materials and concepts for solid oxide fuel cells (SOFCs) in stationary and mobile applications. *Journal of Power Sources, 127*(1–2), 273–283. https://doi.org/10.1016/j.jpowsour.2003.09.024

West, M., Sher, S. J., & Manthiram, A. (2014). Effects of in substitution in $Y_{1-x}In_xBaCo_3ZnO_{7+\delta}$ ($0 \le x \le 0.5$) cathodes for intermediate temperature solid oxide fuel cells. *Journal of Power Sources, 271*, 252–261. https://doi.org/10.1016/j.jpowsour.2014.08.006

Winter, M., & Brodd, R. J. (2004). What are batteries, fuel cells, and supercapacitors? *Chemical Reviews, 104*(10), 4245–4269. https://doi.org/10.1021/cr020730k

Yan, L., Ding, H., Zhu, Z., & Xue, X. (2011). Investigation of cobalt-free perovskite $Ba_{0.95}La_{0.05}FeO_{3-\delta}$ as a cathode for proton-conducting solid oxide fuel cells. *Journal of Power Sources, 196*(22), 9352–9355. https://doi.org/10.1016/j.jpowsour.2011.07.020

Yang, C., Yang, Z., Jin, C., Xiao, G., Chen, F., & Han, M. (2012). Sulfur-tolerant redox-reversible anode material for direct hydrocarbon solid oxide fuel cells. *Advanced Materials, 24*(11), 1439–1443. https://doi.org/10.1002/adma.201104852

Yang, Z., Chen, Y., Xu, N., Niu, Y., Han, M., & Chen, F. (2015). Stability investigation for symmetric solid oxide fuel cell with $La_{0.4}Sr_{0.6}Co_{0.2}Fe_{0.7}Nb_{0.1}O_{3-\delta}$ electrode. *Journal of the Electrochemical Society, 162*(7), F718–F721. https://doi.org/10.1149/2.0551507jes

Yang, Z., Guo, M., Wang, N., Ma, C., Wang, J., & Han, M. (2017). A short review of cathode poisoning and corrosion in solid oxide fuel cell. *International Journal of Hydrogen Energy, 42*(39), 24948–24959. https://doi.org/10.1016/j.ijhydene.2017.08.057

Zhang, K., Ge, L., Ran, R., Shao, Z., & Liu, S. (2008). Synthesis, characterization and evaluation of cation-ordered $LnBaCo_2O_{5+\delta}$ as materials of oxygen permeation membranes and cathodes of SOFCs. *Acta Materialia, 56*(17), 4876–4889. https://doi.org/10.1016/j.actamat.2008.06.004

Zhang, W., Zhang, L., Guan, K., Zhang, X., Meng, J., Wang, H., … Meng, J. (2020). Effective promotion of oxygen reduction activity by rare earth doping in simple perovskite cathodes for intermediate-temperature solid oxide fuel cells. *Journal of Power Sources, 446*(November 2019), 227360. https://doi.org/10.1016/j.jpowsour.2019.227360

Zhao, L., He, B., Zhang, X., Peng, R., Meng, G., & Liu, X. (2010). Electrochemical performance of novel cobalt-free oxide $Ba_{0.5}Sr_{0.5}Fe_{0.8}Cu_{0.2}O_{3-\delta}$ for solid oxide fuel cell cathode. *Journal of Power Sources, 195*(7), 1859–1861. https://doi.org/10.1016/j.jpowsour.2009.09.078

Zhao, X. Y., Yao, Q., Li, S. Q., & Cai, N. S. (2008). Studies on the carbon reactions in the anode of deposited carbon fuel cells. *Journal of Power Sources, 185*(1), 104–111. https://doi.org/10.1016/j.jpowsour.2008.06.061

Zhou, W., Shao, Z., Ran, R., & Cai, R. (2008a). Novel $SrSc_{0.2}Co_{0.8}O_{3-\delta}$ as a cathode material for low temperature solid-oxide fuel cell. *Electrochemistry Communications, 10*(10), 1647–1651. https://doi.org/10.1016/j.elecom.2008.08.033

Zhou, W., Shao, Z., Ran, R., Jin, W., & Xu, N. (2008b). A novel efficient oxide electrode for electrocatalytic oxygen reduction at 400–600°C. *Chemical Communications*, 44, 5791–5793. https://doi.org/10.1039/b813327a

Zhu, T., Fowler, D. E., Poeppelmeier, K. R., Han, M., & Barnett, S. A. (2016). Hydrogen oxidation mechanisms on perovskite solid oxide fuel cell anodes. *Journal of the Electrochemical Society*, *163*(8), F952–F961. https://doi.org/10.1149/2.1321608jes

Zhu, T., Yang, Z., & Han, M. (2015). Performance evaluation of solid oxide fuel cell with in-situ methane reforming. *Fuel*, *161*, 168–173. https://doi.org/10.1016/j.fuel.2015.08.050

Zhu, Y., Sunarso, J., Zhou, W., Jiang, S., & Shao, Z. (2014). High-performance $SrNb_{0.1}Co_{0.9-x}Fe_xO_{3-\delta}$ perovskite cathodes for low-temperature solid oxide fuel cells. *Journal of Materials Chemistry A*, *2*(37), 15454–15462. https://doi.org/10.1039/C4TA03208J

8 Overview on Metal Oxide Perovskite Solar Cells

N. Thejo Kalyani and S. J. Dhoble

CONTENTS

8.1 INTRODUCTION

Human evolution has been an extensive expedition. Discovery of fire with the help of a stone was the supreme accomplishment of primitive man. Soon after, numerous discoveries and inventions absolutely transformed human life. One among them is the generation of electricity—a great invention that made man's life more and more comfortable [1]. This comfortable life led to the consumption of energy to satisfy his own needs, and energy has become the most wanted necessity for everyone.

Since industrialization, the demand for energy increased tremendously, and the era of energy crisis has evolved gradually [2,3]. Because humans are more concerned about their future to conquer the energy crisis, research speaks volumes on sustainable renewable energy resources that are inexhaustible and environment friendly. The technology that can suppress carbon emissions needs aggressive exploration [4]. One way to address this crisis is through exploiting solar energy. This energy is an abundant and natural renewable resource, CO_2-free; it is a source of clean energy with no greenhouse gas emissions released into the atmosphere, and no fuel is required when solar panels are used to create electricity. So, this technology is an appropriate technology to harvest the most abundant solar energy; however, it is possible only during daylight hours and is frequently at the mercy of weather [5]. The energy received from the sun onto the earth can be easily harvested by (1) solar thermal collectors; (2) concentrating the sunrays onto water bodies by lenses to generate steam, which drives a turbine to engender power; and (3) photovoltaic (PV) technology, which converts the light energy to create potential difference [6,7]. To deal with these issues, the journey of PVs started from first-generation silicon solar cells to fourth-generation perovskite solar cells (PSCs). Among various structural materials of perovskites, certainly, metal oxide perovskite solar cells with higher efficiency and stability evolved as a gorgeous substitute to traditional silicon solar cells [8,9]. However, in view of their quantum efficiency, confronts still exist and hence need to be addressed [10]. This chapter mainly focuses on a miniature segment of the gigantic world of metal oxide perovskite solar cells, which improves the stability and effectiveness of solar cells with the preference of electrodes, charge carrier transport materials, and encapsulation approaches based on a path-breaking holistic approach.

8.1.1 Photovoltaic Generations

There has been tremendous progress in improving the efficiency of PV technology by exploring novel materials and relevant structures, which consequently generate electricity by employing solar energy effectively and efficiently. With time, different materials evolved based on their crystalline/amorphous nature, band gap, absorption, and manufacturing complexity.

8.1.1.1 First-Generation Solar Cells

First-generation solar cells are popularly known as traditional solar cells. They are generally made of the basic semiconductor element silicon (Si), whose energy band gap is of the order of 1.12 eV. This silicon wafer-based solar cell technology exploits crystalline silicon, and they encompass a large area single layer *p-n* configuration. They can be further divided into mono or single crystalline silicon solar cells (single crystal of silicon) and poly or multi-crystalline silicon solar cells (silicon fragments are melted together) [11]. These solar cells were found to have high efficiency, longer lifetime, expansive spectral absorption assortment, elevated carrier mobility, and good efficiency [12]. Other compound semiconductors such as GaAs can also be used for solar cells. Such solar cells are made by doping *n*-type and *p*-type impurities, so as to enhance the required number of electrons and holes, respectively.

However, they lag owing to its fragility, rigid structures and expensive production processes, and poor photon to potential difference conversion efficiency.

8.1.1.2 Second-Generation Solar Cells

Second-generation solar cells are conventional thin-film-based solar cells. They include amorphous silicon (hydrogenated amorphous silicone-Si: H) and chalcogenides, which includes (1) cadmium telluride (CdTe) solar panels, which can confine energy at lower wavelengths as compared with silicon panels. Furthermore, their developing costs are relatively low. However, they offer poor efficiency as well employ toxic element (Cd) and scarce element (Te); (2) copper zinc tin sulfide (CZTS) and other class of materials, which belong to I_2-II-IV-VI_4 groups such as copper zinc tin selenide (CZTSe) and the sulfur-selenium alloy copper zinc tin sulfide selenide (CZTSSe), raised the solar cell efficiency to 12.6%. Nevertheless, more work is essential before their commercialization [13]; and (3) copper indium gallium diselenide (CIGS) solar panels, which have higher efficiencies than CdTe panels. It has a high absorption coefficient and even very slender film has the ability to strongly absorb sunlight more than other semiconductor materials [14]. This generation of solar cells are not expensive and even have improvised efficiency. They also employ alternative manufacturing technologies such as vacuum deposition, and electroplating. However, high production costs have averted CIGS panels for commercialization.

8.1.1.3 Third-Generation Solar Cells

Third-generation solar cells are thin film solar cells, and they include (1) photoelectrochemical (PEC) cell, (2) dye-sensitized solar cells (DSSC), (3) nanocrystal solar cells, (4) quantum dot solar cell (QDSC), and (5) organic photovoltaics (OPV).

8.1.1.3.1 Photoelectrochemical Cell

It is a photocurrent-generating device that consists of working electrode (either *n*- or *p*-type photoactive semiconductor) and counterelectrode (metal such as platinum or any semiconductor). When the electrodes are submerged in the electrolyte enclosing appropriate redox couples, at the metal–electrolyte junction, a potential drop on the solution site takes place. However, at the semiconductor–electrolyte junction, a potential drop takes place both at the semiconductor and the solution site. The space charge region is created due to the fact that the charge is dispersed deep in the core of the semiconductor. When the semiconductor–electrolyte junction is illuminated by light with photon energy (E) > optical energy band gap (E_g) of the semiconductor material, photoengendered electrons and holes become alienated in the space charge region, creating a potential difference.

8.1.1.3.2 Dye-Sensitized Solar Cell

DSSCs, popularly known as Gratzel cells, are translucent, power-generating building blocks [15]. They are supported on a semiconductor produced between a photosensitized anode and an electrolyte, which forms a PEC system. Unique features of these DSSC include (1) use of conventional roll-printing techniques, (2) they are partly flexible and partly transparent, and (3) they are mechanically robust [16]. Best semiconducting materials for DSSCs must hold some unique features, such as

(1) broad energy band gap, (2) elevated electron injection capacity, (3) cheap, (4) rich carrier concentration, (5) poor static dielectric constant, and (6) high electron mobility. Materials, generally used in the photoelectrode of the DSC are titanium oxide (TiO_2), zinc oxide (ZnO), tin oxide (SnO_2), and indium (III) oxide (In_2O_3). The electrolytes that can be used in DSSC are liquid electrolytes (LEs) or gel polymer electrolytes (GPEs). LEs are generally ionic liquids such as potassium iodide (KI_2), sodium iodide (NaI_2), or electrolytes. The performance of LEs is effective only at moderate temperatures. At lower and higher temperatures, it may freeze or expand, thereby creating the stability crisis and even physical damage. Recently, GPEs commenced to triumph over the inadequacy of LEs [17]. However, they are bit costly because they use ruthenium (dye), conducting glass or plastic (contact), and platinum (catalyst).

8.1.1.3.3 Nanocrystal Solar Cells

Nanocrystal solar cells are a type of solar cell with nanocrystals coating the substrate. They employ nanocrystals, which are crystalline particles with at least one dimension measuring less than 1000 nm (quantum dot, quantum wire, quantum well). Nanocrystalline silicon is totally different from bulk-silicon in some properties, such as (1) carrier relaxation time (the time taken by the carriers to return back lower energy state) is dramatically reduced because of the established fact that they have large band gap, due to which electrons cannot quickly relax to its initial state because of a weak Coulomb interaction [18]. Hence, these electrons can contribute to the current; (2) formation of more than one electron per absorbed photon, generally known as multiple exciton generation (MEG) is possible. Silicon nanocrystal solar cells are relatively easy to make and hence cheap. By employing nanocrystals in solar cells, their efficiency can be enhanced to a certain extent.

8.1.1.3.4 Quantum Dot Solar Cell

A QDSC employs quantum dots (QDs) as the absorbing PV material. The basic difference between nanoparticles and QDs is that nanoparticles are the particles in the nanometer size regime, while quantum dots are those nanoparticles in the quantum size regime characterized by the discretization of the energy levels inside the material. The particle size is comparable to the exciton bhor radius, which depends on effective mass of the electron/hole and dilectric constant, which is different for different materials. One of the significant striking features of these QDs is their band gaps, which can be finely tuned over a broad assortment of energy levels by manipulating their particle size. This makes a superior option for multi-junction solar cells. In bulk materials, the band gap is fixed for a given material, and hence this band gap tunable property makes QDs gorgeous for multi-junction solar cells [19,20]. A wide range of materials are employed to enhance the efficiency by gathering numerous segments of the solar spectrum. Solar quantum dots (SQDs) have the prospective to augment the power conversion efficiency in PV operation due to enhancement of photoexcitation [21,22]. However, CdSe-based QD solar cells brick bats are (1) extremely toxic and hence need more stable polymer shells that can modify optical properties, (2) degradation increases in aqueous and UV conditions, and (3) difficult to control the particle size.

8.1.1.3.5 Organic Photovoltaics

Instead of employing inorganic materials such as silicon or copper, OPVs make use of slender coatings of organic vapor or solution to produce electricity by exploiting sunlight. Either they employ small organic molecules or conductive organic polymers for light absorption and charge transport to manifest electricity from sunlight [23,24]. The presence of impurities also affects exciton diffusion length, charge separation, and collection. This generation of solar cells are more efficient, highly durable, and non-toxic than the earlier generation PVs. However, their efficiency and life span needs to be improvised before commercialization.

8.1.1.4 Fourth-Generation Solar Cells

Fourth-generation solar cells combine the striking features of polymer thin films as well as novel inorganic nanostructures, and hence they are cheap, flexible, and stable. These features help in enhancing the optoelectronic properties of the low-cost thin film PV. They include emerging hybrid PVs such as PSCs and hybrid inorganic crystals within a polymer matrix. Perovskite materials have materialized as proficient low-cost energy materials for a variety of optoelectronic and photonic device applications. Figure 8.1 highlights the progress journey of solar cells from their first to fourth generation. A comparison of various attributes such as efficiency, toxicity, and payback period is tabulated in Table 8.1. The power conversion efficiency of PSCs has augmented at a phenomenal rate in comparison to other PVs, as depicted in Figure 8.2.

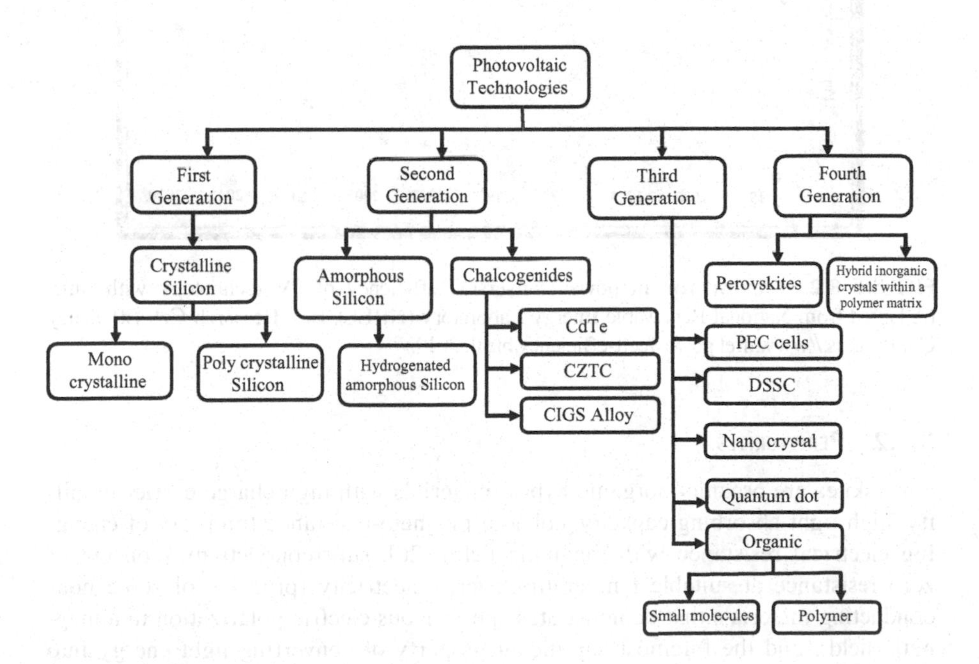

FIGURE 8.1 Evolution of solar cells from 1G to 4G.

TABLE 8.1
Comparison of Photovoltaic Technologies

S.No.	Characteristics	First Generation	Second Generation	Third Generation	Fourth Generation
1.	Abundance	High	Low/Moderate	High	High
2.	Cost of pre-cursors	Low	Low	Low	Low
3.	Fabrication cost	High	Moderate	Low	Low
4.	Efficiency	High	Moderate	High	High
5.	Toxicity	Low	Moderate	Low	High
6.	Durability	Low	Moderate	High	High
7.	Flexibility	Low	Moderate	Moderate	High
8.	Band gap	Low	Moderate	Moderate	Tunable
9.	Payback period	Low	Moderate	Low	Low

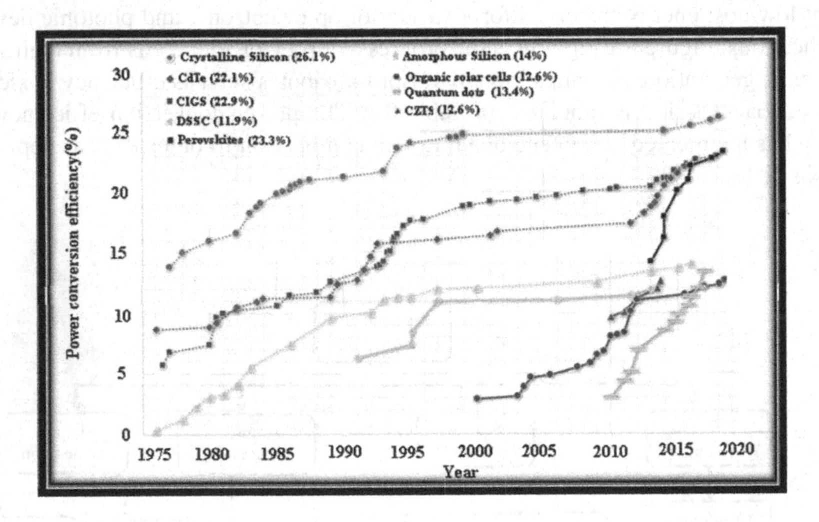

FIGURE 8.2 Improvement in power conversion efficiency of PV technology with time. (Adapted from National Renewable Energy Laboratory (NREL). Best Research-Cell Efficiency Chart, https://www.nrel.gov/pv/cell-efficiency.html, 2019.)

8.1.2 Perovskites

Perovskites are organic-inorganic hybrid materials with high charge carrier mobility, high light absorbing capacity, colossal magneto-resistance (property of changing electrical resistance with magnetic field) [26], superconductivity (concept of zero resistance at suitable temperature), ferro-electricity (property of some non-conducting materials that demonstrates spontaneous electric polarization in a magnetic field), and the foremost significant property of converting light energy into potential difference (photovoltaics) to name a very few. These perovskites came into the lime-light with the discovery of calcium titanate ($CaTiO_3$) in 1839 by a Russian mineralogist Lev von Alekseevich Perovskite. Materials with similar crystal structure are also known as the Perovskite materials. A characteristic example of the metal oxide perovskite structure is barium titanate ($BaTiO_3$). It has 1/8th of Ba atoms at each of the eight corners, one titanium atom at the center of the cube, and 1/2 of oxygen atom at the center of all six faces, resulting in the formula ABO_3 (A and B are cations, and O is the oxygen anion) [27] as shown in Figure 8.3a. Other typical examples include $SrTiO_3$, $BiFeO_3$, $CaTiO_3$, $SrZrO_3$, $CaMnO_3$, $PbTiO_3$ etc. These class of oxides can endure huge levels of dopants with no phase transformations, they are chemically stable even at elevated temperatures and their electronic properties can be tuned by choosing appropriate dopants [28]. The other class of perovskites include compounds with basic chemical formula ABX_3, where "A" and "B" represent cations (size of A > size of B, usually an alkaline earth or rare-earth element), and X is an anion (generally oxides or halogens). A-site ion is located on the corners of the lattice, while B-site ions occupy the center of the lattice as shown in Figure 8.3b.

These perovskites have submetallic to metallic luster, colorless streak, and cube-like structure, along with imperfect cleavage and brittle tenacity. Crystals of perovskite appear as cubes, but are pseudocubic and crystallize in the orthorhombic system. These materials can be considered as versatile class of compounds used as PV materials as a result of the material's excellent light absorption capacity, high charge-carrier motilities, tunable band gap, good lifetime, and many more as depicted in Figure 8.4.

8.1.3 Classification of Perovskites

Perovskites can be classified as inorganic oxide perovskites and halide perovskites. Inorganic oxide perovskites are further divided into two classes, namely, intrinsic inorganic oxide perovskites and doped inorganic oxide perovskites.

Similarly, halide perovskites are categorized as alkali halide perovskites (ABX_3), and organo-metal halide perovskites as shown in Figure 8.5.

8.1.3.1 Transition Metal Oxide Perovskites

Transition metal oxides (TMOs) fall under the class of solids, which exhibit variety of structures and properties. Remarkable properties of these TMOs is due to

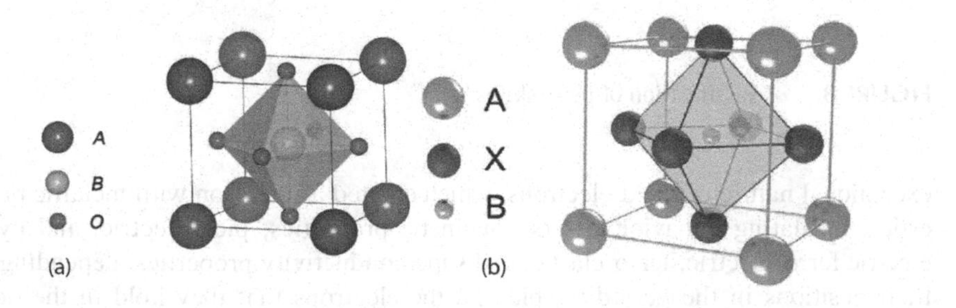

FIGURE 8.3 Structure of (a) inorganic oxide perovskite (ABO_3) and (b) metal halide (ABX_3).

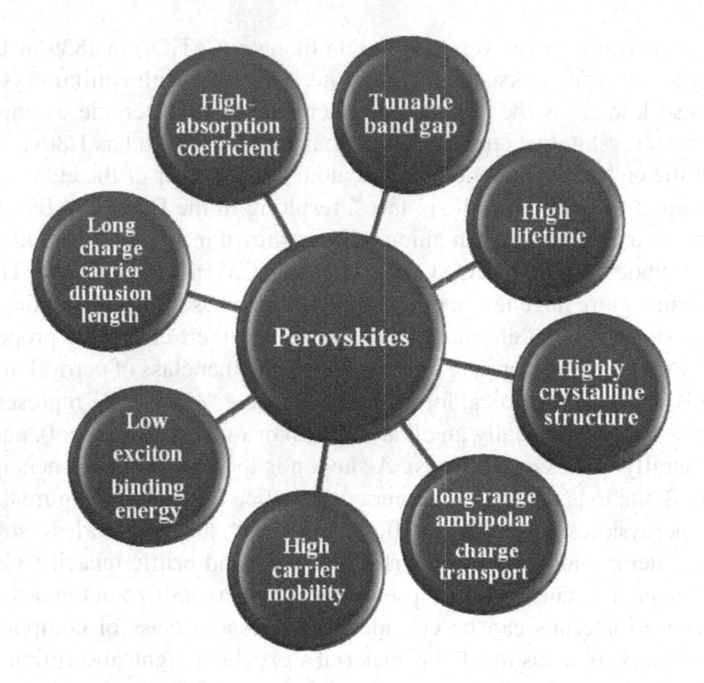

FIGURE 8.4 Superior features of perovskites.

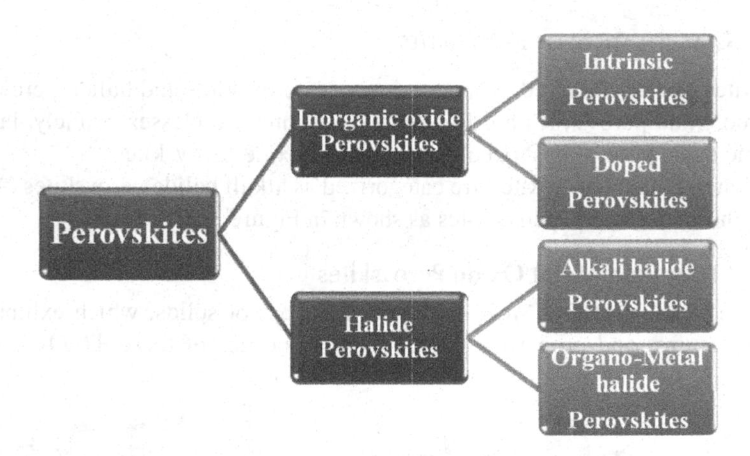

FIGURE 8.5 Classification of perovskites.

exceptional nature of the d-electrons, which created a sensation with metallic properties, insulating behavior, diverse magnetic properties, piezoelectric, and pyroelectric ferro-electric, ferro-elastic, and superconductivity properties, depending on their positions in the periodic table and the electrons that they hold in the outer most orbit [29]. For instance, the compound $SrTiO_3$ has no d electrons; it behaves as semiconductor as well as diamagnetic material under certain conditions. Similarly,

TABLE 8.2
Classification of Transition Metal Oxides

Transition Metal Oxides			
Base Transition Metal Oxides			Nobel Transition Metal Oxides
Monometallic Oxides		Bimetallic Oxides	
Examples	NiO, CuO, TiO_2, CoO, ZnO, MnO_2, Fe_2O_3, Fe_3O_4, V_2O_5, VO	$NiFe_2O_4$, $NiCO_2O_4$, $ZnCO_2O_4$, $BaTiO_3$, $NiMn_2O_4$, $MnCO_2O_4$, NiV_2O_8	RuO_2,IrO_2,MoO_3
Applications	Photovoltaic cells, biological activities, magnetic resonance imaging etc	Photovoltaic cells, drug delivery, water purification and catalysis etc.	Pseudo-capacitors

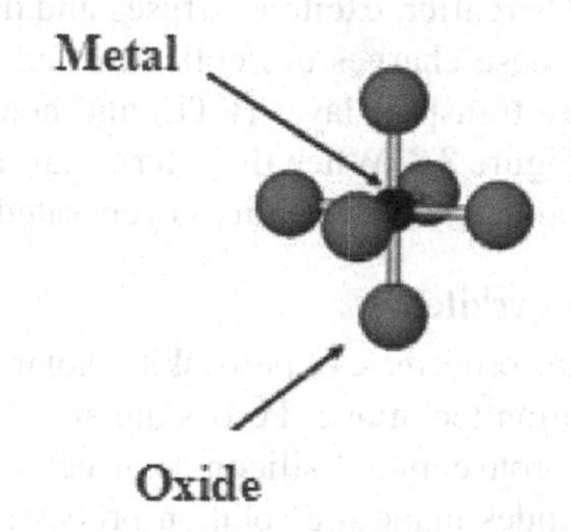

FIGURE 8.6 Generalized structure of transition metal oxide (TMO).

$CaMnO_3$ has four d-electrons, and it behaves as semiconductor as well as antiferromagnetic material under apt conditions. Classification and general structure of TMOs are shown in Table 8.2 and in Figure 8.6, respectively.

Based on crystal structure, bonding, band structure, and their electrical and magnetic properties, TMOs can take three different structures: (1) perovskite (AMO_3/ReO_3), (2) rock salt (MO), and (3) Rutile (MO_2). In all the three structures, M-O interactions rule the properties; however, the latter two structures also consider M-M bonding. These materials find diverse applications in various fields such as electromechanical devices, transducers, capacitors, dielectrics, dynamic random-access memory, field effect transistors (FETs), logic and integrated circuitry, and many more [30].

Monometallic TMOs such as ZnO, TiO_2, and V_2O_5 have special physical and chemical properties; they can be processed as nano-dimensional materials with immensely varied designs and properties. Bimetallic metal oxide perovskites gained momentum as light absorbers for PV applications. In ferroelectric oxide perovskite ($BaTiO_3$)-based solar cells, the photoexcited carriers are separated by the polarization-induced internal electric field, thereby engendering open-circuit voltage (Voc) that exceeds the band gap (E_g) of the absorbing material. However, the power conversion efficiencies of these PV devices are poor due to very low short-circuit current density [28]. Noble transition metal oxides are generally used as the materials, which exhibit pseudocapacitance activity.

8.1.3.2 Basic Anatomy of Planar Perovskite Solar Cells

A PSC is a type of solar cell that includes a perovskite-structured compound, most a commonly hybrid organic-inorganic lead or tin halide/oxide-based material, as the light-harvesting active layer. Basic anatomy of planar perovskite solar cells and material requirements of different layers and commonly used materials for perovskite solar cells are shown in Figure 8.7 and Table 8.3, respectively.

8.1.3.3 Working Mechanism

The basic working mechanism involves absorption of sunlight by the light-absorbing material (perovskite in this case), which then leads to the formation of excitons (a pair of electron-hole). Thereafter, exciton diffuses and dissociates (electron-hole pair gets separated), and these charges are collected and transported to the electrodes through the electron transport layer (ETL) and hole transport layer (HTL), respectively, as shown in Figure 8.8. When these terminals are connected across the load, potential difference and hence electricity is generated [43].

8.1.3.4 Synthesis of Perovskites

The most common synthesis procedure of perovskites solar cells is by either solution technique or vapor deposition technique. Perovskite solar cells hold the advantage of simple processing than conventional silicon solar cells. Methyl ammonium and formamidinium lead trihalides made the solution processing and vapor deposition techniques feasible. Classification of perovskite synthesis processes is detailed in Figure 8.9.

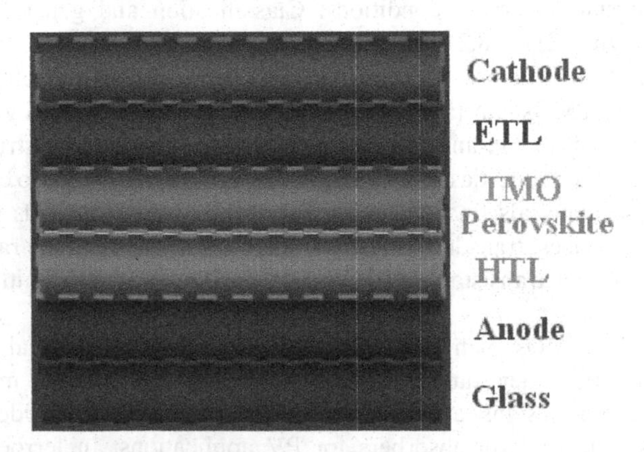

FIGURE 8.7 Basic anatomy of planar perovskite solar cells.

TABLE 8.3
Material Requirements of Different Layers in Perovskite Solar Cell

Materials	Function	Requirements	Examples	References
Anode	To collect positive charge carriers	• Good electrical conductivity • Transparent to visible light • Exceptional linkage to the substrates	ITO ZFO FTO Graphene	[31,32]
HTL	To transport holes from light-absorbing layer to anode	• To prevent Au infiltration in perovskite • To protect perovskite from moisture and oxygen • Block electrons and improve hole transfer efficiency • Physical energetic barrier between anode and perovskite • High thermal and morphological stability • Good solubility in orthogonal solvents • High glass transition temperature >100°C to prevent crystallization	CuPc TIPS-Pentacene NiO, PEDOT:PSS PTAA	[33,34]
Transition metal oxide Perovskite	To absorb the incident light	• Excellent light-absorption capacity • High charge-carrier motility • Tunable band gap • Good lifetime	TiO_2 MnO_2 $SrTiO_3$ $SrZrO_3$ $PbTiO_3$	[35–37]
ETL	To transport electrons from light-absorbing layer to cathode	• Good electron transporting and hole-blocking properties • High electron affinity • High ionization potential	ADN Alq_3	[38,39]
Cathode	To collect negative charge carriers	• Low work function • Less reactive to nature	Ag Al Au	[40–42]

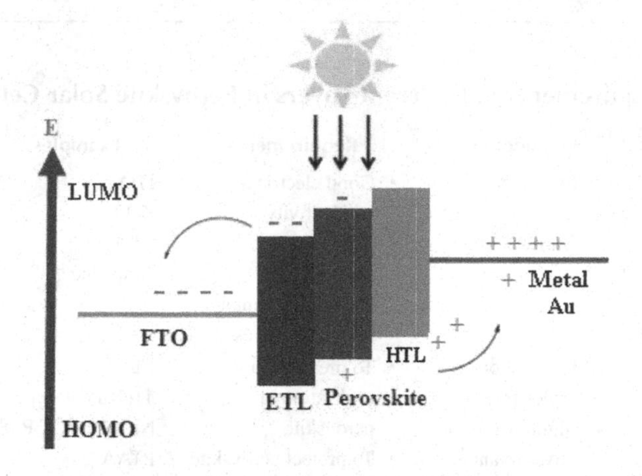

FIGURE 8.8 Basic working mechanism of perovskite solar cells.

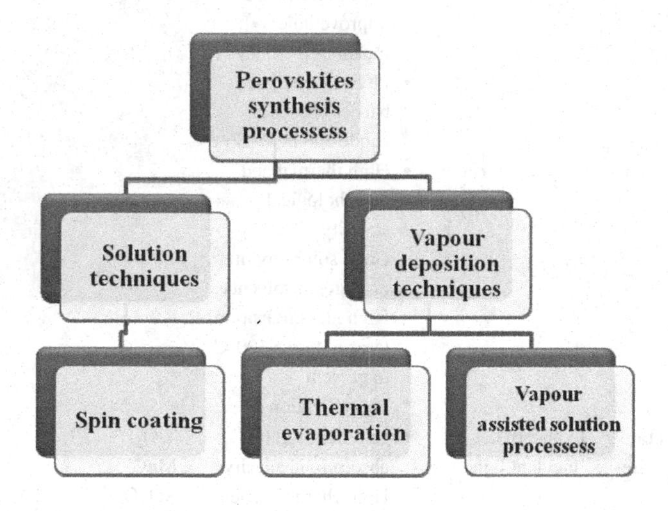

FIGURE 8.9 Classification of perovskite synthesis processes.

Perovskite solar cells can be prepared effortlessly by low-cost solution techniques such as spin coating, either by one step or by two steps. In the former case, perovskite light-absorbing layers, say X and Y, are dissolved in preferred solvents to form a precursor solution, maintaining stoichiometric ratio and then spin coated on a substrate at suitable temperature and atmosphere by using dropper. The substrate is then rotated at very high revolutions per minute (rpm) so that these materials can be homogeneously dispersed onto the substrate, followed by the annealing process. Though this process is easy going, controlling the morphology and size of the synthetic crystals is a big challenge. In the latter case, material of layer X and material of layer Y (say) are mixed by the process of sonication, considered as Step 1. In this case, initially the material of layer X is deposited on the substrate, dried for certain time, and later material of layer Y is deposited and they are allowed to react to form perovskite material on the substrate itself. Finally, the formed perovskite material is subjected for annealing. This method is favorable for the fabrication of perovskite films even under high humidity [44].

The vapor deposition technique is a kind of thermal evaporation process, normally carried out under high vacuum, hence less chances of degradation of perovskite materials. This technique can be subclassified as a co-evaporation process and sequential evaporation process. In the process of co-evaporation, required layers (say X and Y) are deposited on to the precoated substrate, simultaneously at an appropriate temperature and pressure and then crystallized into a perovskite film. This process is carried out at high temperature so as to evaporate the solid into vapor. In case of sequential evaporation, three steps are involved. In the first step, say layer X is deposited on the precoated substrate by thermal evaporation, followed by vapor depositions of layer Y in the second step. In this case, materials of layer X and Y are depositing on one another [45]. Finally, they are subjected to annealing process in Step 3. The PV performance of the devices depends on the temperature of the substrate. The other method, which involves vapor deposition, is a vapor-assisted solution process; the first step is solution processing like the spin-coating technique, while the second step is exposure of organic vapors (all materials in vapor phase) on the top of that materials. The growth of perovskite film is carried out in situ-reaction of the deposited film X with organic vapors Y. This process avoids the drawbacks of the solution method and the vapor deposition method. However, requirement of high vacuum is the major limitation in thermal evaporation method. The technical features of silicon solar cells, and perovskite solar cells are tabulated in Table 8.4.

8.1.4 Characterization Techniques

For the optimization of the solar cells, various analytical methods, which allow the investigators to characterize them on a molecular level and determine their characteristic components, are employed. They include scanning and transmission electron microscopy, X-ray, ultraviolet, synchrotron radiation photoelectron spectroscopy, Raman and UV/Vis spectroscopy, inverse photoelectron spectroscopy and energy loss spectroscopy, X-ray diffraction, quantum efficiency, and I–V characteristics with a solar simulator.

TABLE 8.4
Comparison of Technical Features: Si Versus Perovskite Solar Cell

S.No.	Parameter	Silicon Solar Cells	Perovskite Solar Cells
1.	Processing temperature	High > 1000°C	Moderate/Low
2.	Materials	Inorganic	Inorganic-Organic hybrid
3.	Technology	Wafer technology	Thin-film technology
4.	Scaling	Limited to small areas	Large area coatings in one stretch
5.	Commercialization	Commercialized	In the stage of technology transfer

8.1.5 Prior State-of-the-Art on Metal Oxide Perovskite Solar Cells

Perovskite solar cells have been considered as the rising star in the world of PVs due to swift enhancement in their efficiency and hence creating enormous attention in the scholastic community. However, their outfitted methods are yet new relatively, and hence there is enormous prospect for future research in the field of perovskites. In the past few years, the engineering up-gradation of perovskite formulations and fabrication techniques elevated the power conversion efficiency to around 23%. Prior to their use for solar cells, they have been conventionally employed in diverse areas. In 1993, T. Arima et al. [46] carried out the optical study of trivalent 3d–TMO compounds (RMO_3) with the perovskite-like structure. Their study revealed that electronic structure and properties of perovskites varies with the metal (3d element) and A-site rare-earth element. In 1996, T. Mizokawa et al. [47] employed Hartree-Fock (HF) approximation to study 2p lattice models of transition-metal 3d-oxygen, where complete degeneracy of transition-metal 3d and oxygen 2p orbitals and onsite Coulomb and swap interactions among 3d electrons are considered. They extensively studied the relationship between spin- and orbital-ordered solutions and the Jahn-Teller-type and $GdFeO_3$-type distortions in $RTiO_3$, RVO_3, $RMnO_3$, and $RNiO_3$ (R is a rare earth atom or Y). S. Ishihara et al. in 1997 [48] studied the effect of the degeneracy of energy gap orbitals in Perovskite transition-metal oxides (PTMOs) in the limit of strong repulsive electron-electron interaction. They established isospin field to illustrate the orbital degrees of freedom. Few inconsistent experiments on $La_{1-x}Sr_xMnO_3$ in the low-temperature ferromagnetic phase were also analyzed. The possible implications in coupling orbital degrees of freedom and Jahn-Teller lattice distortion were interpreted.

In 1998, Rodriguez-Martinez [49] studied the structural changes by powder neutron diffraction in a series of 30% hole-doped $AMnO_3$ perovskites. Eng et al. [50] in 2003 investigated on Ti^{4+}, Nb^{5+}, Ta^{5+}, Mo^{6+}, and W^{6+} ions in perovskite and its related structures. They have established the fact that the effective electro-negativity of these ions followed the order: $Mo^{6+} > W^{6+} > Nb^{5+} \sim Ti^{4+} > Ta^{5+}$. Since 1995, hundreds of such new-fangled materials have been demonstrated. Imperative progress has comprised perovskites with complex cation orderings on A and B sites, multi-ferroic bismuth-based perovskites, manganites illustrating colossal magneto-resistance (CMR) and charge ordering properties [51]. The *n*-type TMOs comprising of molybdenum oxide (MoO_x) and vanadium oxide (V_2O_x) are employed as an efficient hole extraction layer (HEL) in heterojunction ZnO/PbS quantum dot solar cells (QDSC) by Gao et al. in 2011 with efficiency improvements [52]. Ferroelectrics have strong inversion symmetry breaking due to spontaneous electric polarization, which improves the bifurcation of photoexcited charge carriers, thereby enabling higher voltages and efficiency ahead of its predecessor conventional solar cells. Hence, these materials are considered to be the potential candidates for coupling light absorption in addition to some peculiar functional properties. They can be fabricated by sol–gel thin-film deposition and sputtering, which are cost effective with an advantage of high mechanical, chemical, and thermal stability. It has been established that trimming down the thickness of ferroelectric layer and cautiously developed domain structures and ferroelectric–electrode interfaces can enormously amplify the harvested current from these ferroelectric absorber materials. This tremendously augments the PCE from 0.0001% to 0.5%. Auxiliary perfections in PV efficiency were hindered by the wide band gaps (2.7 to 4 eV) of ferroelectric oxides, which allocate only 8%–20% utilization of the solar energy. This constrain was prevailed by direct band gap materials with ferro electric properties like $[KNbO_3]_{1-x}[BaNi_{1/2}Nb_{1/2}O_{3-\delta}]_x$ (KBNNO), in the range (1.1 to 3.8 eV). The capability of KBNNO to absorb additional solar energy more than the contemporary ferroelectric materials recommends a direction toward feasible ferroelectric semiconductor-based cells for PVs [53]. The efficiency of the solar cell is calculated by the formula

$$\eta = \frac{V_{oc} \times I_{Sc} \times FF}{P_{in}}$$

where V_{oc} = open circuit voltage: the electrical potential difference between two terminals of a device when disconnected from any circuit, I_{Sc} = short-circuit current: the current through the solar cell when it is short circuited (voltage across the cell is zero), FF = Fill factor = $\frac{P_{max}}{I_{Sc} \times V_{OC}}$.

P_{max} = maximum power that can be obtained from the solar cell, P_{in} = Input power.

The first article published on a perovskite absorber by the research group of Park et al. in 2011 [54] followed an unprecedented rapid improvement in efficiency to more than 19% within less than three years. In 2016, researchers from Korea Research Institute of Chemical Technology (KRICT) and Ulsan National Institute of Science and Technology (UNIST) managed to reach a certified efficiency of over 22%. These results were achieved with the organic-inorganic metal halide, methyl-ammonium lead iodide ($CH_3NH_3PbI_3$), which has a perovskite crystal structure, using solution-based thin-film deposition techniques [55]. Perovskite thin films exhibit exceptional properties for solar cell applications, such as high absorption, a suitable band gap, high charge carrier diffusion length, and low recombination rate. Later, quantum dot-based solar cells have exposed massive prospective with better concert, as economical PVs due to the exceptional ability to generate MEG. The prior state-of-the-art and future perspectives are evidently explored by Etgar in the year 2013 [56].

Zhao et al. reported 11.4% efficient perovskite $CH_3NH_3PbI_3$ solar cell with molybdenum oxide/aluminum (i.e., MoO_x/Al) as the cathode with good hole extraction efficiency. The role of the thickness of cathode materials in the anatomy of solar cells was assessed and concluded that cell performance varies inversely with the cathode layer thickness [57]. Wang et al. in 2014 observed that a *p*-type metal oxide (NiO inorganic metal oxide nanocrystalline) in place of organic hole transport materials proffers best device architecture and efficiency in perovskite-based thin-film solar cells and tandem PVs [58]. In the year 2015, Yeo et al. developed an organometallic hybrid perovskite solar cell with a planar configuration of glass/ITO/RGO/$CH_3NH_3PbI_3$/$PC_{61}BM$/(BCP)/Ag (where ITO = indium tin oxide, RGO = reduced graphene oxide, $CH_3NH_3PbI_3$=methyl aluminum lead halide $PC_{61}BM$ = [6,6]-phenyl C61-butyric acid methyl ester, BCP = bathocuproine and Ag = silver) and attained a maximum PCE of 10.8% [59]. In the same year, Seong Sik Shin fabricated inorganic–organic hybrid PSCs on flexible plastic substrates, which tremendously curtail the costs. They also revealed a new technique to disperse Zn_2SnO_4 (ZSO) nanoparticles at temperatures less than <100°C for the progress of PSCs and attained a PCE of 14.85% under AM 1.5 G 100 $mWcm^{-2}$ illumination [60]. For thin-film based PSCs, Jung et al. exploited Cu-doped NiO_X (Cu:NiO_x) as HTL and attained a PCE of 17.74%. They availed the advantage of synthesizing it by traditional combustion method; also, it can be solution-processed at relatively low temperatures [61]. Xu et al. reported a cross-linked hole-extracting electrical contact for PSCs and established that such configuration improves the stability and also lowers the hysteresis of the cell [62]. In 2016, Werner et al. [63] studied the coloration of molybdenum and tungsten oxide layers, used in fabrication of PVs. Qin et al. in 2017 [64] employed transition metal oxide semiconductors such as NiO_x, CuO_x, CrO_x, MoO_x, WO_3, and V_2O_5 as HTL in an organic semiconductor and hybrid perovskite-based solar cells so as to improve hole-gathering ability, minimize charge recombination, enhance built-in voltage, and hence improved performance and device stability. Hou et al. achieved maximum efficiencies of 21.2% with a Tantalum-doped tungsten oxide ($TaWO_x$)/conjugated polymer multilayer. They achieved illumination stability of about 1000 hours. These findings opened new avenues for the wide range of organics as scalable hole transporting materials for PSCs [65]. This survey concludes that TMO perovskites have grabbed immense curiosity among the scientists and researchers in the development of new class of materials for solar cells, thereby allocating advances of huge significance for contemporary society.

8.1.6 Applications

The superior properties of oxide-based perovskites were comprehensively premeditated for assorted applications as shown in Figure 8.10, and the commonly employed perovskite materials for various applications are tabulated in Table 8.5 [66,67].

8.1.7 Real-World Concerns

Despite numerous striking features, metal oxide transition perovskites still face many challenges with respect to their stability, efficiency, toxicity, and hence real-time perovskite PV applications are yet to be addressed before commercialization.

8.1.7.1 Toxicity

With the inclusion of lead, a toxic metal in perovskite cells rules out the idea of using them under the class of disposable products. So, the researchers are not ready to compromise on lead toxicity. They also tried for different substitutes for lead, such as tin; however, they couldn't even attain the best performance out of them.

8.1.7.2 Scaling Problem

The aforementioned PSCs efficiencies are acquired on small-area devices, less than 1 cm^2 area. However, when accomplished on the large scale, they don't turn out with worthy efficiencies. With the emergence of solution-processed thin-film solar cells technology, the scaling has now raised to an appreciable extent. However, the grounds of the scaling issue are multi-faceted, yet to be explored completely [68,69].

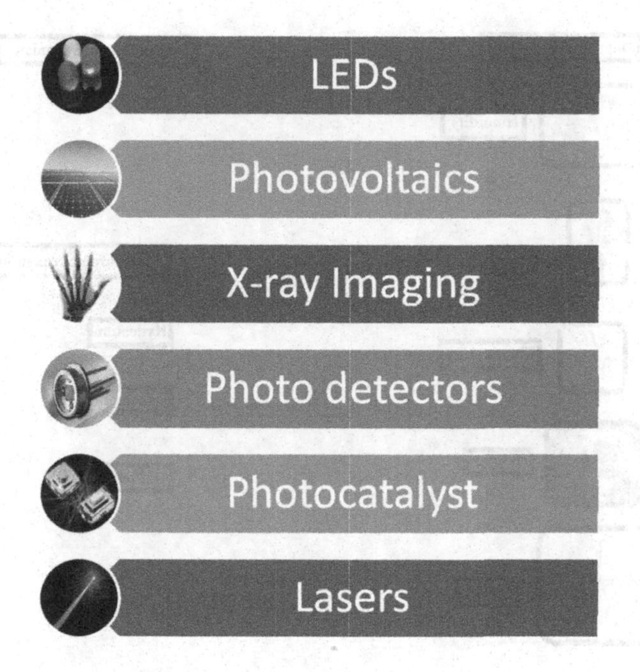

FIGURE 8.10 Prominent fields of perovskite applications.

TABLE 8.5
Commonly Used Perovskite Materials for Different Applications

S.No.	Device	Materials
1.	Light-emitting diode	$Cs_2AgInCl_6$: Mn
		Cs_2SnCl_6:Bi
2.	Photovoltaics	$Cs_2AgBiBr_6$
		Cs_2TiBr_6
3.	X-ray imaging	$Cs_2AgBiBr_6$
4.	Photo detectors	$Cs_2AgInCl_6$
		$Cs_2AgBiBr_6$
5.	Photocatalyst	$Cs_2AgBiBr_6$
6.	Lasers	Cs_2NaGaF_6

8.1.7.3 Au Migration

The most common cathode material used in the metal oxide perovskite solar cells is gold (Au). It has been observed that at high temperatures, gold migrates into the succeeding bottom layers, generally spiro-OMeTAD (2,2′,7,7′-Tetrakis [*N*, *N*-di (4-methoxyphenyl) amino]—9,9′-spirobifluorene) under light [70]. Various performance degradation factors and mechanisms in PSCs are depicted in Figure 8.11.

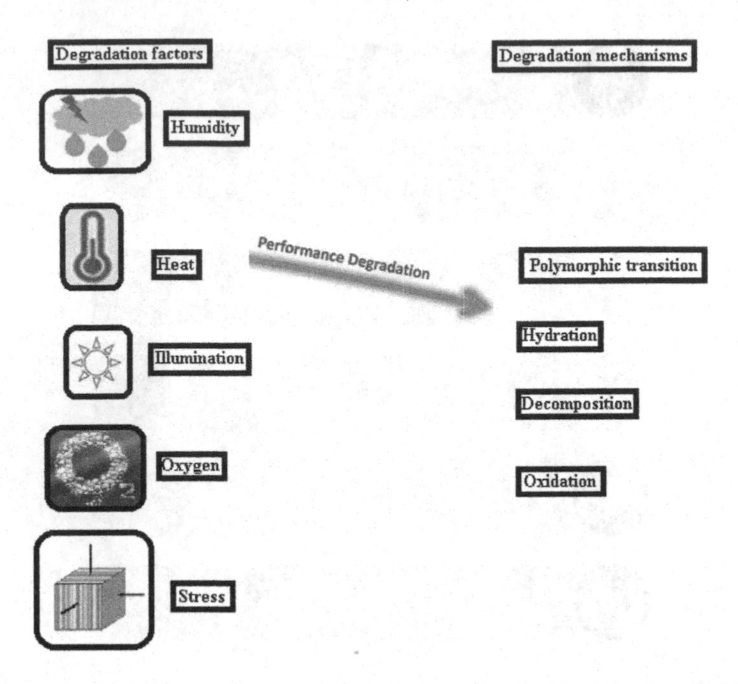

FIGURE 8.11 Depiction of performance degradation factors and mechanisms in PSCs.

8.1.7.4 Degradation

Degradation mechanism in perovskites leads to a device's long-term instability. This mechanism of degradation involves internal as well as external factors. However, this can be overcome by effective encapsulation of the solar cell devices. Internal factors include the change in physical, chemical, and optical properties upon heating or applying stress. These factors result in intrinsic instability [71,72]. The external factors such as moisture, temperature, illumination light, and oxygen also lead to degradation and hence instability. Thus, a drift toward hydrophobic, UV-stable interfacial layers, such as SnO_2, instead of TiO_2, is hydrophilic and susceptible to UV degradation.

8.1.8 Conclusions and Future Outlook

The swiftly rising rate of utilization of natural resources boosts the requirement for clean and never-ending alternative proficient sustainable energy resources. Among many solar cell researchers, a drive in the direction of acquiring electricity from solar energy persists to inspire concentrated research hot spot activity in the up-coming PVs. With this quest, they succeeded in attaining significant improvements by TMO PSCs. The conversion efficiency has propelled to 26.7% from 3.8% with the use of tandem cell perovskite PVs, building new-fangled evidence in the history of PV technology. Astounding developments are expected to achieve superior efficiencies that could yield a better class of non-toxic devices with best robustness and stability. As per the current evolution, it is evident that the next generation high-efficiency perovskite solar cells can be attained by solution-processed approaches with a blended perovskite phase. Perovskites have been proved as the best solar absorbers as compared with other PV technologies. However, these PSCs yet possess immense hurdles for auxiliary enhancements, which need to be addressed before commercialization. Recent advances in the field of nanotechnology make it worthwhile to investigate the nanoscale perovskites so as to enrich the performance of solar cell devices. Numerous research efforts are yet to be invested so as to achieve greater efficiencies even for large-area device efficiency.

REFERENCES

1. Snaith, H.J. Present status and future prospects of perovskite photovoltaics, *Nature Materials*, 17 (2018), 372–376.
2. Jeon, N.J., Noh, J.H., Kim, Y.C., Yang, W.S., Ryu, S., Seok, S.I. Solvent engineering for high-performance inorganic–organic hybrid perovskite solar cells. *Nature Materials*, 13(9) (2014), 897–903.
3. Wang, Z.K., Li, M., Yuan, D.-X., Shi, X.-B., Ma, H., Liao, L.-S. Improved hole interfacial layer for planar perovskite solar cells with efficiency exceeding 15%. *Applied Materials Interfaces*, 7(18) (2015), 9645–9651.
4. Van Le Q., Choi, J.-Y., Kim, S.Y. Recent advances in the application of two-dimensional materials as charge transport layers in organic and perovskite solar cells. *Flat Chemistry*, 2 (2017), 54–66.
5. Samsonov, G.V. (Ed.). *The Oxide Handbook*, IFIi/Plenum, New York (1982).
6. Bergesen, J.D., Heath, G.A., Gibon, T., Suh, S. Thin-film photovoltaic power generation offers decreasing greenhouse gas emissions and increasing environmental co-benefits in the long term. *Environmental Science & Technology*, 48(16) (2014), 9834–9843.
7. Morfa, A.J., MacDonald, B.I., Subbiah, J., Jasieniak, J.J. Understanding the chemical origin of improved thin-film device performance from photodoped ZNO nanoparticles. *Solar Energy Materials and Solar Cells*, 124 (2014), 211–216.
8. Luxová, J., Šulcová, P., Trojan, M. Study of perovskite. *Journal of Thermal Analysis and Calorimetry*, 93(3) (2008), 823–827.
9. Green, M.A., Emery, K., Hishikawa, Y., Warta, W., Dunlop, E.D. Solar cell efficiency tables (version 45). *Progress in Photovoltaics: Research and Applications*, 23(1) (2015), 1–9.
10. Dale, M. Benson, S.M. Energy balance of the global photovoltaic (PV) industry—Is the PV industry a net electricity producer? *Environmental Science & Technology*, 47(7) (2013), 3482–3489.
11. Monkowski, J.R., Bloem, J., Giling, L.J., Graef, M.W.M. Comparison of dopant incorporation into polycrystalline and monocrystalline silicon. *Applied Physics Letters*, 35(5) (1979), 410–412.
12. Winkler, M.T., Wang, W., Gunawan, O., Hovel, H.J., Todorov, T.K., Mitzi, D.B. Optical designs that improve the efficiency of $Cu_2ZnSn(S, Se)_4$ solar cells. *Energy & Environmental Science*, 7(3) (2013), 1029–1036.
13. Green, M.A., Hishikawa, Y., Warta, W., Dunlop, E.D., Levi, D.H., Hohl-Ebinger, J., Ho-Baillie, A.W. Solar cell efficiency tables (version 50). *Progress in Photovoltaics: Research and Applications*, 25(7) (2017), 668–676.
14. Wei, D., Amaratunga, G. Photoelectrochemical cell and its applications in optoelectronics. *International Journal of Electrochemical Science*, 2 (2007), 897–912.
15. O'Regan, B., Grätzel, M. A low-cost, high-efficiency solar cell based on dye-sensitized colloidal TiO_2 films. *Nature*, 353(6346) (1991), 737–740.

16. Conibeer, G. Third-generation photovoltaics. *Materials Today*, 10(11) (2007), 42–50.
17. Aziz, S.B., Woo, T.J., Kadir, M.F.Z., Ahmed, H.M. A conceptual review on polymer electrolytes and ion transport models. *Journal of Science: Advanced Materials and Devices*, 3(1) (2018), 1–17.
18. Zou, C.W., Yan, X.D., Bian, J.M., Gao, W. Enhanced visible photoluminescence of V_2O_5 via coupling ZnO/V_2O_5 composite nanostructures. *Optics Letter*, 35 (2010), 1145.
19. Semonin, O.E., Luther, J.M., Beard, M.C. Quantum dots for next-generation photovoltaics. *Materials Today*, 15(11) (2012), 508–515. doi:10.1016/s1369-7021(12)70220-1.
20. Goodwin, H., Jellicoe, T.C., Davis, N.J., Böhm, M.L. Multiple exciton generation in quantum dot-based solar cells. *Nanophotonics*, 7(1) (2017), 111–126.
21. Sogabe, T., Shen, Q., Yamaguchi, K. Recent progress on quantum dot solar cells: A review. *Journal of Photonics for Energy*, 6(4) (2016), 040901.
22. Zhao, Q., Hazarika, A., Chen, X., Harvey, S.P., Larson, B.W., Teeter, G.R., Liu, J., Song, T., Xiao, C., Shaw, L., Zhang, M., Li, G., Beard, M.C., Luther, J.M. High efficiency perovskite quantum dot solar cells with charge separating heterostructure. *Nature Communications*, 10 (2019), 2842.
23. Brabec, C.J., Sariciftci, N.S., Hummelen, J.K. Plastic solar cells. *Advanced Functional Materials*, 11(1), (2001), 15–26.
24. Mayer, A.C., Scully, S.R., Hardin, B.E., Rowell, M.W., McGehee, M.D. Polymer-based solar cells. *Materials Today*, 10(11) (2007), 28–33.
25. National Renewable Energy Laboratory (NREL). Best Research-Cell Efficiency Chart. 2019. Available at: https://www.nrel.gov/pv/cell-efficiency.html.
26. Huang, Y.H., Karppinen, M., Yamauchi, H., Goodenough, B. Effect of high-pressure annealing on magnetoresistance in manganese perovskites. *Journal of Applied Physics*, 98 (2005), 033911.
27. Rao, C.N.R. Transition metal oxides. *Annual Review of Physical Chemistry*, 40 (1989), 291–326.
28. Torrance, J.B., Lacorro, P., Asavaroengchai, C., Metzger, R.M. Simple and perovskite oxides of transition-metals: Why some are metallic, while most are insulating. *Journal of Solid State Chemistry*, 90 (1991), 168–172
29. Misra, S. Transition metal substituted SrTiO3 perovskite oxides as promising functional materials for oxygen sensor, functional materials. *AIP Conference Proceedings*, 1461 (2012), 379–382.
30. Mao, Y., Zhou, H., Wong, S.S. Synthesis, properties, and applications of perovskite-phase metal oxide nanostructures. *Material Matters*, 5(2) (2010), 50.
31. Kalyani, N.T., Dhoble, S.J. Organic light emitting diodes: Novel energy saving lighting technology—A review. *Renewable and Sustainable Energy Reviews*, 16 (2012), 2696–2723.
32. Calio, L., Kazim, S., Gratzel, M., Ahmad, S. Hole transport materials for perovskite solar cells. *Angewandte Chemie International Edition*, 55(47) (2016), 14522–14545.
33. Wang, Q., Li, H., Zhuang, J., Ma, Z., Wang, F., Zhang, T., Wang, Y., Lei, J. Hole transport materials doped to absorber film for improving the performance of perovskite solar cells, material. *Science in Engineering Processing*, 96 (2019), 113–120.
34. Chen, Y., Zhang, L., Zhang, Y., Gao, H., Yan, H. Large-area perovskite solar cells—A review of recent progress and issues, *RSC Advances*, 8 (2018), 10489–10508.
35. Kalyani, N.T., Dhoble, S.J. Novel materials for fabrication and encapsulation of OLEDs. *Renewable and Sustainable Energy Reviews*, 44 (2015), 319–347.
36. Chitnis, D., Swart, H.C., Dhoble, S.J. Escalating opportunities in the field of lighting. *Renewable and Sustainable Energy Reviews*, 64 (2016), 727–748.
37. Zimmermann, E. Characterization of perovskite solar cells: Towards a reliable measurement protocol. *APL Materials*, 4 (2016), 091901.

38. Ugale, A., Thejo Kalyani, N., Dhoble, S.J. Potential of europium and samarium β-diketonates as red light emitters in organic light-emitting diodes. In *Lanthanide-Based Multifunctional Materials* (2018), pp. 59–97. ISBN: 9780128138403, doi:10.1016/B978-0-12-813840-3.00002-8.
39. Thejokalyani, N., Dhoble, S.J. Novel approaches for energy efficient solid state lighting by RGB organic light emitting diodes—A review. *Renewable and Sustainable Energy Reviews*, 32 (2014), 448–467.
40. Marinova, N., Valero, S., Delgado, J.L. Organic and perovskite solar cells: Working principles, materials and interfaces. *Journal of Colloid and Interface Science*, 488 (2017), 373–389.
41. Haschke, J., Dupré, O., Boccard, M., Ballif, C. Silicon heterojunction solar cells: recent technological development and practical aspects—From lab to industry. *Solar Energy Materials and Solar Cells*, 187 (2018), 140–153.
42. Balasingam, S.K., Kang, M.G., Jun, Y. Metal substrate based electrodes for flexible dyesensitized solar cells: Fabrication methods, progress and challenges. *Chemical Communications*, 49(98) (2013), 11457–11475.
43. Pérez-del-Rey, D., Gil-Escrig, L., Zanoni, K.P., Dreessen, C., Sessolo, M., Boix, P.P., Bolink, H.J. Molecular passivation of MoO_3: Band alignment and protection of charge transport layers in vacuum-deposited perovskite solar cells. *Chemistry of Materials*, 31 (17) (2019), 6945–6949.
44. Zimmermann, Y.-S., Schäffer, A., Hugi, C., Fent, K., Corvini, P.F., Lenz, M. Organic photovoltaics: Potential fate and effects in the environment. *Environment International*, 49 (2012), 128–140.
45. Angmo, D., Hösel, M., Krebs, F.C. All solution processing of ito-free organic solar cell modules directly on barrier foil. *Solar Energy Materials and Solar Cells*, 107 (2012), 329–336.
46. Arima, T., Tokura, Y., Torrance, J.B. Variation of optical gaps in perovskite-type 3d transition-metal oxides. *Physics Review B*, 48 (1993), 17006.
47. Mizokawa, T., Fujimori, A. Electronic structure and orbital ordering in perovskite-type 3d transition-metal oxides studied by hartree-fock band-structure calculations. *Physics Review B*, 54 (1996), 5368.
48. Ishihara, S., Yamanaka, M., Nagaosa, N. Orbital liquid in perovskite transition-metal oxides. *Physics Review B*, 56 (1997), 686.
49. Rodriguez-Martinez, L.M., Attfield, J.P. Cation disorder and the metal-insulator transition temperature in manganese oxide perovskites. *Physics Review B*, 58 (1998), 2426.
50. Eng, H.W., Barnes, P.W., Auer, B.M., Woodward, P.M. Investigations of the electronic structure of d0 transition metal oxides belonging to the perovskite family. *Journal of Solid State Chemistry*, 75(1) (2003), 94–109.
51. Rodgers, J.A., Williams, A.J., Attfield, J.P. High-pressure/high-temperature synthesis of transition metal oxide perovskites. *Naturforsch*, 61b (2006), 1515–1526.
52. Gao, J., Perkins, C.L., Luther, J.M., Hanna, M.C., Chen, H.-Y., Semonin, O.E., Nozik, A.J., Ellingson, R.J., Beard, M.C. *n*-type transition metal oxide as a hole extraction layer in PbS quantum dot solar cells. *Nano Letter*, 11(8) (2011), 3263–3266.
53. Grinberg, I., West, D.V., Torres, M., Gou, G., Stein, D.M., Wu, L., Chen, G., Gallo, E.M., Akbashev, A.R., Davies, P.K., Spanier, J.E., Rappe, A.M. Perovskite oxides for visible-light-absorbing ferroelectric and photovoltaic materials. *Nature*, 503 (2013), 509–512.
54. Im, J.H., Lee, C.R., Lee, J.-W., Park, S.-W., Park, N.-G. 6.5% efficient perovskite quantum-dot-sensitized solar cell. *Nanoscale*, 3 (2011), 4088–4093.
55. Gao, J., Perkins, C.L., Luther, J.M., Hanna, M.C., Chen, H.-Y., Semonin, O.E., Nozik, A.J., Ellingson, R.J., Beard, M.C. *n*-type transition metal oxide as a hole extraction layer in PbS quantum dot solar cells. *Nano Letter*, 11(8) (2011), 3263–3266.

56. Etgar, L. Semiconductor nano-crystals as light harvesters in solar cells. *Materials*, 6(2) (2013), 445–459.
57. Zhao, Y., Nardes, A.M., Zhu, K. Effective hole extraction using MoO_x-Al contact in perovskite $CH_3NH_3PbI_3$ solar cells. *Applied Physics Letters*, 104 (2014), 213906.
58. Wang, K.-C., Jeng, J.-Y., Shen, P.-S., Chang, Y.-C., Diau, E.W., Tsai, C.-H., Chao, T.-Y., Hsu, H.-C., Lin, P.-Y., Chen, P., Guo, T.-F. *p*-type mesoscopic nickel oxide/organometallic perovskite heterojunction solar cells. *Scientific Reports*, 4 (2014), 4756.
59. Yeo, J.-S., Kang, R., Lee, S., Jeon, Y.-J., Myoung, N., Lee, C.Y., Kim, D.-Y., Yun, J.-M., Seo, Y.H., Kim, S.S., Na, S.-I. Highly efficient and stable planar perovskite solar cells with reduced graphene oxide nanosheets as electrode interlayer. *Nano Energy* 12 (2015), 96–104.
60. Shin, S.S. Yang, W.S., Noh, J.H., Suk, J.H., Jeon, N.J., Park, J.H., Kim, J.S., Seong, W.M., Seok, S.I. High-performance flexible perovskite solar cells exploiting Zn_2SnO_4 prepared in solution below 100°C. *Nature Communications*, 6 (2015), 1–8.
61. Jung, J.W., Chueh, C.-C., Jen, A.K. A low-temperature, solution-processable, Cu-doped nickel oxide hole-transporting layer via the combustion method for high-performance thin-film perovskite solar cells. *Advanced Materials*, 27(47) (2015), 7874–7880.
62. Xu, J., Voznyy, O., Comin, R., Gong, X., Walters, G., Liu, M., Kanjanaboos, P., Lan, X., Sargent, E.H. Crosslinked remote-doped hole-extracting contacts enhance stability under accelerated lifetime testing in perovskite solar cells. *Advanced Materials*, 28(14) (2016), 2807–2815.
63. Werner, J., Geissbühler, J., Dabirian, A., Nicolay, S., Morales-Masis, M., Wolf, S.D., Niesen, B., Ballif, C. Parasitic absorption reduction in metal oxide-based transparent electrodes: Application in perovskite solar cells. *ACS Applied Materials & Interfaces*, 27(8) (2016), 17260–17267.
64. Qin, P., He, Q., Ouyang, D., Fang, G., Choy, W.C., Li, G. Transition metal oxides as hole-transporting materials in organic semiconductor and hybrid perovskite based solar cells. *Science China Chemistry*, 60(4) (2017), 472–489.
65. Hou, Y., Du, X., Scheiner, S., McMeekin, D.P., Wang, Z., Li, N., Killian, M.S., Chen, H., Richter, M., Levchuk, I., Schrenker, N. A generic interface to reduce the efficiency-stability-cost gap of Perovskite solar cells. *Science*, 358(6367) (2017), 1192–1197.
66. Robert, F. Service, perovskite LEDs begin to shine. *Science*, 364(6444) (2019), 918.
67. de Figueiredo, A.T., Barrado, C.M., Silva, R.L., Alvarenga, L.D., Motta, F.V., Paskocimas, C.A., Bomio, M.R. Luminescence property of perovskite structure. *International Journal of New Technology and Research*, 1(7) (2015), 22–29.
68. Ramirez, D., Velilla, E., Montoya, J.F., Jaramillo, F. Mitigating scalability issues of perovskite photovoltaic technology through a *p-i-n* meso-superstructured solar cell architecture. *Solar Energy Materials and Solar Cells*, 195 (2019), 191–197.
69. Meredith, P., Armin, A. Scaling of next generation solution processed organic and perovskite solar cells. *Nature Communications*, 9 (2018), 1–4.
70. Correa-Baena, J.P., Saliba, M., Buonassisi, T., Grätzel, M., Abate, A., Tress, W., Hagfeldt, A. Promises and challenges of perovskite solar cells. *Science*, 358(6364) (2017), 739–744.
71. Dharmadasa, I.M., Rahaq, Y., Alam, A.E. Perovskite solar cells: Short lifetime and hysteresis behaviour of current–voltage characteristics. *Journal of Materials Science: Materials in Electronics*, 30(14) (2019), 12851–12859.
72. Ankireddy, K., Ghahremani, A.H., Martin, B., Gupta, G., Druffel, T. Rapid thermal annealing of $CH_3NH_3PbI_3$ perovskite thin films by intense pulsed light with aid of diiodomethane additive. *Journal of Materials Chemistry A*, 6 (2018), 9378–9383.

9 QDs for High Brightness WLEDs and Solar Cell Devices

Vijay B. Pawade, K. N. Shinde, and S. J. Dhoble

CONTENTS

9.1 INTRODUCTION

It is well-known that quantum dots (QDs) are the semiconductor materials that exhibit unique optoelectronics properties due to the presence of quantum confinement effect. QDs are tiny particles or nanocrystals of an inorganic material having small scale size observed in the range of 2–10 nm, and they were first discovered in 1980. Based on their size, it shows unique electronic properties. It means that QDs can emit any emission wavelength of light from the same material simply by changing their particle size. Hence, they require more energy to excite, and they can release more energy when they return to the ground state, thereby showing a color shift from red to blue in the emitted light [1,2]. Usually as the size of the QD decreases, then the energy gap between the valence band (VB) and conduction band (CB) increases as shown in Figure 9.1, and they are classified into different types on the basis of their composition and structure. Currently, QDs gain lot of in the development of LEDs and photovoltaic fields because they have easy machinability, the energy band gap can be adjusted, and depending on their size, they can show tunable properties. QDs exhibit interesting optical properties such as broad excitation band, narrow emission band, size-dependent emission

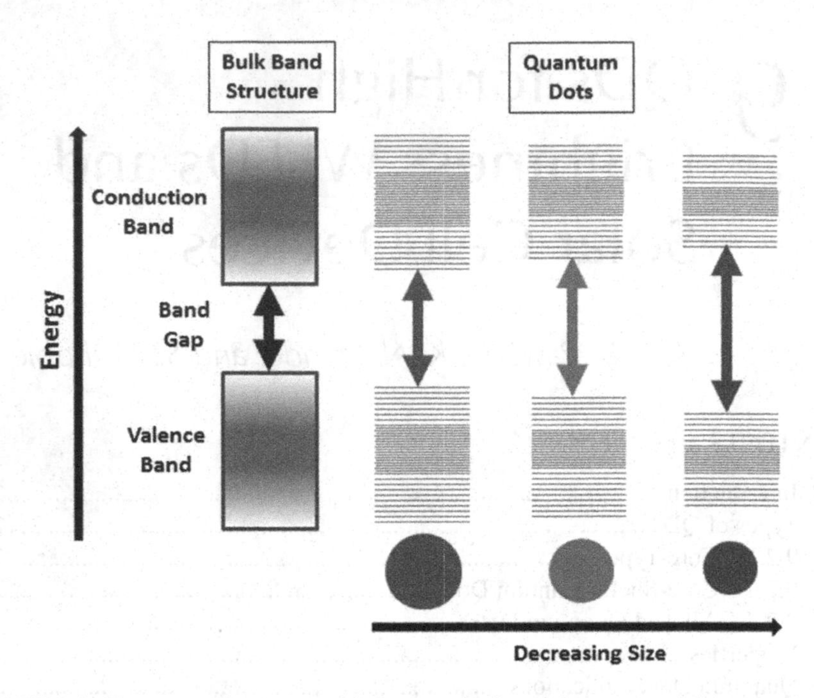

FIGURE 9.1 Energy gap increases with decrease in size of the nanoparticles.

properties, and photobleaching mechanism. Thus, QDs are the promising candidates to replace existing phosphor, which are used in the development of light-emitting diodes (LEDs). In comparison with commercially used phosphor (YAG:Ce^{3+}), QDs shows low light scattering, narrow band emission, high quantum yield (QY), better thermal stability, and good Color Rendering Index (CRI), which are essential parameters to develop QD-based LEDs for the next-generation lighting technology. Also, they have the ability to produce multiple exciton generation (MEG), which makes it excellent materials for the emerging energy saving and conversion technology. Hence, MEG plays an important role in QD system, producing multiple excitations from the VB to the transmission band [3,4]. Recently, many research groups are working on QDs and nanocrystalline solar cell due to its tunable optical properties. Here, the semiconductor QDs materials are used to produce electrons from incident solar energy as photosensitizer materials. It is an ideal way to fabricate the quantum dot-sensitized solar cell (QDSSCs) in which the maximum portion of the solar spectrum can be used by tuning the band gap of the light-absorbing materials. Till now, different kinds of semiconductor QDs materials have been reported for solar cell applications, such as PbS, InAs, Ag_2S, CdS, CdSe, and CdTe, and achieve conversion efficiency up to 13.4%. Therefore, the LEDs, based on the colloidal QD emitting layer, are highly promising for the future display technology, and they are the strong competitor to organic LEDs based on their varying optical properties and cost-effective fabrication techniques [5–7]. Hence, QD-based LEDs can be considered the most excellent candidates for solid-state lighting and display applications. They also show the better response in enhancing the conversion efficiencies when applicable in solar cell devices.

9.2 TYPES OF QDs

QDs can be manufactured in different forms.

9.2.1 Core-Type QDs

Core-type QDs are composed of a single material with uniform internal compositions of metallic chalcogenide, such as cadmium telluride or lead sulfide. Hence, it is a single component material. The photoluminescence and electroluminescence properties of core-type nanocrystals can be tuned by varying the crystallite size.

9.2.2 Core-Shell Quantum Dots

As the size of QDs are reduced, the band gap of the materials are increased. Hence, QDs are small regions of one material buried in another with a larger energy gap. Therefore, they exhibit core–shell structures (e.g., CdSe found in the core and ZnS in the shell form). QDs show excellent luminescent characteristics due to the recombination of electron-hole pairs through a radiative path. However, the exciton decay can also occur through nonradiative methods, thereby reducing the luminescence QY. The quantum efficiency and brightness of nanocrystal materials can be improved by growing shells of another large band gap material around them. Therefore, the QDs with small regions of one material embedded in another with a large band gap are known as core–shell QDs or core–shell nanocrystalline materials.

9.2.3 Alloyed Quantum Dots

It is composed of multiple materials in a homogeneous mixture and not in distinct regions. Therefore, this combination of two distinct semiconductor materials with varying band gaps imparts new and different properties to the particles that are different from the original materials. It well known that QDs have the ability to tune optical as well as electronic properties by varying the crystallite size, and this become a hallmark of QD materials. But this size reduction puts forth restriction on the applications of materials in different fields. But the multicomponent QDs provide alternative techniques to tune properties without varying the crystallite size. Thus, the alloyed semiconductor QDs having homogeneous and gradient internal structures can show tunable optical and electronic properties by merely varying the composition and internal structure without varying the crystallite size. These types of QDs formed by adding two semiconductor materials with different band gap energies, which exhibit interesting properties, differ from the bulk materials. Therefore, the alloyed nanocrystals show new and additional composition-tunable properties, which is different than the properties that arise due to quantum confinement effects.

9.3 PROPERTIES

By controlling the size of the QD, we can "tune" the electrical and optical properties. Nanostructured QDs find potential applications in several areas such as solar cells, semiconductor devices, LEDs, medical imaging, and quantum computing, based on their unique properties. If the radii of spherical-size QDs get reduced, then it results in the blue shift of the characteristic transition energies (such as excitons).

Following are the few points of concern with quantum confinement effects:

- If the radii of the nanostructure materials become very small, then the energy levels for carriers change from continuous bands to discrete energy levels.
- If the radius is reduced, then it enhances the volume-normalized oscillator strength of the exciton, which arises due to the oscillator strength. And it is concentrated over sharp electron-hole transitions, rather than being distributed over a continuum of states as those found in bulk semiconductor materials.
- In QDs, there is an enhancement in exciton nonlinearity and a reduction in optical power, which is required for optical saturation relative to the bulk semiconductor. Hence, there is cooperation between QDs for optical nonlinear and is also an important factor for a practical point of view.
- In a theoretical point, we assume spherical shapes for the crystallite, and this is a reasonable approximation in many cases.

9.4 QUANTUM DOT APPLICATIONS

Currently, the consumption of electricity continuously increases due to the vast increase in the industrial sector. It directly affects the global environment with an increase in level greenhouse gases in the environment. Therefore, to avoid such loss and to protect our planet, there is need to develop environmentally friendly technology products for energy-generation and energy-conversion devices. Thus, to fulfill this requirement, we need energy-efficient and nontoxic materials for said applications. Many types of materials were proposed for this purpose but only few host materials can fulfill this requirement to enhance the device performance of the existing technologies. As per the global concern, only LEDs and solar cells are the best alternative for being environment safety and sustainable. Hence, based on their unique size and composition, QDs show tunable electronic properties. Therefore, QDs are significantly used for optical devices based on their characteristic features, such as their high brightness, pure colors, and ability to emit a rainbow of colors coupled with high quantum efficiencies, longer lifetimes, etc. They have zero-dimensional structures and show sharper densities of states than higher-dimensional structures. Being of very small in size, the electrons do not travel as compared with larger particles, so the electronic devices can also work at a faster rate.

9.4.1 WLEDs

9.4.1.1 Some Reported Materials

9.4.1.1.1 CdS QDs for Blue LED

Currently, colloidal semiconductor QDs have received a great demand in future-generation optoelectronic devices due to their superior photophysical properties, such as size-tunable band gaps, good stability, and better quantum efficiencies. [8–10]. These distinct optoelectronics properties in QDs are observed because of their quantum confinement effect and surface area [11–13]. Hence, considering the demands of energy efficiency for the development of optoelectronic devices, it is good that research and development in the field of colloidal QD-based II–VI semiconductor nanostructured materials have been extensively studied and reported over the last 30 years. Among these, only a few materials fit their criteria for industrial applications, such as CdS QDs, and have been actively studied and referred for many applications, such as nanobarcode, transistors, photocatalysts, LEDs, photodetectors, and solar cells. They also act as down-converters in backlit displays devices [14–21], and at present, many synthesis techniques have been applied for the development of monodispersed and uniform nanostructure QDs. Currently, Ratnesh et al. reported hot injection blended tunable CdS QDs for the production of blue LEDs [22]. However, for the production of QDs, only two techniques are followed, such as hot injection and heating up QDs. Hot injection methodologies are helpful in improving the size, morphology, and stabilization of nanoparticles in a controlled way. Steigerwald et al. [8,11] reported the hot injection colloidal method having a rapid injection of organometallic reagents into hot coordinating solvent in the range of 180°C–300°C. This method was first developed by Murray et al. in 1993 [20,23]. Apart from this study, zero-dimensional binary CdS with a direct band gap of 2.5 eV has received more attention. Also, quantum confinement and confinement properties in CdS QDs has been reported by Rossetti et al. [21,24], and some other researcher who observed it due to the surface defects [25–27]. Hence, QDs show non-radiative decay corresponding to this band gap. Further, Xinmei et al. [25,28] reported the magic size QDs by the same route in which the synthesized QDs exhibit some excellent characteristics, such as having high fluorescence, high QY, stable electronic structure, reduced surface defects, and more biocompatibility. CdS QDs was synthesized by hot injection colloidal techniques, in which oleic acid was used as a capping agent to control the particle size and prevent agglomeration of particles. The ligand exchange technique was used to transform hydrophobic QDs into hydrophilic. Hence, the circular shape and size (2.7–4.0 nm) of QDs was obtained through high-resolution transmission electron microscopy (HRTEM) characterization. Thus, synthesized QDs show high crystallinity and wurtzite structure when characterized by X-ray diffraction (XRD) and HRTEM. Also, the prepared CdS QDs exhibit a strong absorption and better Photoluminescence (PL) properties in the visible region. In Figure 9.2 [22], it is seen that the absorption and PL spectra of the CdS QDs shifted from 366–450 nm and 425–456 nm, respectively, with an increase in time (15–105 min). Thus, the International Commission on Illumination (CIE) color coordinates CdS QDs with varying timelines in the blue region of the spectra, so it acts as a potential blue color component for the production of LEDs.

9.4.1.1.2 Red-Emitting $CuInS_2$/ZnS Quantum Dots for Warm WLEDs

Recently, many researchers took an effort to find an alternative to the microsized red phosphors for white LED (WLED) applications, so in this regards, $CuInS_2$/ZnS (CIS/ZnS) QDs are strongly proposed to be good substitutes. In the past, Dong et al. and coworkers fabricated WLEDs by using yellow-emitting CIS/ZnS QDs with CIE good

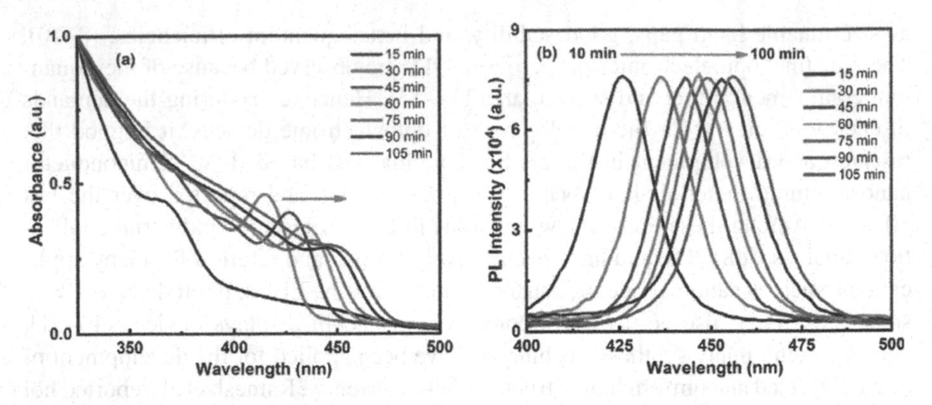

FIGURE 9.2 (a) Absorption and (b) normalized photoluminescence (at excitation 310 nm) spectra of CdS QDs. (Reprinted with permission from Ratnesh, R.K., *Opt. Laser Technol.*, 116, 103–111, 2019.)

color coordinates and color temperature (0.334–0.348, 0.273–0.291, 4447–5380 K [26]. In much literature, CIS/ZnS QDs were synthesized by the non-injection one-pot approach [27–30] and the hot-injection approach, respectively [31–33]. Although the hot-pot method is the simplest one and easy to handle, it cannot observe and monitor the reaction process. However, with the help of the hot-injection method, we can easily monitor the reaction with precise operation and achieve a well-monodispersed and high-purity product. Also, to improve the optical properties of YAG:Ce-based WLEDs, the red-emitting CIS/ZnS QDs were first synthesized by non-injection one-pot techniques, but in this method, QDs with better optical properties could not be achieved.

Therefore, Dong et al. [26] has synthesized CIS/ZnS QDs by hot injection. He found that CIS QDs exhibit lower luminous intensities and QY, but CIS /ZnS QDs manifested high tunable PL emission spectra with a QY of 65.07% after covered by ZnS shells. They measured the fluorescence and absorption spectra at the same time interval. Therefore, they noticed that with increasing reaction time at 220°C, the PL spectra of CIS QDs shifted from 682 nm (5 min) to 738 nm (45 min) as shown in Figure 9.3a. Similarly, the absorption band of CIS QDs also shifted to longer a wavelength as time increased, and the band gap became smaller gradually as shown in Figure 9.3b. CIS QDs and CIS/ZnS QDs exhibit a broad characteristic absorption shoulder. Figure 9.3c shows that the absorption spectra from this absorption edge energy was found to be 3.1209 eV according to Miao's work [34]. Thus, from the measured PL and absorption spectra, it is seen that it may also act as a potential candidate for the fabrication of warm WLEDs.

9.4.2 Solar Cell Devices

Recently, QDs have gained more interest in the development of photovoltaic devices due to their unique features. From metaphoric to artificial atoms, QDs are confined within the three-dimensional box/well in which the electron moves within the bound

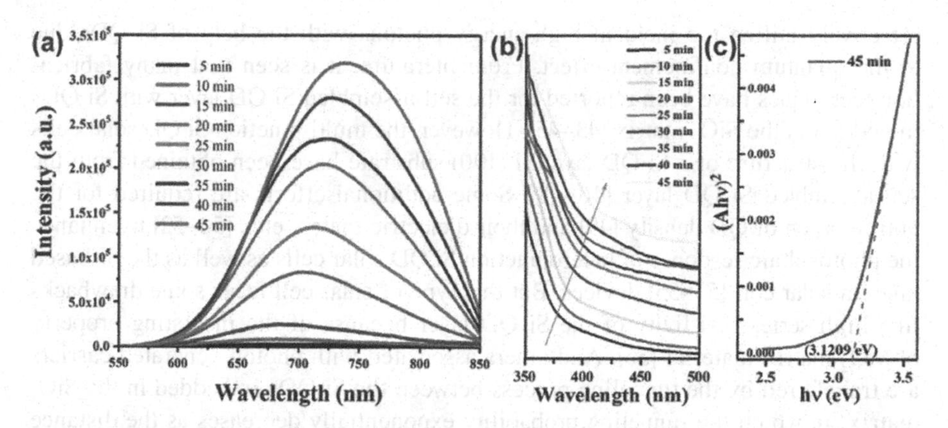

FIGURE 9.3 Evolutions of PL emission spectra (at 450 nm) (a) and UV–vis absorption spectra, (b) of CIS QDs during the synthesis; and (c) absorption edge energy of the sample after 45 min synthesis. (Reprinted with permission from Dong, X. et al., *Dyes Pigments*, 165, 273–278, 2019.)

and distinct energy states. Thus, the electron confinement is due to the presence of an interfacial interaction between semiconductor materials, its surface, and electrostatic potentials due to the addition of impurities, strain, or both factors. Thus, research is needed to identify the problem and solution to control all these related factors to study optoelectronic properties of QDs for the development of solar cell devices used in the third-generation for producing electricity [35,36]. Change in size of the QDs may also be an important factor in concern with the Shockley-Queisser limit of solar cell efficiency [37]. Recently, it was found that the maximum theoretical efficiency of a solar cell by using a single junction QD is reported to be ~42% under AM 1.5 solar illumination having a ~1.4 eV band gap [38]. In the viewpoint of this, researchers are taking more efforts to achieve such high-performance efficiencies of solar devices using novel and energy-efficient QD materials [36]. Nowadays, Si-based solar cells capture the maximum share of the market all over the world because there is no substitute for this type of photovoltaic cell. Also, there is an urgent need to reduce the cost of this solar device to cover the maximum share in underdeveloped countries to explore the environmentally friendly energy source for sustainability at all levels, such as commercial and household purposes. Thus, to develop this kind of device, it is essential to implement and adopt new and ideal ways to explore the scope of nanomaterials in modifying the conversion efficiency of the existing Si-solar cell or to develop new hybrid materials that can replace the existing solar cell devices [39–41].

9.4.2.1 Some Reported Materials

9.4.2.1.1 Silicon QDs

Currently, multiple junction silicon solar cells may be promising next-generation solar cells due to their high conversion efficiencies and low fabrication costs as compared to crystalline Si solar cells. The efficiency of the tandem solar cell has increased up to 42.5%, when top surface of the cell gets coated with a Si QD layer having band gap energy of the order of 1.7–1.8 eV [42]. In this case, solar cells can effectively utilize the incident high-energy photons with the help of Si QDs due to the quantum confinement effect. From literature, it is seen that many fabrication techniques have been reported for the self-assembled Si QD layer with Si QDs imbedded in the SiO_2 matrix [43–45]. However, the multi-junction Si QD solar cells with the structure of a Si QD layer/Si (100) substrate have been obtained from the self-assembled Si QD layer [46–48]. Some additional efforts are required for the optimization of QD density [49] and their dielectric matrix etc., [50–52] to enhance the photovoltaic response of heterojunction Si QD solar cells as well as the Si-based tandem solar cell [53–55] devices. But this type of solar cell faces some drawbacks like high series resistivity of the Si QD layer because of the insulating property of SiO_2 matrix material [56]. And others associated with photon-generated carriers are transferred by the tunneling process between the Si QDs imbedded in the SiO_2 matrix, in which the tunneling probability exponentially decreases as the distance increases [57]. To enhance the conductivity in the Si QD layer, it should be distributed closely and uniformly in the dielectric matrix.

9.4.2.1.2 Perovskite QDs

Recently, perovskite QDs are an emerging class of nanocrystals exhibiting excellent properties over metal chalcogenide QDs. Based on their tunable optical properties and outstanding photovoltaic performance, perovskites have gained a lot of attention from the past few years.

A reduction in the size of perovskite crystals to a few nanometers shows very high photoluminescence QYs and good color quality. Therefore, in display devices, metal chalcogenide QDs already play a role because of their increased PLQY, simple synthesis techniques, synthesis, better color quality, and color tunability, and hence perovskite QDs are promising candidates for this application. Song et al. reported the ITO/PEDOT: PSS/PVK/ $CsPbX_3$/TPBi /LiF/Al structure to fabricate blue, green, and orange LEDs [58]. They observed narrows band emission, high brightness (< 1000 cdm^{-2}), and external quantum efficiencies (EQEs) limited to ~0.1%. Later, Li et al. reported the importance of nanocrystal properties by increasing the EQEs of $CsPbBr_3$ nanocrystal LEDs up to 50× (0.12 to 6.27%) using a mixture of charge-transport layer optimization and by controlling the surface ligand density [59]. Hence, using this approach one can achieve a brightness up to >15000 cdm^{-2}, with excellent color purity (~20 nm emission line width) for nanocrystal sizes up to ~8 nm. So, these are promising QDs for the generation of white light when excited with a blue light. Besides this, there are very few reports available on perovskite QDs for solar cell devices, as compared to bulk and two-dimensional perovskites materials. At present, there are only few materials discovered that show very promising results that indicate the importance of perovskite QDs for the development of photovoltaic devices. In 2011, Im et al. was the first to use perovskite QDs in solar cells. He found that with the use of $MaPbI_3$ nanocrystals as a light-sensitizer in a structure resembles a dye-sensitized solar cell [60]. Then it exhibits a conversion efficiency up to 6.5%. Swarnkar et al. reported use of $CsPbI_3$ QD films [61]. When these materials were fabricated into solar cells, it achieved a power conversion efficiency over 10% and had a large open-circuit voltage up to 1.23 V. And later, a researcher demonstrated the use of coating the nanocrystals in $\mathbf{A^+X^-}$ (**A** is formamidinium, methylammonium or Cs, and **X** = I or Br), which helped in improving perovskite cell efficiency (PCE) ~13.4%; therefore, the researcher reported the highest efficiency in photovoltaic tandem solar cells based on QDs of any kind [62]. In this case, a bulk perovskite film plays the role of the small band-gap absorber, while the perovskite QDs act as wide band-gap absorbers [63].

The quantum confinement effect in perovskite QDs can be elaborated into two parts. In the first part, it will be unremarkable when the particle size of CsPbX3 perovskite QDs is larger than the Bohr radius (7 nm). Hence, band gap is defined between the energy state of Pb 6s–Br 4p hybridized orbitals and the energy state of Pb 6p levels in which the VB and CB are mainly involved by anionic composition [64,65]. Consequently, the emission properties of perovskite QDs can be tuned by varying the anion [66]. In contrast with the tunable properties of QDs, the perovskite QDs are more complex, which shows unique features and advantages in the synthesis and application. If we doped the lanthanide element into perovskite QDs, then it helps in enhancing the QY by changing the energy state or by combining their unique emission wavelength with the original perovskite QDs. Yao et al. [67] reported Ce^{3+} ion-doped CsPbX3 perovskite QDs through the facile hot-injection method, which shows the enhancement in the QY of CsPbX3 QDs. Thus, the doping amount of Ce^{3+} in CsPbBr3 QDs can achieve a higher absolute QY of 89% than undoped perovskite QDs.

9.4.2.1.3 Graphene QDs

It is well known that graphene oxide is a single-layer sheet of graphite oxide, which was discovered a half century ago. Currently, graphene oxide has received great interest due to its potential demand as starting material for the mass production of graphene. Graphene oxide can be a semiconductor or insulator based on its degree of oxidation, as well as its electronic and optical properties, which can be tuned and found in a large scope in the development of energy saving, conversion, and storage devices. Therefore, grapheme consists of a two-dimensional hexagonal structure having a flat monolayer of carbon atoms, and they are attached by strong triangular bonds of the sp^2 hybridized orbitals [68]. Due to this unique structural framework and its excellent properties, it finds diverse applications in engineering, technology, and medical fields [69–73]. Among the different nanostructured such as OD, 1D, 2D, and 3D, graphene nanosheets having the size less than 10 nm are called graphene quantum dots (GQDs) [74]. And this zero-dimensional nanostructure exhibits lower toxicity than those of the two-dimensional graphene materials. The dominant difference in molecular structure between graphene and GQDs is the band gap. The former is a "zero band gap" material, and the latter display a tunable band gap owing to quantum confinement and the edge effect. The tunable band gap, surface defects, and zigzag edges impart the tunable fluorescence property of GQDs, which is the intrinsic drawback of graphene [75,76]. Hence graphene-based inorganic nanostructures are emerging materials in the recent era of science and technology for various application such as development of the solar cell, WLEDs, electrodes for batteries, and fuel cells. Therefore, GQDs exhibit better biocompatibility and easier functionalization than traditional inorganic QDs due to the presence of carbon atoms with multiple functional groups on the surface [77–79]. They also show strong luminescence and good sensors properties [80]. Besides this, they are used as cytotoxic agents in phototherapy because they are photostable. GQDs are used to absorb an appropriate wavelength of photon energy, and then it transfers to specific activation to generate heating effects or reactive oxygen species. Hence, nanosized GQDs with large surface areas may act as potential candidates for effective photochromic carriers similar to other nanomaterials [81].

9.4.2.1.4 Quantum Dot-Sensitized Solar Cells

It is well-known that today energy consumption is gradually increasing day by day due to the growing global demand and vast increases in industries, altering our planet's climate because of an increase in the level of pollutants [8–10,82,83]. Therefore, to protect the environment and to reduce the pollution, clean and eco-friendly sustainable energy sources are urgently required [13,84]. Hence, focusing on this global issue, many universities and national laboratories are working together to utilize natural resources for energy generation, and the best example is sunlight, which is available everywhere. But sunlight alone will not fulfill the criteria of technology. For this purpose, we need energy-efficient materials that have good absorption as well as emission properties in the UV, visible, and near infrared emitting (NIR) ranges so that they can use the maximum portion of sunlight for energy conversion. This conversion mechanism is based on the incident and emitted photon. If we can generate the maximum photon in longer wavelength regions, then we can generate that amount of energy based on the principle of the photovoltaic effect. Many new and advanced materials have been proposed in past few years to improve photovoltaic devices, but at present, the search for a new one has not ended because there is no alternative for the existing solar cell devices. Si-solar cells captured the maximum part of the Asian market. However, the efficiency of this type of device cannot be improved yet. Also, scientists reported the better quantum efficiencies in perovskite solar cells, but due to the presence of toxic materials in these devices, they cannot be used commercialized yet. Some other class of solar cells such as organic, tandem solar cell materials came into existence, but they cannot show a better efficiency. Due to the tremendous growth in materials science and technology, a new class of nanoscale materials have been introduced by researchers, that is, zero-dimensional materials (QDs).

Therefore, these kinds of nonmaterials have been widely utilized for applications, and they fulfill the current requirement of industries [15,16,18,85]. Among the different types low band-gap semiconductors for light-trapping purposes, QD-sensitized solar cells (QDSCs) received great interest for next generation of dye-sensitized solar cells (DSCs) [1]. Besides the quantum confinement effect of the exciton in the absorber, the light conversion efficiency performance of a QDSC can be enhanced further by optimizing the band alignment at the TiO_2/QD-sensitizer interface for an efficient energy transfer mechanism, by increasing the light conversion efficiency and by comparing the absorption spectrum of the QD-sensitizer with the solar spectrum [86]. Therefore, research on the QDSC can be improved, and it finds potential applications by controlling the size of the QD and the band-gap energy of the semiconductor QD-sensitizer.

At present, the reported band-gap value of the II–VI group of compound semiconductors and the mixed metal compounds can be varied with the change in chemical composition of the film in between 1.1 and 3.7 eV. Hence, QDSCs act as promising devices to replace conventional DSCs due to their high light conversion efficiencies, which is reported up to 44% based upon theoretical prediction [87]. But more research and development of QDSCs is required to improve their existing drawback in conversion efficiencies.

9.5 CONCLUSION

Of concern is the global issue related to energy consumption and utilization of natural resources for generation of energy in the current century. There is an urgent need for a sustainable way to generate clean and eco-friendly energy generation plants. Using the traditional source of energy, that is, based on oil or coal, the harmful pollutant level continuously increases in the atmosphere. They also release a great amount of carbon dioxide into the atmosphere and cause global warming and environmental changes. So, to avoid this issue we can propose a better alternative for the sustainable environment for the betterment of human health and other species that live on our lovely planet. The use of LEDs and solar cell devices helps with energy saving and are nonpollutant sources of energy generation. They have the ability to protect the atmosphere and reduce the global effect and changes occurring in the environment. So, as reported in literature, semiconductor inorganic QD-based nonmaterials may fulfill the need and aforementioned gap in the development of LEDs and solar cell devices having high lumen output and high-energy conversion efficiencies.

REFERENCES

1. D. Wang, R. Wang, L. Liu, Y. Qu, G. Wang, Y. Li, Down-shifting luminescence of water soluble $NaYF_4$:Eu^{3+}@Ag core-shell nanocrystals for fluorescence turn-on detection of glucose, *Sci. China Mater.* 60 (2016) 68–74.
2. L. Zeng, L. Xiang, W. Ren, J. Zheng, T. Li, B. Chen, J. Zhang, C. Mao, A. Li, A. Wu, Multifunctional photosensitizer-conjugated core–shell Fe_3O_4@$NaYF_4$:Yb/Er nanocomplexes and their applications in T2-weighted magnetic resonance/upconversion luminescence imaging and photodynamic therapy of cancer cells, *RSC Adv.* 3 (2013) 13915.
3. F. Zhang, G.B. Braun, A. Pallaoro, Y. Zhang, Y. Shi, D. Cui, M. Moskovits, D. Zhao, G.D. Stucky, Stucky mesoporous multifunctional upconversion luminescent and magnetic "Nanorattle" materials for targeted chemotherapy, *Nano Lett.* 12 (2012) 61–67.
4. Y. Zhang, Z. Hong, Synthesis of lanthanide-doped $NaYF_4$@TiO_2 core–shell composites with highly crystalline and tunable TiO_2 shells under mild conditions and their upconversion-based photocatalysis, *Nanoscale* 5 (2013) 8930–8933.
5. L. Zhou, X. Zheng, Z. Gu, W. Yin, X. Zhang, L. Ruan, Y. Yang, Z. Hu, Y. Zhao, Mesoporous $NaYbF_4$@$NaGdF_4$ core-shell up-conversion nanoparticles for targeted drug delivery and multimodal imaging, *Biomaterials* 35 (2014) 7666–7678.
6. R. Wang, L. Zhou, W. Wang, X. Li, F. Zhang, In vivo gastrointestinal drug-release monitoring through second near-infrared window fluorescent bioimaging with orally delivered microcarriers, *Nat. Commun.* 8 (2017) 14702.
7. X. Li, Z. Guo, T. Zhao, Y. Lu, L. Zhou, D. Zhao, F. Zhang, Filtration shell mediated power density independent orthogonal excitations-emissions upconversion luminescence, *Angew. Chem.* 55 (2016) 2464–2469.
8. M.L. Steigerwald, A.P. Alivisatos, J.M. Gibson, T.D. Harris, R. Kortan, A.J. Muller, A.M. Thayer, T.M. Duncan, D.C. Douglass, L.B. Brous, Surface derivatization and isolation of semiconductor cluster molecules, *Molecules* 110 (1988) 3046–3050.

9. J.W. Stouwdam, R.A.J. Janssen, Red, green, and blue quantum dot LEDs with solution processable ZnO nanocrystal electron injection layers, *J. Mater. Chem.* 18 (2008) 1889–1894.
10. S. Coe, W.K. Woo, M. Bawendi, V. Bulovic, Electroluminescence from single monolayers of nanocrystals in molecular organic devices, *Nature* 420 (2002) 3–6.
11. M.L. Steigerwald, L.E. Brus, T.B. Laboratories, M. Hill, Semiconductor crystallites: A class of large molecules, *Acc. Chem. Res.* 23 (1990) 183–188.
12. M.G. Bawendi, M.L. Steigerwald, L.E. Brus, The quantum mechanics of larger semiconductor clusters, *Annu. Rev. Phys. Chem.* 41 (1990) 477–496.
13. A.P. Alivisatos, Perspectives on the physical chemistry of semiconductor nanocrystals, *J. Phys. Chem.* 3654 (1996) 13226–13239.
14. L. Qian, Y. Zheng, J. Xue, P.H. Holloway, Coherent terahertz control, *Nat. Photon.* 5 (2011) 1–6.
15. J.P. Clifford, G. Konstantatos, K.W. Johnston, S. Hoogland, L. Levina, E.H. Sargent, Fast, sensitive and spectrally tuneable colloidal-quantum-dot photodetectors, *Nat. Nanotechnol.* 4 (2009) 3–7.
16. S. Dayal, N. Kopidakis, D.C. Olson, D.S. Ginley, G. Rumbles, Photovoltaic devices with a low band gap polymer and CdSe nanostructures exceeding 3% efficiency, *Nano Lett.* 10 (2010) 239–242.
17. M. Bruchez Jr, M. Moronne, P. Gin, S. Weiss, A.P. Alivisatos, Semiconductor nanocrystals as fluorescent biological labels, *Science* 281 (1998) 2013–2016.
18. B.E. Jang, S. Jun, H. Jang, J. Lim, B. Kim, Y. Kim, White-light-emitting diodes with quantum dot color converters for display backlights, *Adv. Mater.* 712 (2010) 3076–3080.
19. A. Dakka, J. Lafait, C. Sella, S. Berthier, M. Abd-lefdil, J. Martin, M. Maaza, Optical properties of Ag-TiO_2 nanocermet films prepared by cosputtering and multilayer deposition techniques, *Appl. Opt.* 39 (2000) 2745–2753.
20. C.B. Murray, D.J. Noms, M.G. Bawendi, Synthesis and characterization of nearly monodisperse CdE (E = sulfur, selenium, tellurium) semiconductor nanocrystallites, *J. Am. Chem. Soc.* 115 (1993) 8706–8715.
21. R. Rossetti, S. Nakahara, L.E. Brus, Quantum size effects in the redox potentials, resonance Raman spectra, and electronic spectra of CdS crystallites in aqueous solution, *J. Chem. Phys.* 79 (1983) 1086–1088.
22. R.K. Ratnesh, Hot injection blended tunable CdS quantum dots for production of blue LED and a selective detection of Cu^{2+} ions in aqueous medium, *Opt. Laser Technol.* 116 (2019) 103–111.
23. C. Petit, P. Lixon, P. Pileni, Synthesis of cadmium sulfide in situ in reverse micelles. 2. Influence of the interface on the growth of the particles, *J. Phys. Chem.* 94 (1990) 1598–1603.
24. J.D. Levine, P. Mark, Theory and observation of intrinsic surface states on ionic crystals, *Phys. Rev.* 144 (1965) 751–763.
25. X. Liu, Y. Jiang, X. Lan, S. Li, D. Wu, T. Han, H. Zhong, Z. Zhang, Synthesis of high quality and stability CdS quantum dots with overlapped nucleation-growth process in large scale, *J. Colloid Interface Sci.* 354 (2011) 15–22.
26. X. Dong, J. Ren, T. Li, Y. Wang, Synthesis, characterization and application of red-emitting $CuInS_2$/ZnS quantum dots for warm white light-emitting diodes, *Dyes and Pigments* 165 (2019) 273–278.
27. N.T.M. Thuy, T.T.K. Chi, U.T.D. Thuy, N.Q. Liem, Low-cost and large-scale synthesis of $CuInS_2$ and $CuInS_2$/ZnS quantum dots in diesel, *Opt Mater.* 37 (2014) 823–827.
28. Y. Vahidshad, M.N. Tahir, S.M. Mirkazemi, A. Iraji Zad, R. Ghasemzadeh, W. Tremel, One-pot thermolysis synthesis of $CuInS_2$ nanoparticles with chalcopyrite-wurtzite polytypism structure, *J. Mater. Sci. Mater. Electron.* 26 (2015) 8960–8972.

29. Z. Liu, X. Su, One-pot synthesis of strongly fluorescent DNA-$CuInS_2$ quantum dots for label-free and ultrasensitive detection of anthrax lethal factor DNA, *Anal. Chim. Acta* 942 (2016) 86–95.
30. A. Arshad, H. Chen, X. Bai, S. Xu, L. Wang, Coupling reaction of zirconacyclopentadienes with dihalonaphthalenes and dihalopyridines: A new procedure for the preparation of substituted anthracenes, quinolines, and isoquinolines, *Chin. J. Chem.* 34 (2016) 576–582.
31. J. Guo, G. Chang, W. Zhang, X. Liu, Y. He, Facile synthesis of $CuInS_2$ nanoparticles using different alcohol amines as solvent, *Chem. Phys. Lett.* 647 (2016) 51–54.
32. J. Guo, G. Chang, W. Zhang, X. Liu, T. Zhou, Y. He, Mild solution-based method for synthesizing wurtzite CuInS2 nanoplates at low temperature, *Mater. Lett.* 123 (2014) 169–171.
33. T. Zeng, H. Ni, Y. Chen, X. Su, W. Shi, Mild solution-based method for synthesizing wurtzite $CuInS_2$ nanoplates at low temperature, *Mater. Lett.* 172 (2016) 94–97.
34. F. Ran, L. Miao, S. Tanemura, M. Tanemura, Y. Cao, S. Tanaka, Effect of annealing temperature on optical properties of Er-doped ZnO films prepared by sol-gel method, *Mater. Sci. Eng. B* 148(1) (2008) 35–39.
35. F. Priolo, T. Gregorkiewicz, M. Galli, T.F. Krauss, Silicon nanostructures for photonics and photovoltaics, *Nat. Nanotechnol.* 9(1) (2014) 19–32.
36. W.A. Badawy, A review on solar cells from Si-single crystals to porous materials and quantum dots, *J. Adv. Res.* 6(2) (2015) 123–132.
37. M. Samadpour, Efficient CdS/CdSe/ZnS quantum dot sensitized solar cells prepared by ZnS treatment from methanol solvent, *Sol. Energy* 144 (2017) 63–70.
38. Y. Xu, T. Gong, J.N. Munday, The generalized Shockley-Queisser limit for nanostructured solar cells, *Sci. Rep.* 5 (2015) 13536.
39. N. Balaji, H.T.T. Nguyen, C. Park, M. Ju, J. Raja, S. Chatterjee, R. Jeyakumar, J. Yi, Electrical and optical characterization of SiO_xN_y and SiO_2 dielectric layers and rear surface passivation by using SiO_2/SiO_xN_y stack layers with screen printed local Al-BSF for c-Si solar cells, *Curr. Appl. Phys.* 18(1) (2018) 107–113.
40. M. Ju, N. Balaji, C. Park, H.T.T. Nguyen, J. Cui, D. Oh, M. Jeon, J. Kang, G. Shim, J. Yi, The effect of small pyramid texturing on the enhanced passivation and efficiency of single c-Si solar cells, *RSC Adv.* 6(55) (2016) 49831–49838.
41. P. Löper, M. Canino, D. Qazzazie, M. Schnabel, M. Allegrezza, C. Summonte, S.W. Glunz, S. Janz, M. Zacharias, Zacharias, Silicon nanocrystals embedded in silicon carbide: Investigation of charge carrier transport and recombination, *Appl. Phys. Lett.* 102(3) (2013) 033507.
42. M. Green, K. Emery, D. King, S. Igari, W. Warta, *Third Generation Photovoltaics: Advanced Solar Energy Conversion*, Springer, Berlin, Germany, 2003.
43. M. Zacharias, J. Heitmann, R. Scholz, U. Kahler, M. Schmidt, J. Bläsing, Sizecontrolled highly luminescent silicon nanocrystals: A SiO/SiO_2 superlattice approach, *Appl. Phys. Lett.* 80(4) (2002) 661–663.
44. L. Pavesi, L. Dal Negro, C. Mazzoleni, G. Franzo, F. Priolo, Optical gain in silicon nanocrystals, *Nature* 408(6811) (2000) 440–444.
45. V. Vinciguerra, G. Franzo, F. Priolo, F. Iacona, C. Spinella, Quantum confinement and recombination dynamics in silicon nanocrystals embedded in Si/SiO_2 superlattices, *J. Appl. Phys.* 87(11) (2000) 8165–8173.
46. S.H. Hong, J.H. Park, D.H. Shin, C.O. Kim, S.-H. Choi, K.J. Kim, Doping-and size-dependent photovoltaic properties of *p*-type Si-quantum-dot heterojunction solar cells: Correlation with photoluminescence, *Appl. Phys. Lett.* 97(7) (2010) 072108.
47. S. Park, E. Cho, D. Song, G. Conibeer, M.A. Green, *n*-Type silicon quantum dots and p-type crystalline silicon heteroface solar cells, *Sol. Energy Mater. Sol. Cells* 93(6) (2009) 684–690.

48. X. Hao, E. Cho, C. Flynn, Y. Shen, S. Park, G. Conibeer, M. Green, Synthesis and characterization of boron-doped Si quantum dots for all-Si quantum dot tandem solar cells, *Sol. Energy Mater. Sol. Cells* 93(2) (2009) 273–279.
49. S.H. Hong, Y.S. Kim, W. Lee, Y.H. Kim, J.Y. Song, J.S. Jang, J.H. Park, S.-H. Choi, K.J. Kim, Active doping of B in silicon nanostructures and development of a Si quantum dot solar cell, *Nanotechnology* 22(42) (2011) 425203.
50. K.-Y. Kuo, P.-R. Huang, P.-T. Lee, Super-high density Si quantum dot thinfilm utilizing a gradient Si-rich oxide multilayer structure, *Nanotechnology* 24(19) (2013) 195701.
51. G. Conibeer, M.A. Green, D. König, I. Perez-Wurfl, S. Huang, X. Hao, D. Di, L. Shi, S. Shrestha, B. Puthen-Veetil, Silicon quantum dot based solar cells: Addressing the issues of doping, voltage and current transport, *Prog. Photovolt.* 19(7) (2011) 813–824.
52. D. Di, I. Perez-Wurfl, G. Conibeer, M. Green, Formation and photoluminescence of Si quantum dots in SiO_2/Si_3N_4 hybrid matrix for all-Si tandem solar cells, *Sol. Energy Mater. Sol. Cells* 94(12) (2010) 2238–2243.
53. D. Di, H. Xu, I. Perez-Wurfl, M.A. Green, G. Conibeer, Improved nanocrystal formation, quantum confinement and carrier transport properties of doped Si quantum dot superlattices for third generation photovoltaics, *Prog. Photovolt.* 21(4) (2013) 569–577.
54. S. Huang, G. Conibeer, Sputter-grown Si quantum dot nanostructures for tandem solar cells, *J. Phys. D* 46(2) (2013) 024003.
55. S. Yamada, Y. Kurokawa, S. Miyajima, M. Konagai, Silicon quantum dot superlattice solar cell structure including silicon nanocrystals in a photogeneration layer, *Nanoscale Res. Lett.* 9(1) (2014) 1–7.
56. G. Conibeer, I. Perez-Wurfl, X. Hao, D. Di, D. Lin, Si solid-state quantum dot-based materials for tandem solar cells, *Nanoscale Res. Lett.* 7(1) (2012) 1–6.
57. C.-W. Jiang, M.A. Green, Silicon quantum dot superlattices: modeling of energy bands, densities of states, and mobilities for silicon tandem solar cell applications, *J. Appl. Phys.* 99(11) (2006) 114902.
58. J. Song, J. Li, X. Li, L. Xu, Y. Dong, H. Zeng, Quantum dot light-emitting diodes based on inorganic perovskite cesium lead halides ($CsPbX_3$), *Adv. Mater.* 27 (2015) 7162–7167.
59. J. Li, L. Xu, T. Wang, J. Song, J. Chen, J. Xue, Y. Dong, B. Cai, Q. Shan, B. Han, H. Zeng, 50-Fold EQE improvement up to 6.27% of solution-processed all-inorganic perovskite $CsPbBr_3$ QLEDs via surface ligand density control, *Adv. Mater.* 29(5) (2017) 1603885.
60. J.H. Im, C.R. Lee, J.W. Lee, S.W. Park, N.G. Park, 6.5% efficient perovskite quantum-dot-sensitized solar cell, *Nanoscale* 3 (2011) 4088–4093.
61. A. Swarnkar, A.R. Marshall, E.M. Sanehira, B.D. Chernomordik, D.T. Moore, J.A. Christians, T. Chakrabarti, J.M. Luther, Quantum dot–induced phase stabilization of α-CsPbI3 perovskite for high-efficiency photovoltaics, *Science* 354(6308) (2016) 92–95.
62. E.M. Sanehira, A.R. Marshall, J.A. Christians, S.P. Harvey, P.N. Ciesielski, L.M. Wheeler, P. Schulz, L.Y. Lin, M.C. Beard, J.M. Luther, Enhanced mobility CsPbI3 quantum dot arrays for record-efficiency, high-voltage photovoltaic cells, *Sci. Adv.* 3(10) (2017) 4204.
63. J.A. Christians, A.R. Marshall, Q. Zhao, P. Ndione, E.M. Sanehira, J.M. Luther, Perovskite quantum dots: A new absorber for perovskite-perovskite tandem solar cells: Preprint, *NREL.* (2018) NREL/CP-5900–71593.
64. M. Mittal, A. Jana, S. Sarkar, P. Mahadevan, S. Sapra, Size of the organic cation tunes the band gap of colloidal organolead bromide perovskite nanocrystals *J. Phys. Chem. Lett.* 7 (2016) 3270–3277.
65. D.N. Dirin, L. Protesescu, D. Trummer, I.V. Kochetygov, S. Yakunin, F. Krumeich, N.P. Stadie, M.V. Kovalenko, Harnessing defect-tolerance at the nanoscale: Highly luminescent lead halide perovskite nanocrystals in mesoporous silica matrixes, *Nano Lett.* 16 (2016) 5866–5874.

66. G. Nedelcu, L. Protesescu, S. Yakunin, M.I. Bodnarchuk, M.J. Grotevent, M.V. Kovalenko, Fast anion-exchange in highly luminescent nanocrystals of cesium lead halide perovskites (CsPbX3, X = Cl, Br, I), *Nano Lett.* 15 (2015) 5635–5640.
67. J. Yao, J. Ge, B.-N. Han, K.-H. Wang, H.-B. Yao, H.-L. Yu, J.-H. Li, B.-S. Zhu, J. Song, C. Chen, Ce^{3+}-doping to modulate photoluminescence kinetics for efficient CsPbBr3 nanocrystals based light-emitting diodes, *J. Am. Chem. Soc.* 140 (2018) 3626–3634.
68. A.K. Geim, K.S. Novoselov, The rise of graphene, *Nat. Mater.* 6 (2007) 183–191.
69. L. Ma-Hock, V. Strauss, S. Treumann, K. Kuettler, W. Wohlleben, T. Hofmann, S. Groeters, K. Wiench, B. van Ravenzwaay, R. Landsiedel, Comparative inhalation toxicity of multi-wall carbon nanotubes, graphene, graphite nanoplatelets and low surface carbon black, *Part. Fibre Toxicol.* 10 (2013).
70. C. Wei, Z. Liu, F. Jiang, B. Zeng, M. Huang, D. Yu, Cellular behaviours of bone marrow-derived mesenchymal stem cells towards pristine graphene oxide nanosheets, *Cell Prolif.* 50 (2017) 1–10.
71. T.-H. Kim, T. Lee, W. El-Said, J.-W. Choi, Graphene-based materials for stem cell applications, *Materials* 8 (2015) 8674–8690.
72. N. Shadjou, M. Hasanzadeh, Graphene and its nanostructure derivatives for use in bone tissue engineering: Recent advances, *J. Biomed. Mater. Res. A* 104 (2016) 1250–1275.
73. J.-M. Peng, J.-C. Lin, Z.-Y. Chen, M.-C. Wei, Y.-X. Fu, S.-S. Lu, D.-S. Yu, W. Zhao, Enhanced antimicrobial activities of silver-nanoparticle-decorated reduced graphene nanocomposites against oral pathogens, *Mater. Sci. Eng. C Mater. Biol. Appl.* 71 (2017) 10–16.
74. D. Pan, J. Zhang, Z. Li, M. Wu, Hydrothermal route for cutting graphene sheets into blue-luminescent graphene quantum dots, *Adv. Mater.* 22 (2010) 734–738.
75. J. Peng, W. Gao, B.K. Gupta, Z. Liu, R. Romero-Aburto, L. Ge, L. Song, L.B. Alemany, X. Zhan, G. Gao, S.A. Vithayathil, B.A. Kaipparettu, A.A. Marti, T. Hayashi, J.-J. Zhu, P.M. Ajayan, Graphene quantum dots derived from carbon fibers, *Nano Lett.* 12 (2012) 844–849.
76. L.A. Ponomarenko, F. Schedin, M.I. Katsnelson, R. Yang, E.W. Hill, K.S. Novoselov, A.K. Geim, Chaotic dirac billiard in graphene quantum dots, *Science* 320 (2008) 356–358.
77. V.A. Gerard, Y.K. Gun'ko, E. Defrancq, A.O. Govorov, Plasmon-induced CD response of oligonucleotide-conjugated metal nanoparticles, *Chem. Commun. (Camb).* 47 (2011) 7383–7385.
78. C. Wu, J. Hong, X. Guo, C. Huang, J. Lai, J. Zheng, J. Chen, X. Mu, Y. Zhao, Fluorescent core-shell silicananoparticles as tunable precursors: Towards encoding and multifunctional nano-probes, *Chem. Commun.* 6 (2008) 750–752.
79. O. Moudam, B.C. Rowan, M. Alamiry, P. Richardson, B.S. Richards, A.C. Jones, N. Robertson, Europium complexes with high total photoluminescence quantum yields in solution and in PMMA, *Chem. Commun. (Camb)* 43 (2009) 6649–6651.
80. J.-M. Bai, L. Zhang, R.-P. Liang, J.-D. Qiu, Graphene quantum dots combined with europium ions as photoluminescent probes for phosphate sensing, *Chem. Eur. J.* 19 (2013) 3822–3826.
81. Y. Cao, H. Dong, Z. Yang, X. Zhong, Y. Chen, W. Dai, X. Zhang, Aptamer-conjugated graphene quantum dots/porphyrin derivative theranostic agent for intracellular cancer-related MicroRNA detection and fluorescence-guided photothermal/photodynamic synergetic therapy, *ACS Appl. Mater. Interfaces* 9 (2017) 159–166.
82. M.L. Steigerwald, L.E. Brus, T.B. Laboratories, M. Hill, Semiconductor crystallites: A class of large molecules, *Acc. Chem. Res.* 23 (1990) 183–188.
83. Y. Kayanuma, Quantum-size effects of interacting electrons and holes in semiconductor microcrystals with spherical shape, Phys. Rev. B 38 (1988) 9797.

84. L. Qian, Y. Zheng, J. Xue, P.H. Holloway, Stable and efficient quantum dots light emitting diodes based on solution-processed multilayer structures, *Nat. Photon.* 5 (2011) 1–6.
85. L. Spanhel, M. Haase, H. Weller, A. Henglein, Photochemistry of colloidal semiconductors. 20. Surface modification and stability of strong luminescing CdS particles, J. Am. Chem. Soc. 109 (1987) 5649–5655.
86. D. Van-Duong, C. Youngwoo, Y. Kijung, L.L. Liudmila, C. Ho-Suk, Graphene-based nanohybrid materials as the counter electrode for highly efficient quantum-dot-sensitized solar cells, *Carbon* 84 (2015) 383–389.
87. M.C. Hanna, A.J.J. Nozic, Solar conversion efficiency of photovoltaic and photoelectrolysis cells with carrier multiplication absorbers, *J. Appl. Phys.* 100 (2006) 074510.

10 Metals, Metal Oxides, and Their Composites—Safety and Health Awareness

Shrikaant Kulkarni

CONTENTS

10.1 INTRODUCTION

Engineered nanomaterials (NMs) have huge potential on various frontiers of society, namely, medicine. However, the rise in synthesis and application of NMs also is a cause of concern about unknown exposure and its possible evil impacts on the health of human being as well as the environment [1]. This work takes an overview of metal and metal oxide (MO) nanoparticles (NPs), their application areas, and the potential health implications on human exposure [2]. It further talks about the general principles underlying toxicity induced by NPs and different methods employed for testing the toxicity of NPs. A watchful evaluation of the material characteristics is what is often required for a thorough knowledge of the toxicity of NMs; a part of this work refers to testing for their physical and chemical characteristics. The next part is dedicated to a thorough discussion of the state-of-the-art knowledge about 12 of the key metal and MO NPs, for a scientific assessment of in vitro or cell culture and in vivo or animal toxicity studies. However, implications on the human health of exposure to NPs are discussed preferentially over ecotoxicological effects in this chapter [3,4].

10.2 METAL AND METAL OXIDE NANOMATERIALS

10.2.1 Health Implications of Nanoparticles

Nanotechnology is considered a technology that is revolutionary and holds much promise and potential in addressing vital problems confronted by society. Applications in the frontier areas like production of energy and its storage, optical, electronic, mechanical, and ceramic industries are coming to the fore. It also holds much promise and potential on the frontier of nanomedicine in drug-delivery carriers, in medical imaging, and as multifunctional theranostic devices. The application of NPs in these innovative areas is attributed to the unique properties possessed at the nano level. Simultaneously, there has been a growing awareness about the potential hazardous effects that may be caused due to the exposure of the ultrafine particles. NPs by virtue of the large surface area available for reactions to occur, generate reactive oxygen species (ROS) responsible for bringing about further cascading effects [5,6].

Nanotoxicology is aimed at investigating the unison between NMs and biological systems. There are many vital aspects to address in nanotoxicology, such as standard assays and reference materials, as well as proper dose metrices such as mass, number of particles, and surface area. The researchers of today believe that NMs must be examined on an individual basis. The latest study shows that the in-vitro cytotoxic potential of silicon dioxide (SiO_2) NPs, although a thoroughly investigated NM, shows that a more coherent study is required due to knowledge gaps to further the cause of nanotoxicological research [7,8].

NP exposure is not a new phenomenon; for instance, the release of such particles is very common during combustion processes. It is common that air pollutant particles are responsible for causing deleterious health problems [9]. Exposure to particulates in air is instrumental in causing cardiovascular and pulmonary diseases, while exposure to higher levels of indoor airborne particles cause comparatively lower respiratory ailments, which are fatal to young children. For example, exposure to crystalline silica (quartz) has caused severe and lethal fibrotic lung disease (silicosis) to workers like knife grinders and in millstone construction in history. During the construction of a tunnel [10,11], 764 people lost their lives due to acute silicosis, and 1,500 more were infected. Asbestos also causes fatal lung diseases, asbestosis, lung fibrosis, cancer, pleural fibrosis, etc. involving gradually developing malignant tumors. A growing cause of concern today is that a few types of carbon nanotubes (CNTs) have properties resembling with asbestos; however, it is yet to be unraveled about how far CNTs are responsible for causing mesothelioma [12–14].

10.2.2 Applications on Technology and the Consumer Product Front

Engineered NPs are employed to produce products that are high-end technology based, structural, consumer, and cosmetic, like customized imaging and drug delivery systems [15].

Materials are classified into four types, depending upon their chemical composition:

- Metallic or intermetallics, such as semiconductors
- Ceramics (oxides, nitrides, carbides, etc.)
- Polymers/organics
- Composites (organo-inorganic, inorganic-inorganic, metallo-inorganic, etc.) [16]

The very purpose of the work presented here is to provide details about the potential toxicity of NPs of MOs (ceramic) and their composites from the safety and health on the exposure perspective. Metals, namely, Au, Ag, Cu, Pt, and Ti have found widespread uses in the biomedical field by virtue of their antibacterial and antimicrobial [17], optical (plasmonics), and biocompatibility properties. Many MOs (ceramics) find use in consumer products, for example, TiO_2 (additive in sunscreens), Fe_2O_3 (contrasting agents in MRIs or drug-delivery vehicles), SiO_2 (vaccine adjuvants, coatings to increase biocompatibility of materials and protect them from harsh environments, and CeO_2 (catalyst and automobile) [18]. Luminescent NPs, in the form of quantum dots (QDs), may have a host of chemical compositions varying from oxides (ZnO, SnO), metallics/semimetallics like Si and Au, to semiconducting materials like CdSe, CdTe, and InAs. QDs find use in applications such as tracking of cells and fluorescence imaging in animal models [19,20].

10.2.3 Workplace and Consumer Exposure

Increase in the synthesis of NMs has no doubt led to a possibility of more occupational and consumer exposure. Many examinations pertaining to exposure at workplaces have been undertaken so far. Normally, two major approaches are employed to make workplace exposure estimates realistic or based upon simulations [21]. However, in both situations, the problem of particles in the background is a major cause of concern. Exposure of NPs at workplaces and their release, in particular >300 nm were examined when NMs were in dry condition. Furthermore, NMs < 100 nm were found selectively, like the generation of gas in the open during processing. However, the application of protective measures such as fume hoods and ventilation systems can substantially lower potential exposure levels [22,23].

Exposure to NPs may take place during various stages of the life cycle in their manufacturing as in the following:

- Synthesis
- Nanopowder handling
- Nanoproduct dispersion such as spraying
- Abrasion, breakdown of nanoproducts [24].

Some of the examples of studies undertaken are as follows:

- During the production of Ag NPs, the workplace concentration was ranging between 5 and 280 μg/m^3 (approximately 91–1.63 M particles/cm^3). The exposure may be on handling the nanopowders during packaging [25].
- Pigment TiO_2 and nano-TiO_2 exposure during automated packaging at an industrial facility producing TiO_2. The mean exposure varies between 220 to 690 μg/m^3 with more than 90% of the particles <100 nm generated from soot and particulates derived from chemicals used for processing. The highest concentration in mass terms is about 3.0 mg/m^3, above 501 and 201 nm particle size for micro- and nano-TiO_2 when packed, respectively. The handling of powders and the suspensions may result into exposure in a few cases [26].
- Multiwalled carbon nanotubes (MWCNTs) and carbon black are released in the air on sonication of mechanical dispersions, which release droplets of water laced with NPs. The dispersion used during spraying is efficient enough to form liquid aerosols containing NPs. These exercises bring about exposure at a personal level and at distance from the source. Furthermore, there is not much of a difference between a nanospray and a reference spray. The potential exposure study to CNTs during the abrasion of products employed in polymers show that the release takes place on using it in tires, due to tremendous abrasion on application, although the resulting exposure in this case is abysmally low. However, exposure evaluation and prevention at the workplace is quite challenging [27].

There are still several issues concerned with potential human exposure because of the utilization of consumer products in a product inventory because quantitative estimation is a daunting task due to either nonavailability or inadequate data related to concentration levels in various products. Exposure may not be necessarily free NPs but agglomerated ones on several occasions. For example, during the study of exposure on inhaling engineered NPs in cosmetic products, it was observed that users would be vulnerable to exposure to agglomerates of NPs with size greater than 100 nm [28,29].

Moreover, a host of applications in nanomedicine will expose humans to NMs. For example, exposure to paramagnetic iron oxide NPs (SPIONs) are employed as contrast agents in magnetic resonance imaging (MRI) and nanoscale drug delivery systems. Furthermore, the problem that is yet to be addressed is to what degree the concentrations of NPs are put into use; in-vitro and in-vivo toxic potential assessments are in congruence with the exposure situations in reality [30].

10.2.4 Risk Evaluation: Hazard Vis-a-Vis Risk: Regulation of Nanomaterials

The key question confronted by the field of nanotechnology is how far NMs can pose a health risk threat to both human and the ecology at large [31]. The properties most sought after in selective applications, like the capacity to diffuse through biological membranes and surface behavior, are responsible for the rise in toxicity, referred to as NMs paradox. The literature survey showcases that different NPs differ in their toxic effects [32,33]. NMs, like carbon black and TiO_2, have comparatively lower toxicities. It has to be kept in mind that a risk posed by a NM is influenced by the innate toxicity or hazard and the degree of exposure [34]. Therefore, some NPs may exhibit more toxicity, although the risk may be lower due to negligible exposure. More importantly, some NPs may have access to key body locations unlike coarser ones. For example, the olfactory route is amenable to translocation of manganese oxide NPs inhaled to the central nervous system (CNS) in mice. Risk assessment of NMs is a must for ensuring better risk assessment that contains, for instance, long-term studies in individual susceptibility differences, at organ or systems levels, thorough cellular mechanistic studies, and standardization of methods [35,36].

The European Union (EU) provides for, regulation of NPs in terms of Registration, Evaluation, Authorisation and Restriction of Chemicals (REACH) legislation. REACH asks for all chemicals either produced or imported in amounts $\geq$ 1 ton/year to be registered [37]. A single registration may encompass various forms (like both micro- and NPs), and the registrant has to assure the safety in various forms. This, however, doesn't ask for undertaking any typical testing tools for individual form, or to specify the manner in which various forms have been taken care of as a part of registration. Presently, several registrations for NMs were found to possess forms that do not specify in clear terms in the registration protocols. Presently, there is a lot of ambiguity on the way REACH needs to be applied for NMs; however, the European Chemicals Agency (ECHA) is aimed at providing guidelines on whether NMs should be looked upon as different forms of materials [38–40].

In the US, many NMs are considered "chemical substances" under the Toxic Substances Control Act (TSCA). The said act demands producers of fresh NMs to make available typical information to the Environmental Protection Agency (EPA) to examine before producing. From 2005, the EPA has obtained and examined more than 100 fresh registrations for NMs [41]. However, all NMs are not novel, and the EPA has offered to gather information through a voluntary program on all NMs presently in use [42]. This system now demands mandatory reporting and testing due to participation to a limited extent, and concerned legislative authorities are instrumental in the management of NMs in the EU as well as in US. Till now, only a handful of doses as reference or health-related limits have been laid down for NMs. A lot of work is required for having reliable risk assessment of NMs, for metals and MOs in line with the strategic research agenda (SRA) published earlier. It is applicable to research in the safety of NMs in the EU. Traditional risk assessment tool may not be useful to evaluate the potential risks posed by NMs that are engineered due to prevalent method-based constraints and epistemic uncertainties regarding knowledge and its validation [43,44].

This chapter highlights views on the potential risks posed by selective metals and MO NPs, which are examined and mostly manufactured in bulk (e.g., Ag, TiO_2, ZnO). Moreover, further discussion of available scientific literature is required in testimony of the determination of no observed adverse effect concentration (NOAEC) [45], or lowest critical effect dose (CED) against control animals, measured as CED_{05} for metals and their oxide NPs [46].

10.3 TOXICITY INDUCED BY NANOPARTICLES

10.3.1 Effect of Morphology and Surface Charge

Size makes a difference, in particular when it comes to NP uptake efficiency by cells. Moreover, studies in the recent past reveal that the higher the intracellular concentration of NPs, the higher the toxic effect that is generated [47]. A set of experiments show that their NP uptake may be influenced by their optimum size [48]. Further particle size can be regulated by not allowing the agglomeration of NPs to take place. Uptake by cells is not necessary for cytotoxicity to prevail [49]. Therefore, Co and Cr NPs affect human fibroblasts, cutting across a cellular hurdle despite not moving through it. The output varies from that shown by cells exposed directly to NPs. Gold colloidal particles of size $\geq$ 13 nm may behave nontoxic, as against gold particles <2 nm diameter are observed to show toxicity in various cell lines, which brings about a basic change in terms of the response of cells to gold (Au) NPs, particularly in the size varying between individual molecules, their complexes, and the nanoscale [50,51]. The Au clusters with a typical size, 1.4 nm, interact in a distinct way with the main DNA grooves and is attributed to the toxicity of such nanoclusters, unlike somewhat smaller or coarser particles that show substantially low toxicity, as against quite larger Au NPs, which are nontoxic [52]. It is the size, geometry, and morphology of particles that matters. For example, with polystyrene beads (diameter > 0.5 μm) and macrophage cell lines, it is found that geometry in a particular shape decides whether phagocytosis initiates, while the particle size is instrumental in taking the phagocytosis to completion [53,54].

More recently, on examining a number of polyethylene glycol (PEG)-based NPs of varying morphologies having a given chemical composition, it was observed that rod-shaped NPs showcased a comparative advantage in terms of cell internalization rates and were attributed to the response of several rod-shaped bacteria [55]. On the other hand, studies of Au NPs coated with proteins have shown that rod-shaped Au NPs coated with transferrin are less efficient than transferrin-coated spherical NPs when they interact with the HeLa cell line [56]. There is every possibility that the location of the transferrin on the nanorod surface against nanospheres can vary, which may further affect the interaction behavior of the nanostructures with cells. It means that the particle size, shape, and surface attachment with either proteins or other chelating agents are of vital importance when it comes to biological effects of NMs [57]. The toxicity of CNTs ascribed by some researchers to their "needle-like" geometry, with high aspect ratio, make these NMs move through the cell membranes. It can be applicable to few MWCNTs possessing more width and, in turn, rigidity [58].

In several studies, typically theoretical simulations involve computation of aspect ratios for the core of NPs, without taking into account the surface layer with the protein corona attached, which lowers the aspect ratio effectively, thereby neutralizing the probable shape effects [59]. Surface charge also very strongly affects the NP uptake. Normally, NPs possessing a positive charge interact with cells faster than ones with a negative charge, which is by virtue of the net negative charge carried by the cellular surfaces [60]. More toxicity of cationic NPs is attributed to greater cellular uptake efficiency. It is observed that polyamidoamine dendrimers carrying positive charges comparatively show more cytotoxicity than their negative charge-carrying counterparts. Some other studies have demonstrated that the corona of a protein is responsible for acquiring a zeta potential in the order of −10 to −20 mV to the NPs, notwithstanding the chemistry of material, and this "normalization" of zeta potentials is because most plasma proteins are anionic at physiological pH [61–63].

10.3.2 Effect of Solubility and Metal Ion Release

Several metals and their oxide NPs tend to dissolve in acidic environments in the cell, which impart toxicity by following the uptake mechanism of a Trojan horse-type as it overcomes the plasma membrane barrier and lets toxic metal ionic species migrate into cells, as observed in case of oxides of Co, Fe, Mn, and Zn [64,65]. It is particularly important that CuO and NiO NPs show high toxicity and that their cellular uptake efficiency is found to be dependent upon the NP morphology [66,67]. The evaluation of pulmonary inflammogenicity of several metals and their oxide NPs show that toxicity is strongly correlated with either of the two physicochemical factors, namely, the zeta potential in acidic environments with poor solubility and those with rich solubility [68]. Therefore, inflammogenicity of rich-solubility NPs is influenced by the nature of ions, which are formed as a result of solubilization of NPs in the acidic phagolysosomes; toxic ions (e.g., Cu and Zn) destabilize phagolysosomes and thereby cause inflammation. Low-solubility NPs have their zeta potential in an acidic environment correlated with their strength toward the inflammogenicity of lung [69,70]. The possible mechanism when the NPs are inhaled involve adsorption with a protein or lipid corona; therefore, the actual zeta potential is not obtained [71]. The NPs get either phagocytosed or endocytosed within the cells and the lysosomes, then both proteolytic enzymes and the acid medium cumulatively remove the corona from the surface of the particle. If the surface under these conditions is cationic, then the stability of the lysosomal membrane is compromised and inflammation takes place [72,73].

10.3.3 Effect of Nanoparticle Surface with Biocorona

NMs can behave differently upon incorporation into a living organism by virtue of sorption of biomolecules like proteins and lipids by as functional on the surfaces of NPs [74]. The adsorption of biomolecules on NP surfaces is induced by the innate material properties like surface charge, lipophobicity or lipophilicity, and particle size [75]. It is observed that the dextran deposit and iron oxide NPs coated with dextran are made available to typical probes on plasma-induced incubation, showing that surface of the NPs is accessible to recognition by cell lines in spite of having the biocorona [76]. The earlier studies show that adsorbability of cytoplasmic proteins is governed by the surface topography and NP surface chemistry. Thus, NPs of TiO_2 in bare anatase and rutile phase forms readily adsorb proteins alike, while coated rutile with alumina and silicone phases of TiO_2 selectively bind some proteins [77]. The corona of the protein is instrumental in controlling the uptake and toxicity by single-walled CNTs (SWCNTs). Earlier research shows that the adsorption tendency of proteins on iron oxide NP surfaces takes place at varied concentration levels of plasma varying between 1% and 100% [78,79]. It is observed that at lower plasma concentration levels, the NPs show tendency to agglomerate, leading to cluster formations driven by proteins like fibrinogen, while at higher plasma doses other proteins, like apolipoproteins, tend to coat in order to stabilize NPs individually, which reflects upon in-vitro uptake by macrophages [80,81].

An experimental research shows that Au NPs coated with polyacrylic acid readily bind to and bring about uncoiling of fibrinogen, which facilitates better binding with integrin receptors on macrophages in vitro, and thereby cytokines are released, which are inflammatory [82,83]. Neither NPs nor fibrinogen in isolation can show similar responses. All such studies support the presence of a "corona" effect [84]. Furthermore, the corona forms very rapidly on silica and polystyrene NPs, which affects hemolysis, activation of thrombocyte, NP uptake, and endothelial cell death [85,86].

10.3.4 Effect of Translocation

NPs possess the potential to diffuse through the biological barriers in the body, which is a vital and deciding factor in the toxicity of NPs, and thereby the degree to which they can be translocated from the deposition site, like skin or lungs, to circulate in a systemic way [87]. There is not much reliable evidence about translocated NPs (mass fraction > 1% of concentration) given to organs like the lungs in reference to human studies [88]. However, in the study of animals for the translocation of metals, such as Au, Ag, and MOs like, TiO_2 NPs with size ranging from 5 to 100 nm, through the air blood barrier, it is observed that their mass fractions are rarely >5% of the administered lung dose. The recent study shows that the translocation of Au NPs across the air blood barrier does take place [89]; their piling and retention is governed typically by the NPs characteristics, namely, surface charge and area. The accumulation of NPs in secondary organs takes place due to chronic exposure over prolonged time spans [90,91]. However, evil effects may not be only because of NP translocation. Rather, NP inhalation may drive or alter the autonomous nervous system, which will have its own health effects [92]. Furthermore, it should be understood that translocation exercises are carried out by using model NPs, while practical world-engineered NPs have a tendency to agglomerate in aerosols and may not therefore be available systemically [93]. The most probable translocation is through the skin, and studies show no permeation and systemic availability. However, if the skin is damaged by ultraviolet (UV) diffusion of TiO_2 or ZnO NPs in, say, sunscreen formulations available in market, it is marginally increased, although transdermal absorption is not observed [94,95]. NPs getting into the gastrointestinal (GI) tract are most of the times excreted, although absorption to a small extent systemically is observed. Thus, in comparative terms, cationic NPs are absorbed with greater efficiency than anionic and neutral ones. Similarly, the absorbability depends upon the size, for example, polystyrene NPs with size range (50–100-nm) absorb about 250 times more than larger-sized particles [96,97].

10.4 PHYSIOCHEMICAL CHARACTERIZATION

10.4.1 Methods of Characterization

Various methods have been in use for the characterization of a host of characteristics of inorganic NPs. The qualitative analysis of inorganic NMs begins with identification of a crystal lattice and determination of composition [98]. X-ray diffraction (XRD) is a powerful technique to accomplish these objectives that works on the principle of the diffraction of X-rays by the sample in question, producing a characteristic lattice pattern, which is a reflection of its structure [99,100]. The generated pattern for the material under examination is then compared with a library database to identify the crystal structure of the sample concerned [101]. NM occurs naturally, or resembles the engineered one at the laboratory scale, and this matching exercise will come up with many options. However, sometimes the XRD pattern in itself will not help in identifying the material regarding whether it is a solitary phase or a combination of them [102]. Hence, XRD characterization has to be complemented by analysis at the elemental level of the NM under examination. In the event of having a less amount of material to get the XRD pattern, another technique referred to as selected area electron diffraction (SAED) may be carried out by using transmission electron microscopy (TEM) [103,104]. The diffraction spots or rings generated in the form of a diffractogram may be used to compute the lattice spacing, which on matching with the reference library database help identify the crystal structure of the NM under study [105,106].

Composition analysis can be brought about by employing a number of dry or wet chemical analysis methods [107]. A traditional method involves dissolution of NM in an acidic medium and subsequently subjecting it to digestion, followed by atomic absorption spectroscopy (AAS) or inductively coupled plasma (ICP) techniques wherein samples are subjected to atomization either in flame or higher energy plasma [108,109]. On atomizing the sample, elemental analysis can be undertaken with the help of the optical absorption or emission spectroscopy signals obtained by the excitation of atoms [110]. Both AAS and ICP may be employed for qualitative and quantitative analysis. Later, one demands the reference solutions of known concentration for calibration purposes prior to subjecting the sample to analysis [111].

AAS works on the principle of absorption of light radiations of typical wavelengths by atoms of elements in the gaseous phase that present unique and characteristic information using fingerprint analysis [112]. On calibration of the system, the data obtained was subsequently be put to use for quantification of the different elements so as to determine the composition in elemental terms of the sample in powder form [113].

The principle of ICP involves the detection of signals on the emission of radiations by promoted atoms to higher levels of elements in a higher energy plasma medium, which are then either sensed by optical emission spectroscopy (OES) or fragments formed by the plasma medium with the help of mass spectrometry (MS) [114,115]. The analysis is confined to elements, preferably metallic, by AAS while a few nonmetallic elements by ICP [116]. However, if the sample in the dry state is to be tested, then emission of X-ray radiations by promoted atoms is subjected to the elemental analysis, and the tool concerned is referred to as energy-dispersive X-ray spectroscopy (EDXS) [117]. Furthermore, thorough compositional analysis so as to differentiate between various oxidation states of the given element may be carried out with either X-ray photoelectron spectroscopy (XPS) or electron energy loss spectroscopy (EELS) coupled with TEM [118–120].

The characteristics of foremost importance for NPs are morphology, size, and their distribution, as they are related directly with the surface area of the NPs [121]. Particle size may be determined either in the dry or wet state. In the dry state, it is determined by subjecting the nanopowder to analysis by employing electron microscopic techniques such as scanning electron microscopy (SEM) or TEM while sophisticated software for image processing can help in obtaining size distribution profile diagrams. SEM presents information pertaining to the sample surface topography, on the other hand TEM offers a thorough analysis. For example, core-shell, porous, and mesoporous structured NMs can be examined better by TEM analysis only [122–124].

When the NPs are in a dispersion form in a liquid, they show variations in size due to stacked molecular layers of surfactants on the surface or due to agglomeration based on the technique employed for the synthesis of them [125]. The particle size distribution profile in a suspension is determined by the dynamic light scattering (DLS) technique, photon correlation spectroscopy (PCS), or quasi elastic light scattering (QELS) [126]. A difference between dry and wet particle size can help in understanding the thickness of coating or aggregation of NPs [127]. The limitation of the DLS technique is its insensitiveness to particle geometry and gives an approximation about the NP size along the longest axis [128,129].

Surface area of NPs (in m^2/g) may be measured using Brunauer-Emmet-Teller (BET) adsorption isotherm development, which relies upon the adsorption of N_2 gas on NPs [130]. The amount of gas adsorbed is determined as a measure of the surface area; while particle size is determined from the surface area, assuming the NPs to be spherical [131,132].

Surface functionality of NPs refers to functional groups attached and the surface charge. Surface chemistry is studied by Fourier transform infrared spectroscopy

(FTIR) or Raman spectroscopy [133]. The underlying principle is the absorption of infrared (IR) light by the characteristic molecular vibrational modes corresponding to unique functional groups. The characteristic vibrational frequencies of absorption help in identifying the various groups [134]. One of the key aspects of surface chemistry is the charge carried by NPs when dispersed in liquids [135]. The surface charge, type, and magnitude are analyzed using the zeta (ζ) potential technique. The NPs under the influence of an electric field are made to migrate toward oppositely charged electrodes [136]; the direction of migration and mobility are employed for the estimation of surface charge, which are detected by light-scattering techniques. The magnitude of the surface charge is very vital in evaluating NP stability in colloidal dispersions [137]. The stability of a dispersion is assured at a given pH if the surface charge > 25 mV < –25 mV. For example, the net surface charge of iron oxide NPs is 0 $[p.z.c.]_{pH=7}$ [138], at which they begin to agglomerate and subsequently get sedimented/precipitated. By coating the NPs with different functional groups, the p.z.c. is made to shift the pH to either higher or lower values, thereby stabilizing the NPs [139,140].

Optically active NPs are evaluated for their absorption and emission characteristics. Absorption characteristics are specifically studied by using UV-visible spectroscopy [141]. If NPs are luminescent, the characteristic wavelength of emission is detected by photoluminescence spectroscopy. For example, CdSe (semiconductor) NPs is an excellent example that exhibits a spectrum of colors when light is incident upon it [142,143], which can be truncated by maneuvering the NP size [144]. Size determination of QDs can easily be done using UV-Vis spectroscopy from their absorption and emission behavior [145]. NMs are characterized for their magnetic properties using techniques such as vibrating magnetometer or superconducting quantum interference device, which can scan for the magnetism in the NMs in relation to variation in the magnetic field [146]. Novel magnetic characteristics on downsizing of NPs can therefore be determined by studying the nature of magnetization demagnetization diagrams [147,148].

Thermal stability or degradation of material is normally observed using thermogravimetric analysis (TGA), a study of the percent change in the quantity of material against an increase in temperature [149].

TGA is a very powerful technique to

- Test the thermal degradation behavior of inorganic or polymeric NMs [150]
- Determine the surface-coating behavior on a NM surface, which is correlated with calculation of the thickness of the organic modifiers and organic-inorganic nanocomposites and hybrid materials [151]
- Estimate the composition [152].

The extent of loss in mass varies from material to material, which further can help in estimating the weight percent of various constituents [153]. NPs normally have largely perturbed surfaces, which affect their density too [154]. The density of NPs is commonly determined by a pycnometer, by determining the volume of a given mass of nanopowder [155] or the volume of a liquid displaced by it or by using controlled gas chambers wherein the gas confined in a known volume is quantified both in presence and absence of a known mass of sample nanopowder. The volume is thus estimated and used to calculate the density [156,157].

10.4.2 Workplace Exposure

Three metrics are in use for evaluating workplace exposure to NPs; although, each of them has its own innate limitations [158]. The mass of NPs may be measured, but if it is quite small then detection by a gravimetric exercise is difficult to use and thereby may need longer sampling times [159]. The coarser particle mass may easily mask a rise in mass of the NPs. The concentration in number terms at times may have large variations on spatial and temporal basis, which makes it hard to differentiate the synthesized NPs against reference concentrations. The concentration at the surface is not easy to use because of exhibition of either coagulation or aggregation behaviors [160,161]. Simultaneous measurement of all three metrics is vital to use, but there is no solitary instrument in existence that can measure reliably all three parameters. The number of particles can be quantified by using an instrument condensation particle counter (CPC) in real time in the size range varying from 10 to 1000 nm [162,163]. Another instrument of importance is a scanning mobility particle sizer, which consists of a differential mobility analyzer and a CPC, which detects the real-time size-dependent concentration in number terms in the size range varying from 3 to 800 nm, and therefore gives size distribution of an aerosol [164]. The surface area of NPs ranging from 7 nm to 10 μm can be measured, with an electrical low-pressure impactor [165]. The mass of NPs is detected online with either a size-selective sampler or tapered element oscillating microbalance, which enables personal monitoring [166]. The particle count and its size distribution profile measurements at the workplace are considered indicative but without providing any concrete proof about the presence of typical NPs [167]. The latter is done by particle sampling followed by electron microscopic examination. A major limitation of present devices is their inability to different background NPs from engineered ones [168]. The aerosol MS is presently the only tool that has the capability of sizing and chemically characterizing NPs online, although it is a costlier, tedious, and immobile technique [169,170].

10.5 PARTICLES AND THE ENVIRONMENT

10.5.1 Particles in Atmospheric Environment

The atmosphere carries suspended particles with varying size range from nanometers to 10th micron and get added to the atmosphere at the pace of 2.5 btons/year originating both from natural and synthetic sources [171,172]. Naturally occurring salt and soil particles from the sea and the land, respectively, occupy 60% of total number of particles [173]. While, the synthetic particles, namely, nitrates, sulphates, hydrocarbons, and heavy metals that are derived from the anthropological activities occupy only 16%, but due to their higher concentration [174], their effects are detrimental to the ecosystem. It is found that the concentration in number terms of aerosol in the atmosphere that we inhale daily ranging between many hundred and thousand particles/cm^3 in clean area and dusty areas, respectively, having the size range 10 nm to tens of μm [175,176].

The size distribution in the nanoscale is observed based on the causes of particle production, and it is normally bimodal with peaks appearing with the size lying between 10 μ and below [177]. It is observed from the variations in the particle production processes, tiny particles (PM2.5) are formed from molecules, while the NPs are largely complex in their chemical composition than the large-sized ones and many a time have serious and deleterious health effects [178]. The latest epidemiological study reports show that the concentration of PM2.5 is positively correlated with the mortality because of pulmonary diseases [179,180].

Different research tools are employed to explore the mechanism of growth of particles and to identify the cause of emission [181].

10.5.2 Particles in Exhaust Gases

Most of the time, NPs in exhaust gases are examined to study the effects of all particulates in the environment. The concept of "nanoparticles" is put into use only in a handful of cases, while "fine particles" are often used for exploration [182,183]. The causes of exhaust gas generation are mainly immobile (stationary) mass-scale combustion facilities and diesel-run engines for stationary and portable applications [184]. To run such stationary combustion facilities, fossil fuels are normally used [185]. Lighter fuels release particulates at a slower pace, though they contain more fine particles, including NPs [186]. It gives a profile of the frequency distribution when coal and heavy oil are combusted. The distribution profile can be based upon both a number and mass basis [187].

It is observed that the total mass of NPs with a size ≤ 1 mm is quite low, though their number is too high. It is also found that while the number of particulates is far more in coal as against oil on combustion [188], the difference gets narrowed down for particles with diameters ≤1 mm, including NPs [189]. The majority of the particles found in ground coal combustion exhaust gases are generated as particulates derived from ash content originating from coal, and some unburned carbon is part of it; for example, all coarse-sized particles so formed belong to such types of particles [190]. Fine particles are of two types: The first one is generated in the process when a metal with a lower boiling point present in coal ash is subjected to evaporation followed by vaporization in an elevated temperature combustion region, which then settles in the exhaust gas condensation process. The second type is carbon particles in the gas phase (soot) [191] that are formed as a result of the delayed air supply for combustion of volatiles in the beginning. It correlates the metallic content in traces in coal ash to the diameter of particles [192]. For example, high boiling aluminum has a fixed concentration irrespective of the particle diameter [193]. However, it is observed that for lower boiling metals, the lower the particle diameter, the higher the metal content [194]. For particles having size ≤ 1 mm in the nanoscale, the generated content increases rapidly on reducing the combustion air supply in the volatiles combustion zone, for example, when the air supply is cut down in two stages during combustion, which also shows the substantial contribution made by carbon particles generated in the gas phase [195]. For the ash contents generated from heavy oil burning, coarse-sized carbon particles are formed both from liquid sprayed and in the gaseous phase [196]. Further in coal, low-boiling metal in traces from heavy oil settles into tiny particles and gets discharged. While in Liquified Natural Gas (LNG), burning the carbon emissions are in traces in the gaseous phase [197,198].

However, in diesel engines, the fuel is injected in a high-temperature and high-pressure environment by subjecting air to compression such that it ignites spontaneously and subsequently burns the fuel and air mixture in the combustion chamber [199]. Thus, particulates are mainly comprising of unburnt hydrocarbons and are formed because of incomplete combustion [200]. The variations in the particle diameter take place with the alterations in the diesel engine load. The effective particle count increases linearly with the commensurate rise in the engine load rate [201]. SEM images show that tiny particles in diesel engine exhaust emissions are found to consist of fine primary particles of a size from tens of nanometers, and coarser ones with carbon hydride deposited on the surface are covered with secondary particles as aggregates generated from primary particles [202,203].

10.5.3 Particles in Groundwater Environment

Particulate matters exist in water in the form of colloidal solutions or sols. Although the structure and molecular weight of the inorganic colloids, for example, oxides of aluminum, silicon, humic, and fulvic acid, [204] may change and their NPs are normally dispersed in water having particle size < 100 nm [205]. The number concentration of colloidal particles in water may range between 1000 and 1020 (number/m^3) and change substantially based on the geochemical parameters of the water body [206]. In moving water, colloidal particles may either move along with water or faster than water [207].

Safety and migration behavior of colloidal particles have a bearing on the migration of molecules that are ionized in aqueous environments [208]. Since NPs typically acquire stability as a colloidal solution, it is key to understand their characteristics and effect in various application areas [209].

10.5.4 Indoor Environment

People adopting urbanized lifestyles spend lot much of time in closed residential environments rather than outdoor ones [210]. Thus, it is important to analyze exposure to indoor particulates and integrate it with outdoor so as to evaluate the impact of indoor air quality on human health [211]. The origin of indoor NPs lies in many sources namely products obtained in chemical processes, nonvolatile residue from droplets of liquids, printers, bio-aerosols, and outdoor air infiltration [212].

10.6 INDUSTRIAL PROCESSES AND THE PARTICLES

It refers to the causes of NP production at industrial-scale reactions by classifying them into typical reactions and other general ones [213].

10.6.1 Industrial-Scale Reactions

The causes of generation of undesirable NPs at common workplaces are such as vapors emanated from processes carried out at elevated temperatures, like smelting, refining, and welding due to combustion not going to completion [214].

Generation of NPs are favored under conducive conditions at workplaces where there is

- Availability of material vulnerable to vaporization [215],
- Reasonably higher temperature to generate adequate vapor, and then condensation to become an aerosol [216], and
- Fast quenching and a higher temperature variation [217].

Innumerable studies have been undertaken on occupational exposure to NPs on the public health frontier. Normally, higher levels of exposure to NPs are found when operations are underway [218], and then a gradual post-operation decline in it because of coagulation, evaporation, dilution, or deposition [219]. The total number of NPs in general reduces, while submicron particles increase over time and distance from the emission source [220]. To precisely assess exposure, the impacts of spatial and temporal variations will have to be assessed. Hence, it is of much importance to measure the time taken by a typical NP concentration to reduce against the reference ones [221]. For example, in the grinding process, steel is pulverized by using a high-velocity grinder, and the concentration distribution profile of particles generated has a unique bimodality [222], due to particles formed of different sizes by the grinder due to the vaporization or burning of pulverized material/s, the mechanical abrading action, and downsizing. However, the total concentration resulted is about 105 particles/cm^3, which is not so high [223,224].

10.6.2 Industrial Processes with Clean Rooms

For clean rooms with controlled environments, the maximum tolerable airborne particle concentration is <103 particles/m^3 having size ≥0.1 mm, as against the ordinary indoor environments at ≥10^9 particles/m^3, which are normally accepted to prevent particle contamination of products by engineered particles in industrial processes [225], semiconductors, and other electronic or optical gadgets due to the accumulation of particles on the product surfaces, resulting in quality and quantity reduction [225]. The sources of emission may emit trace amounts of nanometer-sized solid substances or NPs, which may not be considered as particulates but as chemical or molecular contaminants deposited on solid surfaces due to a chemical reaction [226,227].

10.6.3 Air Exhaled by Humans

Human body emissions won't contribute much to the indoor environment as particle concentrations in air in these situations are substantial, though the emission in clean-room environments [228] is not negligible. The main emissions generated by humans are in the form of atmospheric dust accumulated on clothes and skin, which are mostly submicron sized. Particles in the exhaled air consist of minute droplets of liquid obtained from spittle, which evaporate to generate NPs of non-volatile residues [229]. Particles are measured in exhaled air by introducing air into a device on drying with the help of a diffusion dryer [230]. Smoking results in the fast increase in concentration of particles ≥0.1 mm by more than 10^4-fold and NPs by twofold, so due care has to be taken about the management of employee's clothes and other protective shields, like face masks, as they make their way into a clean room post-smoking [231].

10.6.4 Ionizer Emissions

Ionizers are normally employed in clean rooms to nullify the electrostatic charge carried by particles using electronic devices precisely [232]. The most commonly used ionizer is a corona discharge type and is classified in three types: AC, DC, and pulsed-DC [233]. There are problems of emission of pollutants, namely, ozone, NOx, and particles, and they are very much applicable to air cleaners employing a corona discharge [234]. The major problem is that the particle emission may contaminate the product surfaces and subsequent reduction in the product recovery [235]. The particle emission under study since the latter part of the nineteenth century is attributed to foreign particle accumulation on the probes, probe poisoning and damage, and condensation of gas to particulate [236]. The problem of electrode erosion may be tackled by using well-designed innovative electrode materials, while the change from gas-to-particle problem may be managed by controlling the airborne molecular contamination [237]. For example, silicon-based materials precipitate on the probes due to the gas-to-particulate transformation of lower molecular weight cyclosiloxane obtained from a silicone sealing agent and induced by a corona discharge [238,239].

10.6.5 Particles Containing Boron

The highly efficient air cleaners comprising of borosilicate glass fiber filters, which can filter particulate matter in air and have ultra-low penetration, are used in most of the clean rooms in the industry [240]. It has been observed that by virtue of the chemical reaction, BF_3 vapor forms from said filters when HF gas is leaked from the wet cleaning equipment and is subjected to filtration. Boron, used for doping semiconductors, has been found to be responsible for failure of the semiconductor devices [241,242]. Furthermore, it has been observed that boron even in traces as boric acid (H_3BO_4) forms from the fibers due to the reaction with humidity in the atmosphere [243]. It shows the variation in vaporized boron mass from different filters as an airborne boron concentration. Particularly on ventilation, the vaporized boron rises with a rise in moisture relatively in the atmosphere. Boric acid, a solid at ambient temperature, is vulnerable to form the particulate matter. Its qualitative analysis is done only by chemical analysis because it is found only in traces [244,245].

$$B_2O_3 + 6HF \rightarrow 2BF_3 + 3H_2O$$

$$B_2O_3 + 3H_2O \rightarrow 2H_3BO_3$$

10.6.6 Haze Deposition on Solid Surfaces

Haze may deposit on the surface of glass lenses and mirrors used in optical devices on their exposure for a prolonged time period to clean-room environments wherein airborne molecular concentrations are not regulated appropriately [246]. The haze acts like a barrier to light incident on exposed surface in photolithography [247]. The major reason is the formation of salt, ammonium sulfate ($(NH_4)_2SO_4$), and deposits on the glass surface due to the chemical reaction between sulfur dioxide (SO_2) and ammonia (NH_3) or amines ($-NH_2$) [248]. While the other reason is hexamethyldisilazane added in resistance coating or low molecular weight siloxanes formed from silicones as sealing agents get adsorbed, followed by its degradation to form SiO_2, which deposits on glass surfaces because of a photoinduced reaction on irradiating with a laser [249], and the subsequent reduction in the laser penetration [250]. A report said that haze, with a size of ≤ 0.2 mm, was generated on surfaces of silicon wafers due to the adsorption of organosilicates forming thin films either by using chemical vapor deposition (CVD) or silica (SiO_2) deposition [251,252].

10.6.7 Solid Surfaces with Watermarks

The surface of a silicon wafer on cleaning by using demineralized water followed by air drying leaves behind a watermark [253]. Oxygen from air dissolves and diffuses into water droplets or the wafer surface adsorbs water, leading to silicate formation by the silanol reaction. The watermark is examined as nanometer-sized particles using electron microscopy [254,255].

10.6.8 Leakage from Nanoparticle Production Processes

There are risks due to leakage from production processes of NPs through processing equipment [256]. The synthesis of NPs can be done commonly by the following two approaches [257]. One is "top down" that initiates with material in bulk, downsizing it to nanoscale using etching, ball milling, sputtering, and laser ablation [258]. The second approach is "bottom up" that involves developing materials from the atomic or molecular scale by growing and assembling so as to design and develop the required NPs using methods like sol-gel, CVD, and flame synthesis laser pyrolysis. The majority of these exercises are carried out in a closed vessel set up in a clean room in regulated environments [259,260]. Humans won't get exposed to engineered NPs while synthesis is underway as long as there is no sudden failure in the system, such as breakdown of a seal [261]. Human exposure takes place following the manufacturing while, say, on opening the reactor, drying, or handling the end products after the reaction. The release of NPs during the cleaning of a reaction vessel with water or some solvent is another key factor [262]. Brushes, sponges, tissue papers, etc. used for cleaning may take NPs to drain or waste streams. Disposal of such waste streams, processing NPs by compression, and coating to develop target products may be responsible for the release of NPs in the environment and the ultimate exposure [263,264].

Earlier studies to assess the aerosol discharge reveal that while handling the carbon nanotubes, the NPs generate under vacuum, although the concentrations are quite low [265]. Moreover, measurements of the concentration level of NPs during packaging of carbon black, typically engineered NM, exhibited that there is no rise in the surrounding air [266,267].

10.7 FOOD SAFETY, PUBLIC HEALTH, AND CONSUMER ISSUES

The state-of-the-art developments in nanotechnology has made it possible to apply the engineered NMs in the food industry, which has led to the creation of delivery systems for encapsulating and protecting a host of edible NPs [268]. Food ingredients like proteins, lipids, polysaccharides, and surfactants are a few of the representative materials that may be used to form edible NPs [269]. Chemical, physical, and enzymatic instability existing within the food ingredients may facilitate the degradation of microingredients. Thus, systems are key to balance the microingredients in the food products by identifying the parameters responsible for the degradation [270]. The delivery property profile and dissolution tendency of the bioactivity of compounds can be enhanced by downsizing the bioactive materials as biological activity of a food nutrient that is influenced by its transfer potential from across intestines to the blood [271]. For example, successful application of nanotechnology to give an impetus to the stability of nutrients like Omega-3 fatty acids and beta-carotene commercially has been exhibited in the isolation of a few useful probiotic bacterial species [272]. However, with the continuous upsurge in the utilization of nanotechnology in the food sector is equally important to evaluate their potential health risks to human. Routes through which various NPs get into the human body include ingestion, inhalation, injection, and permeation through the skin [273]. The potential toxicity exhibited by NMs at the primary exposure site, such as skin, lungs, and the GI tract, is relevant to a large extent [274]. However, the NMs are not only confined to their entry site but get distributed over a large number of secondary locations, such as the kidneys, liver, brain, and spleen that are required for assessing the toxic potential of NMs [275,276].

NPs have been finding a use as dietary supplements, food additives, and drugs and are amenable to move through the GI tract. Ingested NMs will be driven out on account of their instability due to the drastic conditions prevailing in the GI tract [277]. In the event of having a highly acidic environment, it may induce NMs to undergo agglomeration inside the intestinal tract, leading to formation of barriers in the GI tract, resulting in death [278]. However, smaller agglomerates can be expelled through the feces with some absorption. For example, mice administered with zinc NPs (5 g/kg of body mass) acquired diarrhea, vomiting, and anorexia severely and subsequently died, which has been revealed by their autopsy results showing a formation of agglomerates of NPs that choked the intestine [279,280]. Many earlier studies have reported that the ingested NPs are removed to the extent of 98% through the feces within 48 h, while most of the leftover was removed through urine and some transferred to blood for circulation. A few other studies show that NPs pierce through the GI tract and translocate to organs [281]. For example, an oral dose of 25 and 80 nm size TiO_2 NPs, say, 5 g/kg body mass, cause more lesions to the liver and kidneys than NPs of the size 155 nm in mice on exposing them for 2 weeks [282].

Certain NPs have exhibit detrimental effects such as inflammation, oxidative stress, and symptoms of premature tumor growth in human organs [283]. Several studies on animals and humans prescribe that NPs alveolar translocation can bring about the circulatory access, which leads the NPs to disperse into the body organs like the heart, liver, vasculature, spleen, and blood [284]. NMs are instrumental in causing oxidative stress and can facilitate the mitochondrial perturbation, may cause damage to the inner membrane, may bring about energy failure, and enhance cytotoxicity [285]. There are a host of NMs that hold much more potential toxicity in the nanoscale range than in the macroscale. For example, while SWCNTs check proliferation of the human kidney cell, on the other hand [286], it negates the cell growth. Food industries may not be interested in using the NMs of such kind [287]. However, the potential toxic effect has to be identified in the beginning of the technology development [288]. The present testing tools employed to determine the toxicity is the right approach to assess the risks posed by NPs because not much of the research data is available pertaining to food-based nanotechnology toxicity and risk evaluation [289]. There is a lack of inadequate concrete data on the toxic potential and health risk of engineered NMs that can stifle the use of nanotechnology in the food industry in the time to come [290]. Thus, assessing safe and secure manufacturing, utilizing engineered NMs, and ascertaining the nanotoxicity would help contribute substantially in getting positive outcomes [291].

10.8 TESTING TOOLS: IN VITRO VIS-A-VIS IN VIVO

Several nanotoxicological tests have been conducted so far by using modified cell lines, but testing NMs that are engineered in primary cell cultures is of much importance because these model systems resemble that of the in-vivo situation [292]. The limitation of this tool is that primary cell cultures exhibit a better degree of variation among various donors. However, this doesn't only show a technical problem but also applicable to biosystems: a heterogeneity in the human population is not expected to behave in the similar fashion, just like clones of a cancer cell line immortalized one. The comparative edge of using cell lines on the research frontier is as expected because it [293]

- Is easily manipulated [294]
- Yields reproducible results, provided the cell culture conditions are maintained [295]
- Has no in-vitro system that can resemble in totality the intricacies of the in-vivo system [296]

Hence, animal examinations are required to showcase, for example, the biological distribution of NMs throughout the body on taking various exposure modes [297]. However, there are various reasons for restraining the use of animals for experimentation, and use of sophisticated ex-vivo models would strike a balance by providing a mediocre pathway for screening the hazard risk of NPs in a more practical way [298], for example, a "lung-on-a-chip" microdevice that simulates mechanical, functional, and structural properties exhibited by the human alveolar capillary interface [299]. The cyclic mechanical strain hastens the toxicity and inflammatory behavior toward silica NPs. This chapter discusses data obtained from both in-vitro and in-vivo methods for selective but representative metals or MO NPs [300].

10.8.1 Real-Time Modeling and Screening

Real-time screening (HTS) is a scientific study that consists of the screening of a compound activity library database in comparison with the biological targets by way of application of tools such as automation, scale-down, and scale-up data analysis [301]. HTS methods are prescribed as quite meaningful tools to make predictions about the potential hazard risks of NMs [302]. However, NMs interfere with in-vitro assays, and therefore new nanotoxicity assays need the label-free detection of cellular signals. HTS can't be substitute for traditional NMs for further testing [303].

In the recent past, highly quantitative assays have been implemented for potential toxicity evaluations of NMs that are engineered [304]. Earlier studies of about 50 NMs with four doses in four cell types with the help of four different assays have been undertaken [305]. The purpose of the study was to analyze wide pattern profiles of the activity of the NMs among themselves, and not just by extrapolating the findings of a solitary in-vitro assay [306]. The cell types identified represent a host of tissues compatible with the assessment of agents injected intravascularly, such as vascular cells, monocytes, and hepatocytes. Hierarchically clustering the data for selected NMs has a semblance or disturbance in their patterns of biological activity across various cellular lines [307]. Furthermore, when NPs were tested in rats, they were found to exhibit a similar kind of activity pattern in vitro and show effects similar on the monocyte number in vivo, which used a multifactor cytotoxicity assay, which assesses oxidative stress to distinguish the metals and MO NPs effects in bronchial epithelial and macrophage cells [308,309]. With this kind of approach, the tailored ZnO cytotoxicity was obtained by doping with iron that altered the crystal lattice to force the slow release of Zn ions [310]. Further study for 24 MO NPs showed that energy levels in the conduction band can be used to estimate potential toxicity [311]. For the materials, the overlapping conduction band energy levels with the redox potential of cells was strongly related to the MO NP capacity to force oxidative stress and inflammatory responses. CuO and ZnO produced oxidative stress and acute pulmonary inflammatory responses against that shown by conduction-band energy levels, but detrimental biological effects of these materials are attributed to their dissolution tendencies [312]. Therefore, the structure activity relationship depends upon their conduction band-gap energy levels and particle solubility predict assessment of MO NPs [313]. Apart from scientific experimentation, in silico methods may be used for hazard potential assessment to correlate structure-activity relationship with toxicity [314]. The structure activity relationships may be qualitative or quantitative in nature and make predictions about the toxicity due to the physicochemical property profile of the NPs. The cytotoxic potential of

17 MO NPs toward *Escherichia coli* was studied [315]. Furthermore, 109 NPs with the same metal core with varying organic moieties as surface attachment groups showed that the cellular uptake efficiency of NPs can be measured based on the chemical structure of the moieties or functional groups attached. The combination of HTS and mathematical modeling techniques like quantitative nanostructure toxicity relationship models can help in predicting the NP potential toxicity and hazard ranking [316].

10.9 TOXICITY STUDY

10.9.1 Titanium Dioxide (TiO_2)

10.9.1.1 Production and Uses

Titanium dioxide (TiO_2) is used as a pigment in paint formulations, ointments, and toothpaste, as well as it filters off UV radiations when used in sunscreens [317]. TiO_2 NPs are synthesized by employing different methods, namely, co-precipitation, sol-gel, hydrothermal, oxidation, and CVD [318].

10.9.1.2 Toxicity Study: In Vitro

The in-vitro toxicity of TiO_2 NPs has been studied by using several cell models. Reports of earlier studies show that the high doses of TiO_2 NP's exhibit toxic effects [319]. The toxicity of TiO_2 particles in the anatase and rutile phases with different sizes, on exposing A549 cells, is investigated [320]. The particles get internalized in the cells in vacuoles of the cytoplasma or lysosomes, although the cytotoxicity exhibited by these particles is abysmally low [321]. The size, and the type of phase, affect the toxicity. Smaller is TiO_2 particle size and larger is anatase quantity in the sample, the larger is its potential to cause cell mortality [322]. It is observed that TiO_2 NPs of size ranging between 3 and 10 nm, in anatase form, is more toxic than rutile ones when the toxicity of various TiO_2 NPs are compared in epithelial cells like A549 and BEAS-2B and normal human bronchial epithelial (NHBE) cells [323]. TiO_2 NPs having both anatase and rutile phases brought about the maximum release of IL-8, although not much effect in A549 cells in all cell types as against the pure anatase and rutile forms. The effect is medium on cell viability; however, TiO_2 NPs of 15 nm size are lethal to cells in bronchial epithelial cells [324]. The mechanism followed is peroxidation of lipid and destabilization of the lysosomal membrane, which is responsible for causing the release of cathepsin B, activation of caspases, followed by cell fatality due to apoptosis [325].

TiO_2 NPs may damage DNA using what is called the comet assay, although earlier studies either won't corroborate or show effects only when exposed to UV light [326]. TiO_2 NPs within 6 to 8 nm size range in the comet, the hypoxanthine guanine phosporibosyl transferase assay of mutational changes in genes, and the micronucleus test show positive results [327].

10.9.1.3 Toxicity Study: In Vivo

TiO_2 NPs of 21 and 250 nm size are found to show higher toxicity as compared to their heavier counterparts when mice were exposed through the inhalation route when exposed for 3 months, and the effects were assessed for a 16 months post-exposure period [328]. Moreover, mice were injected with TiO_2 particles of varying sizes of 12, 21, 230, and 250 nm, through a single intratracheal induction, and the effects were studied 24 h after exposure [329]. TiO_2 NPs caused a larger pulmonary inflammation accompanied by neutrophil infiltration against larger ones [330]. Small-sized TiO_2 NPs reside within the lung for larger time periods, say, 501 days on inhalation, while large-sized TiO_2 particles reside for 174 days, thereby showing that allowance is poorer for the NPs [331]. The in-vitro studies show that there are other factors than mere size of the particles, which induce toxicity. The crystalline pattern of TiO_2 affects pulmonary toxicity in mice [332]. A sample containing a mixture of 80% anatase and 20% micron-sized rutile with –ve particle control, and an α-quartz with +ve particle control, was analyzed for the pulmonary inflammation to the host of particles 3 months post-exposure [333]. Though quartz has the highest inflammogenicity, it is observed that TiO_2 phases in isolation effected an inflammatory response for a shorter duration, while the combination of the rutile and anatase phases is found to be more potent, having the largest surface area to the extent of 53 m^2/g, which is attributed to the crystallinity, which is further instrumental in the higher inflammatory potential. For example, chronic exposure to TiO_2 NPs shows its vulnerability to tumor formation [334] and thus shows a carcinogenic potential. DNA damage takes place in mice on exposure to TiO_2 NPs as a mixture of anatase and rutile phases in drinking water at a concentration of 500 mg/kg [335]. A rise in the degree of DNA damage is measured by the comet assay, and micronuclei induction in white blood corpuscles on exposure. TiO_2 NPs translocate to the mice brain and induce the oxidative stress and inflammation after inhalation through the nose at higher exposure levels, although the relevance of the results may be questioned [336,337].

According to the European Commission report, the application of TiO_2 NPs without much photocatalytic behavior at a dose of 25% in sunscreens can be taken as harmless with no risk or adverse effects to human beings on application to healthy skin [338]. It is therefore evident that in acute and sub-chronic inhalation, toxicity considerations won't allow the use of TiO_2 NPs in products for spray applications from an overall safety point of view. Therefore, inhalation of TiO_2 NPs must be strictly prohibited [339]. Furthermore, according to the National Institute for Occupational Safety and Health, the exposure limits of fine and ultrafine TiO_2 NPs are 2.4 and 0.3 mg/m^3, respectively, for 10 h/day over a span of a 40-h workup [340].

10.9.2 Gold Nanoparticles

10.9.2.1 Production and Uses

Gold was discovered as a first metal many centuries back. Au NPs have a spectrum of uses on medical and biological frontiers, such as genomics, biosensors, immunoassay, clinical chemistry of microbes and cancer cells, drug delivery, bioimaging, and cell and tissue studies [341]. Au NPs have been in use practically in many medical applications, along with diagnostics, therapy, and hygiene [342]. The host of applications of Au NPs are attributed to their unique physicochemical property profile, in particular the reliance of optical properties on the geometry and morphology of Au NPs. There are key factors in the synthesis and application of gold sols in the fields of biology and medicine [343].

The metal NPs like Au, Ag, and Pt in solution are synthesized by chemical reduction and use organic solvents and chelating agents. In water solutions, Au NPs have been specifically synthesized by the reducing chemically $AuCl_4^-$ ions obtained from a gold salt such as chloroauric acid [344], by employing the Turkevich-Frens method, using reductants such as citric and ascorbic acid. The reduction takes place in the presence of a surfactant or stabilizing (capping) agent and on supplying energy by either photoirradiation or heating. Such methods help control the size and concentration of the NPs. Moreover, the capping or stabilizing agents provide stability to colloidal dispersions and prevent agglomeration of NPs from taking place. Furthermore, the growth of Au NPs can be controlled so as to obtain them in different morphologies. There are various methods for production of Au NPs in solutions using nonaqueous solvents [345]. For example, in the Brust-Schiffrin method, $AuCl_4^-$ is subjected to reduction using sodium borohydride and alkylthiols as capping agents. The Au NPs then are protected with firmly bound ligands that keep them segregated and dispersible due to the same charge on their surfaces in organic solvents that check further change in the surface chemistry, keeping in view target applications [346].

10.9.2.2 Toxicity Study: In Vitro

Au NPs have been widely studied because of their several uses in the field of medicine. Au is considered a material inert in behavior, and few studies emphasize upon the significance of NPs due to the biological effects exhibited by them. So the normal approach is in-vitro studies so as to characterize cytotoxic potential by using a 3-(4,5-dimethythiazol-2-yl)- 2,5-diphenyl tetrazolium bromide (MTT) assay on exposing Au NPs of varying sizes and at different times in the range of 3 to 24 h [347]. Several studies have reported that Au NPs are biocompatible. For example, the cytotoxicity and immunogenicity of Au NPs of 3.5 nm size to RAW264.7 macrophage cells were studied, and it is found that the NPs were found to be noncytotoxic or immunogenic below a concentration of 100 mM. Therefore, the Au NPs are biocompatible and show no cytotoxicity or inflammatory response on lung cell exposure to Au NPs of the size 4 and 13 nm [348]. However, there has been a marked effect on cytokine induction because of the chemically base solvents. The unreacted materials of the synthetic process, impurities like lipopolysaccharide and toxicity due to surface ligands, have to be taken into account [349]. A study was undertaken by modifying the surface of 18-nm Au NPs by attaching different ligands, such as citrate, biotin, and hexadecyltrimethylammonium bromide (CTAB). The findings showed that there was no toxicity due to citrate- and biotin-modified particles with high concentrations, say, 250 mM, while CTAB-coated NPs were comparatively more toxic at even lower of the order 0.05 mM. However, the NPs have no more toxicity when CTAB is removed from the particle surface, showing that the toxic behavior was primarily because of CTAB [350].

There are several examples in the literature about the effect of size on toxicity of Au NPs. It is observed that there is a high toxicity of Au NPs having a size of the order of 1.4 nm in Au55 nanoclusters toward several cancer and healthy cells in humans. The greater toxicity is attributed to the perfect match between the size and the principle groove in DNA. Au NPs with sizes of 0.8, 1.2, 1.4, and 1.8 nm containing clusters of 8, 35, 55, and 150 Au atoms, respectively, and 15-nm NPs are capped with triphenylphosphine and its derivatives and studied. It was observed that the clusters containing NPs with size 1.4 nm were highly cytotoxic with half of maximum inhibitory concentration (IC50) varying between 30 and 46 μM [351]. The 15-nm NPs were noncytotoxic even at concentration 100 times more than IC50 for the smaller clusters. Thus, it can be said that the 1.4 nm NPs led to cell necrosis following 12 h incubation while the 1.2 nm ones led to apoptosis. Furthermore, it is understood that Au NPs may affect autophagy, a lysosome-based degradative pathway, and are instrumental in maintaining cellular homeostasis. Intracellular matter is surrounded by double-membrane vesicles referred to as autophagosomes during this process. Au NPs enhance the piling up of autophagosomes in cells due to stifling autophagy flux than induces autophagy [352].

10.9.2.3 Toxicity Study: In Vivo

Many of the studies have so far been conducted regarding the biodistribution of Au NPs in major organs such as the liver and spleen. The ICP-MS coupled technique has been in use to study the size-dependent biodistribution of Au NPs 24 h after administering them intravenously (i.v.) to mice. Particles with sizes 10, 50, 100, and 250 nm were examined and found to vulnerable to accumulate to the maximum in the major target organs like the liver and spleen. A key observation was selectively 10 nm NPs were accumulated in various organs. However, the question that remains unanswered is whether the deposition of Au NPs in the liver is responsible for toxicity or otherwise. On intravenously injecting (i.v.) 13 nm PEG-coated Au NPs with concentrations 0.17, 0.85, and 4.26 mg/kg to mice, it exhibited longer blood circulation time spans in tandem with earlier research findings [353]. Moreover, TEM images also showed PEG-coated Au NPs in several cytoplasma (vesicles) and liver Kupffer cells (lysomes) and in spleen macrophages, resulting in acute inflammation and apoptosis in the liver. On intraperitoneal injection to mice with of Au NPs clusters (Au25) stabilized by glutathione (GSH) with size 2.1 nm with the help of bovine serum albumin (BSA) having mean size of 8.2 nm, the biodistribution showed that the BSA-stabilized Au NPs were deposited in the liver and spleen, while GSH-stabilized ones accumulated to a lesser extent [354]. The renal clearance for BSA-stabilized Au NPs and GSH-protected ones was 1% and 36%, respectively, after 24 h exposure. Furthermore, after 24 h, the white blood corpuscles of exposed rats increased substantially as well as the kidney function was influenced. GSH-stabilized Au NPs didn't show much observable effects the following 4 weeks, but BSA-stabilized particles did exhibit toxicity. Hence, it is concluded that GSH-stabilized Au NPs hold potential in the area of in-vivo imaging and therapy while BSA-stabilized ones can be employed in liver and cancer therapies.

Effects after inhalation of Au NPs of size from 4 to 5 nm have been studied in mice exposed for 3 months (6 h/day and 5 days/week) in an inhalation chamber. Distribution of Au NPs in tissue showed accumulation of Au NPs depending upon dose preferentially in the lung and kidney [355].

10.9.3 Silver Nanoparticles

10.9.3.1 Production and Uses

Silver (Ag) compounds find widespread use as bactericidal and antimicrobial agents as well as in the medical field in the treatment of burns and various infections. Ag NPs have been used in drug-delivery systems and composites in sterilizing filters and coatings. Ag NPs by virtue of their antibacterial properties find use in checking microbial growth in wound dressing, textile fibers, and wooden floorings. However, the bactericidal property depends upon how far they remain stable in the growth medium because this allows for the larger residence time for the interaction to a greater extent. It is a challenging task to synthesize stable Ag NPs to markedly check the growth of bacteria [356].

Several synthetic routes are used in the preparation of Ag NPs with controlled growth and morphology. The Turkevich and Burst method is one of them. The preparation of monodispersed Ag nanocubes in bulk is brought about by chemically reducing Ag^+ ions from an $AgNO_3$ salt solution with polyalcohol as a reductant and polyvinylpyrrolidone as a surfactant. Ag NPs were also prepared by using condensation in inert atmosphere and co-condensation techniques. The underlying principle for both of these techniques is the evaporation of a metal in an inert atmosphere followed by cooling for allowing the nuclei to form and grow to form NPs [357].

10.9.3.2 Toxicity Study: In Vitro

Ag NPs have been in use for decades as a biocide. A host of in-vitro studies have been undertaken to study various kinds of toxic effects on exposing them to a host of cell types. Several studies showed that Ag NPs are responsible for causing apoptosis, DNA damage, and oxidative stress, although differential toxicity is reported in various studies. The variations in toxicity can be ascribed to changes in size, surface coatings, and the release of Ag^+ ions along with the kind of cell. Some studies have shown size-dependency toxicity. And 15, 30, and 55 nm NPs, following exposure to alveolar macrophages, showed that small-sized particles produced ROS large in number and caused a significant detrimental effect on cell growth as against the coarser ones. Similarly, when toxicity for differently sized Ag NPs was tested, it showed that the small-sized NPs, say, 10 nm, affected cell viability selectively, and the effect wasn't dependent upon the surface coating and can't be due to variations in cellular uptake [358].

Many properties of small-sized Ag NPs increase cytotoxicity. The higher surface areas are responsible for improved release of toxic silver (Ag^+) ions, and the number of Ag^+ ions on the surface are vital for toxicity. The release of Ag^+ ions, because of surface oxidation, may generate peroxide (H_2O_2), which reacts very fast with biomolecules or silver metal, causing the release process to enhance. Moreover, the toxicity is a cumulative effect of release processes like producing ROS and the Ag^+ ions. Free Ag^+ ions chelate with the ligands like chloride ions and sulfhydryl or thiol (–SH) groups. Therefore, proteins containing a thiol group are found to be major moieties responsible for toxicity due to Ag^+ ion [359].

However, there are factors other than the release of Ag^+ ions contributing to toxicity. It is the shape of Ag NPs in A549 cells that is of greater importance for the toxicity over the number of Ag^+ ions released. Spherical particles won't affect much; however, silver wires markedly lower cell viability but rise in lactate dehydrogenase (LDH) generated by human alveolar carcinoma A549 cells. Moreover, a rise in intracellular Ca concentration was noticed within few minutes following wire additions. Samples with the highest ion release haven't caused toxicity, and further, the control experiments of ions release wherein cells processed with pre-incubated medium haven't exhibited any such effects either. The crystal defects in the NMs with varying shapes were found to enhance Ag NPs toxicity other than the Ag^+ ions [360].

Earlier studies have shown DNA damage and the probable mechanism underlying it involves oxidative stress, although there has been predictive evidence of damage by virtue of indirect effects. Indeed, it has been observed that Ag NPs can bring about a rise in the concentration of 8-oxo-2′-deoxyguanosine in cultured cells. Moreover, this can to a certain extent be explained in terms of a reduction in activity of the DNA repair enzyme *N*-glycosylase/DNA lyase (OGG1). The release of Ag^+ ions derived have also been found to deactivate the enzyme FPG, which is the *E. coli* form of OGG1 responsible for DNA repair [361].

10.9.3.3 Toxicity Study: In Vivo

Earlier studies have reported the in-vivo toxic potential of Ag NPs on exposure and inhalation. The inflammatory response and alteration in pulmonary function in mice was studied for three months of inhalation to varying degree of concentration levels of Ag NPs. Lung function analysis was conducted and found that the tidal volume reduced substantially during the exposure to Ag NPs. About inflammatory responses, no marked variations were observed in the cellular counts; however, histopathological investigations showed a rise in lesions depending upon dose, for example, thick alveolar walls and minute granulomatous lesions. Further studies showed similar kinds of dose-dependency, which led to rise in bile duct hyperplasia in mice liver of both sexes and increased erythrocyte agglomeration typically in females. Thus, the organs exposed to Ag NPs were lung and liver in mice of both sexes, and the NOAEL was found to be 100 $\mu g/m^3$ [362].

The distribution of various metals or MO NPs was studied in various organs by estimating the metal content; however, it is still yet to be ascertained whether the tissue contains metal NPs or ions. The ICP-MS technique is used to detect Ag NPs in tissue samples collected after 29, 36, and 84 days on a 4-week oral exposure in mice. NPs were found to be accumulated on the 29th day in the liver, spleen, and lung and showed that particles are derived in vivo from the Ag^+ ions, although no NPs were observed on 36 and 84 days. On the 29th day, Ag was found in all organs checked with the maximum concentration in liver and spleen, which showed that Ag^+ ions to a greater extent and Ag NPs to a far lower extent moved through the intestine membrane of the exposed mice. Neither hepato- nor immune-toxicity was noticed after exposure to Ag NPs. However, an abysmally low rate of removal was noticed in the brain and testis still containing more Ag concentration 2 months after the ultimate exposure. In a further oral study, mice were exposed to Ag NPs of size 56 nm for a time period of 3 months, and the target organ was the liver. A NOAEL and LOAEL of the order of 30 and 125 mg/kg, respectively, were identified from the study. Injection of Ag NPs by intravenous pathway, and thereby bypassing the requirement of crossing the cellular membrane barriers in the lung and GI tract, leads to effects at far lesser doses. The most prominent effects found after intravenous administration of mice to Ag NPs of the size 20 and 100 nm were an increase in spleen weight and lowering of the natural killing cell activity [363]. The minimum critical effect dose for a 5% change as against control animals (CED05) was noticed in the case of the thymus weight of 0.01 mg/kg, and the CED05 for lowered natural killing cell activity was 0.06 mg/kg. In a latest study, humans were orally administered with Ag NPs of about 10 nm size and 32 nm with determined ingestion of 100 and 480 μg/day, respectively, for varying time spans from 3 to 14 days. No clinically vital alterations could be made out in the target organs such as lungs, heart, or abdomen.

The major ill effect of chronic exposure to Ag in humans is an irreversible change in skin (bluish-gray) or eyes color, and the health problems caused are called argyria or argyrosis, respectively. Ag accumulated is not translocated Ag NPs but secondary particles derived from partial dissolution in the GI tract after uptake of ions by systemic circulation as organo-metal (Ag) chelates and immobilization as Ag NPs by photoinduced reduction in irradiated parts of the skin. The results show that Ag NPs are formed as secondary particles in a green environment after exposure to better conventional Ag forms [364].

As far as risks to humans are concerned by the inhalation pathway, the human indicative number effect levels (INELs) for reproducible exposure are recorded. With the 90-day inhalation investigation explained earlier and several evaluation parameters, INELs of 0.1 to 0.7 μg/m^3 with 1200 to 7000 particles/cm^3 were determined for the lung and liver. The assessment shows that repetitive inhalation at the workplace and by consumers poses a threat to health. Furthermore, Ag NPs with uncontrolled growth intake and larger parts subjected to burn treatments of the body having wound dressings can be threatening to health.

10.9.4 Zinc Oxide (ZnO)

10.9.4.1 Production and Uses

ZnO NPs are synthesized in bulk and are largely examined as a semiconductor by virtue of its electronic and optical properties that are influenced by the size of NPs. ZnO NPs are optically transparent because they absorb UV light and are therefore used as active additives to sunscreens. Several techniques are in use for the synthesis of ZnO NPs, such as sol-gel, evaporative decomposition, traditional ceramic fabrication, wet chemistry, hydrothermal, and the gas-phase-based processes. A host of morphologies such as spheres, ellipticals, nanorods, flowers, and belts have been produced using some modifications of the incumbent techniques [365].

10.9.4.2 Toxicity Study: In Vitro

In-vitro examinations about the toxic potential of ZnO NPs are carried out by using airways-derived cells and exposure models, and the commonly used ones are the human bronchial epithelial (BEAS-2B) and alveolar adenocarcinoma (A549) cells. ZnO NPs are the most toxic as compared to other MOs. A cytotoxicity examination with four dyes showed that 20 nm ZnO NPs reduced the mitochondrial membrane potential and the performance in BEAS-2B cells and RAW264.7 murine macrophages. Zn NPs doped with iron lower the rate at which Zn^{2+} dissolves and reduces cytotoxicity, demonstrating the significance of the ion's solubility. It is evident even from other types of cells that the levels of Zn^{2+} ions are of much importance from the toxicity point of view. ZnO NPs with varying sizes of the order of 10 to 30, 30, 100 nm, and 1-μm, imparted cytotoxicity to Ana-1 murine macrophages. The dose of Zn^{2+} that instilled 50% viability in cells was the same as 50% lowering in $ZnCl_2^-$ treated cells on exposure to ZnO nano and microparticles. Moreover, the level of Zn^{2+} in supernatant liquids is related with lowered viability and LDH release. Zn^{2+} therefore plays a vital role in the cytotoxicity induced by nano- and micrometer-sized ZnO. Exposure of Jurkat cells led to apoptosis independent of caspase as well as ROS formation [366]. The effect was the fall out of extracellular solubilization of ZnO NPs. Functionalization of ZnO NPs with mercaptopropyl trimethoxysilane-SiO_2 or methoxyl resulted into lesser cell death. After uptake, ZnO NPs generate intracellular Zn^{2+} chelated with ligands. Thus, ZnO NP toxic potential is dependent upon NP uptake after intracellular solubilization. Furthermore, ZnO NPs are found to be cytotoxic to various primary immunocompetent cells, and transcriptomics analysis proved that the ZnO NPs with a common gene "signature" and regulation of metallothionein genes are affected mainly by the solubilization of the NPs [367,405].

10.9.4.3 Toxicity Study: In Vivo

In-vivo studies on the effects of ZnO NPs to the lung showed that intratracheal instillation of 50–70 nm NPs and <1000 nm micron-sized in mice led to marked but reversible inflammatory response that heightened after 24 hr and was subsided within 1 month. The inflammatory effect of ZnO NPs occurs due to eosinophilic inflammation. After instillation of ZnO NPs in mice lungs, an eosinophilic and cytotoxic inflammatory response was found to be correlated with the rise in concentrations of eotaxin, IL-13, and IFN-γ levels. ZnO NPs retains their identity at pH = 7.4 but dissolve rapidly in an acidic environment at about pH = 4.5 in lysosomes, resulting into destabilization of lysosomes and subsequent cell fatality. Moreover, the soluble portion of ZnO NPs was very much active in all in-vitro inflammation assays in A549 cells, although it showed lower inflammation in vivo, meaning Zn^{2+} ions may generate pseudo-positive in-vitro effects. Since eosinophils are instrumental for causing asthma and other allergies, there is an increase in the risk potential of getting attacks of asthma because of the ZnO NPs infiltration to eosinophils [368].

If the release of Zn^{+2} ions is a critical factor in the toxic potential of ZnO NPs, then it is observed that lower solubility reduces toxicity. A study reported that mice when exposed to Fe-doped ZnO NPs of size 20 nm with lower solubility than undoped ones then lowered lung toxicity as the neutrophil count, IL-6 and heme oxygenase enzyme expression and reduction in inflammation was observed when compared with undoped. NOAEC reported regarding inhalation threshold limits for ZnO, with the particle size of 140 nm as 1.5 and 2.0 mg/m^3 in a 90 and 14-day study in mice, respectively. No acute toxicity was found at a concentration of 500 μg/m^3 on 2-hr exposure in humans.

ZnO NPs in cosmetics have necessitated studying their effects on skin and their diffusion potential through the skin. Many studies in human healthy have shown that

ZnO NPs in sunscreens don't have the ability to penetrate below the stratum corneum. ZnO NPs can penetrate only superficially. However, the depth of penetration in increased in skin damaged by UVB, unlike healthy skin. After the application of sunscreen onto the skin of a healthy human for 5 days, a mere 0.1% of all Zn makes its way into the blood [369].

The Scientific Committee on Consumer Safety noted the no observed adverse effect level (NOAEL) for ZnO NPs is 50 mg/day through the oral route. Moreover, the exposure through lip and sun protection products is confined to the ingestion that is accidental in nature and in small amounts. Thus, the effects of ZnO NPs in sun-protection products are not very adverse in humans on application on the lips and skin. However, this is not applicable to other uses like spray products that lead to their inhalation. Thus, in-vitro and in-vivo studies of ZnO NPs exhibit toxicity in the form of inflammation and cytotoxicity, which are frequently more prevalent than other NPs [370].

10.9.5 Cerium Oxide (CeO_2)

10.9.5.1 Production and Uses

Nanoceria (CeO_2) finds applications in a host of frontier areas such as materials for electrodes in sensors for environmental testing, fuel cells, UV-protecting agents, abrasives, elevated-temperature superconductors, additives to diesel fuel in the automobile industry to cut down the exhaust emissions of particulate matter, and as a catalyst in catalytic converters for the oxidation of contaminants leading to tail pipe emissions.

Various techniques are in use for the synthesis of CeO_2 NPs in traditional, microwave-assisted hydrothermal, co-precipitation, organometallic decomposition, etc. Nanocrystalline CeO_2 with morphologies such as nanospheres, nanowires, and nanorods have been derived by modifying the composition of anions or counter-ions, namely, chloride, nitrate, and phosphate before the hydrothermal technique is brought about [371].

10.9.5.2 Toxicity Study: In Vitro

CeO_2 NPs are quite useful because they exhibit both cytotoxic and protective effects; the latter ones are attributed to the reducing abilities of these NPs. Of course, CeO_2 NPs mimic superoxide dismutase mimetic activity. The enzyme superoxide dismutase is responsible for catalyzing the dis-mutation of the SO^- anion in living cells and exhibiting an activity resembling an enzyme catalase, that is, the ability to bring about decomposition of H_2O_2 to O_2 and H_2O. Many in-vitro studies exhibit toxic potential of CeO_2 NPs to lung cells on inhalation exposure. The cytotoxicity exhibited by CeO_2 NPs in BEAS-2B is studied in comparison with that observed in other cell lines. BEAS-2B cell activity was found to be reduced on treating it with NPs of sizes from 15 to 45 nm. ROS were produced depending upon the dose, GSH levels were reduced, and caspase-3, which is instrumental in following the apoptotic route, increased on treating with CeO_2 NPs. Similarly, CeO_2 NPs show toxicity by virtue of oxidative stress, as do CeO_2 NPs antioxidants. A similar kind of protective and anti-apoptotic behavior is noticed in leukocytes. Furthermore, CeO_2 NPs provided protection to cardiac progenitor cells from cytotoxicity brought about by H_2O_2 [372].

10.9.5.3 Toxicity Study: In Vivo

CeO_2 NPs have been examined for their inflammatory effects on instillation in mice lungs. CeO_2 NPs don't show inflammation in the lung. However, with CeO_2 NPs, if dispersed and when instilled in mice lungs, an acute inflammatory response was evidenced on much less surface area. Furthermore, if rats are intratracheally exposed to CeO_2 NPs, then inflammation, air/blood barrier effect, cytotoxicity, and phospholipidosis with inflated alveolar macrophages are found to take place. Similarly, CeO_2 brings about lung inflammation and damage, which may end up in fibrosis. Ce can cause heart fibrosis, while CeO_2 is found to cause myocardial fibroblast and collagen accumulation in mice. Moreover, CeO_2 NPs have exhibited cardio-protective effects with the help of animal models following intravenous injection, showing that they tone up myocardial oxidative stress as well as inflammatory response mechanisms by way of their autoregenerative antioxidant behavior [373].

10.9.6 Silicon Dioxide (SiO_2)

10.9.6.1 Production and Uses

Silica (SiO_2) is used in cosmetics, drugs, toners, varnishes, and food products as an additive. Moreover, silica NPs are synthesized for applications on biomedicine and biotechnology frontiers such as diagnosis, cancer treatment, drug delivery, and immobilization of enzymes. SiO_2 NPs are normally produced by the Stöber method, that is, by bringing about hydrolysis followed by condensation of alkyl silicates. This method allows for the regulated growth of spherical-shaped SiO_2 NPs of uniform size by hydrolyzing alkyl silicates followed by condensation of alcoholic silicic acid. Many modified versions of this pathway have been in use to synthesize SiO_2 NPs. An innovative procedure used to obtain monodispersed SiO_2 by using the two-step hydrolysis of Si in water as a solvent was developed recently, and the method has successfully helped in synthesizing size-controlled SiO_2 NPs.

Mesoporous Si is a family of synthetically designed silica in colloidal form wherein an array of highly ordered and mesoscale (2–50 nm) pores are generated by way of the self-assembling (templating) technique of amphiphiles, especially micelles derived from emulsifying agents. Decomposition of silicates thermally around the surfactant molecules by templating them facilitates the formation of mesoporous SiO_2 with a highly ordered network of pores. Mesoporous silica is used preferentially as drug-delivery carriers, by virtue of its distinct properties attributed to a highly ordered pore network, controllable pore sizes and volumes, large and functional surface area, which holds a huge potential as a candidate for drug loading, delivery, and its controlled release. Moreover, mesoporous-SiO_2 is used to deposit several nanocrystals to get rid of the core material toxicity. Mesoporous SiO_2-based materials with an array of highly ordered pores have been synthesized and characterized thoroughly so far [374].

10.9.6.2 Toxicity Study: In Vitro

Several in-vitro investigations studying the cytotoxic potential of silica NPs have so far been undertaken and reported. Although every study proved to be conclusive, for the want of availability of standard methods, still they hinted at some properties that drive the cytotoxic effects. Most studies pertaining to investigation of toxicity, depending upon size, have shown that the cytotoxicity of SiO_2 NPs increases reciprocally with their size. For example, the effect of amorphous SiO_2 NPs of sizes between 15 and 30 nm and micrometer-sized ones on cell viability, cycle, apoptosis, etc. has been studied. Modifications in the morphology were noticed on exposing for 24 h, and subsequently the human epidermal keratinocyte (HaCaT) cell viability was found to be substantively reduced. SiO_2 NPs hold greater potential than the micrometer-sized particles as far as cytotoxicity is concerned, which further induces a greater apoptotic rate. Proteomic analysis showed variability in the induction of as many as 16 protein expressions, such as oxidative stress, cytoskeleton, tumor-related proteins, energy metabolism molecular chaperones, and apoptosis. In a further study, 15-nm SiO_2 NPs and micrometer-sized particles were used to investigate the DNA methylation in HaCaT cells. NPs affect hypomethylation and reduce the concentration of methyltransferases (DNMT1, DNMT3a) based on the concentration levels of the mRNA and protein [375].

Many studies have reported cellular uptake efficiency of SiO_2 NPs, irrespective of whether the uptake is essential or not, for instilling cytotoxicity is yet to be unraveled. It was found that particles of sizes tested move through the cytoplasm, leading to localization as nuclei typically in cells exposed to NPs ranging from 40 to 70 nm. Furthermore, the uptake by the clusters of topoisomerase I and aggregate of proteins in the nucleoplasm, which is responsible for resistance to cell growth without modifying the viability of cells unlike cells exposed to 0.5 to 5 μm particles, exhibited similar replication and transcription behavior like untreated control cells. However, some investigations have reported the translocation of NPs to the cell nucleus [376].

The underlying mechanisms by virtue of which SiO_2 exposure tends to silicosis and chronic inflammatory response, however, is yet to be ascertained. However, a latest study showed differential toxicity of amorphous SiO_2 depending upon the synthetic pathway and subsequent physicochemical changes. The unusual toxicity of SiO_2 compared to that of nano-SiO_2 is ascribed to its framework and surface topography, as well as its chain structure by virtue of elevated temperature synthesis and fast thermal cooling. The Silanol groups on the surface preferentially facilitate interaction with cell membranes and are instrumental in the hemolytic and cytotoxic activity of SiO_2. Silanols donate protons to quaternary and phosphatic ester groups in phospholipid membranes and cause the denaturation of cell membrane proteins, which can be checked by serum proteins, attributed to shielding of surface functionality because of complete protection offered by cultured cells against SiO_2 than ZnO NPs. Another potential parameter responsible for surface reactivity to SiO_2 toxicity is ROS generation. Plasma membranes and red blood cells are vulnerable to oxidative stress, leading to lipid peroxidation and hemolysis.

Another factor that is deciding the toxicity determination is particle uptake; for example, the fumed SiO_2 particles were attached to the cell membrane surface and not taken up by cells, which is dangerous to $NLRP_3$ inflammasome, and its subsequent activation tends to IL-1β secretion. The fumed SiO_2 aggregates cause damage to cell membranes and induce ROS production, both of which drive inflammasome activation unlike the $NLRP_3$ inflammasome activation routes for silica in crystalline form, wherein phagocytosed particles show lysosomal breakdown and release of cathepsin B [377].

10.9.6.3 Toxicity Study: In Vivo

Amorphous SiO_2 is normally "safe" according to the Food and Drug Administration (FDA). While, the International Agency for Research on Cancer (IARC) has categorized crystalline silica as a Group 1 carcinogen, with evidenced carcinogenicity in experimental animals and in humans. Carcinogenicity may be due to intrinsic properties of the crystalline SiO_2 or external agents influencing its biological activity or distribution of its polymorphic forms. Genotoxicity may be noticed following cell exposure to crystalline SiO_2 in vitro, though at very large doses, thereby evidencing secondary genotoxicity driven by inflammation as an underlying major mechanism. Various studies have explored the toxicity of SiO_2 NPs in vivo, after lung exposure, showing an inflammation at higher doses. Histopathological studies noticed acute neutrophils and chronic granulomatous inflammatory responses. The expression of cytokines and chemokines was substantially hiked in the beginning, without any changes after the first week. When rats were exposed through inhalation to newly produced aerosolized amorphous SiO_2 NPs of the size ranging between 37 and 83 nm for 1- or 3-day exposure, and a smaller dose, the particle count was considered as the dose metric. Pulmonary toxicity in bronchoalveolar lavage (BAL) fluid and genotoxicity endpoints (micronuclei induction) were evaluated from 1 day to 2 months following exposure. No reasonable pulmonary inflammatory, genotoxic, or detrimental lung histopathological effects in mice exposed to very large particle counts in a range of concentration levels (1.8 or 86 mg/m^3) may be noticed. Reports of good tolerance to mesoporous SiO_2 materials on oral exposure in mice were obtained. The findings hold promise for using the mesoporous SiO_2 as a drug-delivery capsule for oral injection [378].

10.9.7 Iron Oxide (Fe_2O_3)

10.9.7.1 Production and Uses

Iron oxides are very important transition MOs for a host of uses in frontier areas. As many as 16 forms of iron oxides exist, such as oxides, hydroxides, and oxyhydroxides. They preferentially have iron in the oxidation state +3 (ferric state), and possess lower solubility but brilliant colors. The iron oxides have applications in various frontier technological areas such as catalysis, sorption, pigments, flocculation, coatings, gas sensors, ion exchanging agents, and lubricants. Iron oxide nanocomposites hold much promise and potential in magnetic storage of data, resonance imaging, toners, inks for printers, wastewater treatment, bioseparation, and medicine.

Magnetic NPs are a family of very widely explored because they present potential applications in the areas like diagnostics and therapies. The application of SPION as an MRI contrast agent is very well-known, and a host of SPIONs are listed out in the approval list of US-based Food and Drug Administration (FDA). However, for suitability in biomedical applications, SPIONs ought to possess well-defined geometry, monodispersed, and show a better magnetization ability [379].

Iron oxides are prepared by well-established wet-chemistry-based methods, but tuning the particle size to nanoscale and morphology in order to meet given applications is a great challenge. Iron oxides, either Fe_3O_4 or γ-Fe_2O_3, can be synthesized from Fe^{2+}/Fe^{3+} aqueous salt solutions in basic conditions by bubbling inert gas in it at either ambient or higher temperature. A thermal decomposition method is designed typically to synthesize highly qualitative semiconductor and oxide NPs with controlled geometry and morphology. Organometallic materials as starting materials, namely, $Fe(Cup)_3$, $Fe(CO)_5$, or iron(III) acetylacetonate [$Fe(acac)_3$], are subjected to decomposition in nonaqueous media having higher boiling points in the presence of stabilizing agents (surfactants). Fe_2O_3 NPs obtained by elevated temperature decomposition produce monodispersed particles with great size control. Furthermore, methods like microemulsion, hydrothermal, and sol-gel are explored to prepare magnetic iron oxide NPs. Various phase-transfer agents are used to get lyophobic magnetic NPs as aqueous colloidal dispersions essential in biomedical applications [380].

10.9.7.2 Toxicity Study: In Vitro

The cytotoxicity of different types of Fe_2O_3 NPs, either functionalized or otherwise, show generally poor or non-cytotoxicity, which is related with their doses. However, some studies show toxicity but not causing cell death, due to exposure of Fe_2O_3 NPs. For example, the effect of citrate-coated SPION (diameter 5 nm) exposure on mice macrophages shows oxidative stress, predicted from a substantial rise in the concentration of malonyldialdehyde and protein carbonyls with no cytotoxicity. The rise in oxidative stress was checked by an iron chelating agent, showing that Fe released from NPs in the beginning of incubation may have been instrumental in reflecting upon the observed effects. Furthermore, on comparing the toxicity of monodispersed SiO_2-coated Fe_2O_3 NPs of sizes 30, 50, and 120 nm with dextran-coated Fe_2O_3 ones of sizes 20 and 50 nm show that the former ones were nontoxic to primary human monocyte-derived macrophages used for testing. Uptake was evaluated in quantitative terms using ICP-MS and testified with TEM. Moreover, toxicity of the small-sized SiO_2-coated NPs (30 and 50 nm size) was examined at higher concentrations towards primary monocyte-derived dendritic cells against dextran-coated Fe_2O_3 NPs of similar size and it was observed that they won't differentiate in their behaviors. No macrophage or dendritic cell secretion of proinflammatory cytokines was noticed on introduction of the NPs [381].

10.9.7.3 Toxicity Study: In Vivo

When Fe_2O_3 NPs were administered into rats it was noticed that there was a small but uncertain rise in liver markers like aspartate transaminase (AST), alanine transaminase (ALT), and alkaline phosphatase, despite the concentrations return to normalcy in 3 days post-administration. Furthermore, alterations in oxidative stress that quantified as lipid hydroperoxide levels were found in various organs such as the spleen, kidney, and liver. Histological studies of samples obtained on Days 1 and 7 have shown apparently no abnormal effects. Similar effects were observed on injecting PEGylated Fe_2O_3 NPs of varied size to rats. The quantity of Fe was prominently on the rise in the liver and spleen, and substantial amounts were left over for 1 month after injection. Blood reports show increasing AST and ALT concentrations 1 month after injection. As the particles are preferentially taken up by Kupffer cells (liver) and AST and ALT enzymes (hepatocytes), it is considered that the rise in levels may be ascribed to the formation of toxic substances by the degradation of the macrophages.

A 4-week reproducible dose oral toxicity examination on Fe_2O_3 NPs of 30 nm size and the bulk material particles was done. Biochemical enzymes and histopathology-based modifications were studied in various organs of female albino Wistar mice. Dose-related toxic effects of Fe_2O_3 NPs for fine particles were observed. Similarly, it was observed that intratracheal administration of both nano and submicrometer-sized Fe_2O_3 particles led to oxidative stress. Fe_2O_3 NPs brought about an increase in cell lysis in the lung epithelium, which has influenced the blood coagulation mechanism to a greater degree than submicrometer-sized Fe_2O_3. Mice with Fe_2O_3 NPs (γ-Fe_2O_3), with sizes varying between 20 and 140 nm at a concentration level of 57 to 90 $\mu g/m^3$; 6 hr/day for 3 days' exposure, showed no toxicity, though a rise in IL-1β was observed at maximum concentration. On exposing mice to particles of size 3.5 mg/m^3; 4 hr/day; 5 days/week for 2 weeks, no cytotoxicity was noticed, but a rise in macrophages, neutrophils, and lymphocytes was observed in the BAL fluid. The inflammatory response was observed by an evaluation of the BAL fluid for 3 weeks after exposure [382].

An in-vivo study in humans observed that Ferumoxtran-10, Fe_2O_3 coated with dextran is responsible for inducing side effects like diarrhea, urticaria, and nausea, although the effects were milder and short-lived. The particles are considered to be subjected to degradation and taken off from circulation by endogenous Fe metabolic routes. Fe derived from the NPs is metabolized in the liver and further made use of in the development of red blood cells or excreted by the kidney. An increase in concentration of free Fe^{+2} ions in exposed organs can bring about a loss of homeostasis with indicators in the form of oxidative stress, epigenetic effects, and DNA damage. Fe has been, over time, related to cancer and different mechanisms underlying Fe-induced carcinogenesis, such as the generation of ROS that can directly affect DNA, proteins, and lipid peroxidation.

10.9.8 Nickel Oxide (NiO)

10.9.8.1 Production and Uses

Nickel oxide (NiO) NPs are in use for a host of applications in frontier technology areas like smart window, electrochemical supercapacitors in the form of a transparent semiconducting antiferromagnetic films, and dye-sensitized photocathodes. A few fine NiO NPs exhibiting superparamagnetic behavior are employed for the separation and purification of polyhistidine tag or recombinant proteins obtained from a multi-component solution containing a cell lysate. Different techniques have been reported so far for the production of NiO NPs, such as magnetron sputtering, evaporation, sol-gel, oxalate and hydroxides decomposition, and wet-chemistry [383].

10.9.8.2 Toxicity Study: In Vitro

Some of the in-vitro examinations have reported the toxic potential of Ni and NiO NPs. In general, the NPs are more toxic than coarser particles, and moreover, Ni^{+2} ions released within cells is of vital importance. NiO NPs show cytotoxicity at high levels through cellular uptake and Ni^{2+} release within cells. As against this the cellular effect on fine NiO particles exposure was constrained because of a very small quantity of Ni^{2+} release. The investigation of toxicity observed in human lung epithelial A549 cells exposed to Ni NPs of size 65 nm with a concentration level of 25 μg/mL for a period of 24 and 48 hr was done. Different assays pertaining to oxidative stress and cell vitality were obtained. Ni NPs reduce cell vitality and enhance the oxidative stress to 2 μg/mL, which is a comparatively small level as against those employed in other in-vitro examinations. As far as the toxic potential of Ni NPs and that of Ni micrometer-sized particles is concerned, it was observed that soluble $NiCl_2$ and NiO NPs almost had the same cytotoxicity; however, Ni metal NPs exhibited low toxicity while Ni micrometer-sized particles were not toxic on exposure to lung cells in humans. Both Ni and NiO NPs induced stability as well as nuclear translocation of the hypoxia-inducible transcription factor 1α (HIF-1α). Such an effect is not showcased by the micro-sized particles. HIF-1α is observed often in human cancers and help in the identification of cells showing resistance to hypoxia and tumor formation [384].

10.9.8.3 Toxicity Study: In Vivo

Nickel compounds are categorized as Group 1 carcinogens by the International Agency for Research on Cancer (IARC); however, the carcinogenicity is shown by different nickel compounds. Micrometer-sized NiO is weakly soluble and exhibits biopersistence in the lung, but is less carcinogenic as compared to nickel sulfide. According to the nickel ion bioavailability model, the carcinogenicity of a nickel compound is dependent upon the biological availability of Ni^{+2} ions and then the uptake by cells, intracellular solubility, and finally the presence of Ni^{+2} ions. Therefore, if Ni and NiO NPs show better uptake and solubility within cells, then it results in a higher carcinogenicity. The size-dependency of the toxicity of NiO particles is investigated after intratracheal instillation into Wistar mice. Histopathological results predicted a significant infiltration of macrophages and of alveolitis in mice exposed to NiO NPs over time. When the inflammatory effect was studied over a 6-month time period, it was observed that micrometer-sized NiO exhibited only a marginal effect. The effects of intratracheal induction in mice demonstrated that exposure to NiO NPs for 24 h brought about neutrophilic inflammation and mild cytotoxicity. Furthermore, 4 weeks post-instillation, neutrophilic/lymphocytic inflammation and drastic cytotoxicity was noticed. Ni^{2+} ions having intracellular solubility are expected to be instrumental in the immuno-inflammation. On 4 weeks post-instillation of NiO NPs in the lung, mice showed chronic inflammatory responses characterized by lymphocytes in the BAL fluid and higher doses of interferon gamma (IFN-γ) signaling the Th1 kind of immunity. NiO NPs work as haptens, that is, lower molecular weight moieties aren't antigenic in themselves but on linking with a protein (host) the protein undergoes a modification in structure to induce an immunity. On induction into mice, NiO NPs were found to demonstrate a delayed hypersensitive behavior in Th1 or Th17 CD4 T-cells. Moreover, the particles brought about pulmonary alveolar proteinosis in the form of piling up of protein and lipid surfactants in the alveoli. It signifies that a solitary higher exposure of NiO NPs induces a gradual lung disease characterized by delayed hypersensitivity, which is attributed to the slow release of Ni^{2+} ions from NPs, leading to reproducible cycles of host protein haptenization, causing what is called as chronic autoimmune inflammation. An inhalation investigation of NiO aerosol NPs with average diameter of the order of 59 nm by mice in a complete-body chamber to a dose of 200 μg/m^3 containing 9.2 × 104 particles/cm^3 for time period of 6 h/day; 5 days/week for 4 weeks showed an inflammatory response detected after 3 days and 1 month of exposure, which is considered as the level at which effects are detectable (i.e., "LOAEL") [385].

There is a report wherein a 38-year old healthy man was exposed to nickel sprayed in an arc process. He lost his life after 13 days of exposure, and the reason for death was attributed to respiratory disease syndrome. Ni NPs of size <25 nm were found in lung macrophage on examination with TEM. Moreover, a higher concentration of Ni was observed in the urine, meaning that inhaled NPs can journey systemically and influence the kidneys and therefore develop an acute tubular necrosis, implying that the kidney is vulnerable to damage. Furthermore, in regard to health risk evaluation in humans, a World Health Organization (WHO) reports that nickel salts in general, irrespective of their size, are carcinogens via inhalation exposure, and since there is a linear dose response, there are no safety limits as such for nickel compounds worth recommending. The concentrations responsible for an excessive lifetime risk potential of cancer are of the order of 0.25 and 0.025 μg/m_3, respectively.

10.9.9 Copper Oxide (Cu_2O/CuO)

10.9.9.1 Production and Uses

Copper has been used as a biocide for years. Copper ions either in isolation or as copper complexes find use as disinfectants for human organs. Copper even in traces plays an important role in several human physiology and metabolism-based processes like wound repairing. Cu NPs are major ingredients of conductive printer inks and pastes and numerous electronic components. Cu_2O and CuO are semiconductors that have dragged the attention of the research community because of a host of potentially meaningful physical properties possessed by them. These semiconductors are put into use in diverse application areas like gas sensing, batteries, elevated-temperature superconducting materials, field emission emitters, heat exchanging fluids, and catalysts.

Cu NPs are synthesized either in aqueous or nonaqueous solutions along with a plethora of surface active agents. A host of methods have been in use for their synthesis, for example, thermal decomposition, microemulsion, UV irradiation, chemical reduction, and the polyol process. A comparatively mass-scale, high-yield process is the chemical reduction of sulfate, nitrate, acetate of Cu in presence of a reductant like sodium borohydride or sodium hypophosphite and polyethylene glycol, polymer surfactant, sodium do-decyl, or lauryl sulfate as surfactants. CuO NPs are developed by precipitation method involving the use of dissolved copper salts in an alkaline environment or through elevated temperature thermal decomposition of organometallic materials, namely, cuprous acetate [386].

10.9.9.2 Toxicity Study: In Vitro

Cu metal and CuO NPs exhibit higher toxicity as observed in many of the earlier studies, as against their other metal and MO counterparts. CuO NPs are found to have the maximum strength in bringing about DNA damage and cell mortality as against many other simple and mixed MOs, such as TiO_2, Fe_2O_3, ZnO, and $CuZnFe_2O_4$ on exposing them to A549 cells. Furthermore, it was observed that the toxicity of Cu and CuO NPs were dependent upon size; micrometer-sized Cu and CuO show less toxicity when on exposure of cells to the same concentration. Moreover, the cytotoxicity of 24 synthesized NPs in A549 and macrophage (THP-1) cell lines was tested in two different time spans, 3 and 24 hr. It was revealed that Cu NPs and Zn NPs were highly toxic. The toxicity of various MO NPs when compared on exposing them to three different cell lines showed that CuO was highly toxic, whereas SiO_2 was modest in terms of its effect on cell viability as against other MO NPs, which were found to be nontoxic [387].

The toxicity of CuO NPs can be described using a Trojan-horse-based mechanism according to which structure of solid particles is amenable to higher uptake efficiency for Cu, and the corresponding intracellular release of Cu ions tends to have a higher degree of toxicity and oxidative stress. Oxidative stress is signaled by the generation of enzymes like catalase and superoxide dismutase, intracellular ROS, and oxidative DNA lesions, as well as a reduction in glutathione on cellular exposure to CuO NPs. On exposure of A549 cells to CuO NPs, Cu ions were released in cell media while 594 genes were induced by the NPs in themselves, 648 were upregulated by CuO NPs, and 54 were ascribed to the Cu ions.

The mechanism underlying the cytotoxicity effect of Cu NPs seems to be slightly different from that of CuO NPs and causes rapid damage to the membrane performance. Cu NPs may get oxidized to ions on interaction with the cell membrane because the latter has more O_2 concentration as compared to intracellular media. The metal release mechanism can further generate H_2O_2 at the cell membrane, leading to its rapid damage. The cell membrane loss may not be straightway brought about by Cu ions because very little damage has been noticed on exposure of A549 cells to Cu NPs or ions released and may be dependent upon the metal release at the cell surface. This mechanism is supported by an investigation exhibiting higher cell membrane loss due to Cu NPs than carbon-coated Cu NPs [388,400,403].

10.9.9.3 Toxicity Study: In Vivo

In-vivo studies similar to in-vitro ones also show the higher toxicity of 16 differently sized particles of CuO NPs to lung compared following intratracheal instillation in mice, which further revealed that CuO particles have high toxic potential typically on the nanoscale. Histopathology-based modification of the lung on Day 1 following exposure was more marked on exposure to both CuO and Cu metal NPs as compared to quartz (positive control). Lungs exposed to CuO NPs showed acute edema. On the 28th day, alterations in the lung because of exposure to Cu NPs couldn't be evaluated because of cell mortality. Further studies involving intratracheal instillation of mice have observed severe cytotoxic inflammatory responses [389].

Two different investigations showed the toxic effect of Cu NPs having 23.5 nm average diameter and 17 μm average diameter Cu, following exposure of rats through the GI tract. Another observation was that Cu NPs instilled sudden spleen atrophy and brought about modifications in color and morphological features of the kidney. The amount of Cu in many organs was checked with the help of ICP-MS, which showed that Cu NPs get preferentially accumulated in kidney. These findings were in tandem with size dependency of toxicity. Cu NPs on investigation at of 3.5 and 3.7 mg/m^3 inhalation level studies in rats for 4 h/day; 5 days/week, and for 14 days caused acute inflammation with a rise in macrophages, lymphocytes, and neutrophil levels in the BAL fluid, which lasted for 3 weeks after exposure. Furthermore, a reduction in bacteria was observed, implying that on exposure to Cu NPs, there is an increased level of pulmonary infection risk. Cu NPs, when inducted intranasally in rats every alternate day for 21 days, have altered the secretion extents of several neurotransmitters in various brain domains (olfactory bulb) at which NPs were accumulated in preference. Therefore, it is noticed that significant exposure through the respiratory pathway to Cu NPs poses a threat to pulmonary cells and neurotransmitters in the CNS. It is not yet ascertained however whether the accumulation of NPs in the brain domain olfactory bulb is applicable to humans as well or otherwise. CuO NP exposure led to eosinophilia while Cu ions didn't [390].

10.9.10 Aluminum Oxide (Al_2O_3)

10.9.10.1 Production and Uses

Aluminum oxide (Al_2O_3) (or alumina) is a ceramic material used as a catalyst, catalyst support, absorbent, and wear-resistant coatings. Alumina NPs find widespread use in a host of application areas such as electronics in general and optoelectronics in particular. Many methods like vapor deposition or liquid-solid phase synthesis are in use for their production and processing in the nanoscale range. The sol-gel method is commonly used to synthesize several nanoscale MO particles because of their cost effectiveness, easy fabrication, and lower processing temperatures. The process involves solubilization of starting materials in the aqueous phase with gelling and chelating agents added to it. By suitably adjusting the pH, the metal ions are chelated to form metal chelates, and raising the temperature permits the gradual vaporization of water, forming a gel-like mass, which traps all the ingredients in its pores, which is then subjected further to calcination for the breakdown and remediation of all organic compounds so as to form the desired phase. The same method has also been employed for the fabrication of alumina NPs [391].

10.9.10.2 Toxicity Study: In Vitro

Al_2O_3 NPs seem to be nontoxic from a few studies. Al_2O_3 NPs with average particle size 40 nm won't show marked effects on cytotoxic potential, cell permeability, and inflammatory responses when exposed to endothelial cells in humans characterized over concentration levels from 0.001 to 100 μg/mL. Cellular exposure and toxicity of aluminum-based NMs in various forms, for example, Al_2O_3 NPs of the size between 30 and 40 nm, and Al NPs coated with 2–3 nm of oxide of particle size, say, 50, 80, and 120 nm were examined. Cell vitality assays showed that a mild effect on macrophages takes place on exposure to Al_2O_3 NPs at higher concentrations, say, 100 µg/mL for 1 day, while Al NPs generated substantially lower cell viability. The effects of aluminum and alumina particles were studied in an alveolar coculture in conjunction with a respiratory pathogen such as *Staphylococcus aureus*. Both Al and Al_2O_3 NPs haven't shown any influence on cell vitality, though the response of the cells to the pathogen was changed. Thus, although the NPs exhibited lower toxicity, still their very presence altered the cells' natural tendency to respond to pathogens [392].

10.9.10.3 Toxicity Study: In Vivo

The oxidative stress caused the following oral administration of high concentration levels of Al_2O_3 NPs and Al_2O_3 (bulk) in Wistar mice. It was investigated and found to be reduced with time. A histopathology-based examination exhibited lesions in the mice liver on exposure to Al_2O_3 NPs at the concentrations tested. The study therefore shows that effects associated with the oxidative stress are observed in mice at higher concentrations of Al_2O_3 NPs. In the case of exposure of lung intratracheal induction of Al_2O_3 NPs in mice having a particle surface area of the order of 150 cm^2 amounting to 68 mg per mouse led to lung inflammation indicated in the form of a rise in granulocyte count in the BAL fluid for 24 hr post-instillation. A mouse on inhalation of boehmite (AlOOH) with particle size between 10 to 40 nm for 4 weeks at 6 hr/day and 5 days/week showed marked pathological modifications in BAL factors and lung at the highest exposure dose of the order of 28 mg/m^3. The macrophage count in the BAL fluid was found to be doubled in animals exposed for 2 weeks and six times than those for 4 weeks as against control animals. However, there were no indications of any inflammation and rise in concentration of LDH or cytokines in the BAL fluid [393,400,401].

10.10 CONCLUSION

Metal and MO NPs have been finding use on a large scale in innumerable medical, consumer, and industrial applications. Metal and MO toxicology is a matter of examination over time as shown by the research community. However, with the reduction in particle size, some metallic materials may exhibit a rise in toxicity as against the material in bulk. For some metal or MO NPs like Cu/CuO or Ni/NiO, particle solubility and the release of toxic metal ions within cells are major driving forces causing toxicity [394,404]. However, for few NPs, like QDs, the constituent elements like Cd are found to have toxic potential. However, the investigation of synthesized NP-induced toxicity is still in its infancy stage and is confronted by many challenges. The major problem of proper concentrations of NPs demands a careful study and attention as suggested by the toxicological studies [395]. Furthermore, a demand for calibrated test procedures, meticulously characterized standard materials, and validated in-vitro assays in tandem with in-vivo endpoints such as inflammation and genotoxicity are what is asked for. The most appropriate dose metric in nanotoxicological studies in the form of mass, particle count, or surface area is yet to be quantified, and a comprehensive understanding of the way NMs interact with biological systems and the system biology approach may prove quite productive [396,397]. Further higher throughput and development of mathematical models might give an impetus to the evaluation of the host of NMs that are being synthesized and could prove to be toxic depending upon a thorough knowledge of the structure toxicity relationship. This chapter has brought to the fore state-of-the-art in-vitro and in-vivo toxicology studies of the key metals and MO NPs [398,399]. Regarding exposure evaluation at the workplace, further investigation and more field data is required, accompanied by sound measurement tools [400–403]. Better understanding and knowledge about detrimental health effects of NMs, namely, metals and MOs, has been evolutionary in the field of science, and the incumbent data is not adequate enough to make a reliable health risk evaluation of a host of such materials [404–406].

REFERENCES

1. Adamcakova-Dodd, A., Stebounova, L.V., O'Shaughnessy, P.T., et al., 2012. *Part. Fibre Toxicol.* 9, 22.
2. Aghababazadeh, R., Mirhabibi, A.R., Javadpour, J., et al., 2007. *J. Surf. Sci.* 601 (13), 2864–2867.
3. Ahamed, M., 2011. *Toxicol. Vitro.* 25 (4), 930–936.
4. Ahamed, M., Siddiqui, M.A., Akhtar, M.J., et al., 2010. *Biochem. Biophys. Res. Commun.* 396 (2), 578–583.
5. Akhtar, M.K., Xiong, Y., Pratsinis, S.E., 1991. *AIChE J.* 37, 1561–1570.
6. Alexandridis, P., 2011. *Chem. Eng. Technol.* 34, 15–28.
7. Andon, F.T., Fadeel, B., 2013. *Acc. Chem. Res.* 46 (3), 733–742.
8. Anzai, Y., Piccoli, C.W., Outwater, E.K., et al., 2003. *Radiology* 228 (3), 777–788.
9. AshaRani, P.V., Low Kah Mun, G., Hande, M.P., et al., 2009. *ACS Nano* 3 (2), 279–290.
10. AshaRani, P.V., Xinyi, N., Hande, M.P., et al., 2010. *Nanomedicine Lond.* 5 (1), 51–64.
11. Aslam, M., Gopakumar, G., Shoba, T.L., et al., 2002. *J. Colloid Interf. Sci.* 255, 79–90.
12. Baker, C., Pradhan, A., Pakstis, L., et al., 2005. *J. Nanosci. Nanotech.* 5, 244–249.
13. Bee, A., Massart, R., Neveu, S., 1995. *J. Magn. Magn. Mater.* 149, 6–9.
14. Bekker, C., Brouwer, D.H., van Duuren-Stuurman, B., et al., 2014. *J. Expo. Sci. Environ. Epidemiol.* 24 (1), 74–81.
15. Bekyarova, E., Fornasiero, P., Kašpar, J., Graziani, M., 1998. *Catal. Today* 45, 179–183.
16. Bergamaschi, A., Iavicoli, I., Savolainen, K., 2012. *Adverse Effects of Engineered Nanomaterials*, Elsevier, Boston, MA.
17. Bhabra, G., Sood, A., Fisher, B., et al., 2009. *Nat. Nanotechnol.* 4, 876–883.
18. Bondioli, F., Corradi, A.B., Manfredini, T., et al., 2000. *Chem. Mater.* 12, 324–330.
19. Borkow, G., Gabbay, J., 2009. *Curr. Chem. Biol.* 3, 272–278.
20. Borkow, G., Gabbay, J., Zatcoff, R.C., 2008. *Med. Hypotheses* 70, 610–613.
21. Borm, P.J., Tran, L., Donaldson, K., 2011. *Crit. Rev. Toxicol.* 41 (9), 756–770.
22. Bottini, M., D'Annibale, F., Magrini, A., et al., 2007. *Int. J. Nanomed.* 2, 227–233.
23. Braun, J.H., Baidins, A., Marganski, R.E., 1992. *Prog. Org. Coat* 20, 105–138.
24. Brook, R.D., Rajagopalan, S., Pope 3rd, C.A., et al., 2011. *Circulation* 121 (21), 2331–2378.
25. Braydich-Stolle, L.K., Speshock, J.L., Castle, A., et al., 2010. *ACS Nano* 4 (7), 3661–3670.
26. Broekaert, J.A.C., 1998. *Analytical Atomic Spectrometry with Flames and Plasmas*, 3rd ed., Wiley-VCH, Weinheim, Germany.

27. Brouwer, D., 2010. *Toxicology* 269, 120–127.
28. Brus, L.E., 1984. *J. Chem. Phys.* 80 (9), 4403–4409.
29. Brust, M., Walker, M., Bethell, D., et al., 1994. *J. Chem. Soc. Chem. Commun.* 7, 801–802.
30. Buerki-Thurnherr, T., Xiao, L., Diener, L., et al., 2013. *Nanotoxicology* 7 (4), 402–416.
31. Carlson, C., Hussain, S.M., Schrand, A.M., et al., 2008. *J. Phys. Chem. B* 112 (43), 13608–13619.
32. Caruso, R.A., Ashokkumar, M., Grieser, F., 2002. *Langmuir* 18 (21), 7831–7836.
33. Celardo, I., De Nicola, M., Mandoli, C., et al., 2011. *ACS Nano* 5 (6), 4537–4549.
34. Chakrabarty, S., Chatterjee, K., 2009. *J. Phys. Sci.* 13, 245–250.
35. Champion, J.A., Mitragotri, S., 2006. *Proc. Natl. Acad. Sci.* 103, 4930–4934.
36. Chithrani, B.D., Ghazan, A.A., Chan, C.W., 2006. *Nano Lett.* 6, 662–668.
37. Chithrani, B.D., Chan, W.C.W., 2007. *Nano Lett.* 7, 1542–1550.
38. Chen, M., von Mikecz, A., 2005. *Exp. Cell Res.* 305 (1), 51–62.
39. Chen, Z., Meng, H., Xing, G., et al., 2006. *Toxicol. Lett.* 163 (2), 109–120.
40. Cho, W.S., Choi, M., Han, B.S., 2007. *Toxicol Lett.* 175 (1–3), 24–33.
41. Cho, W.S., Cho, M., Jeong, J., et al., 2009. *Toxicol. Appl. Pharmacol.* 236, 16–24.
42. Cho, W.S., Duffin, R., Poland, C.A., et al., 2010. *Environ. Health Perspect.* 118 (12), 1699–1706.
43. Cho, W.S., Duffin, R., Howie, S.E., et al., 2011. *Part Fibre Toxicol.* 8, 27.
44. Cho, W.S., Duffin, R., Thielbeer, F., et al., 2012a. *Toxicol. Sci.* 126 (2), 469–477.
45. Cho, W.S., Duffin, R., Poland, C.A., et al., 2012b. *Nanotoxicology* 6 (1), 22–35.
46. Cho, W.S., Duffin, R., Bradley, M., et al., 2012c. *Eur. Respir. J.* 39 (3), 546–557.
47. Choi, A.O., Brown, S.E., Szyf, M., et al., 2008. *J. Mol. Med.* 86, 291–302.
48. Choo, K.H., Kang, S.-K., 2003. *Desalination* 154, 139–146.
49. Christensen, F.M., Johnston, H.J., Stone, V., et al., 2010. *Nanotoxicology* 4 (3), 284–295.
50. Chu, M., Wu, Q., Yang, H., et al., 2010. *Small* 6, 670–678.
51. Coats, A.W., Redfern, J.P., 1963. *Analyst* 88, 906–924.
52. Connor, E.E., Mwamuka, J., Gole, A., et al., 2005. *Small* 1 (3), 325–327.
53. Conroy, J., Byrne, S.J., Gun'ko, Y.K., et al., 2008. *Small* 4 (11), 2006–2015.
54. Cornell, R.M., Schwertmann, U. (Eds.), 1996. *The Iron Oxides: Structure, Properties, Reactions, Occurrence and Uses.* VCH Verlagsgesellschaft, Weinheim, Germany.
55. Cronholm, P., Karlsson, H.L., Hedberg, J., et al., 2013. *Small* 9 (7), 970–982.
56. Damoiseaux, R., George, S., Li, M., et al., 2011. *Nanoscale* 3, 1345–1360.
57. De Jong, W.H., Hagens, W.I., Krystek, P., et al., 2008. *Biomaterials* 29, 1912–1919.
58. De Jong, W.H., Van Der Ven, L.T., Sleijffers, A., et al., 2013. *Biomaterials* 34 (33), 8333–8343.
59. Del Monte, F., Ferrer, M.L., Mateo, C.R., et al., 1997. *Langmuir* 13, 3627–3634.
60. Deng, H., Li, X., Peng, Q., et al., 2005. *Angew. Chem. Int. Ed.* 44, 2782–2785.
61. Deng, Z.J., Liang, M., Monteiro, M., et al., 2011. *Nat. Nanotechnol.* 6 (1), 39–44.
62. Denizli, A., Say, R., 2001. *J. Biomater. Sci. Polym. Edn.* 12, 1059–1073.
63. Dikmen, S., Machamer, J., Temkin, N., et al., 2002. *Solid State Sci.* 4, 585–590.
64. Donaldson, K., Seaton, A., 2012. *Part Fibre Toxico* 9, 13.
65. Donaldson, K., Poland, C.A., 2012. *Swiss Med. Wkly.* 142, 135–147.
66. Dong, Y., Li, Y., Wang, C., et al., 2001. *J. Colloid Interf. Sci.* 243, 85–89.
67. Dutta, D., Sundaram, S.K., Teeguarden, J.G., et al., 2007. *Toxicol. Sci.* 100, 303–315.
68. Dykman, L., Khlebtsov, N., 2012. *Chem. Soc. Rev.* 41, 2256–2282.
69. Ekstrand-Hammarström, B., Akfur, C.M., Andersson, P.O., et al., 2012. *Nanotoxicology* 6, 623–634.
70. Elder, A., Gelein, R., Silva, V., et al., 2006. *Environ. Health Perspect.* 114 (8), 1172–1178.
71. Eom, H.J., Choi, J., 2009. *Toxicol. Lett.* 187 (2), 77–83.
72. European Commission, 2012a. COM (2012) 572.

73. European Commission, 2012b. Scientific Committee on Consumer Safety. SCCS/1489/12.
74. European Commission, 2013. Scientific Committee on Consumer Safety. SCCS/1516/13.
75. Fadeel, B., 2012. *Swiss. Med. Wkly.* 142, w13609.
76. Fadeel, B., 2013. *J. Intern. Med.* 274 (6), 578–580.
77. Fadeel, B., Garcia-Bennett, A.E., 2010. *Adv. Drug Deliv. Rev.* 62 (3), 362–374.
78. Fadeel, B., Feliu, N., Vogt, C., et al., 2013. *Wiley Interdiscip. Rev. Nanomed. Nanobiotechnol.* 5, 111–129.
79. Farokhzad, O.C., Langer, R., 2009. *ACS Nano* 3 (1), 16–20.
80. Feliu, N., Fadeel, B., 2010. *Nanoscale* 2, 2514–2520.
81. Feliu, N., Walter, M.V., Montañez, M.I., et al., 2012. *Biomaterials* 33 (7), 1970–1981.
82. Feng, X.D., Sayle, D.C., Wang, Z.L., et al., 2006. *Science* 312, 1504–1508.
83. Ferin, J., Oberdörster, G., Penney, D.P., 1992. *Am. J. Respir. Cell Mol. Biol.* 6, 535–542.
84. Fitzpatrick, J.A., Andreko, S.K., Ernst, L.A., et al., 2009. *Nano Lett.* 9 (7), 2736–2741.
85. Foner, S., 1959. *Rev. Sci. Instrum.* 30 (7), 548–557.
86. Fourches, D., Pu, D., Tassa, C., et al., 2010. *ACS Nano* 4, 5703–5712.
87. Frens, G., 1973. *Nat. Phys. Sci.* 241 (105), 20–22.
88. Pope, C.A., Thun, J.M., Namboodiri, M.M., et al., 1995. *Am. J. Respir. Crit. Care Med.* 151, 669e674.
89. Kasahara, M., 1994. *J. Jpn. Soc. Air Pollut.* 29 (6), A100.
90. Kasahara, M., 1990. *J. Jpn. Soc. Air Pollut.* 25 (2), 115.
91. Nagasaki, S., 1998. *J. Jpn. Soc. Irrig. Drain. Reclam. Eng.* 66 (12), 1261e1269.
92. Makino, H., 1982. Denchuukenhoukoku, 281040.
93. Makino, H., 1989. *J. Aerosol Res. Jpn.* 4 (3) 90–95.
94. Saito, K., 2001. *J. Soc. Powder Technol. Jpn.* 38 (7), 493e502.
95. Pan, J.R., Huang, C., Jiang, W. et al., 2005. *Desalination* 179, 31e40.
96. Umezawa, H., Iseki, M., Takaoka, D., et al., 2003. *Sanyo Tech. Rev.* 35 (2), 22e30.
97. Moriya, M., 1999. *J. Jpn. Soc. Water Environ.* 22, 346e351.
98. Skipperud, L., Salbu, B., Hagebø, E., 1998. *Sci. Total Environ.* 217, 251e256.
99. Karakulski, K., Morawski, A.W., 2002. *Desalination* 149, 163e167.
100. Kang, S.K., Choo, K.H., 2003. *J. Membr. Sci.* 223, 89e103.
101. Higashi, K., 1997. *J. Jpn. Soc. Water Environ.* 20, 210e214.
102. Shinozuka, N., 1995. *J. Jpn. Soc. Water Environ.* 18, 261e265.
103. Homma, S., 1991. Denpun Kagaku 38, 73e79.
104. Nagano, A., Nakamoto, C., Suzuki, M., 1999. *J. Jpn. Soc. Water Environ.* 22, 498e504.
105. Rektor, A., Vatai, G., 2004. *Desalination* 162, 279e286.
106. Rideal, G., 2005. *Filtr. Sep.* 42 (7), 30e33.
107. Yoshimura, K., 1993. *J. Jpn. Soc. Water Environ.* 16, 294e301.
108. Kiso, Y., Li, H.-D., Kitao, T., 1996. *J. Jpn. Soc. Water Environ.* 19, 648e656.
109. Madaeni, S.S., Fane, A.G., Grohmann, G.S., 1995. *J. Membr. Sci.* 102, 65e75.
110. Urase, T., Sato, K., 2005. *J. Jpn. Soc. Water Environ.* 28, 657e662.
111. Kim, S., Suzuki, Y., 2002. *J. Jpn. Soc. Water Environ.* 25, 349e354.
112. Weschler, C.J., Shields, H.C., 1999. *Atmos. Environ.* 33, 2301.
113. Hatakeyama, S., 1991. *J. Aerosol Res. Jpn.* 6, 106 (in Japanese).
114. Iinuma, Y., Böge, O., Gnauk, T. et al., 2004. *Aerosol Environ.* 38, 761.
115. Wolkoff, P., Clausen, P.A., Wilkins, C.K., et al., 2000. *Indoor Air* 10, 82.
116. Ito, T., Nishimura, K., 2002. *J. Soc. Heat. Aircond. Sanit. Eng. Jpn.* 76, 817 (in Japanese).
117. Tanimura, Y., 2003. *J. Aerosol Res. Jpn.* 18, 20 (in Japanese).
118. Jan, B., Lenggoro, I.W., Choi, M., Okuyama, K., 2003. *Anal. Sci.* 19, 843.
119. RAL German Institute for Quality Assurance and Certification, 2004. Basic Criteria for the Award of the Environmental Label, Printer RAL-UZ 85, p. 35.

120. Namiki, N., Otani, Y., Emi, H., Kagi, N., Fujii, S., 2003. *Proc. Air Clean. Contam. Control* 118 (in Japanese).
121. Wallace, L.A., Emmerich, S.J., Howard-Reed, C., 2004. *Environ. Sci. Technol.* 38, 2304.
122. Li, C.-S., Jenq, F.-T., Lin, W.-H., 1992. *J. Aerosol Sci.* 23, S547.
123. Venkataraman, C., Habib, G., Eiguren-Fernandez, A., Miguel, A.H., Friedlander, S.K., 2005. *Science* 307, 1454.
124. Katayama, K., Kitao, S., Shimada, M., Okuyama, K., 2004. *J. Aerosol Res. Jpn.* 19, 50 (in Japanese).
125. Otani, Y., Namiki, N., 2004. *Annu. Res. Rep. Smok. Res. Found.* 795 (in Japanese).
126. Hogan, C.J., Jr., Lee, M.-H., Biswas, P., 2004. *Aerosol Sci. Technol.* 38, 475.
127. Zimmer, A.T., Maynard, A.D., 2002. *Ann. Occup. Hyg.* 46, 663.
128. Otani, Y., Namiki, N., 2005. *Annu. Res. Rep. Smok. Res. Found.* 747 (in Japanese).
129. Suzuki, M., Yamaji, Y., 1989. *J. Jpn. Air Clean. Assoc.* 32, 218 (in Japanese).
130. Namiki, N., Otani, Y., Emi, H., Fujii, S., 1996. *J. Inst. Environ. Sci.* 39, 26.
131. Kato, K., Tanaka, A., Saiki, A., Hirata, J., 1995. *Proc. Air Clean. Contam. Control* 5, 70–72 (in Japanese).
132. Ushio, Y., Nakamura, T., Shimizu, S., et al., 1998. *Proc. Air Clean. Contam. Control* 335, 100–102 (in Japanese).
133. Saiki, A., Ro, S., Fujimoto, T., 1997. Chemical Contamination in Semiconductor Processing Environments and Its Countermeasures, vol. 426, Realize, Inc., Tokyo (in Japanese).
134. The Japan Society of Industrial Machinery Manufactures, 1994. Report on Behavior Control of Individual Sort of Contaminants e 1993 Report on Introduction of Advanced Technologies to Environmental Equipment Industry, vol. 171 (in Japanese).
135. Luther, W., 2004. Industrial Application of Nanomaterials e Chance and Risks, Future Technologies, Division of VDI Technologiezentrum, Düsseldorf, Germany, p. 112.
136. Maynard, A.D., Baron, P.A., Foley, M., et al., 2004. *J. Toxicol. Environ. Health A* 67, 87.
137. Kuhlbusch, T.A.J., Neumann, S., Fissan, H., 2004. *J. Occup. Environ. Hyg.* 1, 660.
138. Fukuoka, A., Araki, H., Kimura, J., et al., 2004. *J. Mater. Chem.* 14, 752–756.
139. Gallas, M.R., Hochey, B., Pechenik, A., et al., 1991. *J. Am. Ceram. Soc.* 77, 2107–2112.
140. Ge, C., Du, J., Zhao, L., et al., 2011. *Proc. Natl. Acad. Sci. U S A* 108, 16968–16973.
141. Gehrke, H., Pelka, J., Hartinger, C.G., et al., 2011. *Arch. Toxicol.* 85 (7), 799–812.
142. Geiser, M., Kreyling, W.G., 2010. *Part. Fibre. Toxicol.* 7, 2.
143. George, S., Pokhrel, S., Xia, T., et al., 2010. *ACS Nano* 4, 15–29.
144. George, S., Lin, S., Ji, Z., et al., 2012. *ACS Nano* 6, 3745–3759.
145. Gilbert, B., Fakra, S.C., Xia, T., et al., 2012. *ACS Nano* 6, 4921–4930.
146. Gliga, A.R., Skoglund, S., Odnevall Wallinder, I., et al., 2014. *Part. Fibre Toxicol.* 11 (1), 11.
147. Godinho, M.J., Gonçalves, R.F., Santos, L.P.S., et al., 2007. *Mater. Lett.* 61, 1904–1907.
148. Goldstein, J., Newbury, D., Joy, D., et al. (Eds.), 2003. *Scanning Electron Microscopy and X-Ray Microanalysis.* Kluwer Academic/Plenum Publishers, New York.
149. Gong, C., Akfur, C.M., Andersson, P.O., et al., 2010. *Biochem. Biophys. Res. Commun.* 397 (3), 397–400.
150. Gong, C., Tao, G., Yang, L., et al., 2012. *Toxicol. Lett.* 209 (3), 264–269.
151. Goodman, J.E., Prueitt, R.L., Thakali, S., et al., 2011. *Crit. Rev. Toxicol.* 41 (2), 142–174.
152. Gratton, S.E., Ropp, P.A., Pohlhaus, P.D., et al., 2008. *Natl. Acad. Sci. U. S. A* 105, 11613–11618.
153. Grzelczak, M., Pérez-Juste, J., Mulvaney, P., et al., 2008. *Chem. Soc. Rev.* 37, 1783–1791.
154. Gu, L., Fang, R.H., Sailor, M.J., et al., 2012. *ACS Nano* 6 (6), 4947–4954.
155. Gulson, B., McCall, M., Korsch, M., et al., 2009. *Toxicol. Sci.* 118 (1), 140–149.

156. Guo, J.J., Liu, X.H., Cheng, Y.C., et al., 2008. *J. Colloid. Interf. Sci.* 326, 138–142.
157. Gupta, A., Silver, S., 1998. *Nat. Biotechnol.* 16, 888–890.
158. Gwinn, M.R., Tran, L., 2010. *Wiley Interdiscip. Rev. Nanomed. Nanobiotechnol.* 2 (2), 130–137.
159. Haas, V., Birringer, R., Gleiter, H., et al., 1997. *Aerosol. Sci.* 28, 1443–1453.
160. Häfeli, U., Schutt, W., Teller, J., Zborowski, M., 1997. *Scientific and Clinical Applications of Magnetic Carriers.* Plenum Press, New York.
161. Haile, S.M., Jonhagon, D.W., Wiserm, G.H., 1989. *J. Am. Ceram. Soc.* 72, 2004–2008.
162. Hämeri, K., Lähde, T., Hussein, T., et al., 2009. *Inhal. Toxicol.* 21, 17–24.
163. Hamm, B., Staks, T., Taupitz, M., et al., 1994. *J. Magn. Reson. Imaging* 4, 659–668.
164. Han, K.R., Lim, C.S., Hong, M.J., et al., 2000. *J. Am. Ceram. Soc.* 83, 750–754.
165. Hanagata, N., Zhuang, F., Connolly, S., et al., 2011. *ACS Nano* 5 (12), 9326–9338.
166. Hansen, S.F., Michelson, E.S., Kamper, A., et al., 2008. *Ecotoxicology* 17 (5), 438–447.
167. Hauck, T.S., Anderson, R.E., Fischer, H.C., et al., 2010. *Small* 6, 138–144.
168. Heinrich, U., Fuhst, R., Rittinghausen, S., et al., 1995. *Inhal. Toxicol.* 7, 533–556.
169. Hirst, S.M., Karakoti, A.S., Tyler, R.D., et al., 2009. *Small* 5 (24), 2848–2856.
170. Horie, M., Nishio, K., Fujita, K., et al., 2008. *Chem. Res. Toxicol.* 22 (8), 1415–1426.
171. Hu, X., Zrazhevskiy, P., Gao, X., 2009. *Ann. Biomed. Eng.* 37, 1960–1966.
172. Huh, D., Matthews, B.D., Mammoto, A., et al., 2010. *Science* 328 (5986), 1662–1668.
173. Hunter, R.J., 1989. *Foundations of Colloid Science.* Oxford University Press, Oxford.
174. Hussain, S., Thomassen, L.C., Ferecatu, I., et al., 2010. *Part Fibre Toxicol.* 7, 10.
175. Hyeon, T., Lee, S.S., Park, J., et al., 2001. *J. Am. Chem. Soc.* 123, 12798–12801.
176. Inaba, H., Tagawa, H., 1996. *Solid State Ionics* 83, 1–16.
177. Ivers-Tiffee, E., Seitz, K., 1987. *Am. Ceram. Soc. Bull.* 66, 1384–1388.
178. Jain, T.K., Reddy, M.K., Morales, M.A., et al., 2008. *Mol. Pharm.* 5 (2), 316–327.
179. Jain, M.P., Vaisheva, F., Maysinger, D., 2012. *Nanomedicine* 7, 23–37.
180. Jasinski, P., Suzuki, T., Anderson, H.U., 2003. *Sens. Actuators B* 95, 73–77.
181. Jiang, L., Yubai, P., Changshu, X., et al., 2005. *Ceramics Int.* 32 (5), 587–591.
182. Johnson, D.R., Methner, M.M., Kennedy, A.J., et al., 2010. *Environ. Health Perspect.* 118 (1), 49–54.
183. Johnston, H.J., Hutchison, G.R., Christensen, F.M., et al., 2009. *Particle Fibre Toxicol.* 6, 33.
184. Joint Research Center, 2011. JRC Reference Report EUR 24847 EN.
185. Jugan, M.L., Barillet, S., Simon-Deckers, A., et al., 2012. *Nanotoxicology* 6 (5), 501–513.
186. Kain, J., Karlsson, H.L., Möller, L., 2012. *Mutagenesis* 27, 491–500.
187. Kapoor, S., Palit, D.K., Mukherjee, T., 2002. *Chem. Phys. Lett.* 355, 383–387.
188. Karlsson, H.L., 2010. *Anal. Bioanal. Chem.* 398 (2), 651–666.
189. Karlsson, H.L., Cronholm, P., Gustafsson, J., et al., 2008. *Chem. Res. Toxicol.* 21 (9), 1726–1732.
190. Karlsson, H.L., Gustafsson, J., Cronholm, P., et al., 2009. *Toxicol. Lett.* 188 (2), 112–118.
191. Karlsson, H.L., Cronholm, P., Hedberg, Y., et al., 2013. *Toxicology* 313 (1), 59–69.
192. Khlebtsov, N., Dykman, L., 2011. *Chem. Soc. Rev.* 40 (3), 1647–1671.
193. Kim, H.R., Lähde, T., Hussein, T., et al., 2011. *Mutat. Res.* 726 (2), 129–135.
194. Kim, J., Lee, J.E., Lee, J., et al., 2006. *J. Am. Chem. Soc.* 128, 688–689.
195. Kim, J., Kim, H.S., Lee, N., et al., 2008. *Angew. Chem. Int.* 47, 8438–8441.
196. Kim, J.S., Yoon, T.J., Yu, K.N., et al., 2006. *Toxicol. Sci.* 89, 338–347.
197. Kim, J.S., Adamcakova-Dodd, A., O'Shaughnessy, A., et al., 2011. *Part. Fibre Toxicol.* 8, 29.
198. Kim, S., Kim, H.J., 2006. *Int. Biodeterior. Biodegrad.* 57, 155–162.
199. Kim, S., Ryu, D.Y., 2013. *J. Appl. Toxicol.* 33 (2), 78–89.
200. Kim, Y.S., Song, M.Y., Park, J.D., et al., 2010. *Part. Fibre Toxicol.* 6 (7), 20.

201. Klein, C.L., Wiench, K., Wiemann, M., et al., 2012. *Arch. Toxicol.* 86 (7), 1137–1151.
202. Koch, U., Fojtik, A., Weller, H., et al., 1985. *Chem. Phys. Lett.* 122, 507–510.
203. Koivisto, A.J., Lyyränen, J., Auvinen, A., et al., 2012. *Inhal. Toxicol.* 12, 839–849.
204. Kuhlbusch, T.A., Asbach, C., Fissan, H., et al., 2011. *Part. Fibre Toxicol.* 8, 22.
205. Kumari, M., Rajak, S., Singh, S.P., et al., 2012. *J. Nanosci. Nanotechnol.* 12 (3), 2149–2159.
206. Kunzmann, A., Andersson, B., Thurnherr, T., et al., 2011a. *Biochim. Biophys. Acta* 1810 (3), 361–373.
207. Kunzmann, A., Andersson, B., Vogt, C., et al., 2011b. *Toxicol. Appl. Pharmacol.* 253, 81–93.
208. Kupferschmidt, N., Xia, X., Labrador, R.H., et al., 2013. *Nanomed. Lond.* 8 (1), 57–64.
209. Kreyling, W.G., Semmler-Behnke, M., Takenaka, S., et al., 2013. *Acc. Chem. Res.* 46 (3), 714–722.
210. Kreyling, W.G., Hirn, S., Möller, W., et al., 2014. *ACS Nano* 8 (1), 222–233.
211. Krug, H.F., Wick, P., 2011. *Angew. Chem. Int. Ed. Engl.* 50, 1260–1278.
212. Kuriahra, L.K., Chow, G.M., Schoen, P.E., 1995. *Nanostruct. Mater.* 5, 607–613.
213. Landsiedel, R., Fabian, E., Ma-Hock, L., et al., 2012. *Arch. Toxicol.* 86 (7), 1021–1060.
214. Lanone, S., Rogerieux, F., Geys, J., et al., 2009. *Part. Fibre Toxicol.* 6, 14.
215. Lartigue, L., Hugounenq, P., Alloyeau, D., et al., 2012. *ACS Nano* 6, 2665–2678.
216. Lauf, R.J., Bond, W.D., 1984. *Am. Ceram. Soc. Bull.* 63, 278–281.
217. Laurent, S., Boutry, S., Mahieu, I., et al., 2009. *Curr. Med. Chem.* 16, 4712–4727.
218. Lee, I.I.S., Lee, N., Park, J., et al., 2006. *J. Am. Chem. Soc.* 128, 10658–10659.
219. Lee, K., Prakash, I., Chen, Y., et al., 2012. *J. Nanopart. Res.* 14, 134–138.
220. Lee, Y., Choi, J.R., Lee, K.J., et al., 2008. *Nanotechnology* 19, 415604–415607.
221. Li, R., Ji, Z., Chang, C.H., et al., 2014. *ACS Nano* 8 (2), 1771–1783.
222. Li, W., Shah, S.I., Huang, C.P., et al., 2002. *Mater. Sci. Eng. B* 96, 247–253.
223. Li, W.J., Shi, E.W., Zheng, Y.Q., et al., 2001. *J. Mater. Sci. Lett.* 20, 1381–1383.
224. Li, Y., Yang, R.T., 2007. *Ind. Eng. Chem. Res.* 46, 8277–8281.
225. Li, Y., Wu, Y., Ong, B., 2008. Stabilized silver nanoparticle composition. US patent 2008 US2008000382. http://www.rexresearch. com/alchemy5/nanogold.htm (accessed March 23, 2014).
226. Lin, Y.S., Haynes, C.L., 2010. *J. Am. Chem. Soc.* 132 (13), 4834–4842.
227. Litter, M.I., Navio, J.A., 1994. *J. Photochem. Photobiol. A Chem.* 84, 183–193.
228. Liu, J., Wang, Z., Liu, F.D., et al., 2012. *ACS Nano* 6 (11), 9887–9899.
229. Lu, C.H., Yeh, C.H., 2000. *Ceram. Int.* 26, 351–357.
230. Lu, S., Duffin, R., Poland, C., et al., 2009. *Environ. Health Perspect.* 117 (2), 241–247.
231. Lundqvist, M., Stigler, J., Elia, G., et al., 2008. *Proc. Natl. Acad. Sci.* 105, 14265–14270.
232. Ma, J.Y., Zhao, H., Mercer, R.R., et al., 2011. *Nanotoxicology* 5 (3), 312–325.
233. Ma, X., Wu, Y., Jin, S., et al., 2011. *ACS Nano* 5 (11), 8629–8639.
234. Mandal, M., Ghosh, S.K., Kundu, S., et al., 2002. *Langmuir* 18 (21), 7792–7797.
235. Maneerung, T., Tokura, S., Rujiravanit, R., 2008. *Carbohydr. Polym.* 72, 43–51.
236. Meng, H., Chen, Z., Xing, G., et al., 2007. *Toxicol. Lett.* 175 (1–3), 102–110.
237. Michalet, X., Pinaud, F.F., Bentolila, L.A., et al., 2005. *Science* 307, 538–544.
238. Midander, K., Cronholm, P., Karlsson, H.L., et al., 2009. *Small* 5 (3), 389–399.
239. Mikhaylova, M., Kim, D.K., Berry, C.C., et al., 2004. *Chem. Mater.* 16 (12), 2344–2354.
240. Milt, V., Querini, C.A., Miro, E.E., et al., 2003. *J. Catal.* 220, 424–432.
241. Minocha, S., Mumper, R.J., 2012. *Small* 8 (21), 3289–3299.
242. Miyata, T., Ishino, Y., Hirashima, T., 1978. *Synthesis*, 834–835.
243. Mizuguchi, Y., Myojo, T., Oyabu, T., et al., 2013. *Inhal. Toxicol.* 25 (1), 29–36.
244. Monteiro-Riviere, N.A., Wiench, K., Landsiedel, R., et al., 2011. *Toxicol. Sci.* 123 (1), 264–280.
245. Morris, J., Willis, J., De Martinis, D., et al., 2011. *Nat. Nanotechnol.* 6 (2), 73–77.

246. Munger, M.A., Radwanski, P., Hadlock, G.C., et al., 2014. *Nanomedicine* 10 (1), 1–9.
247. Murray, A.R., Kisin, E.R., Tkach, A.V., et al., 2012. *Part. Fibre Toxicol.* 9, 10.
248. Murray, C.B., Norris, D.J., Bawendi, M.G., 1993. *J. Am. Chem. Soc.* 115, 8706–8715.
249. Musa, A.O., Akomolafe, T., Carter, M.J., 1998. *Sol. Energy Mater. Sol. Cells* 51, 305–316.
250. Nandiyanto, A.B.D., Kim, S.-G., Iskandar, F., et al., 2009. *Microporous Mesoporous Mater.* 120, 447–453.
251. Nagai, H., Okazaki, Y., Chew, S.H., et al., 2011. *Proc. Natl. Acad. Sci. U S A* 108, 1330–1338.
252. Napierska, D., Thomassen, L.C., Lison, D., et al., 2010. *Part. Fibre Toxicol.* 7 (1), 39.
253. Natile, M.M., Boccaletti, G., Glisenti, A., 2005. *Chem. Mater.* 17, 6272–6286.
254. Nazarenko, Y., Zhen, H., Han, T., et al., 2012. *Environ. Health Perspect.* 120 (6), 885–892.
255. Nel, A., Xia, T., Meng, H., et al., 2013. *Acc. Chem. Res.* 46 (3), 607–621.
256. Nemamacha, A., Rehspringer, J.L., Khatmi, D., 2006. *J. Phys. Chem. B* 110, 383–387.
257. Neves, M.C., 2001. *Mater. Res. Bull.* 36, 1099–1108.
258. NIOSH (National Institute for Occupational Safety and Health), 2011. Current Intelligence Bulletin 63. Occupational exposure to titanium dioxide. Department of Health and Human Services, Public Health, Cincinnati, OH.
259. NIOSH., 2013. Current Intelligence Bulletin 65. Occupational exposure to carbon nanotubes and nanofibers. Department of Health and Human Services, Public Health, Cincinnati, OH.
260. Niu, J., 2007. *Cardiovasc. Res.* 73 (3), 549–559.
261. Nomiya, K., Yoshizawa, A., Tsukagoshi, K., et al., 2004. *J. Inorg. Biochem.* 98, 46–60.
262. Nowack, B., David, R.M., Fissan, H., et al., 2013. *Environ. Int.* 59, 1–11.
263. O'Brien, S., Brus, L., Murray, C.B., 2001. *J. Am. Chem. Soc.* 123, 12085–12086.
264. Oberdörster, G., Oberdörster, E., Oberdörster, J., 2005. *Environ. Health Perspect.* 13 (7), 823–839.
265. Oberdörster, G., 2010. *J. Intern. Med.* 267 (1), 89–105.
266. Ogami, A., Morimoto, Y., Myojo, T., et al., 2009. *Inhal. Toxicol.* 21 (10), 812–818.
267. Oh, W.K., Jeong, Y.S., Kim, S., et al., 2012. *ACS Nano* 4, 5301–5313.
268. Olivares, M., Uauy, R., 1996. *Am. J. Clin. Nutr.* 63, 791S–796S.
269. Pach, L., Roy, R., Komarneni, S., 1990. *J. Mater. Res.* 5, 278–285.
270. Padua, G.W., Wang, Q., 2012. *Nanotechnology Research Methods for Foods and Bioproducts.* Wiley, Hoboken, NJ.
271. Pagliari, F., Mandoli, C., Forte, G., et al., 2012. *ACS Nano* 6, 3767–3775.
272. Palmisano, L., Auggliaro, V., Sclafani, A., et al., 1988. *J. Phys. Chem.* 92, 6710–6713.
273. Park, E.J., Choi, J., Park, Y.K., et al., 2008. *Toxicology* 245 (1–2), 90–100.
274. Park, E.J., Tak, T.H., Na, D.H., et al., 2010. *Arch. Pharm. Res.* 33 (5), 727–735.
275. Park, J., An, K.J., Hwang, Y.S., et al., 2004. *Nat. Mater.* 3, 891–895.
276. Park, Y.K., Tadd, E.H., Zubris, M., et al., 2005. *Mat. Res. Bull.* 40 (9), 1506–1512.
277. Pecharsky, V., Zavalij, P., 2009. *Fundamentals of Powder Diffraction and Structural Characterization of Materials.* Springer Science, LLC, Boston, MA.
278. Peng, Z.A., Peng, X., 2001. *J. Am. Chem. Soc.* 123, 183–184.
279. Pettibone, J.M., Adamcakova-Dodd, A., Thorne, P.S., et al., 2008. *Nanotoxicology* 2 (4), 189–204.
280. Pfaff, G., Reynders, P., 1999. *Chem. Rev.* 99, 1963–1981.
281. Phillips, J.I., Green, F.Y., Davies, J.C., et al., 2010. *Am. J. Ind. Med.* 53 (8), 763–767.
282. Piao, M.J., Kim, K.C., Choi, J.Y., et al., 2011. *Toxicol. Lett.* 207 (2), 143–148.
283. Pietroiusti, A., Campagnolo, L., Fadeel, B., 2013. *Small* 9 (9–10), 1557–1572.
284. Pintar, A., Batista, J., Levec, J., 2001. *Catal. Today* 66, 503–510.
285. Prabhakar, P.V., Reddy, U.A., Singh, S.P., et al., 2012. *J. Appl. Toxicol.* 32 (6), 436–445.

286. Prakash, K.B., Shivakumar, K., 1998. *Biol. Trace Elem. Res.* 63 (1), 73–79.
287. Qi, L.M., Ma, J.M., Shen, J.L., 1997. *J. Colloid. Interf. Sci.* 186, 498–500.
288. Qin, J., Laurent, S., Jo, Y.S., et al., 2007. *Adv. Mater.* 19, 1874–1878.
289. Rai, M., Yadav, A., Gade, A., 2009. *Biotechnol. Adv.* 27, 76–83.
290. Raj, K., Moskovitz, R., 1990. *J. Magn. Magn. Mater.* 85, 233–245.
291. Ram, S., Mitra, C., 2001. *Mater. Sci. Eng. A* 304–306, 805–809.
292. Pan, Y., Neuss, S., Leifert, A., et al., 2007. *Small* 3, 1941–1949.
293. Pauluhn, J., 2009. *Toxicol. Sci.* 109 (1), 152–167.
294. Pelka, H., Gehrke, H., Esselen, M., et al., 2010. *Chem. Res. Toxicol.* 22 (4), 649–659.
295. Pfaller, T., Colognato, R., Nelissen, I., et al., 2010. *Nanotoxicology* 4 (1), 52–72.
296. Pietruska, J.R., Liu, X., Smith, A., 2011. *Toxicol. Sci.* 124 (1), 138–148.
297. Pope 3rd, C.A., Burnett, R.T., Thun, M.J., et al., 2002. *JAMA* 287 (9), 1132–1141.
298. Puzyn, T., Rasulev, B., Gajewicz, A., et al., 2011. *Nat. Nanotechnol.* 6 (3), 175–178.
299. Recalenda, A., M.Sc. Thesis, 2009. *Synthesis and Characterizations of Multifunctional Nanoparticles for Simultaneous drug Delivery and Visualization.* KTH, Sweden.
300. Redl, F.X., Black, C.T., Papaefthymiou, G.C., et al., 2004. *J. Am. Chem. Soc.* 126, 14583–14599.
301. Reimer, P., Balzer, T., 2003. *Eur. Radiol.* 13, 1266–1276.
302. Reiss, P., Bleuse, J., Pron, A., 2002. *Nano Lett.* 2, 781–784.
303. Rocha, R.A., Muccillo, E.N.S., 2003. *Adv. Powder Technol.* 416 (4), 711–717.
304. Rouxoux, A., Schulz, J., Patin, H., 2002. *Chem. Rev.* 102, 3757–3778.
305. Salvador, A., Pascual-Martí, M.C., Adell, J., et al., 2000. *J. Pharm. Biomed. Anal.* 22, 301–306.
306. Savolainen, K., Backman, U., Brouwer, D., et al., 2013. Nanosafety in Europe 2015–2025: Towards Safe and Sustainable Nanomaterials and Nanotechnology Innovations. Published by: Finnish Institute of Occupational Health, Helsinki, Finland.
307. Sayes, C.M., Wahi, R., Kurian, P.A., et al., 2006. *Toxicol. Sci.* 92, 174–185.
308. Sayes, C.M., Reed, K.L., Warheit, D.B., 2007. *Toxicol. Sci.* 97 (1), 163–180.
309. Sayes, C.M., Reed, K.L., Glover, K.P., et al., 2010. *Inhal. Toxicol.* 4, 348–354.
310. Schneider, T., Brouwer, D.H., Koponen, I.K., et al., 2011. *J. Expo. Sci. Env. Epid.* 21, 450–463.
311. Schrurs, F., Lison, D., 2012. *Nat. Nanotechnol.* 7, 546–548.
312. Schulte, P.A., Murashov, V., Zumwalde, R., et al., 2010. *J. Nanopart. Res.* 12, 1971–1987.
313. Schutt, W., Gruttner, C., Hafeli, U., et al., 1997. *Hybridoma* 16, 109–117.
314. Sharama, P.K., Jilavi, M.H., Bugard, D., et al., 1998. *J. Am. Ceram. Soc.* 81, 2732–2734.
315. Sharrock, M.P., Bodnar, R.E., 1985. *J. Appl. Phys.* 57, 3919–3924.
316. Shaw, S.Y., Westly, E.C., Pittet, M.J., et al., 2008. *Proc. Natl. Acad. Sci. USA* 105, 7387–7392.
317. Shi, J., Karlsson, H.L., Johansson, K., et al., 2012. *ACS Nano* 6 (3), 1925–1938.
318. Shukla, R., Bansal, V., Chaudhary, M., et al., 2005. *Langmuir* 21, 10644–10654.
319. Shvedova, A.A., Pietroiusti, A., Fadeel, B., et al., 2012. *Toxicol. Appl. Pharmacol.* 261, 21–33.
320. Simberg, D., Park, J.H., Karmali, P.P., et al., 2009. *Biomaterials* 30, 3926–3933.
321. Simon-Deckers, A., Gouget, B., Mayne-L'hermite, M., et al., 2008. *Toxicology* 253, 137–146.
322. Singh, M., Singh, S., Prasad, S., et al., 2008. *Digest J. Nanomater. Biostruct.* 3, 115–122.
323. Singh, N., Jenkins, G.J., Asadi, R., et al., 2010. *Nano Rev.* 1, 5358.
324. Skoog, D.A., Holler, F.J., Crouch, S.R., 2007. *Principles of Instrumental Analysis*, 6th ed. Thomson Brooks/Cole, Belmont, CA.
325. Slowing, I.I., Wu, C.W., Vivero-Escoto, J.L., et al., 2009. *Small* 5 (1), 57–62.
326. Smith, K.R., Samet, J.M., Romieu, I., et al., 2000. *Thorax* 55 (6), 518–532.
327. Song, H.Z., 2003. *Solid State Ionics* 156, 249–254.

328. Song, W., Zhang, J., Guo, J., et al., 2010. *Toxicol. Lett.* 199, 389–397.
329. Stroh, A., Zimmer, C., Gutzeit, C., et al., 2004. *Free. Radic. Biol. Med.* 36, 976–984.
330. Stoehr, L.C., Gonzalez, E., Stampfl, A., et al., 2011. *Part. Fibre Toxicol.* 8, 36.
331. Sugunan, A., Guduru, V.K., Uheida, A., et al., 2010. *J. Am. Ceram. Soc.* 93, 3740–3744.
332. Stark, W.J., 2011. *Angew. Chem. Int. Ed. Engl.* 50, 1242–1258.
333. Stöber, W., Fink, A., Bohn, E., 1968. *J. Coll. Interf. Sci.* 26, 62–69.
334. Sun, J., Wang, S., Zhao, D., et al., 2011. *Cell. Biol. Toxicol.* 27 (5), 333–342.
335. Sun, S., Zeng, H., Robinson, D.B., et al., 2004. *J. Am. Chem. Soc.* 126, 273–279.
336. Sun, T., Yan, Y., Zhao, Y., et al., 2012. *PLoS One* 7 (8), 43442.
337. Sun, X., Jiang, X., Dong, S., et al., 2003. *Macromol. Rapid Commun.* 24 (17), 1024–1028.
338. Sun, Y., Xia, Y., 2002. *Science* 298, 2176–2179.
339. Sund, J., Alenius, H., Vippola, M., et al., 2011. *ACS Nano* 5, 4300–4309.
340. Sung, J.H., Ji, J.H., Yoon, J.U., et al., 2008. *Inhal. Toxicol.* 20 (6), 567–574.
341. Sung, J.H., Ji, J.H., Park, J.D., et al., 2009. *Toxicol. Sci.* 108 (2), 452–461.
342. Sung, J.H., Ji, J.H., Park, J.D., et al., 2011. *Part. Fibre. Toxicol.* 1 (14), 8–16.
343. Talapin, D.V., Rogach, A.L., Shevchenko, E.V., et al., 2001. *Nano Lett.* 1, 207–211.
344. Tenzer, S., Docter, D., Rosfa, S., et al., 2011. *ACS Nano* 5, 7155–7167.
345. Tenzer, S., Docter, D., Kuharev, J., et al., 2013. *Nat. Nanotechnol.* 8 (10), 772–781.
346. Toyokuni, S., 2009. *Cancer Sci.* 100 (1), 9–16.
347. Trouiller, B., Reliene, R., Westbrook, A., et al., 2009. *Cancer. Res.* 69 (22), 8784–8789.
348. Trovarelli, A., Zamar, F., Llorca, J., et al., 1997. *J. Catal.* 169, 490–502.
349. Tschopp, J., Schröder, K., 2010. *Nat. Rev. Immunol.* 10 (3), 210–215.
350. Tsoli, M., Kuhn, H., Brandau, W., et al., 2005. *Small* 1, 841–844.
351. Tsunekawa, S., Fukuda, T., Kasuya, A., 2000. *J. Appl. Phys.* 87, 1318–1321.
352. Tuomela, S., Autio, R., Buerki-Thurnherr, T., et al., 2013. *PLoS One* 8 (7), e68415.
353. Turkevich, J., 1985. *Gold Bull.* 18, 86–91.
354. Turkevich, J., Stevenson, P.C., Hillier, J., 1951. *Discuss. Faraday Soc.* 11, 55–75.
355. Uauy, R., Olivares, M., Gonzalez, M., 1998. *Am. J. Clin. Nutr.* 67, 952S–959S.
356. Vallet-Regi, M., Balas, F., Arcos, D., 2007. *Angew. Chem. Int. Ed.* 46, 7548–7558.
357. van Broekhuizen, P., 2011. *J. Biomed. Nanotechnol.* 7 (1), 15–17.
358. van der Zande, M., Vandebriel, R.J., Van Doren, E., et al., 2012. *ACS Nano* 6 (8), 7427–7442.
359. van der Zande, M., Vandebriel, R.J., Groot, M.J., et al. (2014), *Part. Fibre Toxic* 01.11:8.
360. Vanwinkle, B.A., de Mesy Bentley, K.L., Malecki, J.M., et al., 2009. *Nanotoxicology* 3, 307–318.
361. Veiseh, O., Gunn, J.W., Zhang, M., 2009. *Adv. Drug Deliv. Rev.* 62, 284–304.
362. Vissokov, G.P., Pirgov, P.S., 1996. *J. Mater. Sci.* 31, 4007–4010.
363. Vogt, C., Toprak, M.S., Muhammed, M., et al., 2010. *J. Nanopart. Res.* 12 (4), 1137–1147.
364. Wagner, A.J., Bleckmann, C.A., Murdock, R.C., et al., 2007. *J. Phys. Chem. B* 111 (25), 7353–7359.
365. Walkenhorst, A., Schmitt, M., Adrian, H., et al., 1994. *Appl. Phys. Lett.* 64, 1871–1873.
366. Walkey, C.D., Chan, W.C., 2012. *Chem. Soc. Rev.* 41, 2780–2799.
367. Wan, Y., Zhao, D., 2007. *Chem. Rev.* 107, 2821–2860.
368. Wang, C.C., Zhang, Z., Ying, J.Y., 1997. *Nanostruct. Mater.* 9, 583–586.
369. Wang, J., Liu, Y., Jiao, F., et al., 2008. *Toxicology* 254, 82–90.
370. Wang, J.J., Sanderson, B.J., Wang, H., 2007. *Mutat. Res.* 628 (2), 99–106.
371. Wang, W., Howe, J.Y., Li, Y., et al., 2010. *J. Mater. Chem.* 20, 7776–7781.
372. Wang, Y., Hao, Y., Cheng, H., et al., 1999. *J. Mater. Sci.* 34, 2773–2779.
373. Warheit, D.B., Webb, T.R., Reed, K.L., et al., 2007. *Toxicology* 230, 90–104.
374. Wen, Z., Liu, J., Li, J., 2008. *J. Adv. Mater.* 20, 743–747.

375. Williams, D.B., Carter, C.B., 2009. *Transmission Electron Microscopy.* Springer Science, LLC, New York.
376. Williams, K.R., Burstein, G.T., 1997. *Catal. Today* 38, 401–410.
377. Wilkinson, K.E., Palmberg, L., Witasp, E., et al., 2011. *ACS Nano* 5 (7), 5312–5324.
378. Winkler, D.A., Mombelli, E., Pietroiusti, A., et al., 2013. *Toxicology* 313 (1), 15–23.
379. Winnik, F.M., Maysinger, D., 2013. *Acc. Chem. Res.* 46 (3), 672–680.
380. Witasp, E., Kupferschmidt, N., Bengtsson, L., et al., 2009. *Toxicol. Appl. Pharmacol.* 239, 306–319.
381. Xia, T., Li, N., Nel, A.E., 2009. *Annu. Rev. Public Health* 30, 137–150.
382. Xia, T., Zhao, Y., Sager, T., 2011. *ACS Nano* 5 (2), 1223–1235.
383. Xie, R., Kolb, U., Mews, A., et al., 2005. *J. Am. Chem. Soc.* 127, 7480–7488.
384. Yahiro, H., Baba, Y., Eguchi, Y., et al., 1988. *J. Electrochem. Soc.* 135, 2077–2080.
385. Yan, M., Zhang, Y., Xu, K., et al., 2011. *Toxicology* 282, 94–103.
386. Yang, X., Liu, J., He, H., et al., 2010. *Part. Fibre Toxicol.* 19 (7), 1.
387. Ye, L., Yong, K.T., Liu, L., et al., 2012. *Nat. Nanotechnol.* 7 (7), 453–458.
388. Yin, M., Wu, C.K., Lou, Y., et al., 2005. *J. Amer. Chem. Soc.* 127 (26), 9506–9511.
389. Yokohira, M., Kuno, T., Yamakawa, K., et al., 2008. *Toxicol. Pathol.* 36 (4), 620–631.
390. Yong, K.T., Law, W.C., Hu, R., et al., 2013. *Chem. Soc. Rev.* 42 (3), 1236–1250.
391. Yu, W.W., Chang, E., Drezek, R., et al., 2006. *Nanotechnology* 17, 4483–4487.
392. Yuan, S.A., Chen, W.H., Hu, S.S., 2005. *Mater. Sci. Eng. C* 25, 479–485.
393. Zallen, R., Moret, M.P., 2006. *Solid State Commun.* 137, 154–157.
394. Zawadzki, M., 2008. *J Alloys Compd.* 454, 347–351.
395. Zhang, H., Ji, Z., Xia, T., et al., 2012a. *ACS Nano* 6 (5), 4349–4368.
396. Zhang, H., Dunphy, D.R., Jiang, X., et al., 2012b. *J. Am. Chem. Soc.* 134 (38), 15790–57804.
397. Zhang, J., Sun, L., Pan, H., et al., 2002. *New J. Chem.* 26, 33–34.
398. Zhang, L., Bai, R., Liu, Y., et al., 2012. *Nanotoxicology* 6 (5), 562–575.
399. Zhang, X.D., Wu, D., Shen, X., et al., 2012. *Biomaterials* 33 (18), 4628–4638.
400. Zhou, Y.M., Zhong, C.Y., Kennedy, I.M., et al., 2003. *Environ. Toxicol.* 18, 227–235.
401. Zhu, M.T., Feng, W.Y., Wang, B., et al., 2008. *Toxicology* 247 (2–3), 102–111.
402. Kulkarni, S., 2015. *Size Controlled Biosynthesis of Ag Metal Nanoparticles Using Carrot Extract. Chemical and Bioprocess Engineering—Trends and Developments*, 1st ed., Chapter 33, Apple Academic Press.
403. Kulkarni, S., 2015. *Chem. Sci.*, 4 (4), 922–926.
404. Kulkarni, S.R., Saptale, S.P., Borse, D.B., et al., 2011. Green synthesis of Ag nanoparticles using Vitamin C (Ascorbic Acid) in a batch process, International Conference on Nanoscience, Engineering and Technology (ICONSET 2011), IEEE, pp. 88–90.
405. Kokate, K.K., Kulkarni, S., Bhandarkar, S.E., 2018. *Asian J. Res. Chem.*, 11 (1), 91–102.
406. Kulkarni, S.R., 2015. *J. Chem. Chem. Sci.*, 5 (7), 377–383.

11 Future Scope of Various Nanostructured Metal Oxides for Sustainable Energy Sources

Jaykumar B. Bhasarkar, Bharat A. Bhanvase, and Vijay B. Pawade

CONTENTS

11.1 INTRODUCTION

The clean energy and water demand requirements have drastically increased as the global population increases. Production of clean energy through convectional processes, that is, combustion of fossil fuel, is widely used. However, these processes lead to releasing a large amount of toxic and greenhouse gases, that is, NOx and SOx, to the environment as their final product. Moreover, the nature of non-renewable sources, inadequate availability, and adverse impact on the environment caused by processing these fossil fuels urges one to search for a new source of sustainable energy. Sustainable energy is one of the keys to overcoming the many drawbacks of fossil and non-renewable energy sources. Production of clean energy from sustainable energy has more potential to produce energy in large excess without an adverse impact on the environment [1,2].

A large number of metal oxides have been designed and developed due to them having more chemical stability, simple method of production, and low cost. The bonding between metal and oxygen in oxide formation, which is typical, forms the more chemically stable rock-forming mineral [3]. The various groups of metal oxides have received keen interest; out of these, many oxides have large band gaps, such as magnesium oxide (MgO) and aluminum oxides (Al_2O_3). Such types of metal oxides are more useful for electrical insulators and not very suitable as semiconductors. Although limited group oxides like tin (IV) oxide (SnO_2), indium (III) oxides (In_2O_3), and gallium (III) oxide (Ga_2O_3) have n-type wide band-gap transparent oxides. Due to n-type semiconductors in nature, it is very difficult to convert them to become p-type semiconductors [4]. Another property of metal oxides such as surface properties has to be considered because of their greater influence on the chemistry of the reaction. Heterogeneous reaction systems such as solid-gas or solid-liquid can be mostly limited to the surface of the solid.

Nanostructured metal oxides have shown their beneficial effects on energy saving and energy storage devices such as batteries (lithium-ion), fuel and solar cells, light-emitting diodes (LEDs), production of hydrogen using photolysis of water, and the degradation of many hazardous organic and inorganic pollutants. Apart from these applications, metal oxides are greatly used in biological treatment and the medical field. In the medical field, metal oxides are widely used for drug delivery systems, treatment of cancer, etc. Recently, metal-oxide-based nanomaterials (NMs) like ZnO and TiO_2 have been developed as novel NMs due to easy availability and simple synthesis method [5,6]. One of the prominent features of these nanostructured metal oxides is biocompatibility, which helps encourage the interdisciplinary research to have better tie-up between chemist, physicist, and biotechnologist [7]. In this chapter, we highlight the different noble nanostructured metal oxides and their applications in various fields. Finally, we present their current and future scope as a source of sustainable energy.

11.2 STRUCTURE OF NANOSTRUCTURED METAL OXIDE

Among the number of nanostructured material developed, controllable incorporation of different materials such as metal oxides (e.g., TiO_2, ZnO, CeO_2, ZrO_2, etc.) into sole nanostructures has lately become one of the modern and hottest topics for research in various fields, due to their functionality, synergetic and collective catalyst properties compared with individual NMs. As per the geometrical configuration of nanostructured material, they can be categorized into four types: (1) NM-decorated metal oxide nanoarrays, (2) core–shell nanostructures, (3) yolk/shell nanostructures, and (4) Janus noble metal-metal oxide nanostructures.

11.2.1 NM—Decorated Metal Oxide Nanoarrays

More recently, metal oxide NMs have been used to decorate various types of metal oxide nanoarrays, which including nanotubes, nanorods, nanosheets, and nanowires. Many studies have focused on the loading of NMs on metal oxide nanoarrays [8,9]. The deposition of NMs on the surface of metal oxides can be done by using chemical and photoreduction techniques. Similarly, these techniques can also be used for attaching the NMs with metal oxide nanoarrays. Many NMs can be modified with these techniques, such as the synthesis of Au–Pd nanotubes modified with TiO_2 films that have been done by simultaneous photoreduction techniques. The acute task of NM deposition on nanotube arrays is the struggle to fill cylindrical pores. This leads to the clogging of a huge quantity of NMs in the nanotubes. The growth of NMs on metal oxide nanotube arrays is much simpler and cost-effective with the electroless deposition method [10]. This method provides the uniform covering of NMs on the outer and inner surfaces of the metal oxides.

11.2.2 Core–Shell Nanostructures

Encompassing metal oxide NMs, a shell of semiconductor is an appropriate technique to ease accumulation and deterioration of metal oxides nanoparticles (NPs) due to their contact with harsh chemicals. Figure 11.1 shows the mechanism and preparation route for Au/TiO_2 core/shell NPs. Many authors have reported the preparation and synthesis of core–shell nanostructured metal oxides by the solvothermal and/or hydrothermal method [11–13]. Preparation of Au–TiO_2 core–shell structures have been thoroughly studied by Bian and his co-workers [14]. They have prepared Au-TiO_2 core–shell nanostructures by dissolving the mixture of known quantity of glycerol and ethyl ether. This mixture was further developed with an autoclave

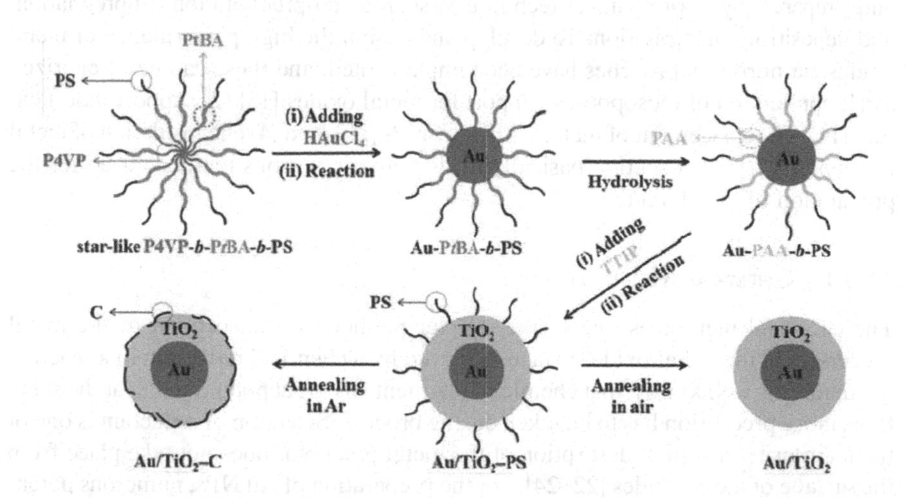

FIGURE 11.1 Schematic illustration of the preparation route for Au/TiO_2 core–shell NPs. (Reprinted with permission from Zheng, D. et al., *Chem. Mater.*, 27, 5271–5278, 2015. Copyright 2015 American Chemical Society.)

followed by the solvothermal treatment. In another study, the preparation of Au/TiO_2 core–shell NPs was done by exploiting the copolymers (star-like) and metal precursors.

11.2.3 Yolk–Shell Nanostructures

The yolk–shell structure is also known as a rattle-like structure shell with a solid core of NMs. The solid core NM helps to improve the free movement of metal oxides in the shell. Yolk–shell nanostructures (YSNs) are made up of a hollow cavity core surrounded by a porous shell. The yolk–shell structure has received more attention in current research fraternity due to their exceptional nanostructural, physical, and chemical properties, and extensive applications in various fields. The production of YSNs has been very easy for SiO_2 metal oxides due to the hollow shell structure. The only one limitation for SiO_2 metal oxide YSNs is they can be used as stabilizers. Producing the other metal oxides, YSNs are more difficult as compared to SiO_2 because of corrosive nature during the etching process. Consequently, SiO_2 is extensively used to produce the YSN shell. YSNs cannot be obtained by the single direct method; a combination of various methods can be adopted for the YSN shell. The application of YSNs is wider in the biomedical field, which includes cancer therapy, bio-sensing, bio-imaging, gene delivery, and drug delivery systems [16,17]

11.3 METHODS FOR SYNTHESIS OF METAL OXIDES NANOSTRUCTURED

Various NMs have been developed with governable incorporation of different metals (e.g., Pt, Ag, Au, and Pd) and metal oxides (e.g., ZnO, TiO_2, CeO_2, and ZrO_2, etc.) into single nanostructures. Conventionally, these metal oxides are easily synthesized and prepared by various novel techniques, such as co-precipitation, impregnation, and deposition–precipitation. To develop and design the high performance of metal oxides, numerous approaches have been implemented, and these can be categorized as (1) application of mesoporous support for metal oxides [18] (2) support modification [19], (3) replacement of metal with alloy NPs [20], and (4) encapsulation of metal to support [21]. This section basically deals with the various methods used for the preparation of metal oxides.

11.3.1 Chemical Methods

The chemical method is one of the simpler methods for adsorption of the metal precursor on the metal oxide surface, followed by a chemical reduction in a reaction medium. It is well known that chemical treatment has great potential, but at the same time, more precaution has to be taken during process. Selection of reductant is one of the main criteria so that desorption of the metal precursor does not take place from the surface of metal oxides [22–24]. For the preparation of Au NPs, numerous potential methods are available to prepare the colloidal mixture with the combination of the proper solvent. The Tuekevich (citrate reduction) method [25], Brust method, and $NaBH_4$ reduction methods are the most appropriate and well-known processes to produce Au NPs [26]. Au NPs can be adsorbed on the surface of the metal oxide semiconductors by simple agitation of the colloidal solution. The one disadvantage observed during these processes is aggregation of Au NPs during adsorption, thus leading to a reduction in the efficiency of the anticipated applications [27]. The sonochemical-assisted processes have wider applications in advanced oxidation processes, synthesis of NPs, and oxidation of organic and inorganic compounds [28,29]. The ultrasound method is another potential method to prepare Au NPs doped on metal oxide semiconductor surfaces. Basically, two main effects have observed during sonochemical-assisted processes are cavitation and ultrasound (intense mixing). Both physical and chemical effects of ultrasound have important impacts on the synthesis of NPs. Many researchers have done tremendous work of preparation of photocatalysts through sonochemical or photochemical routes. The preparation of Au NPs using the deposition precipitation method has been studied. In this study, a TiO_2 particle prepared by the sol–gel method and Au NPs were deposited on the surface of TiO_2 particles [30].

11.3.2 Thermal Methods

As mentioned in the previous section, solvothermal and hydrothermal methods are commonly used for preparation of metal oxide NPs, and these methods fall in the thermal method category. In the hydrothermal process, the metal oxides are mixed with a precursor of metal with water. This mixture has been treated under autoclave conditions, that is, the reaction mixture boils at the above-boiling temperature of the water or solvent. During this process (high temperature), the precursors are abridged and start adsorbing on the surface of metal oxide semiconductors [31]. In the case of the solvothermal process, the same mechanism is applied as the hydrothermal process except the aqueous solution (water). The metal oxides are added in metal precursors in the presence of an appropriate solvent other than water [32].

11.4 APPLICATION OF NANOSTRUCTURED MATERIALS

The applications of metal oxides have a large range of applications in the energy sector, environmental remediation, medical, and sustainable energy. In recent years, many researchers have particularly focused only on the performance of metal oxides that explain the properties of NPs. Therefore, with this information, the NPs can ideally be synthesize/prepared with only preferred material properties that do not include any hazard or adverse impact on environment. The few applications of nanostructured metal oxides are listed in following section.

11.4.1 Application of Zinc Oxide (ZnO)

Many researchers have shown their focus on ZnO as a metal oxide, due to the low cost and it is a good replacement of Si and GaN. ZnO is one of the most promising materials of the future with n-type and wide band-gap material. ZnO metal oxide has several applications in wastewater treatment and gas-sensor applications. ZnO, present in its polycrystalline structure with 3.40 eV of direct band-gap energy and higher activity, was observed in the UV blue region. The possible applications of ZnO are discussed in the following sections.

11.4.1.1 Solar Cells and Light Detectors

Research of renewable energy sources is highly required because of the limited sources and generation of pollution due to the burning of fossil fuel. Fabrication of new solar cells and modification of existing solar cells is much needed in recent years. Generally, solar cells are made up of p-n junction diodes. These diodes are photocarriers, which are generated by sunrays/sunlight. The more efficient, high-charge mobility and low hole-trapping level material is required to fabricate more efficient solar cells. Commercially crystalline silicon solar cells have been used with 20% maximum efficiency, which is occupies 90% of the market, but it is very cost effective. Therefore, other material has to develop for high efficiency solar cells with the low cost of fabrication. Because ZnO is a wide band-gap semiconductor with higher electron mobility, it leads to reducing the photocarrier loss. Apart from these physical and electronic properties of ZnO, it can be easily grown in various nanostructures with a low electron trap level, which makes it more potential material for the future alternative of silicon-based solar cells.

11.4.1.2 Hydrogen Generation and Storage

Hydrogen is one of the major sources of renewable energy and has more potential to replacing combustion of fossil fuel. Production of hydrogen from water is the cheapest process, and it can be broken into its molecules by the photolysis method. ZnO nanostructured metal oxide can be used for the production of hydrogen by electrochemical photolysis of water. Recently, the production of hydrogen with ZnO/TiO_2 metal oxide system has been studied by César et al. [33]. Another study proved that the combination of ZnO and TiO_2 show good composite for production of hydrogen by the partial combustion of methanol. Electrical charges of ZnO nanowires get accumulated on the material/metal surface, and this leads to a decrease in the flow of current. However, this phenomenon can be utilized to construct small pressure sensors, which can identify the flow rate of blood generated in veins. Other applications of ZnO-based biosensors are used to detect glucose and urea content available.

11.4.1.3 Water and Air Purification

ZnO metal oxide has extensively been used for degradation or removal of many toxic compounds such as arsenic from wastewater. ZnO can be used in porous form, which is more suitable for wastewater treatment and air purification due to the high absorption rate of toxic gases such as CO, CO_2, SOx, and NO on the surface of metal oxide. ZnO can be a better agent in degradation of organic pollutants, such as organic dyes and other organic chemicals released from industries in the presence of UV light [34,35]. Therefore, ZnO has a remarkable application in the field of water treatment and air purification. Apart from these, ZnO is widely used in the rubber industry, concrete industry, cigarette filters, coating, and pigments.

11.4.2 Applications of Titanium Dioxide (TiO_2)

TiO_2 powders have been commercially used as white pigments from early times. The stability of TiO_2 has held under dark condition, whereas it is more active in the presence of UV light. Due to this, TiO_2 is widely used as a photocatalyst in wastewater treatment and the degradation of toxic chemicals. In this section, we have discussed the application of TiO_2 in various fields.

11.4.2.1 Production of Hydrogen

As discussed in the previous section, the production of clean and sustainable energy is the major issue today. Many studies have shown that hydrogen is a better replacement for fossil fuel. Production of hydrogen can be done through various methods such as the photochemical, photoelectrochemical, and photocatalytic splitting of water. There have been many technical reports and studies done in the past decade on the production of hydrogen using nanostructured TiO_2. Recently, numerous research groups have prepared TiO_2 nanotubes as a thin film for the production of hydrogen. The material which have lower band-gap, the dimensions of nanotubes (i.e., length of the tube), and doping of species play a major role in the production of hydrogen using TiO_2. Several researchers have reported the modification of TiO_2 nanotubes that can improve the hydrogen production rate. From these studies, it was concluded that the rate of hydrogen production is highly influenced in the TiO_2 nanotube compared to the TiO_2 NP system.

11.4.2.2 Wastewater Treatment

TiO_2 is the most widely used as photocatalyst for energy and environmental applications. However, the usage of TiO_2 has some technical difficulties, and proper utilization of TiO_2 is hampered due to its wide band gap. Several attempts have been made to lower the band gap of TiO_2. To maximize the efficiency of the TiO_2-based system, TiO_2 NPs should be irradiated under UV light. The activity of the TiO_2 (photocatalytic activation) nanostructure is strongly enhanced under the UV light irradiation. Commercial TiO_2, containing 20:80 rutile and anatase with 30 mm of average particle size, has an average particle size of 30 nm and is extensively applicable for the degradation of toxic contaminants. However, the average particle size of 6–9 nm (nanostructured TiO_2) is used as a prominent photocatalyst for wastewater treatment and the purification process. Singh et al. [36] have studied the degradation of phenol (organic compound) using TiO_2 as a photocatalyst in a photoreactor. They have suggested the possible reaction mechanism for the degradation of phenol with nanostructured TiO_2 photocatalyst as described below:

$$TiO_2(ns) + \vartheta \rightarrow e^- + h^+$$

$$2H_2O + h^+ \rightarrow H_2O^+ + H_2O \rightarrow OH^{\cdot} + H_3O^+$$

$$OH^{\cdot} + OH^{\cdot} \rightarrow H_2O_2$$

$$\text{Phenol} + H_2O_2 \rightarrow \text{Product}$$

From the results, it was observed that the concentration of phenol was reduced from 100 to ~68 mg/L with total yield of 32% in 1 h of only UV illumination. The hybrid system (nanostructured TiO_2 + UV light) has to be applied on the same reaction mixture for 1 hr, and the maximum yield was observed of 68%. On the other hand, many researchers have found that the degradation efficiency of the photocatalyst can be increased with doped TiO_2 nanostructures for wastewater treatment.

11.4.2.3 Other Applications

In other applications of nanostructured metal oxides (TiO_2), apart from those mentioned above, TiO_2 NMs are widely being used for the manufacture of electrochromic devices. Different nanostructures of TiO_2 such as nanotubes or nanowires are also used for the storage of hydrogen. Irradiated TiO_2 nanofilm electrode suspensions (colloidal) can be effectively used for killing HeLe cells for cancer treatment. Primarily, the nanostructured TiO_2 photocatalysts were widely used for water or degradation of toxic/organic pollutants in aqueous systems. But, in recent days, it has been revealed that the nanostructured TiO_2 or ZnO can be used for photocatalytic volatile organic detoxification.

11.5 FUTURE PROSPECTS

Due to the higher performance of nanostructured metal oxides, wide application, being economically cheap, easily available, and easily synthesized, the preparations of nanostructured metal oxides have more attention nowadays. Nanostructured metal oxides can be more efficient for semiconductor industries due to the high demand in semiconductor devices. Metal oxides may be effective for the manufacture of the electronic devices and electronic chips. Intense developments and demand in the energy sector has been observed by many international agencies as well as researchers. Therefore, future growth in the energy sector, that is, photovoltaics, should be centered on the low-cost materials with easy synthesis process in order to persist in the aggressive competition with the silicon-based technology. Again, recent advancement in the application of organic and inorganic hybrid-type material in the energy sector has boosted greatly. To cater to the requirement of materials and costs of the process, development of low-cost material compared to silicon technology has been much needed in recent years. The best plausible solution for these is to produce/synthesize low-cost nanostructured metal oxides that can be a potential replacement of silicon technology. Metal oxide nanostructured material is expected to be the consistent alternative material in the agricultural, biological, and medical sector.

11.6 CONCLUSION

This chapter focuses on emerging nanostructured metal oxides, synthesis processes, their application, and current and future prospect. In this chapter, we highlight the actual mechanistic aspect of nanostructure metal oxide materials that provide the exceptional physical and chemical properties with extensive benefits toward sustainable energy and the environment. Many of the metal oxides have shown their great potential toward wastewater treatment, energy storage devices, biological application, medical treatment, and the rubber, cosmetic, and paint industries. However, all nanostructure metal oxides materials are effective for the above-mentioned applications; the activity of these metal oxides can be boosted after modification through doping on surface of other material. Therefore, the nanostructured metal oxides have a more accessible and economically realistic way for technological upgradation. Meanwhile, the principle design and application of nanostructure metal oxides also helps to solve the critical energy and environmental challenges.

CONFLICT OF INTEREST

The authors declare that there is no conflict of interest.

REFERENCES

1. Garca-Melchor, M., Braga, A.A.C., Lledós, A., Ujaque, G., Maseras, F. 2013. Computational perspective on Pd-catalyzed C–C cross-coupling reaction mechanisms. *Acc. Chem. Res.* 46: 2626–34.
2. Zhang, H., Jin, M.S., Xiong, Y.J., Lim, B., Xia, Y.N. 2013. Shape-controlled synthesis of Pd nanocrystals and their catalytic applications. *Acc. Chem. Res.* 46: 1783–94.
3. Rao, C.N.R. 1989. Transition metal oxides. *Annu. Rev. Phys. Chem.* 40: 291–26.
4. Ray, C., Pal, T. 2017. Recent advances of metal-metal oxide nanocomposites and their tailored nanostructures in numerous catalytic applications. *J. Mater. Chem. A.* 5: 9465–87.
5. Ahmad, J., Majid, K. 2018. In-situ synthesis of visible-light responsive Ag_2O/graphene oxide nanocomposites and effect of graphene oxide content on its photocatalytic activity. *Adv. Compos. Hyb. Mater.* 1: 374–88.
6. Li, H., Zhou, Y., Tu, W., Ye, J., Zou, Z. 2015. State-of-the-art progress in diverse heterostructured photocatalysts toward promoting photocatalytic performance. *Adv. Funct. Mater.* 25: 998–1013.
7. Zhu, Y.P., Ren, T.Z., Yuan, Z.Y. 2015. Mesoporous phosphorus-doped g-C3N4 nanostructured flowers with superior photocatalytic hydrogen evolution performance. *ACS Appl. Mater. Interfaces.* 7: 16850–56.
8. Roy, P., Berger, S., Schmuki, Angew, P. 2011. TiO_2 nanotubes: Synthesis and applications. *Chem. Int. Ed.* 50: 2904–39.
9. Gao, Z.D., Liu, H.F., Li, C.Y., Song, Y.Y. 2015. Carbon cladded TiO_2 nanotubes: Fabrication and use in 3D-RuO_2 based supercapacitors. *Chem. Commun.* 49: 7614–17.
10. Wang, T., Jin, B. Jiao, Z., Lu, G., Ye, J., Bi, Y. 2015. Electric field-directed growth and photoelectrochemical properties of cross-linked Au–ZnO hetero-nanowire arrays *Chem. Commun.* 51: 2103–06.
11. Burda, C., Chen, X., Narayanan, R., El-Sayed, M.A. 2005. Chemistry and properties of nanocrystals of different shapes. *Chem. Rev.* 105: 1025–1102.
12. Garnett, E.C., Brongersma, M.L., Cui, Y., McGehee, M.D. 2011. Nanowire solar cells. *Annu. Rev. Mater. Res.* 41: 269–95.
13. Law, M., Greene, L.E., Johnson, J.C., Saykally, R., Yang, P. 2005. Nanowire dye-sensitized solar cells. *Nat. Mater.* 4: 455–59.
14. Bian, Z., Zhu, J., Cao, F., Lu, Y., Li, H. Solvothermal synthesis of well-defined TiO_2 mesoporous nanotubes with enhanced photocatalytic activity. *Chem. Commun.* 44: 8451–53.

15. Zheng, D., Pang, X., Wang, M., He, Y., Lin, C., Lin, Z. 2015. Unconventional route to hairy plasmonic/semiconductor core/shell nanoparticles with precisely controlled dimensions and their use in solar energy conversion. *Chem. Mater.* 27: 5271–78.
16. Mura, S., Nicolas, J., Couvreur, P. 2013. Stimuli-responsive nanocarriers for drug delivery. *Nat. Mater.* 12: 991–1003.
17. Robinson, I., Tung, L.D., Maenosono, S., Wälti, C., Thanh, N.T.K. 2010. Synthesis of core-shell gold coated magnetic nanoparticles and their interaction with thiolated DNA. *Nanoscale.* 2: 2624–30.
18. Chen, C., Nan, C., Wang, D., Su, Q., Duan, H., Liu, X., Zhang, L., Chu, D., Song, W., Peng Q., Li, Y. 2011. Angew mesoporous multicomponent nanocomposite colloidal spheres: Ideal high-temperature stable model catalysts. *Chem. Int. Ed. Engl.* 50:3725–29.
19. Zhao, K., Qiao, B., Wang, J., Zhang, Y., Zhang, T. 2011. A highly active and sintering-resistant Au/FeOx–hydroxyapatite catalyst for CO oxidation. *Chem. Commun.* 47: 1779–81.
20. Liu, X., Wang, A., Wang, X., Mou, C., Zhang, T. 2008. Au–Cu alloy nanoparticles confined in SBA-15 as a highly efficient catalyst for CO oxidation. *Chem. Commun.* 27: 3187–89.
21. Qi, J., Chen, J., Li, G., Li, S., Gao, Y., Tang, Z. 2012. Facile synthesis of core–shell Au@ CeO_2 nanocomposites with remarkably enhanced catalytic activity for CO oxidation. *Energy Environ. Sci.* 5: 8937–41.
22. Kochuveedu, S.T., Oh, J.H., Do, Y.R., Kim, D.H. Surface-plasmon-enhanced band emission of ZnO nanoflowers decorated with Au nanoparticles. *Chem. Eur. J.* 18: 7467–71.
23. Zhang, Q., Lima, D.Q., Lee, I., Zaera, F., Chi, M.F., Yin, Y.D. Angew. 2011. A highly active titanium dioxide based visible-light photocatalyst with nonmetal doping and plasmonic metal decoration. *Chem. Int. Ed. Engl.* 50: 7088–92.
24. Mallik, K., Mandal, M., Pradhan, N., Pal, T. 2001. Seed mediated formation of bimetallic nanoparticles by UV irradiation: A photochemical approach for the preparation of "core–shell" type structures. *Nano Lett.* 1: 319–22.
25. Turkevich, J., Stevenson, P.C., Hillier, J. 1953. The formation of colloidal gold. *J. Phys. Chem.* 57: 670–73.
26. Brust, M. Walker, M., Bethell, D., Schiffrin D.J., Whyman, R. 1994. Synthesis of thiol-derivatised gold nanoparticles in a two-phase liquid–liquid system. *J. Chem. Soc. Chem. Commun.* 7: 801–02.
27. Patwa, A., Labille, J., Bottero, J., Thiéry, A., Barthélémy, P. 2015. Decontamination of nanoparticles from aqueous samples using supramolecular gels. *Chem. Commun.* 51: 2547–50.
28. Bhasarkar, J.B., Chakma, S., Moholkar, V.S. 2013. Mechanistic features of oxidative desulfurization using sono–fenton– peracetic acid (Ultrasound/Fe_2+ –CH_3COOH^- H_2O_2) system. *Ind. Eng. Chem. Res.* 52: 9038–47.
29. Chakma, S., Moholkar, V.S. 2013. Physical mechanism of sono-fenton process. *AIChE. J.* 59: 4303–13.
30. Murdoch, M., Waterhouse, G., Nadeem, M.A., Metson, J.B., Keane, M.A., Howe, R.F., Llorca, J., Idriss, H. 2011. The effect of gold loading and particle size on photocatalytic hydrogen production from ethanol over Au/TiO_2 nanoparticles. *Nat. Chem.* 3(6): 489–92.
31. Varghese, N., Panchakarla, L.S., Hanapi, M., Govindaraj, A., Rao, C.N.R. 2007. Solvothermal synthesis of nanorods of ZnO, N-doped ZnO and CdO. *Mater. Res. Bull.* 42: 2114–24.

32. Tonto, P., Mekasuwandumrong, O., Phatanasri, S., Pavarajarn, V., Praserthdam, P. 2008. Preparation of ZnO nanorod by solvothermal reaction of zinc acetate in various alcohols. *Ceramic Inter.* 34: 57–62.
33. Cesar, I., Sivula, K., Kay, A., Zboril, R., Grätzel, M. 2009. Influence of feature size, film thickness, and silicon doping on the performance of nanostructured hematite photoanodes for solar water splitting. *J. Phys. Chem. C.* 113: 772–82.
34. Yu, J., Yu, X. 2008. Hydrothermal synthesis and photocatalytic activity of zinc oxide hollow spheres. *Environ. Sci. Technol.* 42: 4902–07.
35. Quintana, M., Ricra, E., Rodriguez, J., Estrada, W. 2002. Spray pyrolysis deposited zinc oxide films for photo-electrocatalytic degradation of methyl orange: Influence of the pH. *Catalys. Today,* 76: 141–48.
36. Singh, D., Neti, N.R., Sinha, A.S.K., Srivastava, O.N. 2007. Growth of different nanostructures of Cu_2O (Nanothreads, Nanowires, and Nanocubes) by simple electrolysis based oxidation of copper. *J. Phys. Chem. C,* 111: 1638–45.

Index

Note: Page numbers in italic and bold refer to figures and tables, respectively.

I

L

M

N

O

P

Q

U

V

W

X

Y

Z

Printed in the United States
by Baker & Taylor Publisher Services

Printed in the United States
by Baker & Taylor Publisher Services